BUILDING CONSTRUCTION
HANDBOOK

BUILDING CONSTRUCTION HANDBOOK

Fourth edition

R. Chudley
MCIOB
and
R. Greeno
BA (Hons) MCIOB

AMSTERDAM BOSTON HEIDELBERG LONDON NEW YORK OXFORD
PARIS SAN DIEGO SAN FRANCISCO SINGAPORE SYDNEY TOKYO

Butterworth-Heinemann
An imprint of Elsevier Science
Linacre House, Jordan hill, Oxford OX2 8DP
200 Wheeler Road, Burlington MA 01803

First published 1988
Reprinted 1988, 1989 (twice), 1990 (three times), 1991
Revised and reprinted 1992
Reprinted 1993, 1994
Second edition 1995
Revised and reprinted 1996, 1997
Third edition 1998
Reprinted 1999, 2000
Fourth edition 2001
Reprinted 2002, 2003 (twice)

© R. Chudley 1988, 1995, 1996
© R. Chudley and R. Greeno 1998, 2001

Illustrations by the authors

British Library Cataloguing in Publication Data
A catalogue record for this book is available from the British Library

ISBN 0 7506 3753 6

For information on all Butterworth-Heinemann publications
visit our website at www.bh.com

Composition by Keyword Typesetting Services Ltd.
Printed and bound in Great Britain by MPG Books Ltd, Bodmin, Cornwall

CONTENTS

Preface ix

Part One   General

Built environment 3
The structure 6
Primary and secondary elements 13
Construction activities 19
Construction documents 20
Construction drawings 22
Planning application 35
Modular coordination 40
Construction regulations 42
CDM regulations 43
Safety signs and symbols 44
Building Regulations 46
British Standards 52
European Standards 53
CPI System of Coding 54
CI/SfB system of coding 55

Part Two   Site Works

Site investigations 59
Soil investigation 61
Soil assessment and testing 68
Site layout considerations 75
Site security 78
Site lighting and electrical supply 81
Site office accommodation 85
Materials storage 88
Materials testing 93
Setting out 102
Levelling 106
Road construction 108
Tubular scaffolding and scaffolding systems 116
Shoring systems 128

Part Three   Builders Plant

General considerations 139
Bulldozers 142
Scrapers 143
Graders 144
Tractor shovels 145
Excavators 146
Transport vehicles 151
Hoists 154
Rubble chutes and skips 156
Cranes 157
Concreting plant 169

Part Four   Substructure

Foundations–function, materials and sizing 177
Foundation beds 185
Short bored pile foundations 190
Foundation types and selection 192
Piled foundations 197
Retaining walls 215
Basement construction 228
Waterproofing basements 235
Excavations 241
Concrete production 247
Cofferdams 253
Caissons 255
Underpinning 257
Ground water control 266
Soil stabilization and improvement 276

Part Five   Superstructure

Choice of materials 283
Brick and block walls 284
Gas resistant membranes 303
Arches and openings 308
Windows, glass and glazing 314
Domestic and industrial doors 334
Timber frame construction 345
Reinforced concrete framed structures 347
Formwork 360
Precast concrete frames 365
Structural steelwork 376
Composite timber beams 396
Timber pitched and flat roofs 400
Long span roofs 442

Shell roof construction 448
Rainscreen cladding 468
Panel walls and curtain walling 470
Concrete claddings 473
Thermal insulation 478
Thermal bridging 497
Sound insulation 500
Access for the disabled 505

Part Six   Internal Construction and Finishes

Internal elements 511
Internal walls 512
Construction joints 517
Partitions 518
Plasters and plastering 523
Dry lining techniques 525
Wall tiling 529
Domestic floors and finishes 531
Large cast insitu ground floors 537
Concrete floor screeds 539
Timber suspended floors 541
Timber beam design 545
Reinforced concrete suspended floors 547
Precast concrete floors 551
Raised access floors 554
Timber, concrete and metal stairs 555
Internal doors 578
Fire resisting doors 584
Plasterboard ceilings 586
Suspended ceilings 587
Paints and painting 591
Joinery production 595
Composite boarding 600
Plastics in building 602

Part Seven   Domestic Services

Drainage effluents 605
Subsoil drainage 606
Surface water removal 608
Road drainage 611
Rainwater installations 613
Drainage systems 615
Drainage pipe sizes and gradients 624
Water supply 625
Cold water installations 627
Hot water installations 629

Cisterns and cylinders 633
Sanitary fittings 636
Single and ventilated stack systems 639
Domestic hot water heating systems 642
Electrical supply and installation 646
Gas supply and gas fires 654
Services–fire stops and seals 658
Open fireplaces and flues 659
Telephone installations 667

Index   669

PREFACE

This book presents the basic concepts of techniques of building construction, mainly by means of drawings illustrating typical construction details, processes and concepts. I have chosen this method because it reflects the primary means of communication on site between building designer and building contractor – the construction drawing or detail. It must be stressed that the drawings used here represent typical details, chosen to illustrate particular points of building construction or technology; they do not constitute the alpha and omega of any buildings design, detail or process. The principles they illustrate must therefore, in reality, be applied to the data of the particular problem or situation encountered. This new edition has been revised by Roger Greeno, in line with current building regulations.

Readers who want to pursue to greater depth any of the topics treated here will find many useful sources of information in specialist textbooks, research reports, manufacturer's literature, codes of practice and similar publications. One such subject is building services, which are dealt with here only in so far as they are applicable to domestic dwellings. A comparable but much wider treatment of services is given in *Building Services Handbook* by F. Hall and R. Greeno, also published by Butterworth-Heinemann.

In conclusion, I hope that this book will not only itself prove useful and helpful to the reader, but will act as a stimulus to the observation of actual buildings and the study of works in progress. In this way the understanding gained here will be continually broadened and deepened by experience.

R.C.

PREFACE TO FOURTH EDITION

This fourth edition incorporates numerous legislative changes, with new applications to design and construction. Roy Chudley's original illustrations are retained throughout, with regard to existing construction and reference to contemporary practice. Particular attention is afforded to current directives, which require modern buildings to be more fuel-efficient. A reduction in carbon gases and other atmospheric pollutants from buildings is shown, by attaining higher insulative standards in all elements of construction. In keeping with the earlier editions, additions and supplements are presented in illustrative format with comprehensive text as guidance to the many aspects of the building process.

R.G.

1 GENERAL

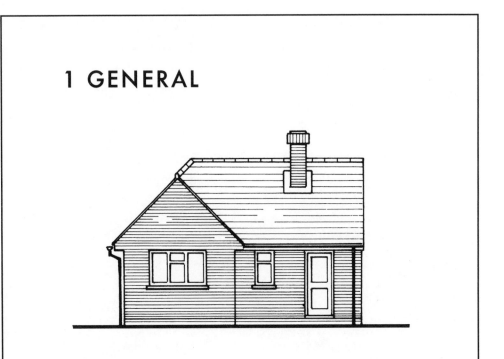

BUILT ENVIRONMENT

THE STRUCTURE

PRIMARY AND SECONDARY ELEMENTS

CONSTRUCTION ACTIVITIES

CONSTRUCTION DOCUMENTS

CONSTRUCTION DRAWINGS

CDM REGULATIONS

SAFETY SIGNS AND SYMBOLS

PLANNING APPLICATION

MODULAR COORDINATION

CONSTRUCTION REGULATIONS

BUILDING REGULATIONS

BRITISH STANDARDS

EUROPEAN STANDARDS

CPI SYSTEM OF CODING

CI/SFB SYSTEM OF CODING

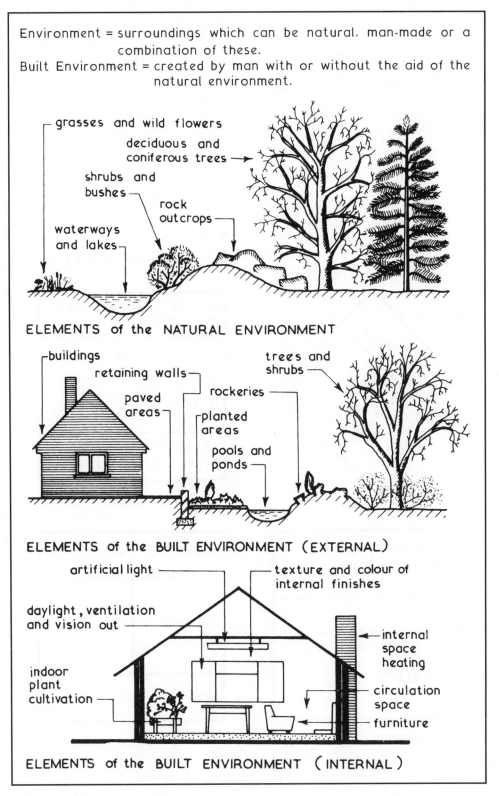

Environment = surroundings which can be natural, man-made or a combination of these.

Built Environment = created by man with or without the aid of the natural environment.

grasses and wild flowers

deciduous and coniferous trees →

shrubs and bushes

rock outcrops

waterways and lakes

ELEMENTS of the NATURAL ENVIRONMENT

buildings

retaining walls

trees and shrubs

paved areas

rockeries

planted areas

pools and ponds

ELEMENTS of the BUILT ENVIRONMENT (EXTERNAL)

artificial light

texture and colour of internal finishes

daylight, ventilation and vision out

internal space heating

indoor plant cultivation

circulation space

furniture

ELEMENTS of the BUILT ENVIRONMENT (INTERNAL)

Built Environment

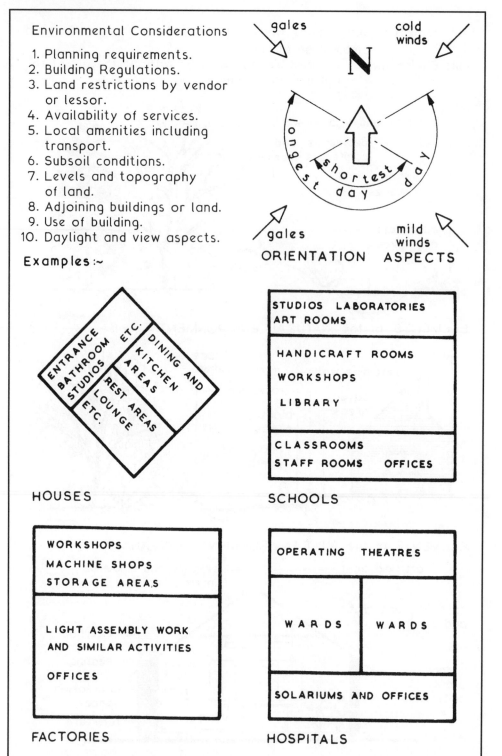

Environmental Considerations

1. Planning requirements.
2. Building Regulations.
3. Land restrictions by vendor or lessor.
4. Availability of services.
5. Local amenities including transport.
6. Subsoil conditions.
7. Levels and topography of land.
8. Adjoining buildings or land.
9. Use of building.
10. Daylight and view aspects.

Examples :~

gales

cold winds

N

longest day   shortest day

gales   mild winds

ORIENTATION   ASPECTS

ENTRANCE BATHROOM STUDIOS ETC.

DINING AND KITCHEN AREAS

REST AREAS LOUNGE ETC.

HOUSES

STUDIOS LABORATORIES ART ROOMS

HANDICRAFT ROOMS

WORKSHOPS

LIBRARY

CLASSROOMS
STAFF ROOMS   OFFICES

SCHOOLS

WORKSHOPS
MACHINE SHOPS
STORAGE AREAS

LIGHT ASSEMBLY WORK
AND SIMILAR ACTIVITIES

OFFICES

FACTORIES

OPERATING   THEATRES

WARDS   WARDS

SOLARIUMS AND OFFICES

HOSPITALS

Physical considerations

1. Natural contours of land.
2. Natural vegetation and trees.
3. Size of land and/or proposed building.
4. Shape of land and/or proposed building.
5. Approach and access roads and footpaths.
6. Services available.
7. Natural waterways, lakes and ponds.
8. Restrictions such as rights of way; tree preservation and ancient buildings.
9. Climatic conditions created by surrounding properties, land or activities.
10. Proposed future developments.

Examples:~

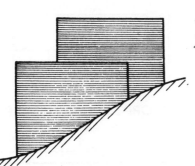

Split level construction to form economic shape.

Shape determined by existing trees.

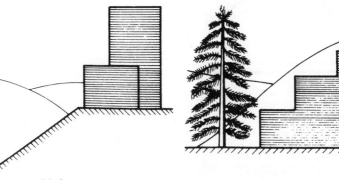

Plateau or high ground solution giving dry site conditions on sloping sites.

Stepped elevation or similar treatment to blend with the natural environment.

The Structure—Basic Types

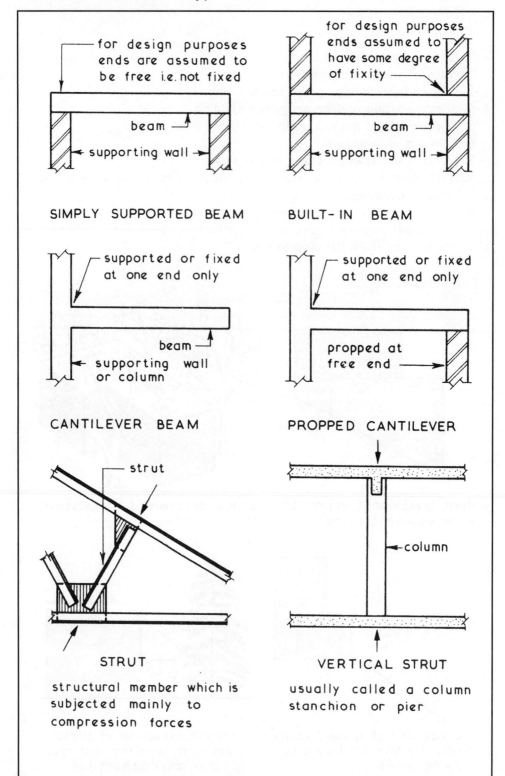

for design purposes ends are assumed to be free i.e. not fixed

beam →

← supporting wall →

SIMPLY SUPPORTED BEAM

for design purposes ends assumed to have some degree of fixity —

beam →

← supporting wall →

BUILT-IN BEAM

supported or fixed at one end only

beam →

← supporting wall or column

CANTILEVER BEAM

supported or fixed at one end only

propped at free end →

PROPPED CANTILEVER

strut

STRUT

structural member which is subjected mainly to compression forces

← column

VERTICAL STRUT

usually called a column stanchion or pier

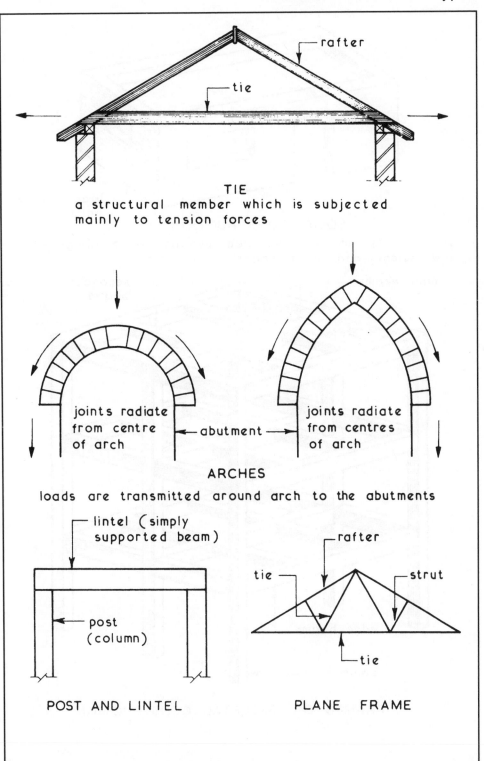

TIE
a structural member which is subjected
mainly to tension forces

joints radiate from centre of arch — abutment — joints radiate from centres of arch

ARCHES
loads are transmitted around arch to the abutments

lintel (simply supported beam)

post (column)

rafter

tie

strut

tie

POST AND LINTEL

PLANE FRAME

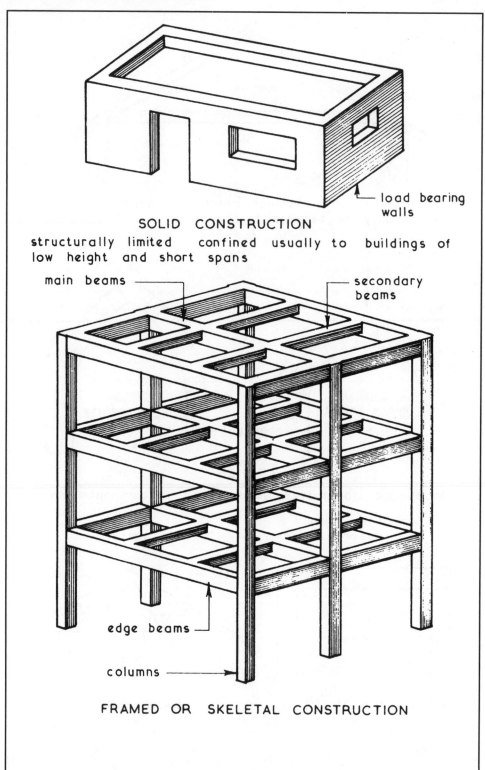

SOLID CONSTRUCTION

structurally limited confined usually to buildings of
low height and short spans

FRAMED OR SKELETAL CONSTRUCTION

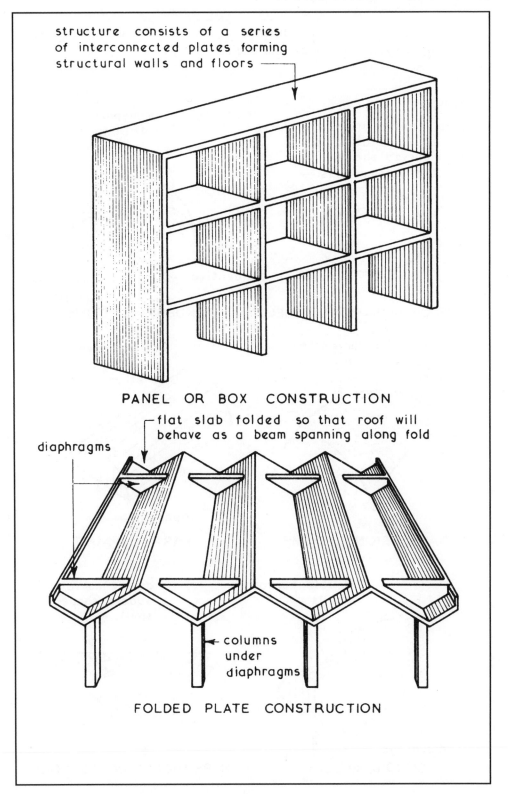

structure consists of a series
of interconnected plates forming
structural walls and floors

PANEL OR BOX CONSTRUCTION

flat slab folded so that roof will
behave as a beam spanning along fold

diaphragms

columns
under
diaphragms

FOLDED PLATE CONSTRUCTION

9

Shell Roofs ~ these are formed by a structural curved skin covering a given plan shape and area.

Examples ~

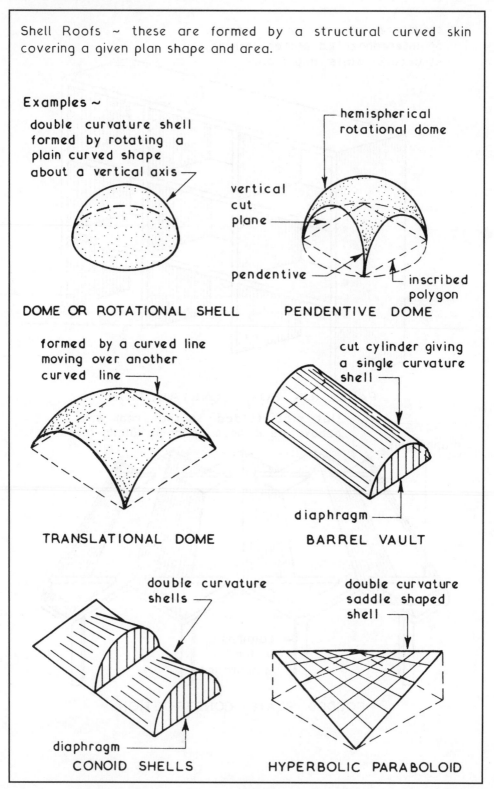

double curvature shell formed by rotating a plain curved shape about a vertical axis

DOME OR ROTATIONAL SHELL

hemispherical rotational dome

vertical cut plane

pendentive

inscribed polygon

PENDENTIVE DOME

formed by a curved line moving over another curved line

TRANSLATIONAL DOME

cut cylinder giving a single curvature shell

diaphragm

BARREL VAULT

double curvature shells

diaphragm

CONOID SHELLS

double curvature saddle shaped shell

HYPERBOLIC PARABOLOID

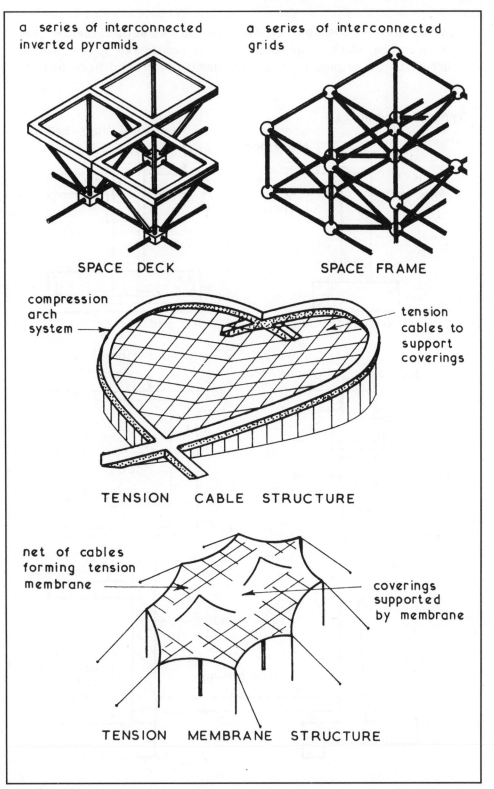

a series of interconnected inverted pyramids

a series of interconnected grids

SPACE DECK

SPACE FRAME

compression arch system

tension cables to support coverings

TENSION CABLE STRUCTURE

net of cables forming tension membrane

coverings supported by membrane

TENSION MEMBRANE STRUCTURE

Substructure

Substructure ~ can be defined as all structure below the superstructure which in general terms is considered to include all structure below ground level but including the ground floor bed.

Typical Examples~

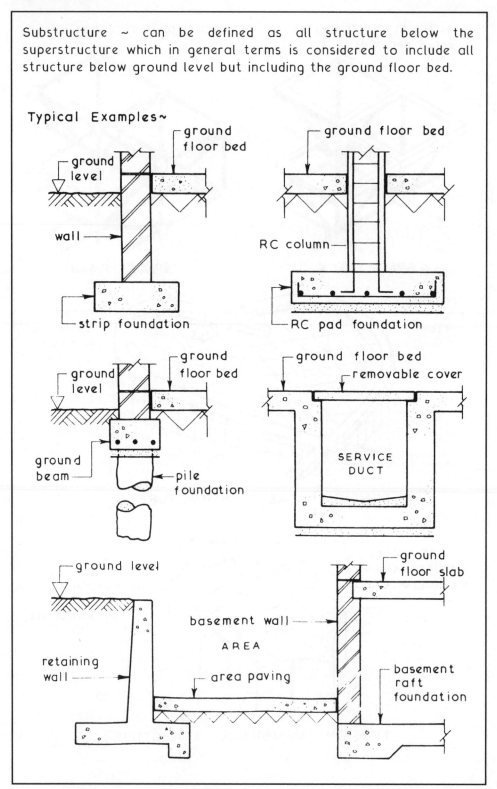

Superstructure ~ can be defined as all structure above substructure both internally and externally.

Primary Elements ~ basically components of the building carcass above the substructure excluding secondary elements, finishes, services and fittings.

Typical Examples~

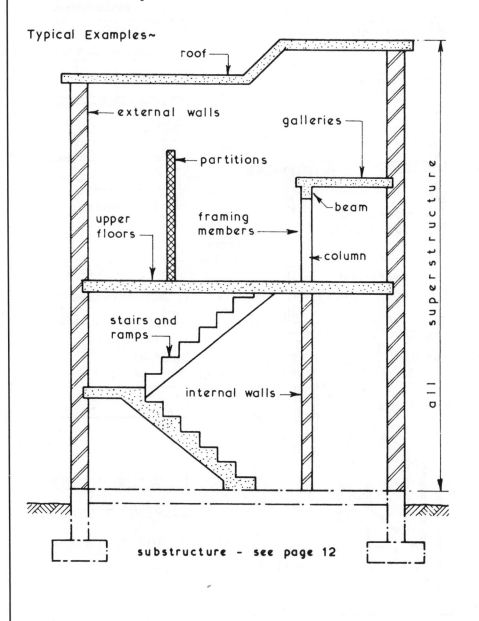

Secondary Elements ~ completion of the structure including completion around and within openings in primary elements.

Typical Examples~

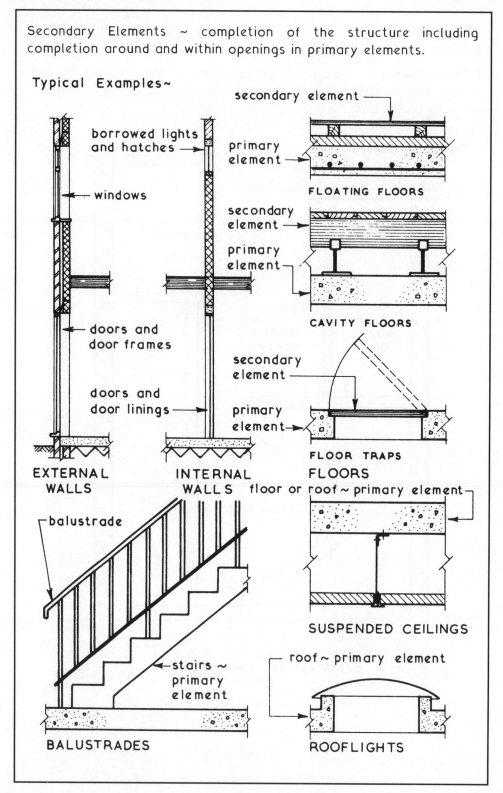

borrowed lights and hatches →

windows

doors and door frames

doors and door linings →

EXTERNAL WALLS

INTERNAL WALLS

secondary element

primary element

FLOATING FLOORS

secondary element

primary element

CAVITY FLOORS

secondary element

primary element →

FLOOR TRAPS

FLOORS   floor or roof ~ primary element

balustrade

stairs ~ primary element

BALUSTRADES

SUSPENDED CEILINGS

roof ~ primary element

ROOFLIGHTS

Finish ~ the final surface which can be self finished as with a trowelled concrete surface or an applied finish such as floor tiles.

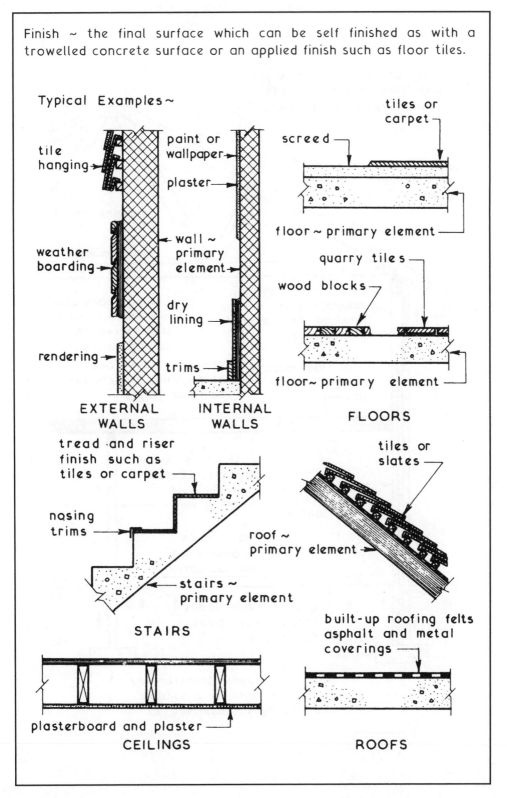

Typical Examples ~

EXTERNAL WALLS
- tile hanging
- weather boarding
- rendering

INTERNAL WALLS
- paint or wallpaper
- plaster
- wall ~ primary element
- dry lining
- trims

FLOORS
- tiles or carpet
- screed
- floor ~ primary element
- quarry tiles
- wood blocks
- floor ~ primary element

STAIRS
- tread and riser finish such as tiles or carpet
- nosing trims
- stairs ~ primary element

ROOFS
- tiles or slates
- roof ~ primary element
- built-up roofing felts asphalt and metal coverings

CEILINGS
- plasterboard and plaster

Domestic Structures :~

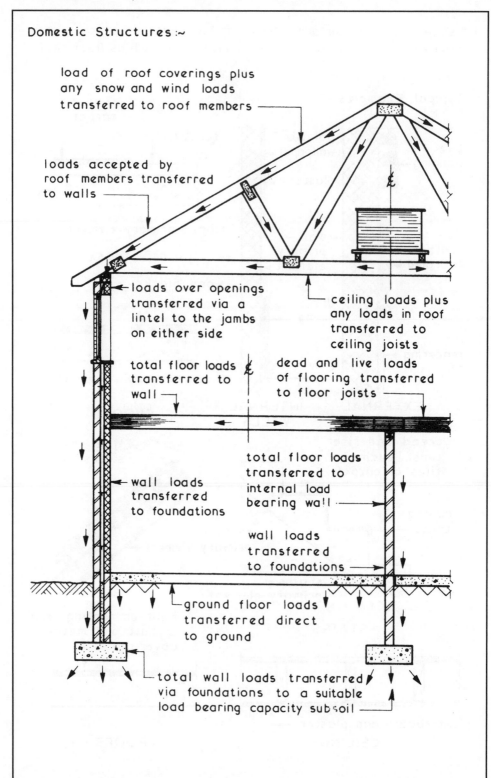

load of roof coverings plus any snow and wind loads transferred to roof members

loads accepted by roof members transferred to walls

loads over openings transferred via a lintel to the jambs on either side

ceiling loads plus any loads in roof transferred to ceiling joists

total floor loads transferred to wall

dead and live loads of flooring transferred to floor joists

total floor loads transferred to internal load bearing wall

wall loads transferred to foundations

wall loads transferred to foundations

ground floor loads transferred direct to ground

total wall loads transferred via foundations to a suitable load bearing capacity subsoil

Framed Structures:~

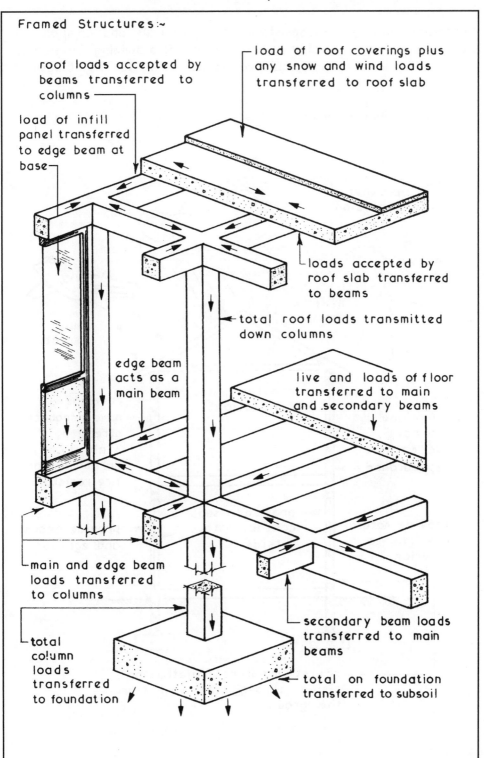

roof loads accepted by beams transferred to columns

load of roof coverings plus any snow and wind loads transferred to roof slab

load of infill panel transferred to edge beam at base

loads accepted by roof slab transferred to beams

total roof loads transmitted down columns

edge beam acts as a main beam

live and loads of floor transferred to main and secondary beams

main and edge beam loads transferred to columns

secondary beam loads transferred to main beams

total column loads transferred to foundation

total on foundation transferred to subsoil

External Envelope ~ consists of the materials and components which form the external shell or enclosure of a building. These may be load bearing or non-load bearing according to the structural form of the building.

Primary Functions :~

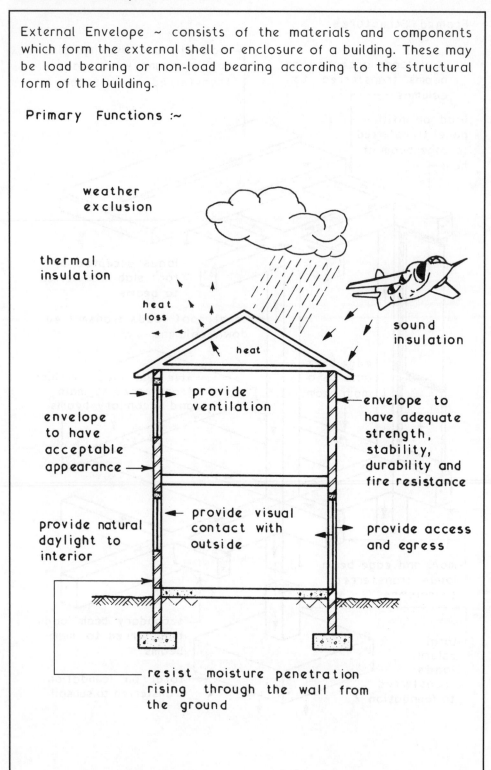

weather exclusion

thermal insulation

heat loss

heat

sound insulation

provide ventilation

envelope to have acceptable appearance

envelope to have adequate strength, stability, durability and fire resistance

provide natural daylight to interior

provide visual contact with outside

provide access and egress

resist moisture penetration rising through the wall from the ground

A Building or Construction Site can be considered as a temporary factory employing the necessary resources to successfully fulfil a contract.

Manpower:~
in the form of
managerial and
supervisory staff.

Manpower:~
in the form of
craftsmen and general
site operatives.

Materials:~
for temporary
works, access
provisions,
security and
final structure.

Plant:~
from the simple
hand held tools
to large items
such as tower
cranes.

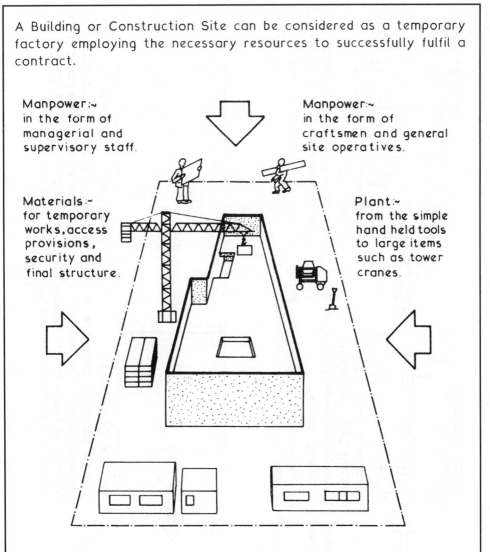

Money :~
in the form of capital investment from the building
owner to pay for the land, design team fees and a
building contractor who uses his money to buy materials,
buy or hire plant and hire labour to enable the
project to be realised.

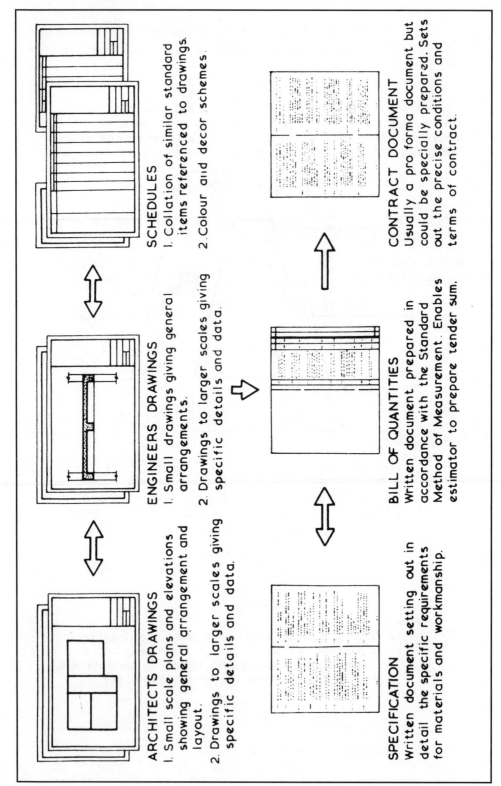

SCHEDULES
1. Collation of similar standard items referenced to drawings.
2. Colour and decor schemes.

CONTRACT DOCUMENT
Usually a pro forma document but could be specially prepared. Sets out the precise conditions and terms of contract.

ENGINEERS DRAWINGS
1. Small drawings giving general arrangements.
2. Drawings to larger scales giving specific details and data.

BILL OF QUANTITIES
Written document prepared in accordance with the Standard Method of Measurement. Enables estimator to prepare tender sum.

ARCHITECTS DRAWINGS
1. Small scale plans and elevations showing general arrangement and layout.
2. Drawings to larger scales giving specific details and data.

SPECIFICATION
Written document setting out in detail the specific requirements for materials and workmanship.

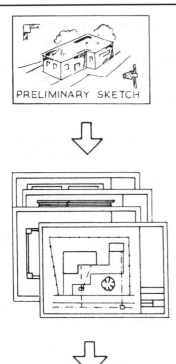

PRELIMINARY SKETCH

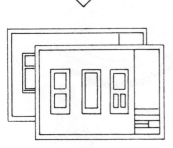

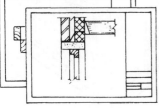

Location Drawings ~

Site Plans — used to locate site, buildings, define site levels, indicate services to buildings, identify parts of site such as roads, footpaths and boundaries and to give setting out dimensions for the site and buildings as a whole. Suitable scale not less than 1:2500

Floor Plans — used to identify and set out parts of the building such as rooms, corridors, doors, windows, etc., Suitable scale not less than 1:100

Elevations — used to show external appearance of all faces and to identify doors and windows. Suitable scale not less than 1:100

Sections — used to provide vertical views through the building to show method of construction. Suitable scale not less than 1:50

Component Drawings ~

used to identify and supply data for components to be supplied by a manufacturer or for components not completely covered by assembly drawings. Suitable scale range 1:100 to 1:1

Assembly Drawings ~

used to show how items fit together or are assembled to form elements. Suitable scale range 1:20 to 1:5

All drawings should be fully annotated, fully dimensioned and cross referenced.

Ref. BS EN ISO 7519: Technical drawings — construction drawings — general principles of presentation for general arrangement and assembly drawings.

Sketch ~ this can be defined as a draft or rough outline of an idea, it can be a means of depicting a three-dimensional form in a two-dimensional guise. Sketches can be produced free-hand or using rules and set squares to give basic guide lines.

All sketches should be clear, show all the necessary detail and above all be in the correct proportions.

Sketches can be drawn by observing a solid object or they can be produced from conventional orthographic views but in all cases can usually be successfully drawn by starting with an outline 'box' format giving length, width and height proportions and then building up the sketch within the outline box.

Example~ Square Based Chimney Pot.

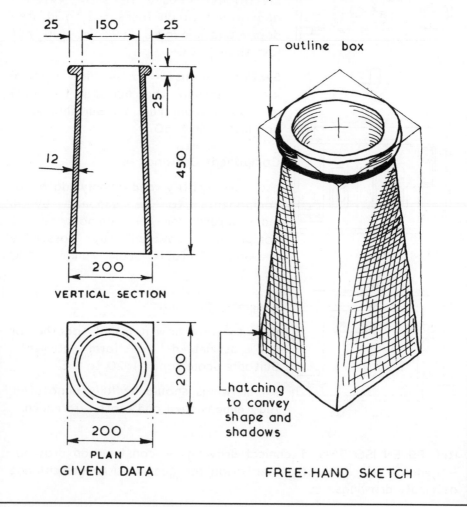

VERTICAL SECTION

PLAN

GIVEN DATA

outline box

hatching
to convey
shape and
shadows

FREE-HAND SKETCH

Return this card today and enter £100 book draw

Select the subjects you'd like to receive information about, enter your email and mail address and freepost it back to us.

TECHNOLOGY
Architecture and Design:
- Architecture and Design: ○
- History of architecture ○
- Landscape ○
- Urban design ○
- Sustainable architecture ○
- Planning and design ○

☐ **Building and Construction**

Computing: Professional:
- Communications ○
- Data Management ○
- Enterprise Computing ○
- IT Management ○
- Operating Systems ○

☐ **Computing: Beginner:**
- Computing ○
- Programming ○

☐ **Conservation and Museology**

☐ **Engineering:**
- Aeronautical Engineering ○
- Automotive Engineering ○
- Chemical Engineering ○

- Health & Safety ○
- Environmental Engineering ○
- Plant / Maintenance / Manufacturing ○
- Marine Engineering ○
- Materials Science & Engineering ○
- Mechanical Engineering ○
- Petroleum Engineering ○
- Quality ○

☐ **Electronics and Electrical Engineering:**
- Electrical Engineering ○
- Electronic Engineering ○
- Radio, Audio and TV Technology ○
- Computer Technology ○

☐ **Film, Television, Video & Audio:**
- Audio/Radio ○
- Post Production ○
- Lighting ○
- Theatre Performance ○
- Photography/Imaging ○
- Radio ○

- TV ○
- Film/TV/Video Production ○
- Journalism ○
- Multimedia ○
- Computer Graphics/ Animation ○
- Broadcast Management & Theory ○
- Broadcast & Communications Technology ○

☐ **Security**

MANAGEMENT
- ☐ Finance and Accounting
- ☐ Hospitality, Leisure and Tourism
- ☐ HR and Training
- ☐ Pergamon Flexible Learning
- ☐ Knowledge Management
- ☐ Management
- ☐ Marketing
- ☐ IT Management

Name:

Email address:

Mail address:

Postcode _____ Date _____

Please keep me up to date by ☐ email ☐ post ☐ both

Jo Blackford
Data Co-ordinator
Elsevier
FREEPOST - SCE5435
Oxford
Oxon
OX2 8BR

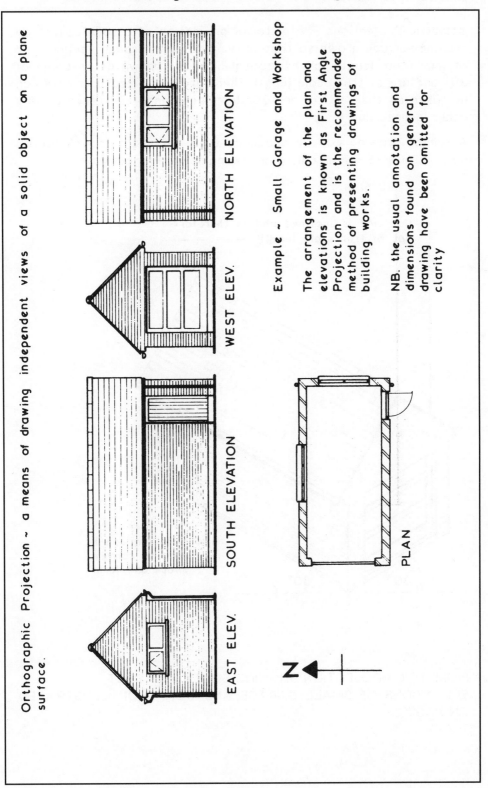

Orthographic Projection ~ a means of drawing independent views of a solid object on a plane surface.

NORTH ELEVATION

WEST ELEV.

SOUTH ELEVATION

EAST ELEV.

PLAN

Example ~ Small Garage and Workshop

The arrangement of the plan and elevations is known as First Angle Projection and is the recommended method of presenting drawings of building works.

NB. the usual annotation and dimensions found on general drawing have been omitted for clarity

Communicating Information—Isometric Projections

Isometric Projections ~ a pictorial projection of a solid object on a plane surface drawn so that all vertical lines remain vertical and of true scale length, all horizontal lines are drawn at an angle of 30° and are of true scale length therefore scale measurements can be taken on the vertical and 30° lines but cannot be taken on any other inclined line.

A similar drawing can be produced using an angle of 45° for all horizontal lines and is called an Axonometric Projection

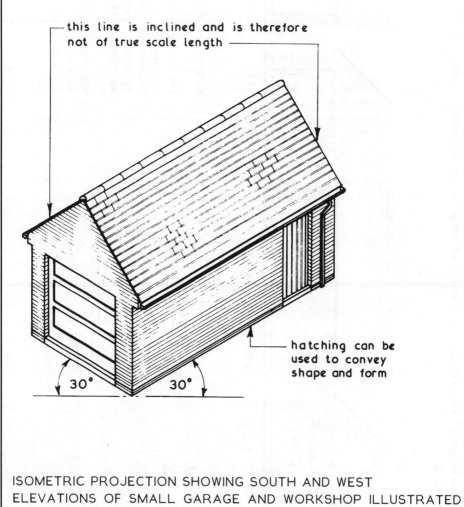

this line is inclined and is therefore not of true scale length

hatching can be used to convey shape and form

30°　　30°

ISOMETRIC PROJECTION SHOWING SOUTH AND WEST ELEVATIONS OF SMALL GARAGE AND WORKSHOP ILLUSTRATED ON PAGE 23

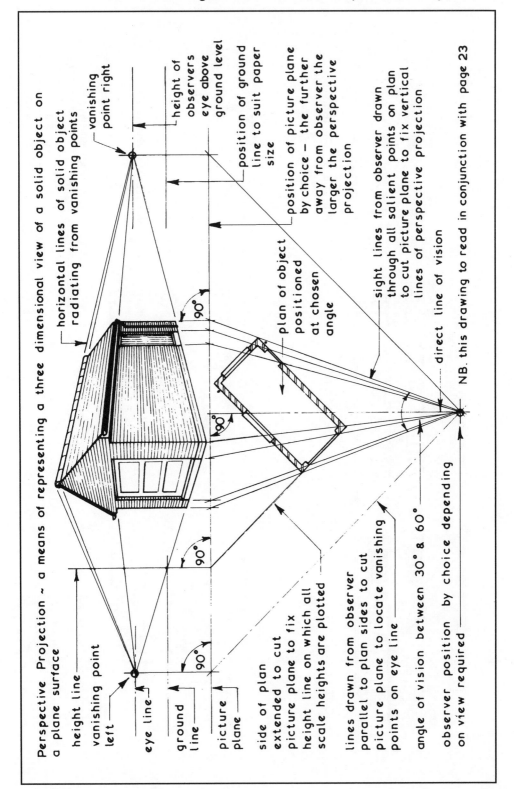

Perspective Projection ~ a means of representing a three dimensional view of a solid object on a plane surface

horizontal lines of solid object radiating from vanishing points

vanishing point right

height of observers eye above ground level

position of ground line to suit paper size

position of picture plane by choice — the further away from observer the larger the perspective projection

plan of object positioned at chosen angle

sight lines from observer drawn through all salient points on plan to cut picture plane to fix vertical lines of perspective projection

direct line of vision

NB. this drawing to read in conjunction with page 23

vanishing point left

eye line

ground line

picture plane

side of plan extended to cut picture plane to fix height line on which all scale heights are plotted

lines drawn from observer parallel to plan sides to cut picture plane to locate vanishing points on eye line

angle of vision between 30° & 60°

observer position by choice depending on view required

height line

90°

90°

90°

90°

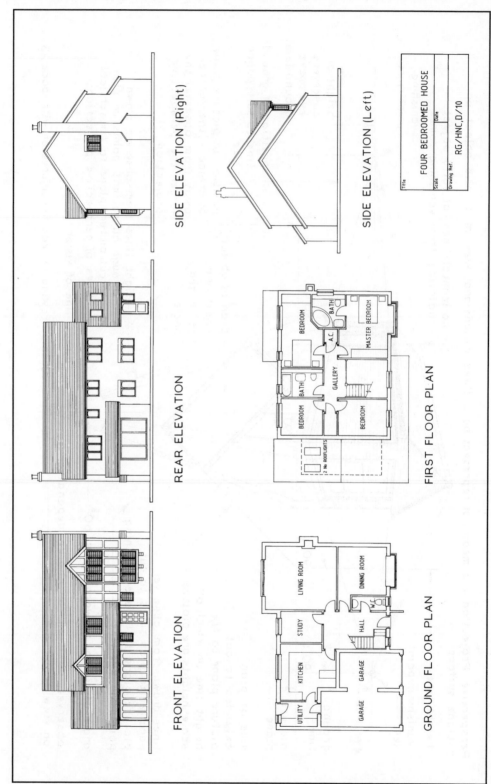

SIDE ELEVATION (Right)

SIDE ELEVATION (Left)

FOUR BEDROOMED HOUSE

Title

Scale | Date

Drawing Ref. | RG/HNC,D/10

REAR ELEVATION

FIRST FLOOR PLAN

BEDROOM

BATH

A.C.

MASTER BEDROOM

BATH

GALLERY

BEDROOM

BEDROOM

2 No ROOFLIGHTS

FRONT ELEVATION

GROUND FLOOR PLAN

LIVING ROOM

DINING ROOM

STUDY

HALL

W.C.

KITCHEN

GARAGE

UTILITY

GARAGE

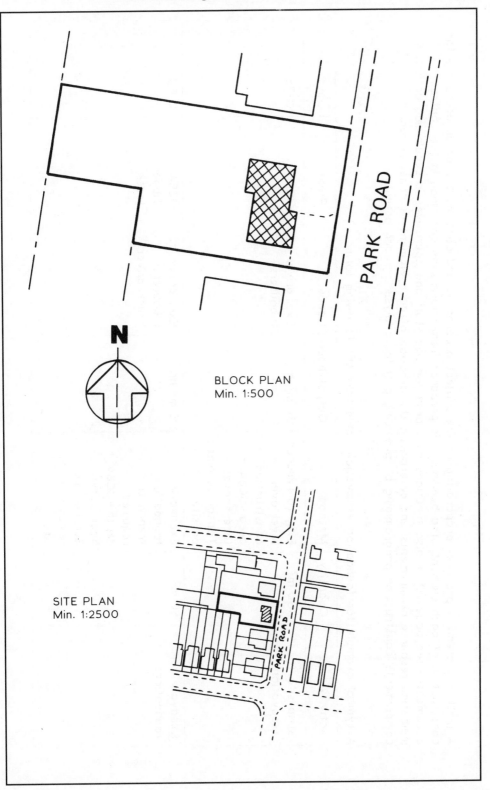

PARK ROAD

N

BLOCK PLAN
Min. 1:500

SITE PLAN
Min. 1:2500

PARK ROAD

A method statement precedes preparation of the project programme and contains the detail necessary for construction of each element of a building. It is prepared from information contained in the contract documents - see page 20. It also functions as a brief for site staff and operatives in sequencing activities, indicating resource requirements and determining the duration of each element of construction. It complements construction programming by providing detailed analysis of each activity.

A typical example for foundation excavation could take the following format:

Activity	Quantity	Method	Output/hour	Labour	Plant	Days
Strip site for excavation	300 m²	Exc. to reduced level over construction area - JCB4 + face shovel. Topsoil retained on site.	50 m²/hr	Exc. driver +2 labourers	JCB4	0·75
Excavate for foundations	60 m³	Excavate foundation trench to required depth - JCB4 + back acter. Surplus spoil removed from site.	15 m³/hr	Exc. driver +2 labourers. Truck driver.	JCB4. Tipper truck.	0·50

Material	Weight (kg/m^2)
BRICKS, BLOCKS and PAVING —	
Clay brickwork — 102·5 mm	
low density	205
medium density	221
high density	238
Calcium silicate brickwork — 102·5 mm	205
Concrete blockwork, aerated	78
.. .. lightweight aggregate	129
Concrete flagstones (50 mm)	115
Glass blocks (100 mm thick) 150×150	98
.. .. 200×200	83
ROOFING —	
Thatching (300 mm thick)	40·00
Tiles — plain clay	63·50
.. — plain concrete	93·00
.. single lap, concrete	49·00
Tile battens (38×20) and felt underlay	5·00
Bituminous felt underlay	1·00
Bituminous felt, sanded topcoat	2·70
3 layers bituminous felt	4·80
SHEET MATERIALS —	
Aluminium (0·9 mm)	2·50
Copper (0·9 mm)	4·88
Cork board (standard) per 25 mm thickness	4·33
.. .. (compressed)	9·65
Hardboard (3.2 mm)	3·40
Glass (3 mm)	7·30
Lead (1·32 mm — code 3)	14·97
.. .. (3·15 mm — code 7)	35·72
Particle board/chipboard (12 mm)	9·26
.. .. (22 mm)	16·82
Planking, softwood strip flooring (ex. 25 mm)	11·20
.. .. hardwood	16·10
Plasterboard (9·5 mm)	8·30
.. .. (12·5 mm)	11·00
.. .. (19 mm)	17·00
Plywood per 25 mm	1·75
PVC floor tiling (2·5 mm)	3·90
Strawboard (25 mm)	9·80
Weatherboarding (20 mm)	7·68
Woodwool (25 mm)	14·50

Typical Weights of Building Materials and Densities

Material	Weight (kg/m²)
INSULATION	
Glass fibre thermal (100 mm)	2·00
.. .. acoustic	4·00
APPLIED MATERIALS -	
Asphalte (18 mm)	42
Plaster, 2 coat work	22
STRUCTURAL TIMBER -	
Rafters and Joists (100×50 @ 400 c/c)	5·87
Floor joists (225×50 @ 400 c/c)	14·93

Densities -

Material	Approx. Density (kg/m³)
Cement	1440
Concrete (aerated)	640
.. .. (broken brick)	2000
.. .. (natural aggregates)	2300
.. .. (no-fines)	1760
.. .. (reinforced)	2400
Metals -	
Aluminium	2770
Copper	8730
Lead	11325
Steel	7849
Timber (softwood/pine)	480 (average)
.. .. .. (hardwood, eg. maple, teak, oak)	720 .. .. ..
Water	1000

Refs. BS 648: Schedule of Weights of Building Materials.
   BS 6399: Pt.1: Code of Practice for Dead
   and Imposed Loads.

Drawings ~ these are the major means of communication between the designer and the contractor as to what, where and how the proposed project is to be constructed.

Drawings should therefore be clear, accurate, contain all the necessary information and be capable of being easily read.

To achieve these objectives most designers use the symbols and notations recommended in BS 1192 and BS 308 to which readers should refer for full information.

Typical Examples~

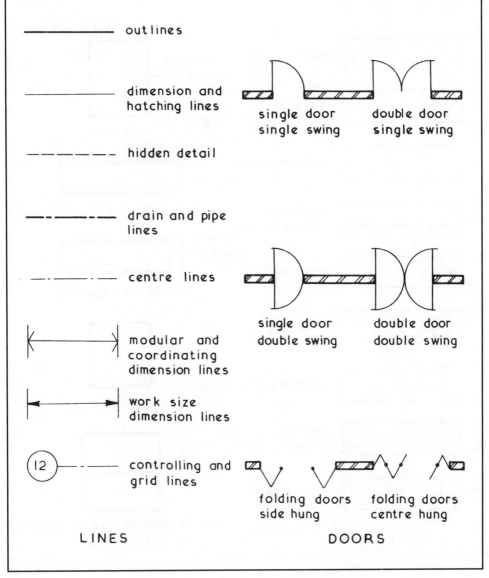

———————	outlines
———————	dimension and hatching lines
— — — — — —	hidden detail
—·——·——	drain and pipe lines
—··———··—	centre lines
↦←————→↤	modular and coordinating dimension lines
↤————→	work size dimension lines
(12)——·——·—	controlling and grid lines

single door single swing   double door single swing

single door double swing   double door double swing

folding doors side hung   folding doors centre hung

LINES   DOORS

Drawings—Hatchings, Symbols and Notations

Hatchings ~ the main objective is to differentiate between the materials being used thus enabling rapid recognition and location. Whichever hatchings are chosen they must be used consistently throughout the whole set of drawings. In large areas it is not always necessary to hatch the whole area.

Symbols ~ these are graphical representations and should wherever possible be drawn to scale but above all they must be consistent for the whole set of drawings and clearly drawn.

Typical Examples~

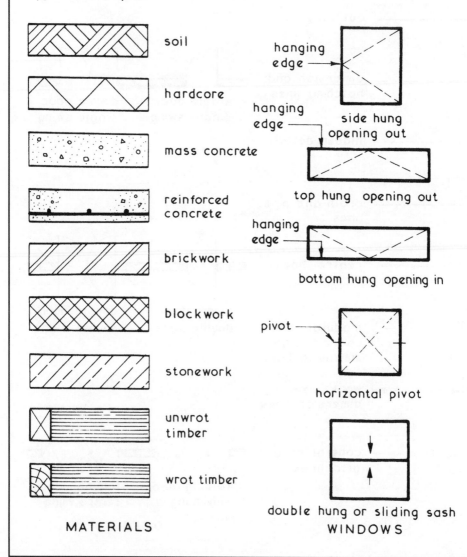

MATERIALS

- soil
- hardcore
- mass concrete
- reinforced concrete
- brickwork
- blockwork
- stonework
- unwrot timber
- wrot timber

WINDOWS

- hanging edge → side hung opening out
- hanging edge → top hung opening out
- hanging edge → bottom hung opening in
- pivot → horizontal pivot
- double hung or sliding sash

Name	Symbol	Name	Symbol
Rainwater pipe	○ R W P	Distribution board	□
Gully	□ G	Electricity meter	⊞
Inspection chambers	IC ┤□┤ soil or foul IC ┤○┤ surface water	Switched socket outlet	◗•
Boiler	□ B	Switch	●
Sink	[S •]	Two way switch	●
Bath	[•]	Pendant switch	●
Wash basin	[W•B]	Filament lamp	○
Shower unit	□ S	Fluorescent lamp	■○■
Urinal	stall bowl	Bed	□
Water closet	▭ ○	Table and chairs	⊡
TYPICAL COMPONENT, FITMENT AND ELECTRICAL SYMBOLS			

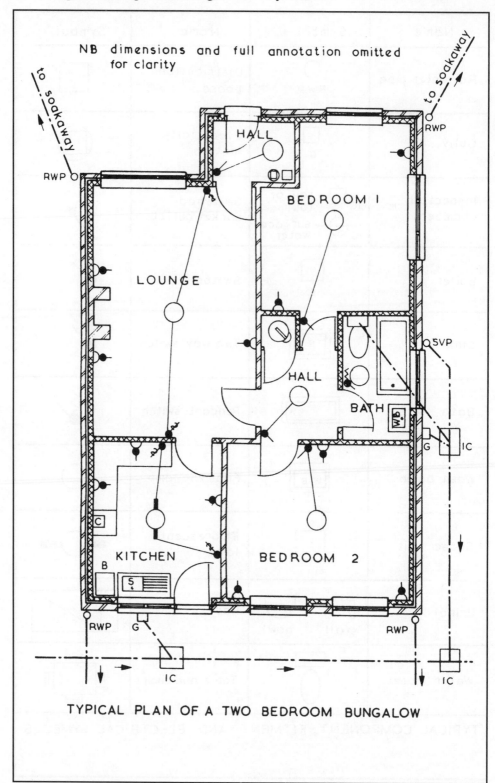

NB dimensions and full annotation omitted for clarity

TYPICAL PLAN OF A TWO BEDROOM BUNGALOW

Principal legislation :~

The Town & Country Planning Act 1990 — Effects control over volume of development, appearance and layout of buildings. The Public Health Acts 1936 to 1961 — limits development with regard to emission of noise, pollution and public nuisance. The Highways Act 1980 — Determines layout and construction of roads and pavements.

The Building Act 1984 — Effects the Building Regulations 1991. which enforce minimum material and design standards. The Civic Amenities Act 1967 — Establishes conservation areas, providing local authorities with greater control of development. The Town & Country Amenities Act 1974 — Local authorities empowered to prevent demolition of buildings and tree felling.

Procedure :~
Outline Planning Application — This is necessary for permission to develop a proposed site. The application should contain :

An application form describing the work.
A site plan showing adjacent roads and buildings (1 : 2500).
A block plan showing the plot, access and siting (1 : 500).
A certificate of land ownership.

Detail or Full Planning Application — This follows outline permission and is also used for proposed alterations to existing buildings.

It should contain : details of the proposal, to include trees, materials, drainage and any demolition.

Site and block plans (as above). A certificate of land ownership. Building drawings showing elevations, sections, plans, material specifications, access, landscaping, boundaries and relationship with adjacent properties (1 : 100).

Permitted Developments — Small developments may be exempt from formal application. These include house extensions <15% of the original volume (vol, in 1948 for older houses), <10% for terraced properties. Other exceptions include porches <2m² floor area, <3m in height and over 2m from the boundary.

Note : All developments are subject to Building Regulation approval.

Certificates of ownership — Article 7 of the Town & Country Planning (General Development Procedure) Order 1995 :

Cert. A — States the applicant is sole site freeholder.

Cert. B — States the applicant is part freeholder and the other owners of the site know of the application.

Cert. C — As Cert. B, but the applicant is only able to ascertain some of the other land owners.

Cert. D — As Cert. B, but the applicant cannot ascertain any owners of the site other than him/herself.

PLANNING APPLICATION

		APPLICATION No
Use this form to apply for Planning Permission for:- • an Extension • a High Wall or Fence • a Loft Conversion • a Garage or Outbuilding • a New or Altered Access • a Satellite Dish	Please return:- • 6 copies of the Form • 6 copies of the Plans • a Certificate under Article 7 • the correct fee	DATE RECEIVED

1. NAME AND ADDRESS OF APPLICANT

Post Code _____

Tel. No. _____

2. NAME AND ADDRESS OF AGENT (If Used)

Post Code _____

Tel. No. _____

3. ADDRESS OF PROPERTY TO BE ALTERED OR EXTENDED

4. OWNERSHIP

Please indicate applicants interest in the property and complete the appropriate Certificate under Article 7.

Freeholder ☐   Other ☐

Leaseholder ☐

Purchaser ☐

5. BRIEF DESCRIPTION OF WORKS (include any demolition work)

6. DESCRIPTION OF EXTERNAL MATERIALS

7. ACCESS AND PARKING

Will your proposal affect?   Please tick appropriate boxes

Vehicular Access   Yes ☐   No ☐

A Public Right of Way   Yes ☐   No ☐

Existing Parking   Yes ☐   No ☐

8. DRAINAGE

a. Please indicate method of Surface Water Disposal

b. Please indicate method of Foul Water Disposal
Please tick one box

Mains Sewer ☐   Septic Tank ☐

Cesspit ☐   Other ☐

9. TREES

Does the proposal involve the felling of any trees?

Please tick box   Yes ☐   No ☐

If yes, please show details on plans

10. PLEASE SIGN AND DATE THIS FORM BEFORE SUBMITTING

I/We hereby apply for Full Planning Permission for the development described above and shown on the accompanying plans.

Signed _____   Date _____

Date

On behalf of (if agent) _____

Use this form to apply for **Planning Permission for:-**
Outline Permission
Full Permission
Approval of Reserved Matters
Renewal of Temporary Permission
Change of Use

Please return:-
   * 6 copies of the Form
   * 6 copies of the Plans
   * a Certificate under
     Article 7
   * the correct fee

DATE RECEIVED

DATE VALID

1. NAME AND ADDRESS OF APPLICANT

_____
_____
_____
_____
_____

Post Code _____

Day Tel. No. _____ Fax No. _____

Email: _____

2. NAME AND ADDRESS OF AGENT (If Used)

_____
_____
_____
_____
_____

Post Code _____

Tel. No. _____ Fax No. _____

Email:_____

3. ADDRESS OR LOCATION OF LAND TO WHICH APPLICATION RELATES.

_____
_____
_____

State Site Area _____ Hectares
This must be shown edged in Red on the site plan

4. OWNERSHIP

Please indicate applicants interest in the property and complete the appropriate Certificate under Article 7.

Freeholder ☐   Other ☐

Leaseholder ☐   Purchaser ☐

Any adjoining land owned or controlled and not part of application must be edged Blue on the site plan

5. WHAT ARE YOU APPLYING FOR?   Please tick one box and then answer relevant questions.

☐ **Outline Planning Permission** Which of the following are to be considered?

☐ Siting ☐ Design ☐ Appearance ☐ Access ☐ Landscaping

☐ **Full Planning Permission/Change of use**

☐ **Approval of Reserved Matters following Outline Permission.**

O/P No. _____ Date_____ No. of Condition this application refers to: _____

☐ **Continuance of Use without complying with a condition of previous permission**

P/P No. _____ Date_____ No. of Condition this application relates to: _____

☐ **Permission for Retention of works.**

Date of Use of land or when buildings or works were constructed: _____ Length of temporary permission: _____

Is the use temporary or permanent? _____ No. of previous temporary permission if applicable: _____

6. BRIEF DESCRIPTION OF PROPOSED DEVELOPMENT.

Please indicate the purpose for which the land or buildings are to be used. _____
_____
_____

Planning Application—New Build (2)

7. NEW RESIDENTIAL DEVELOPMENTS. Please answer the following if appropriate:

What type of building is proposed? _____

No. of dwellings: _____ No. of storeys: _____ No. of Habitable rooms: _____

No. of Garages: _____ No. of Parking Spaces: _____ Total Grass Area of all buildings: _____

How will surface water be disposed of? _____

How will foul sewage be dealt with? _____

8. ACCESS.

Does the proposed development involve any of the following? Please tick the appropriate boxes.

New access to a highway	☐ Pedestrian	☐ Vehicular
Alteration of an existing highway	☐ Pedestrian	☐ Vehicular
The felling of any trees	☐ Yes	☐ No

If you answer Yes to any of the above, they should be clearly indicated on all plans submitted.

9. BUILDING DETAIL

Please give details of all external materials to be used, if you are submitting them at this stage for approval.

_____

_____

List any samples that are being submitted for consideration. _____

_____

10. LISTED BUILDINGS OR CONSERVATION AREA.

Are any Listed buildings to be demolished or altered? ☐ Yes ☐ No

If Yes, then Listed Building Consent will be required and a separate application should be submitted.

Are any non listed buildings within a Conservation Area to be demolished? ☐ Yes ☐ No

If Yes, then Conservation Area consent will be required to demolish. Again, a separate application should be submitted.

11. NOTES.

A special Planning Application Form should be completed for all applications involving Industrial, Warehousing, Storage, or Shopping development.

An appropriate Certificate must accompany this application unless you are seeking approval to Reserved Matters. A separate application for Building Regulation approval is also required.

Separate applications may also be required if the proposals relate to a Listed Building or non-listed building within a Conservation Area.

12. PLEASE SIGN AND DATE THIS FORM BEFORE SUBMITTING.

I/We hereby apply for Planning Permission for the development described above and shown on the accompanying plans.

Signed _____

TOWN AND COUNTRY PLANNING ACT – Article 7

CERTIFICATE A   **For Freehold Owner (or his/her Agent)**

I hereby certify that:-

1.   No person other than the applicant was an owner of any part of the land to which the application relates at the beginning of the period of 21 days before the date of the accompanying application.

2.   ***Either (i)** None of the land to which the application relates constitutes or forms part of an agricultural holding:

       *or (ii) *(I have) (the applicant has) given the requisite notice to every person other than *(myself) (himself) (herself) who, 21 days before the date of the application, was a tenant of any agricultural holding any part of which was comprised in the land to which the application relates, viz:-

Name and Address of Tenant..

....

.... Signed .... Date.........................

Date of Service of Notice.. *On Behalf of ....

CERTIFICATE B   **For Part Freehold Owner or Prospective Purchaser (or his/her Agent) able to ascertain all the owners of the land**

I hereby certify that:-

1   *(I have) (the applicant has) given the requisite notice to all persons other than (myself) (the applicant) who, 21 days before the date of the accompanying application were owners of any part of the land to which the application relates, viz:-

Name and Address of Owner ....

....

.... Date of Service of Notice ....

2.   ***Either (i)** None of the land to which the application relates constitutes or forms part of an agricultural holding;

       *or (ii) *(I have) (the applicant has) given the requisite notice to every person other than *(myself) (himself) (herself) who, 21 days before the date of the application, was a tenant of any agricultural holding any part of which was comprised in the land to which the application relates, viz:-

Name and Address of Tenant..

....

.... Signed .... Date.........................

Modular Coordination ~ a module can be defined as a basic dimension which could for example form the basis of a planning grid in terms of multiples and submultiples of the standard module.

Typical Modular Coordinated Planning Grid ~

Let M = the standard module

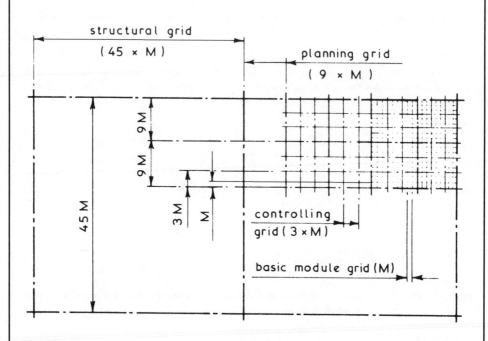

Structural Grid ~ used to locate structural components such as beams and columns.

Planning Grid ~ based on any convenient modular multiple for regulating space requirements such as rooms.

Controlling Grid ~ based on any convenient modular multiple for location of internal walls, partitions etc.

Basic Module Grid ~ used for detail location of components and fittings.

All the above grids, being based on a basic module, are contained one within the other and are therefore interrelated. These grids can be used in both the horizontal and vertical planes thus forming a three dimensional grid system. If a first preference numerical value is given to M dimensional coordination is established — see page 41.

Dimensional Coordination ~ the practical aims of this concept are to :-

1. Size components so as to avoid the wasteful process of cutting and fitting on site.
2. Obtain maximum economy in the production of components.
3. Reduce the need for the manufacture of special sizes.
4. Increase the effective choice of components by the promotion of interchangeability.

BS 6570 specifies the increments of size for coordinating dimensions of building components thus :-

Preference	1st	2nd	3rd	4th
Size (mm)	300	100	50	25

the 3rd and 4th preferences having a maximum of 300 mm

Dimensional Grids — the modular grid network as shown on page 40 defines the space into which dimensionally coordinated components must fit. An important factor is that the component must always be undersized to allow for the joint which is sized by the obtainable degree of tolerance and site assembly :-

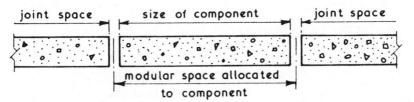

Controlling Lines, Zones and Controlling Dimensions - these terms can best be defined by example :-

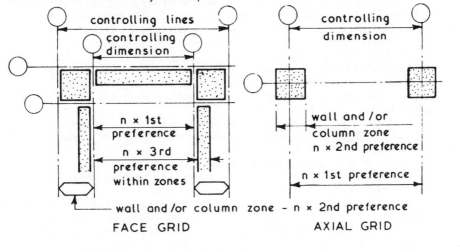

FACE GRID   AXIAL GRID

Construction Regulations

Construction Regulations ~ these are Statutory Instruments made under the Factories Acts of 1937 and 1961 and come under the umbrella of the Health and Safety at work etc., Act 1974. They set out the minimum legal requirements for construction works and relate primarily to the health, safety and welfare of the work force. The requirements contained within these documents must therefore be taken into account when planning construction operations and during the actual construction period. Reference should be made to the relevant document for specific requirements but the broad areas covered can be shown thus:-

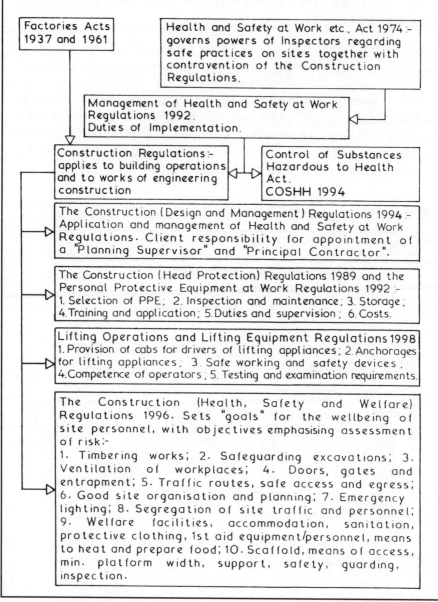

Factories Acts 1937 and 1961

Health and Safety at Work etc., Act 1974 :- governs powers of Inspectors regarding safe practices on sites together with contravention of the Construction Regulations.

Management of Health and Safety at Work Regulations 1992. Duties of Implementation.

Construction Regulations :- applies to building operations and to works of engineering construction

Control of Substances Hazardous to Health Act. COSHH 1994

The Construction (Design and Management) Regulations 1994 :- Application and management of Health and Safety at Work Regulations. Client responsibility for appointment of a "Planning Supervisor" and "Principal Contractor".

The Construction (Head Protection) Regulations 1989 and the Personal Protective Equipment at Work Regulations 1992 :- 1. Selection of PPE; 2. Inspection and maintenance; 3. Storage; 4. Training and application; 5. Duties and supervision; 6. Costs.

Lifting Operations and Lifting Equipment Regulations 1998 1. Provision of cabs for drivers of lifting appliances; 2. Anchorages for lifting appliances; 3. Safe working and safety devices; 4. Competence of operators; 5. Testing and examination requirements.

The Construction (Health, Safety and Welfare) Regulations 1996. Sets "goals" for the wellbeing of site personnel, with objectives emphasising assessment of risk:-
1. Timbering works; 2. Safeguarding excavations; 3. Ventilation of workplaces; 4. Doors, gates and entrapment; 5. Traffic routes, safe access and egress; 6. Good site organisation and planning; 7. Emergency lighting; 8. Segregation of site traffic and personnel; 9. Welfare facilities, accommodation, sanitation, protective clothing, 1st aid equipment/personnel, means to heat and prepare food; 10. Scaffold, means of access, min. platform width, support, safety, guarding, inspection.

Objective — To create an all-party integrated and planned approach to health and safety throughout the duration of a construction project.

Administering Body — The Health and Safety Executive (HSE).

Scope — The CDM Regulations are intended to embrace all aspects of construction, with the exception of very minor works.

Responsibilities — The CDM Regulations apportion responsibility for health and safety issues to all parties involved in the construction process, ie. client, designer, planning supervisor and principal contractor.

Client — Appoints a planning supervisor and the principal contractor. Provides the planning supervisor with information on health and safety matters and ensures that the principal contractor has prepared an acceptable health and safety plan for the conduct of work. Ensures that a health and safety file is available.

Designer — Establishes that the client is aware of their duties. Considers the design implications with regard to health and safety issues, including an assessment of any perceived risks. Co-ordinates the work of the planning supervisor and other members of the design team.

Planning Supervisor — Ensures that:

* a pre-tender, health and safety plan is prepared.
* the HSE are informed of the work.
* designers are liaising and conforming with their health and safety obligations.
* a health and safety file is prepared.
* contractors are of adequate competance with regard to health and safety matters and advises the client and principal contractor accordingly.

Principal Contractor — Develops a health and safety plan, collates relevant information and maintains it as the work proceeds. Administers day-to-day health and safety issues. Co-operates with the planning supervisor and designers, preparing risk assessments as required.

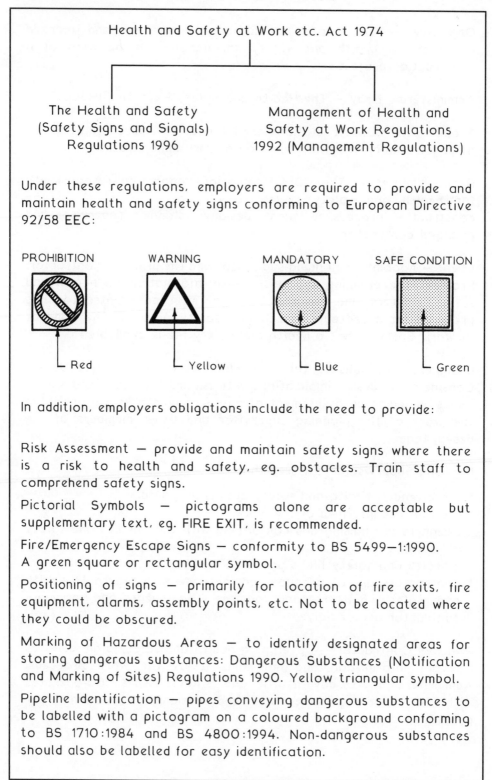

Health and Safety at Work etc. Act 1974

The Health and Safety (Safety Signs and Signals) Regulations 1996

Management of Health and Safety at Work Regulations 1992 (Management Regulations)

Under these regulations, employers are required to provide and maintain health and safety signs conforming to European Directive 92/58 EEC:

PROHIBITION — Red

WARNING — Yellow

MANDATORY — Blue

SAFE CONDITION — Green

In addition, employers obligations include the need to provide:

Risk Assessment — provide and maintain safety signs where there is a risk to health and safety, eg. obstacles. Train staff to comprehend safety signs.

Pictorial Symbols — pictograms alone are acceptable but supplementary text, eg. FIRE EXIT, is recommended.

Fire/Emergency Escape Signs — conformity to BS 5499—1:1990.
A green square or rectangular symbol.

Positioning of signs — primarily for location of fire exits, fire equipment, alarms, assembly points, etc. Not to be located where they could be obscured.

Marking of Hazardous Areas — to identify designated areas for storing dangerous substances: Dangerous Substances (Notification and Marking of Sites) Regulations 1990. Yellow triangular symbol.

Pipeline Identification — pipes conveying dangerous substances to be labelled with a pictogram on a coloured background conforming to BS 1710:1984 and BS 4800:1994. Non-dangerous substances should also be labelled for easy identification.

Typical Examples on Building Sites ~

PROHIBITION (Red)

Authorised personnel only

Children must not play on this site

Smoking prohibited

Access not permitted

WARNING (Yellow)

Dangerous substance

Flammable liquid

Danger of electric shock

Compressed gas

MANDATORY (Blue)

Safety helmets must be worn

Protective footwear must be worn

Use ear protectors

Protective clothing must be worn

SAFE CONDITIONS (Green)

FIRE EXIT

A

Emergency escapes

Treatment area

Safe area

Building Regulations

The Building Regulations ~ this is a Statutory Instrument which sets out the minimum performance standards for the design and construction of buildings and where applicable to the extension of buildings. The regulations are supported by other documents which generally give guidance on how to achieve the required performance standards. The relationship of these and other documents is set out below :-

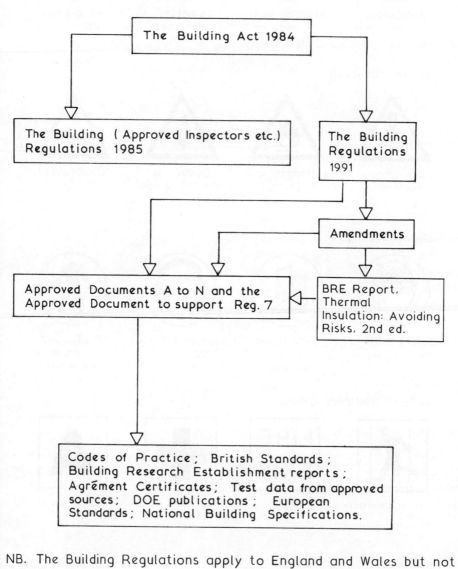

NB. The Building Regulations apply to England and Wales but not to Scotland and Northern Ireland which have separate systems of control.

Approved Documents ~ these are non-statutory publications supporting the Building Regulations prepared by the Department for Transport, Local Government and Regions, approved by the Secretary of State and issued by The Stationery Office. The Approved Documents (ADs) have been compiled to give practical guidance to comply with the performance standards set out in the various regulations. They are not mandatory but in the event of a dispute they will be seen as tending to show compliance with the requirements of the Building Regulations. If other solutions are used to satisfy the requirements of the Regulations the burden of proving compliance rests with the applicant or designer. The various Approved Documents and their titles are set out below :-

Approved Document A - STRUCTURE

Approved Document B - FIRE SAFETY

Approved Document C - SITE PREPARATION AND RESISTANCE TO MOISTURE

Approved Document D - TOXIC SUBSTANCES

Approved Document E - RESISTANCE TO THE PASSAGE OF SOUND

Approved Document F - VENTILATION

Approved Document G - HYGIENE

Approved Document H - DRAINAGE AND WASTE DISPOSAL

Approved Document J - HEAT PRODUCING APPLIANCES

Approved Document K - PROTECTION FROM FALLING, COLLISION AND IMPACT

Approved Document L - CONSERVATION OF FUEL AND POWER

Approved Document M - ACCESS AND FACILITIES FOR DISABLED PEOPLE

Approved Document N - GLAZING - SAFETY IN RELATION TO IMPACT, OPENING AND CLEANING

Approved Document to support Regulation 7 MATERIALS AND WORKMANSHIP

Example in the Use of Approved Documents

Problem :- the sizing of suspended upper floor joists to be spaced at 400 mm centres with a clear span of 3·600 m for use in a two storey domestic dwelling.

Building Regulation A1 :- states that the building shall be constructed so that the combined dead, imposed and wind loads are sustained and transmitted by it to the ground —

(a) safely, and

(b) without causing such deflection or deformation of any part of the building, or such movement of the ground, as will impair the stability of any part of another building.

Approved Document A :- this gives a series of tables the sizing of certain timber members for single family houses of not more than three storeys high. The table which can be applied to the above problem is Table A1 assuming a strength class 3 softwood timber is to be used.

Solution :-

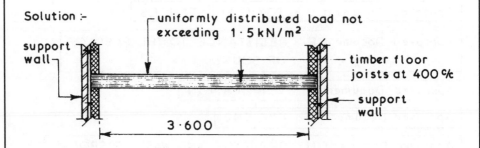

Dead load (kN/m²) supported by joist excluding mass of joist :-

Floor finish — carpet	— 0·03	⎫
Flooring — 20 mm thick particle board	— 0·15	⎬ weights of materials
Ceiling — 9·5 mm thick plasterboard	0·08	⎬ from BS648
Ceiling finish — 3 mm thick plaster	— 0·04	⎭
total dead load —	<u>0·30</u> kN/m³	

Dead loading is therefore in the 0·25 to 0·50 kN/m² band

From table A1 of AD 'A' suitable joist sizes are :- 38×195; 47×195; 50×170; 63×170 and 75×170.

Final choice of section to be used will depend upon cost; availability; practical considerations and/or personal preference.

Building Control ~ unless the applicant has opted for control by a private approved inspector under The Building (Approved Inspectors etc.) Regulations 1985 the control of building works in the context of the Building Regulations is vested in the Local Authority. There are two systems of control namely the Building Notice and the Deposit of Plans. The sequence of systems is shown below : -

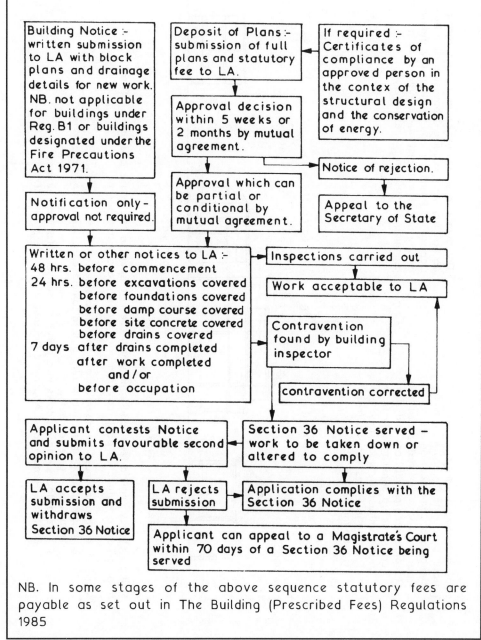

Building Notice :- written submission to LA with block plans and drainage details for new work. NB. not applicable for buildings under Reg. B1 or buildings designated under the Fire Precautions Act 1971.

Deposit of Plans :- submission of full plans and statutory fee to LA.

If required :- Certificates of compliance by an approved person in the contex of the structural design and the conservation of energy.

Approval decision within 5 weeks or 2 months by mutual agreement.

Notification only - approval not required.

Approval which can be partial or conditional by mutual agreement.

Notice of rejection.

Appeal to the Secretary of State

Written or other notices to LA :-
48 hrs. before commencement
24 hrs. before excavations covered
before foundations covered
before damp course covered
before site concrete covered
before drains covered
7 days after drains completed
after work completed
and / or
before occupation

Inspections carried out

Work acceptable to LA

Contravention found by building inspector

contravention corrected

Applicant contests Notice and submits favourable second opinion to LA.

Section 36 Notice served – work to be taken down or altered to comply

LA accepts submission and withdraws Section 36 Notice

LA rejects submission

Application complies with the Section 36 Notice

Applicant can appeal to a Magistrate's Court within 70 days of a Section 36 Notice being served

NB. In some stages of the above sequence statutory fees are payable as set out in The Building (Prescribed Fees) Regulations 1985

Building Regulations Exemptions

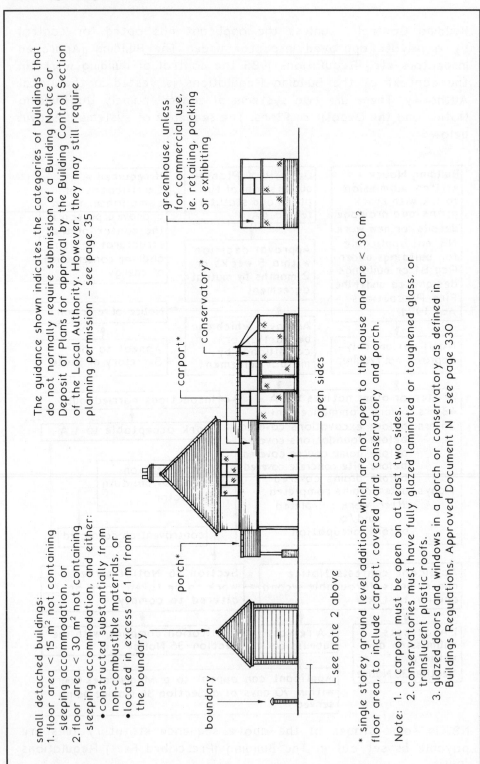

The guidance shown indicates the categories of buildings that do not normally require submission of a Building Notice or Deposit of Plans for approval by the Building Control Section of the Local Authority. However, they may still require planning permission – see page 35

small detached buildings:
1. floor area < 15 m² not containing sleeping accommodation, or
2. floor area < 30 m² not containing sleeping accommodation, and either:
   - constructed substantially from non-combustible materials, or
   - located in excess of 1 m from the boundary

greenhouse, unless for commercial use, ie. retailing, packing or exhibiting

conservatory*

carport*

open sides

porch*

boundary

see note 2 above

* single storey ground level additions which are not open to the house and are < 30 m² floor area, to include carport, covered yard, conservatory and porch.

Note:
1. a carport must be open on at least two sides.
2. conservatories must have fully glazed laminated or toughened glass, or translucent plastic roofs.
3. glazed doors and windows in a porch or conservatory as defined in Buildings Regulations, Approved Document N – see page 330

BUILDING REGULATIONS APPLICATION

APPLICATION No

Use this form to give notice of intention to erect, extend, or alter a building, install fittings or make a material change of use of the building.	Unless specified differently overleaf, Please return:- • 2 copies of the Form • 4 copies of the Plans • the correct fee

DATE RECEIVED

1. NAME AND ADDRESS OF APPLICANT
Applicant will be invoiced on commencement of work.

Post Code _____

Tel. No. _____

2. NAME AND ADDRESS OF AGENT (If Used)

Post Code _____

Tel. No. _____

3. ADDRESS OR LOCATION OF PROPOSED WORK.

4. DESCRIPTION OF PROPOSED WORKS

5. IF NEW BUILDING OR EXTENSION PLEASE STATE PROPOSED USE.

6. IF EXISTING BUILDING PLEASE STATE PRESENT USE.

7. DRAINAGE

Please state means of:-

Water Supply _____

Foul Water Disposal _____

Storm Water Disposal _____

8. CONDITIONS

Do you consent to the Plans being passed subject to conditions where appropriate?   Yes ☐   No ☐

Do you agree to an extension of time if this is required by the Council?   Yes ☐   No ☐

9. COMPLETION CERTIFICATE

Do you wish the Council to issue a Completion Certificate upon satisfactory completion of the work?

Yes ☐   No ☐

10. FIRE PRECAUTIONS ACT 1971

Is the building intended to be used for commercial purposes?

Yes ☐   No ☐

11. FEE
Please state estimated cost of the work (at current market value) £.......................... Amount of Fee submitted £......................

Has Planning Permission been sought?   Yes ☐   No ☐   If Yes, please give Application No _____

12. PLEASE SIGN AND DATE THIS FORM BEFORE SUBMITTING

I/We hereby give notice of intention to carry out the work set out above and deposit the attached drawings and documents in accordance with the requirements of Regulations 11 (1) (b). Also enclosed is the appropriate Plan Fee and I understand that a further Fee will be payable when the first inspection of work on site is made by the Local Authority.

Signed _____ Date _____ On behalf of (if agent) _____

British Standards

British Standards ~ these are publications issued by the British Standards Institution which give recommended minimum standards for materials, components, design and construction practices. These recommendations are not legally enforceable but some of the Building Regulations refer directly to specific British Standards and accept them as deemed to satisfy provisions. All materials and components complying with a particular British Standard are marked with the British Standards kitemark thus :- together with the appropriate BS number.

This symbol assures the user that the product so marked has been produced and tested in accordance with the recommendations set out in that specific standard. Full details of BS products and services can be obtained from, Customer Services, BSI, 389 Chiswick High Road, London, W4 4BR. Standards applicable to building may be purchased individually or as a complete set, under the International Classification for Standards (ICS) ref. 91, CONSTRUCTION MATERIALS AND BUILDING. British Standards are constantly under review and are amended, revised and rewritten as necessary, therefore a check should always be made to ensure that any standard being used is the current issue. There are over 1500 British Standards which are directly related to the construction industry and these are prepared in four formats :-

1. British Standards — these give recommendations for the minimum standard of quality and testing for materials and components. Each standard number is prefixed BS.

2. Codes of Practice — these give recommendations for good practice relative to design, manufacture, construction, installation and maintenance with the main objectives of safety, quality, economy and fitness for the intended purpose. Each code of practice number is prefixed CP or BS.

3. Draft for Development — these are issued instead of a British Standard or Code of Practice when there is insufficient data or information to make firm or positive recommendations. Each draft number is prefixed DD. Sometimes given a BS number and suffixed DC, ie. Draft for public Comment.

4. Published Document — these are publications which cannot be placed into any one of the above categories. Each published document is numbered and prefixed PD.

European Standards — since joining the European Union (EU), trade and tariff barriers have been lifted. This has opened up the market for manufacturers of construction-related products, from all EU and European Economic Area (EEA) member states. The EU is composed of 15 countries; Austria, Belgium, Denmark, Finland, France, Germany, Greece, Ireland, Italy, Luxemburg, Netherlands, Portugal, Spain, Sweden and the United Kingdom. The EEA extends to; Iceland, Liechtenstein and Norway. Nevertheless, the wider market is not so easily satisfied, as regional variations exist. This can create difficulties where product dimensions and performance standards differ. For example, thermal insulation standards for masonry walls in Mediterranean regions need not be the same as those in the UK. Also, preferred dimensions differ across Europe in items such as bricks, timber, tiles and pipes.

European Standards are prepared under the auspices of Comité Européen de Normalisation (CEN), of which the BSI is a member. European Standards that the BSI have not recognized or adopted, are prefixed EN. These are EuroNorms and will need revision for national acceptance.

For the time being, British Standards will continue and where similarity with other countries' standards and EN's can be identified, they will run side by side until harmonisation is complete and approved by CEN.

eg. BS EN 295, replaces the previous national standard
    BS 65 — Vitrefied clay pipes . . . . . for drains and sewers.

European Pre-standards are similar to BS Drafts for
Development. These are known as ENV's.

Some products which satisfy the European requirements for safety, durability and energy efficiency, carry the CE mark. This is not to be assumed a mark of performance and is not intended to show equivalence to the BS kitemark. However, the BSI is recognized as a Notified Body by the EU and as such is authorised to provide testing and certification in support of the CE mark.

International Standards — these are prepared by the International Organisation for Standardisation and are prefixed ISO. Many are compatible with and complement BS's, eg. the ISO 9000 series relates very closely to BS 5750 on quality assurance procedures.

CPI System of Coding

CPI System of Coding ~ the Co-ordinated Project Information initiative originated in the 1970s in response to the need to establish a common arrangement of document and language communication, across the varied trades and professions of the construction industry.

However, it has only been effective in recent years with the publication of the Standard Method of Measurement 7th edition (SMM 7), the National Building Specification (NBS) and the Drawings Code. (Note : The NBS is also produced in CI/SfB format.)

The arrangement in all documents is a co-ordination of alphabetic sections, corresponding to elements of work, the purpose being to avoid mistakes, omissions and other errors which have in the past occurred between drawings, specification and bill of quantities descriptions.

The coding is a combination of letters and numbers, spanning 3 levels : -

Level 1 has 24 headings from A to Z (omitting I and O). Each heading relates to part of the construction process, such as groundwork (D), Joinery (L), surface finishes (M), etc.

Level 2 is a sub-heading, which in turn is sub-grouped numerically into different categories. So for example, Surface Finishes is sub-headed; Plaster, Screeds, Painting, etc. These sub-headings are then extended further, thus Plaster becomes; Plastered/Rendered Coatings, Insulated Finishes, Sprayed Coatings etc.

Level 3 is the work section sub-grouped from level 2, to include a summary of inclusions and omissions.
As an example, an item of work coded M21 signifies : -

> M — Surface finishes
> 2 — Plastered coatings
> 1 — Insulation with rendered finish

The coding may be used to : -

(a) simplify specification writing

(b) reduce annotation on drawings

(c) rationalise traditional taking-off methods

CI/SfB System ~ this is a coded filing system for the classification and storing of building information and data. It was created in Sweden under the title of Samarbetskommittën för Byggnadsfrågor and was introduced into this country in 1961 by the RIBA. In 1968 the CI (Construction Index) was added to the system which is used nationally and recognised throughout the construction industry. The system consists of 5 sections called tables which are subdivided by a series of letters or numbers and these are listed in the CI/SfB index book to which reference should always be made in the first instance to enable an item to be correctly filed or retrieved.

Table 0 — Physical Environment
This table contains ten sections 0 to 9 and deals mainly with the end product (i.e. the type of building.) Each section can be further subdivided {e.g. 21, 22, et seq.) as required.

Table 1 — Elements
This table contains ten sections numbered (—–) to (9—) and covers all parts of the structure such as walls, floors and services. Each section can be further subdivided (e.g. 31, 32 et seq.) as required.

Table 2 — Construction Form
This table contains twenty five sections lettered A to Z (O being omitted) and covers construction forms such as excavation work, blockwork, cast insitu work etc., and is not subdivided but used in conjunction with Table 3.

Table 3 — Materials
This table contains twenty five sections lettered a to z (l being omitted) and covers the actual materials used in the construction form such as metal, timber, glass etc., and can be subdivided (e.g. n1, n2 et seq.) as required.

Table 4 — Activities and Requirements
This table contains twenty five sections lettered (A) to (Z), (O being omitted) and covers anything which results from the building process such as shape, heat, sound, etc. Each section can be further subdivided ((M1), (M2) et seq.) as required.

2 SITE WORKS

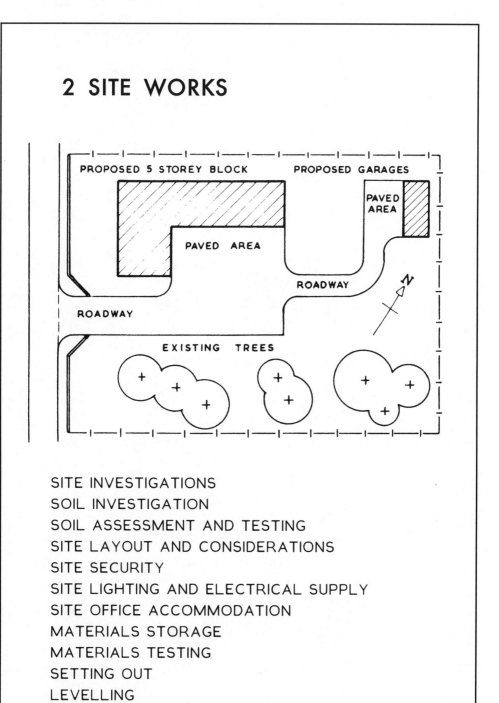

SITE INVESTIGATIONS

SOIL INVESTIGATION

SOIL ASSESSMENT AND TESTING

SITE LAYOUT AND CONSIDERATIONS

SITE SECURITY

SITE LIGHTING AND ELECTRICAL SUPPLY

SITE OFFICE ACCOMMODATION

MATERIALS STORAGE

MATERIALS TESTING

SETTING OUT

LEVELLING

ROAD CONSTRUCTION

TUBULAR SCAFFOLDING AND SCAFFOLDING SYSTEMS

SHORING SYSTEMS

Site Investigation For New Works ~ the basic objective of this form of site investigation is to collect systematically and record all the necessary data which will be needed or will help in the design and construction processes of the proposed work. The collected data should be presented in the form of fully annotated and dimensioned plans and sections. Anything on adjacent sites which may affect the proposed works or conversely anything appertaining to the proposed works which may affect an adjacent site should also be recorded.

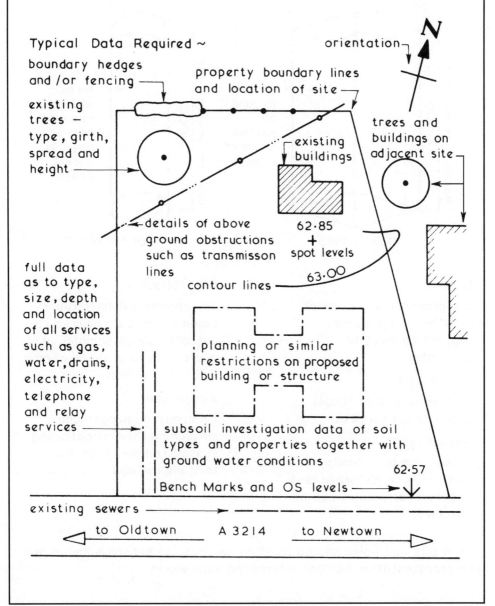

Typical Data Required ~

orientation

boundary hedges and /or fencing

property boundary lines and location of site

existing trees — type, girth, spread and height

existing buildings

trees and buildings on adjacent site

details of above ground obstructions such as transmisson lines

62·85 + spot levels

full data as to type, size, depth and location of all services such as gas, water, drains, electricity, telephone and relay services

contour lines

63·00

planning or similar restrictions on proposed building or structure

subsoil investigation data of soil types and properties together with ground water conditions

62·57

Bench Marks and OS levels

existing sewers

to Oldtown   A 3214   to Newtown

Trial Pits and Hand Auger Holes

Purpose ~ primarily to obtain subsoil samples for identification, classification and ascertaining the subsoil's characteristics and properties. Trial pits and augered holes may also be used to establish the presence of any geological faults and the upper or lower limits of the water table.

Typical Details ~

minimum plan size to provide access for operatives 1·200 x 1·200

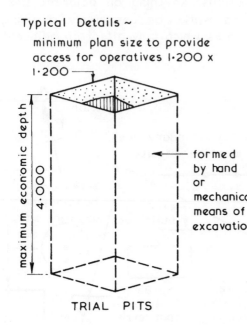

diameter range 50 to 150mm

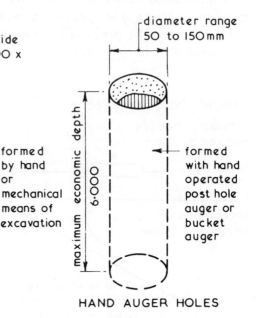

maximum economic depth 4·000 — formed by hand or mechanical means of excavation

maximum economic depth 6·000 — formed with hand operated post hole auger or bucket auger

TRIAL PITS

HAND AUGER HOLES

General use ~

dry ground which requires little or no temporary support to sides of excavation.

Subsidiary use~

to expose and/or locate underground services.

Advantages ~

subsoil can be visually examined insitu — both disturbed and undisturbed samples can be obtained.

General use ~

dry ground but liner tubes could be used if required to extract subsoil samples at a depth beyond the economic limit of trial holes.

Advantages ~

generally a cheaper and simpler method of obtaining subsoil samples than the trial pit method.

Trial pits and holes should be sited so that the subsoil samples will be representative but not interfering with works.

Site Investigation ~ this is an all embracing term covering every aspect of the site under investigation.

Soil Investigation ~ specifically related to the subsoil beneath the site under investigation and could be part of or separate from the site investigation.

Purpose of Soil Investigation ~

1. Determine the suitability of the site for the proposed project.
2. Determine an adequate and economic foundation design.
3. Determine the difficulties which may arise during the construction process and period.
4. Determine the occurrence and/or cause of all changes in subsoil conditions.

The above purposes can usually be assessed by establishing the physical, chemical and general characteristics of the subsoil by obtaining subsoil samples which should be taken from positions on the site which are truly representative of the area but are not taken from the actual position of the proposed foundations. A series of samples extracted at the intersection points of a 20.000 square grid pattern should be adequate for most cases.

Soil Samples ~ these can be obtained as disturbed or as undisturbed samples.

Disturbed Soil Samples ~ these are soil samples obtained from boreholes and trial pits. The method of extraction disturbs the natural structure of the subsoil but such samples are suitable for visual grading, establishing the moisture content and some laboratory tests. Disturbed soil samples should be stored in labelled air tight jars.

Undisturbed Soil Samples ~ these are soil samples obtained using coring tools which preserve the natural structure and properties of the subsoil. The extracted undisturbed soil samples are labelled and laid in wooden boxes for dispatch to a laboratory for testing. This method of obtaining soil samples is suitable for rock and clay subsoils but difficulties can be experienced in trying to obtain undisturbed soil samples in other types of subsoil.

The test results of soil samples are usually shown on a drawing which gives the location of each sample and the test results in the form of a hatched legend or section.

Soil Investigation

Depth of Soil Investigation ~ before determining the actual method of obtaining the required subsoil samples the depth to which the soil investigation should be carried out must be established. This is usually based on the following factors —

1. Proposed foundation type.
2. Pressure bulb of proposed foundation.
3. Relationship of proposed foundation to other foundations.

Typical Examples ~

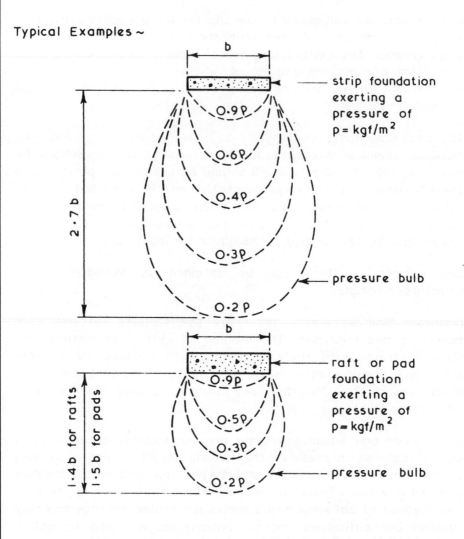

Pressure bulbs of less than 20% of original loading at foundation level can be ignored — this applies to all foundation types.

For further examples see page 63.

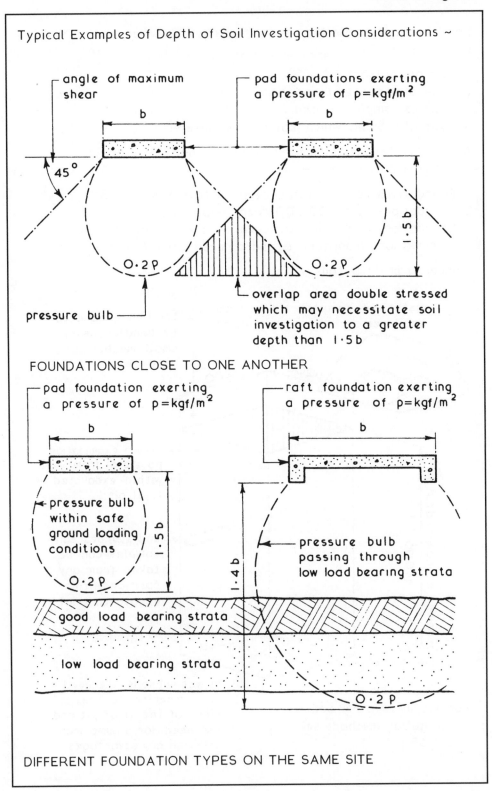

Typical Examples of Depth of Soil Investigation Considerations ~

angle of maximum shear

pad foundations exerting a pressure of $p=kgf/m^2$

b

b

45°

1.5 b

0.2p

0.2p

pressure bulb

overlap area double stressed which may necessitate soil investigation to a greater depth than 1.5 b

FOUNDATIONS CLOSE TO ONE ANOTHER

pad foundation exerting a pressure of $p=kgf/m^2$

raft foundation exerting a pressure of $p=kgf/m^2$

b

b

pressure bulb within safe ground loading conditions

1.5 b

0.2p

pressure bulb passing through low load bearing strata

1.4 b

good load bearing strata

low load bearing strata

0.2p

DIFFERENT FOUNDATION TYPES ON THE SAME SITE

63

Soil Investigation

Soil Investigation Methods ~ method chosen will depend on several factors —

1. Size of contract
2. Type of proposed foundation.
3. Type of sample required.
4. Type of subsoils which may be encountered.

As a general guide the most suitable methods in terms of investigation depth are —

1. Foundations up to   3·000 deep   — trial pits.
2. Foundations up to 30·000 deep   — borings.
3. Foundations over   30·000 deep   — deep borings and insitu examinations from tunnels and/or deep pits.

Typical Trail Pit Details ~

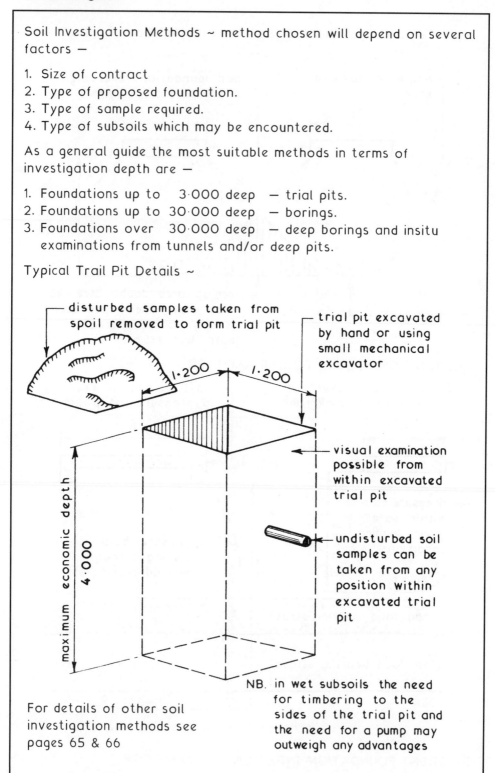

disturbed samples taken from spoil removed to form trial pit

trial pit excavated by hand or using small mechanical excavator

1·200   1·200

maximum economic depth   4·000

visual examination possible from within excavated trial pit

undisturbed soil samples can be taken from any position within excavated trial pit

NB. in wet subsoils the need for timbering to the sides of the trial pit and the need for a pump may outweigh any advantages

For details of other soil investigation methods see pages 65 & 66

Boring Methods to Obtain Disturbed Soil Samples ~

1. Hand or Mechanical Auger — suitable for depths up to 3·000 using a 150 or 200mm diameter flight auger.
2. Mechanical Auger — suitable for depths over 3·000 using a flight or Cheshire auger — a liner or casing is required for most granular soils and may be required for other types of subsoil.
3. Sampling Shells — suitable for shallow to medium depth borings in all subsoils except rock.

Typical Details ~

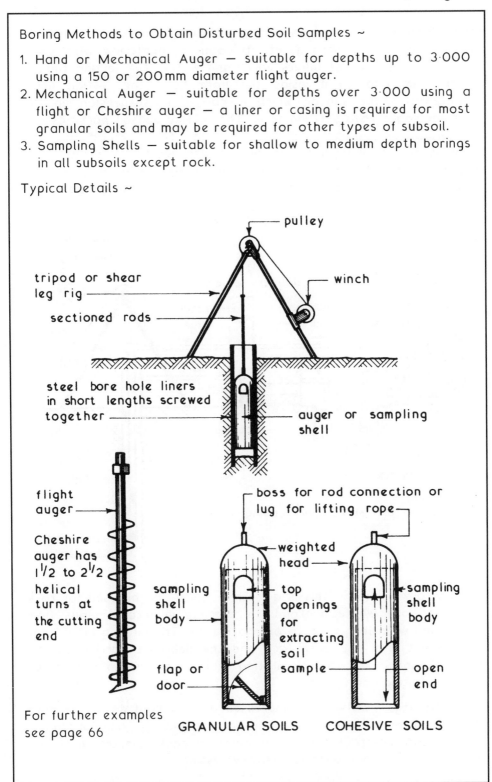

pulley

tripod or shear leg rig

winch

sectioned rods

steel bore hole liners in short lengths screwed together

auger or sampling shell

flight auger

Cheshire auger has 1½ to 2½ helical turns at the cutting end

boss for rod connection or lug for lifting rope

weighted head

sampling shell body

top openings for extracting soil sample

sampling shell body

flap or door

open end

For further examples see page 66

GRANULAR SOILS   COHESIVE SOILS

Wash Boring ~ this is a method of removing loosened soil from a bore hole using a strong jet of water or bentonite which is a controlled mixture of fullers earth and water. The jetting tube is worked up and down inside the bore hole, the jetting liquid disintegrates the subsoil which is carried in suspension up the annular space to a settling tank. The settled subsoil particles can be dried for testing and classification. This method has the advantage of producing subsoil samples which have not been disturbed by the impact of sampling shells however it is not suitable for large gravel subsoils or subsoils which contain boulders.

Typical Wash Boring Arrangement ~

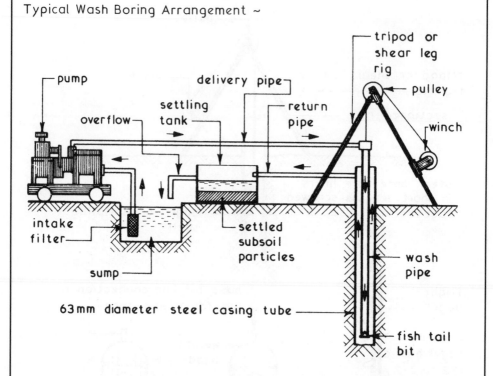

Mud-rotary Drilling ~ this is a method which can be used for rock investigations where bentonite is pumped in a continuous flow down hollow drilling rods to a rotating bit. The cutting bit is kept in contact with the bore face and the debris is carried up the annular space by the circulating fluid. Core samples can be obtained using coring tools.

Core Drilling ~ water or compressed air is jetted down the bore hole through a hollow tube and returns via the annular space. Coring tools extract continuous cores of rock samples which are sent in wooden boxes for laboratory testing.

Bore Hole Data ~ the information obtained from trial pits or bore holes can be recorded on a pro forma sheet or on a drawing showing the position and data from each trial pit or bore hole thus:-

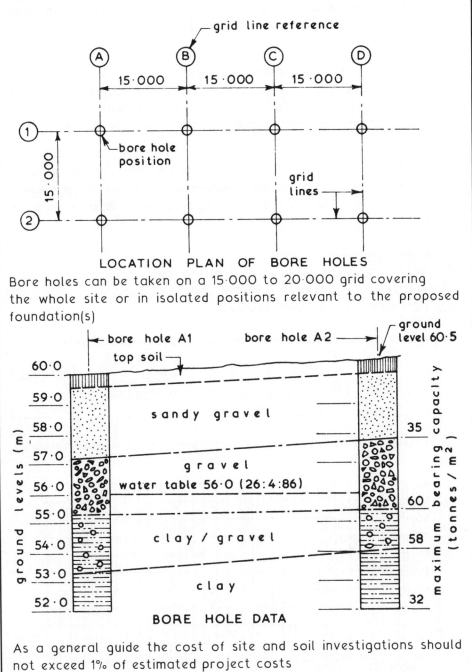

LOCATION PLAN OF BORE HOLES

Bore holes can be taken on a 15·000 to 20·000 grid covering the whole site or in isolated positions relevant to the proposed foundation(s)

BORE HOLE DATA

As a general guide the cost of site and soil investigations should not exceed 1% of estimated project costs

Soil Assessment ~ prior to designing the foundations for a building or structure the properties of the subsoil(s) must be assessed. These processes can also be carried out to confirm the suitability of the proposed foundations. Soil assessment can include classification, grading, tests to establish shear strength and consolidation. The full range of methods for testing soils is given in BS 1377.

Classification ~ soils may be classified in many ways such as geological origin, physical properties, chemical composition and particle size. It has been found that the particle size and physical properties of a soil are closely linked and are therefore of particular importance and interest to a designer.

Particle Size Distribution ~ this is the percentages of the various particle sizes present in a soil sample as determined by sieving or sedimentation. BS 1377 divides particle sizes into groups as follows:-

Gravel particles — over 2mm
Sand particles — between 2mm and 0·06mm
Silt particles — between 0·06mm and 0·002mm
Clay particles — less than 0·002mm

The sand and silt classifications can be further divided thus:-

CLAY	SILT			SAND			GRAVEL
	fine	medium	coarse	fine	medium	coarse	
0·002	0·006	0·02	0·06	0·2	0·6	2	

The results of a sieve analysis can be plotted as a grading curve thus:-

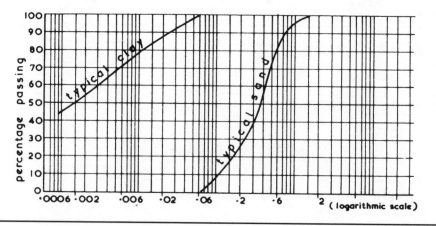

Triangular Chart ~ this provides a general classification of soils composed predominantly from clay, sand and silt. Each side of the triangle represents a percentage of material component. Following laboratory analysis, a sample's properties can be graphically plotted on the chart and classed accordingly.

e.g. Sand — 70%, Clay — 10% and Silt — 20% = Sandy Loam.

Note:

Silt is very fine particles of sand, easily suspended in water.
Loam is very fine particles of clay, easily dissolved in water.

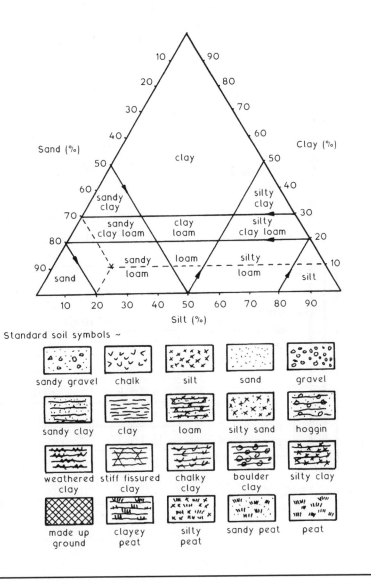

Standard soil symbols ~

sandy gravel	chalk	silt	sand	gravel
sandy clay	clay	loam	silty sand	hoggin
weathered clay	stiff fissured clay	chalky clay	boulder clay	silty clay
made up ground	clayey peat	silty peat	sandy peat	peat

Site Soil Tests ~ these tests are designed to evaluate the density or shear strength of soils and are very valuable since they do not disturb the soil under test. Three such tests are the standard penetration test, the vane test and the unconfined compression test all of which are fully described in BS 1377: Methods of test for soils for civil engineering purposes.

Standard Penetration Test ~ this test measures the resistance of a soil to the penetration of a split spoon or split barrel sampler driven into the bottom of a bore hole. The sampler is driven into the soil to a depth of 150 mm by a falling standard weight of 65 kg falling through a distance of 760 mm. The sampler is then driven into the soil a further 300 mm and the number of blows counted up to a maximum of 50 blows. This test establishes the relative density of the soil.

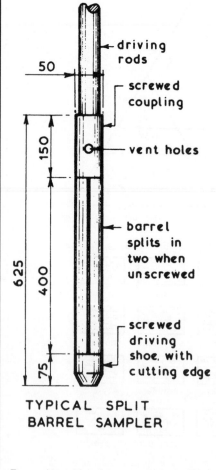

TYPICAL SPLIT BARREL SAMPLER

TYPICAL RESULTS
Non cohesive soils:-

No. of Blows	Relative Density
0 to 4	very loose
4 to 10	loose
10 to 30	medium
30 to 50	dense
50+	very dense

Cohesive soils:-

No of Blows	Relative Density
0 to 2	very soft
2 to 4	soft
4 to 8	medium
8 to 15	stiff
15 to 30	very stiff
30+	hard

The results of this test in terms of number of blows and amounts of penetration will need expert interpretation.

For other tests see pages 71 & 72

Vane Test ~ this test measures the shear strength of soft cohesive soils. The steel vane is pushed into the soft clay soil and rotated by hand at a constant rate. The amount of torque necessary for rotation is measured and the soil shear strength calculated as shown below.

This test can be carried out within a lined bore hole where the vane is pushed into the soil below the base of the bore hole for a distance equal to three times the vane diameter before rotation commences. Alternatively the vane can be driven or jacked to the required depth, the vane being protected within a special protection shoe, the vane is then driven or jacked a further 500 mm before rotation commences.

Calculation of Shear Strength —

$$\text{Formula} :- \qquad S = \frac{M}{K}$$

where S = shear value in kN/m²

M = torque required to shear soil

K = constant for vane

= $3.66\, D^3 \times 10^{-6}$

D = vane diameter

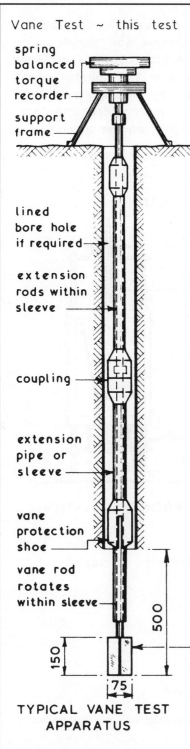

spring balanced torque recorder

support frame

lined bore hole if required

extension rods within sleeve

coupling

extension pipe or sleeve

vane protection shoe

vane rod rotates within sleeve

500

150

75

TYPICAL VANE TEST APPARATUS

stainless steel vane blades forming a cruciform in plan - height of vane to be equal to twice vane diameter

Unconfined Compression Test ~ this test can be used to establish the shear strength of a non-fissured cohesive soil sample using portable apparatus either on site or in a laboratory. The 75mm long×38mm diameter soil sample is placed in the apparatus and loaded in compression until failure occurs by shearing or lateral bulging. For accurate reading of the trace on the recording chart a transparent viewfoil is placed over the trace on the chart.

Typical Apparatus Details~

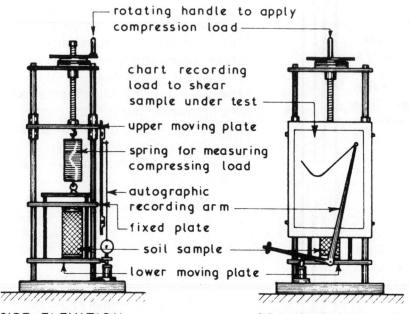

rotating handle to apply compression load

chart recording load to shear sample under test

upper moving plate

spring for measuring compressing load

autographic recording arm

fixed plate

soil sample

lower moving plate

SIDE ELEVATION

FRONT ELEVATION

Typical Results ~ showing compression strengths of clays:-

Very soft clay	— less than 25 kN/m^2
Soft clay	— 25 to 50 kN/m^2
Medium clay	— 50 to 100 kN/m^2
Stiff clay	— 100 to 200 kN/m^2
Very stiff clay	— 200 to 400 kN/m^2
Hard clay	— more than 400 kN/m^2

NB. The shear strength of clay soils is only half of the compression strength values given above.

Laboratory Testing ~ tests for identifying and classifying soils with regard to moisture content, liquid limit, plastic limit, particle size distribution and bulk density are given in BS 1377.

Bulk Density ~ this is the mass per unit volume which includes mass of air or water in the voids and is essential information required for the design of retaining structures where the weight of the retained earth is an important factor.

Shear Strength ~ this soil property can be used to establish its bearing capacity and also the pressure being exerted on the supports in an excavation. The most popular method to establish the shear strength of cohesive soils is the Triaxial Compression Test. In principle this test consists of subjecting a cylindrical sample of undisturbed soil (75mm long × 38mm diameter) to a lateral hydraulic pressure in addition to a vertical load. Three tests are carried out on three samples (all cut from the same large sample) each being subjected to a higher hydraulic pressure before axial loading is applied. The results are plotted in the form of Mohr's circles.

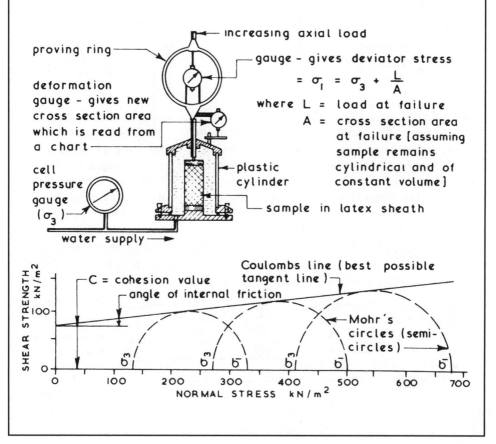

increasing axial load

proving ring

deformation gauge - gives new cross section area which is read from a chart

cell pressure gauge (σ_3)

water supply →

gauge - gives deviator stress

$= \sigma_1 = \sigma_3 + \dfrac{L}{A}$

where L = load at failure
A = cross section area at failure [assuming sample remains cylindrical and of constant volume]

plastic cylinder

sample in latex sheath

C = cohesion value
angle of internal friction

Coulombs line (best possible tangent line)

Mohr's circles (semi-circles)

SHEAR STRENGTH kN/m²

NORMAL STRESS kN/m²

Shear Strength ~ this can be defined as the resistance offered by a soil to the sliding of one particle over another. A simple method of establishing this property is the Shear Box Test in which the apparatus consists of two bottomless boxes which are filled with the soil sample to be tested. A horizontal shearing force (S) is applied against a vertical load (W) causing the soil sample to shear along a line between the two boxes.

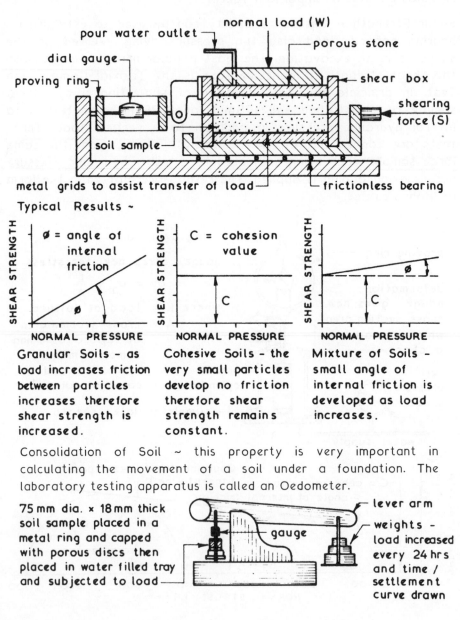

Typical Results ~

Granular Soils – as load increases friction between particles increases therefore shear strength is increased.

Cohesive Soils – the very small particles develop no friction therefore shear strength remains constant.

Mixture of Soils – small angle of internal friction is developed as load increases.

Consolidation of Soil ~ this property is very important in calculating the movement of a soil under a foundation. The laboratory testing apparatus is called an Oedometer.

75 mm dia. × 18 mm thick soil sample placed in a metal ring and capped with porous discs then placed in water filled tray and subjected to load

General Considerations ~ before any specific considerations and decisions can be made regarding site layout a general appreciation should be obtained by conducting a thorough site investigation at the pre-tender stage and examining in detail the drawings, specification and Bill of Quantities to formulate proposals of how the contract will be carried out if the tender is successful. This will involve a preliminary assessment of plant, materials and manpower requirements plotted against the proposed time scale in the form of a bar chart.

Access Considerations ~ this must be considered for both on and off site access. Routes to and from the site must be checked as to the suitability for transporting all the requirements for the proposed works. Access on site for deliveries and general circulation must also be carefully considered.

Typical Site Access Considerations ~

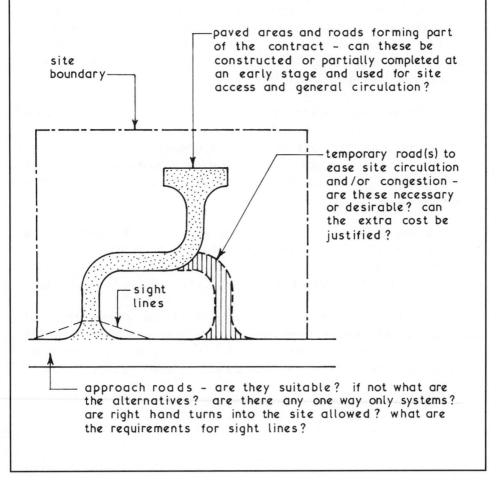

site boundary

paved areas and roads forming part of the contract - can these be constructed or partially completed at an early stage and used for site access and general circulation?

temporary road(s) to ease site circulation and/or congestion - are these necessary or desirable? can the extra cost be justified?

sight lines

approach roads - are they suitable? if not what are the alternatives? are there any one way only systems? are right hand turns into the site allowed? what are the requirements for sight lines?

Storage Considerations ~ amount and types of material to be stored, security and weather protection requirements, allocation of adequate areas for storing materials and allocating adequate working space around storage areas as required, siting of storage areas to reduce double handling to a minimum without impeding the general site circulation and/or works in progress.

Accommodation Considerations ~ number and type of site staff anticipated, calculate size and select units of accommodation and check to ensure compliance with the minimum requirements of the Construction (Health, Safety and Welfare) Regulations 1996, select siting for offices to give easy and quick access for visitors but at the same time giving a reasonable view of the site, select siting for messroom and toilets to reduce walking time to a minimum without impeding the general site circulation and/or works in progress.

Temporary Services Considerations ~ what, when and where are they required? Possibility of having permanent services installed at an early stage and making temporary connections for site use during the construction period, coordination with the various service undertakings is essential.

Plant Considerations ~ what plant, when and where is it required ? static or mobile plant? If static select the most appropriate position and provide any necessary hard standing, if mobile check on circulation routes for optimum efficiency and suitability, provision of space and hard standing for on site plant maintenance if required.

Fencing and Hoarding Considerations ~ what is mandatory and what is desirable? Local vandalism record, type or types of fence and/or hoarding required, possibility of using fencing which is part of the contract by erecting this at an early stage in the contract

Safety and Health Considerations ~ check to ensure that all the above conclusions from the considerations comply with the minimum requirements set out in the various Construction Regulations and in the Health and Safety at Work etc., Act 1974.

For a typical site layout example see page 77.

Typical Site Layout Example ~

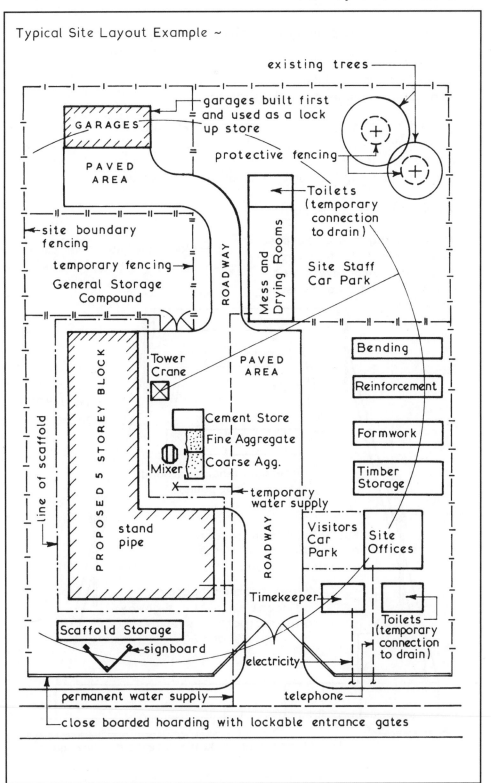

existing trees

garages built first and used as a lock up store

GARAGES

PAVED AREA

protective fencing

site boundary fencing

temporary fencing

General Storage Compound

ROADWAY

Mess and Drying Rooms

Toilets (temporary connection to drain)

Site Staff Car Park

Bending

Reinforcement

Tower Crane

PAVED AREA

Cement Store

Fine Aggregate

Coarse Agg.

Mixer

Formwork

Timber Storage

line of scaffold

PROPOSED 5 STOREY BLOCK

stand pipe

temporary water supply

ROADWAY

Visitors Car Park

Site Offices

Scaffold Storage

signboard

Timekeeper

Toilets (temporary connection to drain)

electricity

permanent water supply

telephone

close boarded hoarding with lockable entrance gates

Site Security ~ the primary objectives of site security are —

1. Security against theft.
2. Security from vandals.
3. Protection from innocent trespassers.

The need for and type of security required will vary from site to site according to the neighbourhood, local vandalism record and the value of goods stored on site. Perimeter fencing, internal site protection and night security may all be necessary.

Typical Site Security Provisions ~

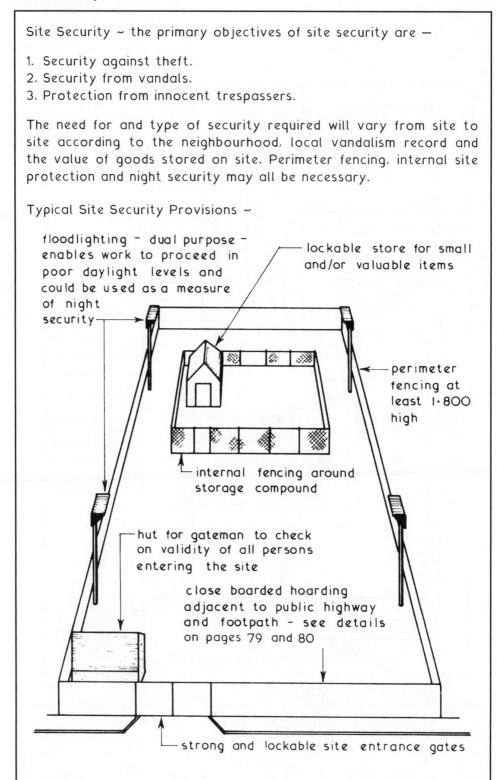

floodlighting — dual purpose — enables work to proceed in poor daylight levels and could be used as a measure of night security

lockable store for small and/or valuable items

perimeter fencing at least 1·800 high

internal fencing around storage compound

hut for gateman to check on validity of all persons entering the site

close boarded hoarding adjacent to public highway and footpath — see details on pages 79 and 80

strong and lockable site entrance gates

Hoardings ~ under the Highways Act 1980 a close boarded fence hoarding must be erected prior to the commencement of building operations if such operations are adjacent to a public footpath or highway. The hoarding needs to be adequately constructed to provide protection for the public, resist impact damage, resist anticipated wind pressures and adequately lit at night. Before a hoarding can be erected a licence or permit must be obtained from the local authority who will usually require 10 to 20 days notice. The licence will set out the minimum local authority requirements for hoardings and define the time limit period of the licence.

Typical Hoarding Details ~

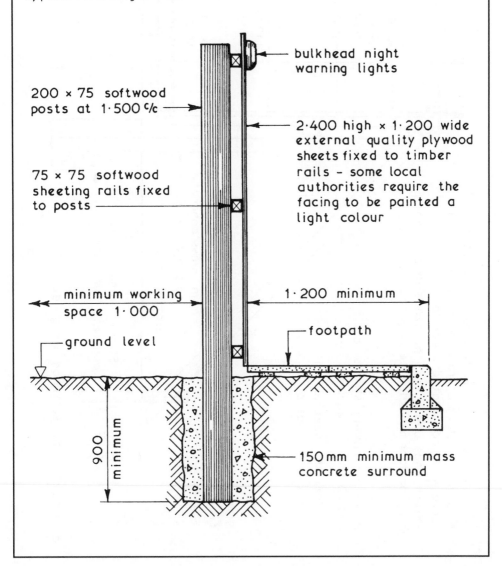

bulkhead night warning lights

200 × 75 softwood posts at 1·500 c/c

2·400 high × 1·200 wide external quality plywood sheets fixed to timber rails – some local authorities require the facing to be painted a light colour

75 × 75 softwood sheeting rails fixed to posts

minimum working space 1·000

1·200 minimum

ground level

footpath

900 minimum

150 mm minimum mass concrete surround

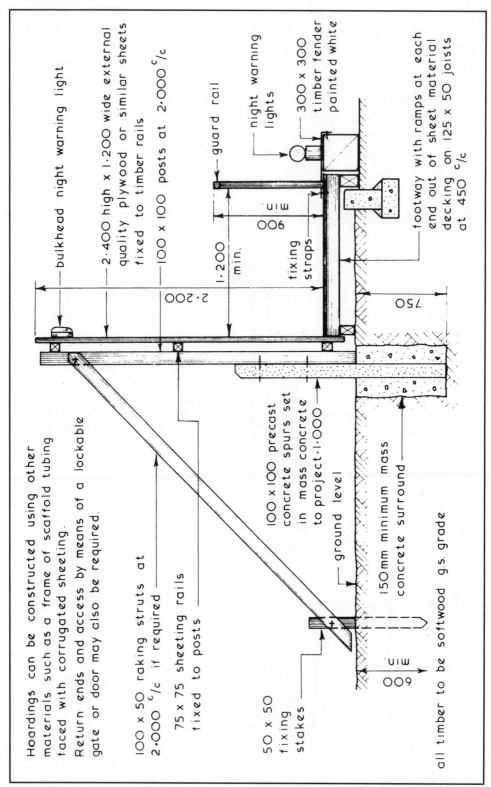

Hoardings can be constructed using other materials such as a frame of scaffold tubing faced with corrugated sheeting.

Return ends and access by means of a lockable gate or door may also be required.

100 x 50 raking struts at 2·000 c/c if required

75 x 75 sheeting rails fixed to posts

50 x 50 fixing stakes

100 x100 precast concrete spurs set in mass concrete to project·1·000

ground level

150mm minimum mass concrete surround

all timber to be softwood g.s. grade

600 min

bulkhead night warning light

2·400 high x 1·200 wide external quality plywood or similar sheets fixed to timber rails

100 x 100 posts at 2·000 c/c

guard rail

night warning lights

300 x 300 timber fender painted white

footway with ramps at each end out of sheet material decking on 125 x 50 joists at 450 c/c

fixing straps

2·200

1·200 min.

900 min

750

Site Lighting ~ this can be used effectively to enable work to continue during periods of inadequate daylight. It can also be used as a deterrent to would-be trespassers. Site lighting can be employed externally to illuminate the storage and circulation areas and internally for general movement and for specific work tasks. The types of lamp available range from simple tungsten filament lamps to tungsten halogen and discharge lamps. The arrangement of site lighting can be static where the lamps are fixed to support poles or mounted on items of fixed plant such as scaffolding and tower cranes. Alternatively the lamps can be sited locally where the work is in progress by being mounted on a movable support or hand held with a trailing lead. Whenever the position of site lighting is such that it can be manhandled it should be run on a reduced voltage of 110 V single phase as opposed to the mains voltage of 230 V.

To plan an adequate system of site lighting the types of activity must be defined and given an illumination target value which is quoted in lux (lx). Recommended minimum target values for building activities are:-

External lighting   —   general circulation ⎫   10 lx
                           materials handling ⎭

Internal lighting   —   general circulation   5 lx
                           general working areas   15 lx
                           concreting activities   50 lx
                           carpentry and joinery ⎫
                           bricklaying   ⎬   100 lx
                           plastering   ⎭
                           painting and decorating ⎫
                           site offices   ⎬   200 lx
                           drawing board positions   300 lx

Such target values do not take into account deterioration, dirt or abnormal conditions therefore it is usual to plan for at least twice the recommended target values. Generally the manufacturers will provide guidance as to the best arrangement to use in any particular situation but lamp requirements can be calculated thus:-

$$\text{Total lumens required} = \frac{\text{area to be illuminated (m}^2\text{)} \times \text{target value (lx)}}{\text{utilisation factor } 0 \cdot 23 \text{ [dispersive lights } 0 \cdot 27\text{]}}$$

After choosing lamp type to be used :-

$$\text{Number of lamps required} = \frac{\text{total lumens required}}{\text{lumen output of chosen lamp}}$$

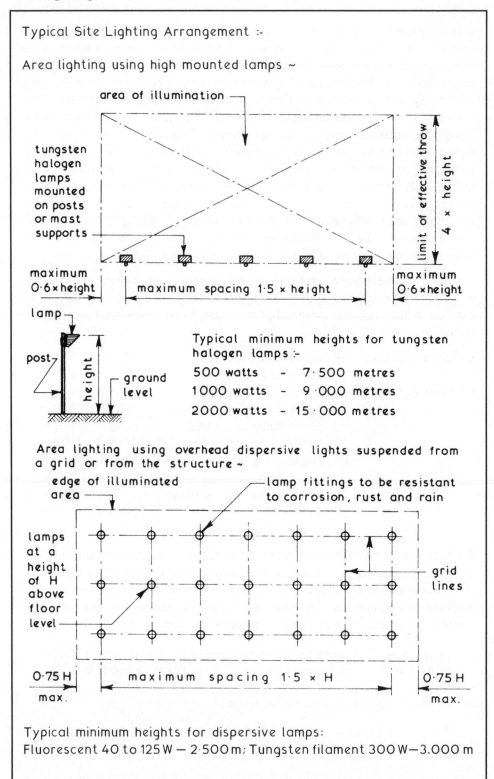

Typical Site Lighting Arrangement :-

Area lighting using high mounted lamps ~

area of illumination

tungsten halogen lamps mounted on posts or mast supports

limit of effective throw 4 × height

maximum 0·6 × height

maximum spacing 1·5 × height

maximum 0·6 × height

lamp

post

height

ground level

Typical minimum heights for tungsten halogen lamps :-

500 watts – 7·500 metres
1000 watts – 9·000 metres
2000 watts – 15·000 metres

Area lighting using overhead dispersive lights suspended from a grid or from the structure ~

edge of illuminated area

lamp fittings to be resistant to corrosion, rust and rain

lamps at a height of H above floor level

grid lines

0·75 H max.

maximum spacing 1·5 × H

0·75 H max.

Typical minimum heights for dispersive lamps:
Fluorescent 40 to 125 W — 2·500 m; Tungsten filament 300 W—3·000 m

Walkway and Local Lighting ~ to illuminate the general circulation routes bulkhead and/or festoon lighting could be used either on a standard mains voltage of 230 V or on a reduced voltage of 110 V. For local lighting at the place of work hand lamps with trailing leads or lamp fittings on stands can be used and positioned to give the maximum amount of illumination without unacceptable shadow cast.

Typical Walkway and Local Lighting Fittings ~

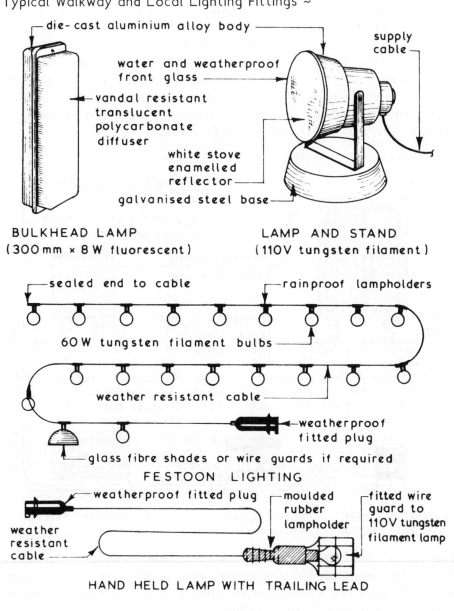

die-cast aluminium alloy body

supply cable

water and weatherproof front glass

vandal resistant translucent polycarbonate diffuser

white stove enamelled reflector

galvanised steel base

BULKHEAD LAMP
(300 mm × 8 W fluorescent)

LAMP AND STAND
(110V tungsten filament)

sealed end to cable

rainproof lampholders

60 W tungsten filament bulbs

weather resistant cable

weatherproof fitted plug

glass fibre shades or wire guards if required

FESTOON LIGHTING

weatherproof fitted plug

moulded rubber lampholder

fitted wire guard to 110V tungsten filament lamp

weather resistant cable

HAND HELD LAMP WITH TRAILING LEAD

Electrical Supply to Building Sites

Electrical Supply to Building Sites ~ a supply of electricity is usually required at an early stage in the contract to provide light and power to the units of accommodation. As the work progresses power could also be required for site lighting, hand held power tools and large items of plant. The supply of electricity to a building site is the subject of a contract between the contractor and the local area electricity company who will want to know the date when supply is required; site address together with a block plan of the site; final load demand of proposed building and an estimate of the maximum load demand in kilowatts for the construction period. The latter can be estimated by allowing $10\,W/m^2$ of the total floor area of the proposed building plus an allowance for high load equipment such as cranes. The installation should be undertaken by a competent electrical contractor to ensure that it complies with all the statutory rules and regulations for the supply of electricity to building sites.

Typical Supply and Distribution Equipment ~

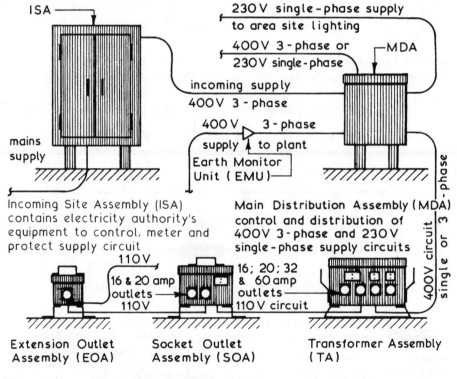

Incoming Site Assembly (ISA) contains electricity authority's equipment to control, meter and protect supply circuit

Main Distribution Assembly (MDA) control and distribution of 400V 3-phase and 230V single-phase supply circuits

Extension Outlet Assembly (EOA)

Socket Outlet Assembly (SOA)

Transformer Assembly (TA)

The units must be strong, durable and resistant to rain penetration with adequate weather seals to all access panels and doors. All plug and socket outlets should be colour coded :- 400V — red; 230V — blue; 110V — yellow

84

Office Accommodation ~ the type of office accommodation to be provided on site is a matter of choice for each individual contractor who can use timber framed huts, prefabricated cabins, mobile offices or even caravans. Generally separate offices would be provided for site agent, clerk of works, administrative staff and site surveyors.

The minimum requirements of such accommodation is governed by the Offices, Shops and Railway Premises Act 1963 unless they are ~

1. Mobile units in use for not more then 6 months.
2. Fixed units in use for not more than 6 weeks.
3. Any type of unit in use for not more than 21 man hours per week.
4. Office for exclusive use of self employed person.
5. Office used by family only staff.

Sizing Example ~

Office for site agent and assistant plus an allowance for 3 visitors. Assume an internal average height of 2·400.
Allow 3·7 m² minimum per person and 11·5 m³ minimum per person.
Minimum area = $5 \times 3·7 = 18·5$ m²
Minimum volume = $5 \times 11·5 = 57·5$ m³

Assume office width of 3·000 then minimum length required is
$$= \frac{57·5}{3 \times 2·4} = \frac{57·5}{7·2} = 7·986 \text{ say } 8·000$$

Area check $3 \times 8 = 24$ m² which is $> 18·5$ m² ∴ satisfactory

Typical Examples ~

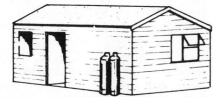

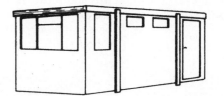

| Timber framed site hut with insulated walls and roof. Wall cladding – painted weatherboard. Hut supplied unequipped. Sizes based on 1·500 module. | Portable cabin with four steel tube jacking legs. Shell – timber framing with plywood facing. Cabin insulated and fully equipped. Wide range of sizes available. |

Site Health and Welfare Requirements

The requirements for health and wellbeing of persons on construction sites are enforced by the Health and Safety Executive, through the Health and Safety at Work etc. Act 1974 and the Construction (Health, Safety and Welfare) Regulations 1996. The following minimum standards were established by the superseded Construction Regulations of 1966.

Provision	Requirement	No of persons employed on site
FIRST AID	Box to be distinctively marked and in charge of responsible person.	5 to 50 — first aid boxes 50 + first aid box and a person trained in first aid
AMBULANCES	Stretcher(s) in charge of responsible person	25 + notify ambulance authority of site details within 24 hours of employing more than 25 persons
FIRST AID ROOM	Used only for rest or treatment and in charge of trained person	If more than 250 persons employed on site each employer of more than 40 persons to provide a first aid room
SHELTER AND ACCOMMODATION FOR CLOTHING	All persons on site to have shelter and a place for depositing clothes	Up to 5 where possible a means of warming themselves and drying wet clothes 5 + adequate means of warming themselves and drying wet clothing
MEALS ROOM	Drinking water, means of boiling water and eating meals for all persons on site	10 + facilities for heating food if hot meals are not available on site
WASHING FACILITIES	Washing facilities to be provided for all persons on site for more than 4 hours	20 to 100 if work is to last more than 6 weeks — hot and cold or warm water, soap and towel. 100 + work lasting more than 12 months — 4 wash places + 1 for every 35 persons over 100
SANITARY FACILITIES	To be maintained, lit and kept clean. Separate facilities for female staff	Up to 100 — 1 convenience for every 25 persons 100 + –1 convenience for every 35 persons

Site Storage ~ materials stored on site prior to being used or fixed may require protection for security reasons or against the adverse effects which can be caused by exposure to the elements.

Small and Valuable Items ~ these should be kept in a secure and lockable store. Similar items should be stored together in a rack or bin system and only issued against an authorised requisition.

Large or Bulk Storage Items ~ for security protection these items can be stored within a lockable fenced compound. The form of fencing chosen may give visual security by being of an open nature but these are generally easier to climb than the close boarded type of fence which lacks the visual security property.

Typical Storage Compound Fencing ~

Close boarded fences can be constructed on the same methods used for hoardings — see pages 79 & 80

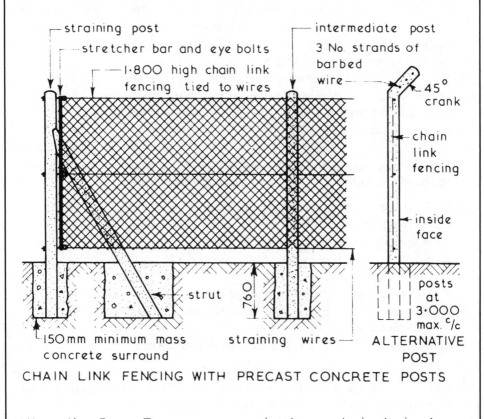

CHAIN LINK FENCING WITH PRECAST CONCRETE POSTS

Alternative Fence Types ~ woven wire fence, strained wire fence, cleft chestnut pale fence, wooden palisade fence, wooden post and rail fence and metal fences — see BS 1722 for details.

Storage of Materials ~ this can be defined as the provision of adequate space, protection and control for building materials and components held on site during the construction process. The actual requirements for specific items should be familiar to students who have completed studies in construction technology at an introductory level but the need for storage and control of materials held on site can be analysed further :-

1. Physical Properties — size, shape, weight and mode of delivery will assist in determining the safe handling and stacking method(s) to be employed on site, which in turn will enable handling and storage costs to be estimated.

2. Organisation — this is the planning process of ensuring that all the materials required are delivered to site at the correct time, in sufficient quantity, of the right quality, the means of unloading is available and that adequate space for storage or stacking has been allocated.

3. Protection — building materials and components can be classified as durable or non-durable, the latter will usually require some form of weather protection to prevent deterioration whilst in store.

4. Security — many building materials have a high resale and/or usage value to persons other than those for whom they were ordered and unless site security is adequate material losses can become unacceptable.

5. Costs — to achieve on economic balance of how much expenditure can be allocated to site storage facilities the following should be taken into account :-

   a. Storage areas, fencing, racks, bins, etc.,
   b. Protection requirements.
   c. Handling, transporting and stacking requirements.
   d. Salaries and wages of staff involved in storage of materials and components.
   e. Heating and/or lighting if required.
   f. Allowance for losses due to wastage, deterioration, vandalism and theft.
   g. Facilities to be provided for sub-contractors.

6. Control — checking quality and quantity of materials at delivery and during storage period, recording delivery and issue of materials and monitoring stock holdings.

Site Storage Space ~ the location and size(s) of space to be allocated for any particular material should be planned by calculating the area(s) required and by taking into account all the relevant factors before selecting the most appropriate position on site in terms of handling, storage and convenience. Failure to carry out this simple planning exercise can result in chaos on site or having on site more materials than there is storage space available.

Calculation of Storage Space Requirements ~ each site will present its own problems since a certain amount of site space must be allocated to the units of accommodation, car parking, circulation and working areas, therefore the amount of space available for materials storage may be limited. The size of the materials or component being ordered must be known together with the proposed method of storage and this may vary between different sites of similar building activities. There are therefore no standard solutions for allocating site storage space and each site must be considered separately to suit its own requirements.

Typical Examples ~

Bricks — quantity = 15,200 to be delivered in strapped packs of 380 bricks per pack each being 1100 mm wide × 670 mm long × 850 mm high. Unloading and stacking to be by forklift truck to form 2 rows 2 packs high.

Area required :- number of packs per row $= \dfrac{15,200}{380 \times 2} = 20$

length of row = 10 × 670 = 6·700
width of row = 2 × 1100 = 2·200

allowance for forklift approach in front of stack = 5·000
∴ minimum brick storage area = 6·700 long × 7·200 wide

Timber — to be stored in open sided top covered racks constructed of standard scaffold tubes. Maximum length of timber ordered = 5·600. Allow for rack to accept at least 4 No. 300 mm wide timbers placed side by side then minimum width required = 4 × 300 = 1·200
Minimum plan area for timber storage rack = 5·600 × 1·200
Allow for end loading of rack equal to length of rack
∴ minimum timber storage area = 11·200 long × 1·200 wide
Height of rack to be not more than 3 × width = 3·600

Areas for other materials stored on site can be calculated using the basic principles contained in the examples above.

Site Allocation for Materials Storage ~ the area and type of storage required can be determined as shown on pages 88 and 89 but the allocation of an actual position on site will depend on :-

1. Space available after areas for units of accommodation have been allocated.
2. Access facilities on site for delivery, vehicles.
3. Relationship of storage area(s) to activity area(s) — the distance between them needs to be kept as short as possible to reduce transportation needs in terms of time and costs to the minimum. Alternatively storage areas and work areas need to be sited within the reach of any static transport plant such as a tower crane.
4. Security — needs to be considered in the context of site operations, vandalism and theft.
5. Stock holding policy — too little storage could result in delays awaiting for materials to be delivered, too much storage can be expensive in terms of weather and security protection requirements apart from the capital used to purchase the materials stored on site.

Typical Example ~

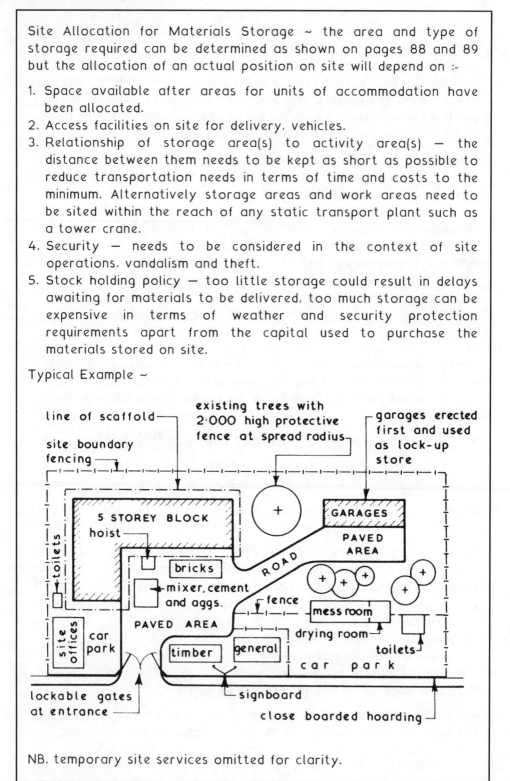

NB. temporary site services omitted for clarity.

Bricks ~ may be supplied loose or strapped in unit loads and stored on timber pallets

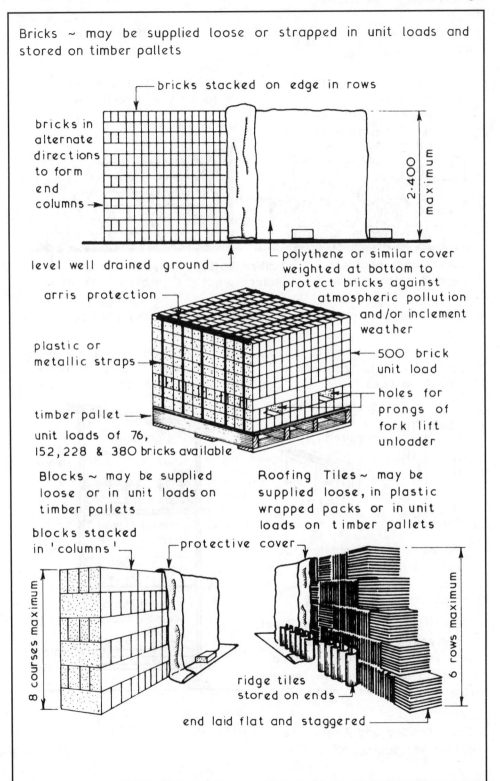

— bricks stacked on edge in rows

bricks in alternate directions to form end columns →

level well drained ground —

2·400 maximum

polythene or similar cover weighted at bottom to protect bricks against atmospheric pollution and/or inclement weather

arris protection —

plastic or metallic straps —

timber pallet —

unit loads of 76, 152, 228 & 380 bricks available

500 brick unit load

holes for prongs of fork lift unloader

Blocks ~ may be supplied loose or in unit loads on timber pallets

blocks stacked in 'columns' —

— protective cover —

8 courses maximum

Roofing Tiles ~ may be supplied loose, in plastic wrapped packs or in unit loads on timber pallets

6 rows maximum

ridge tiles stored on ends —

end laid flat and staggered —

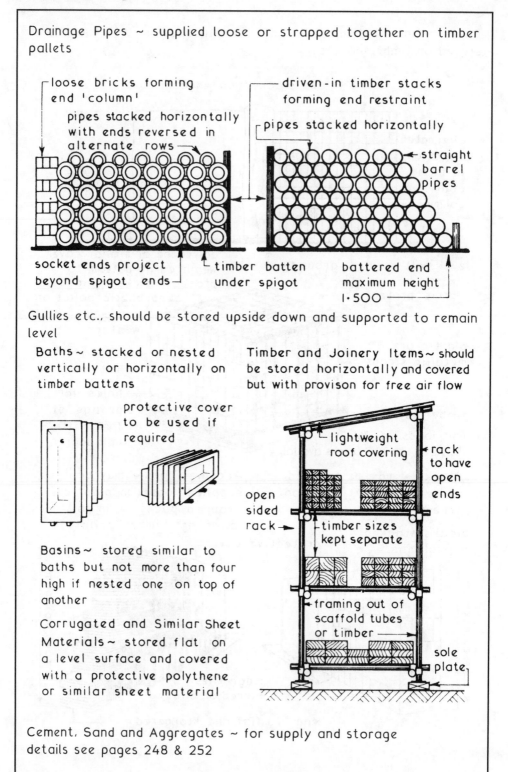

Drainage Pipes ~ supplied loose or strapped together on timber pallets

loose bricks forming end 'column'

pipes stacked horizontally with ends reversed in alternate rows

driven-in timber stacks forming end restraint

pipes stacked horizontally

straight barrel pipes

socket ends project beyond spigot ends

timber batten under spigot

battered end maximum height 1·500

Gullies etc., should be stored upside down and supported to remain level

Baths ~ stacked or nested vertically or horizontally on timber battens

protective cover to be used if required

Timber and Joinery Items ~ should be stored horizontally and covered but with provison for free air flow

lightweight roof covering

rack to have open ends

open sided rack

timber sizes kept separate

framing out of scaffold tubes or timber

sole plate

Basins ~ stored similar to baths but not more than four high if nested one on top of another

Corrugated and Similar Sheet Materials ~ stored flat on a level surface and covered with a protective polythene or similar sheet material

Cement, Sand and Aggregates ~ for supply and storage details see pages 248 & 252

Site Tests ~ the majority of materials and components arriving on site will conform to the minimum recommendations of the appropriate British Standard and therefore the only tests which need be applied are those of checking quantity received against amount stated on the delivery note, ensuring quality is as ordered and a visual inspection to reject damaged or broken goods. The latter should be recorded on the delivery note and entered in the site records. Certain site tests can however be carried out on some materials to establish specific data such as the moisture content of timber which can be read direct from a moisture meter. Other simple site tests ore given in the various British Standards to ascertain compliance with the recommendations such as the test for compliance with dimensional tolerance given in BS 3921 which covers clay bricks. This test is carried out by measuring a sample of 24 bricks taken at random from a delivered load thus :-

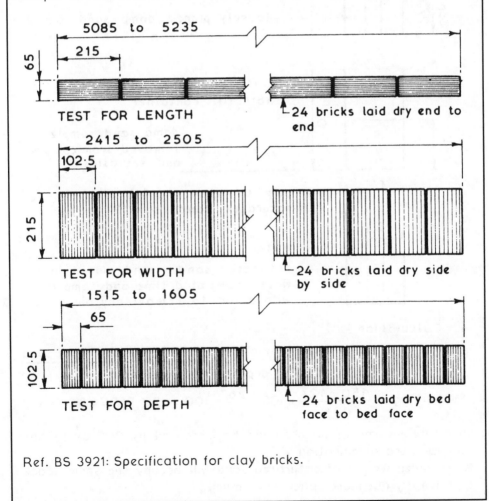

5085 to 5235
215
65
TEST FOR LENGTH

24 bricks laid dry end to end

2415 to 2505
102·5
215
TEST FOR WIDTH

24 bricks laid dry side by side

1515 to 1605
65
102·5
TEST FOR DEPTH

24 bricks laid dry bed face to bed face

Ref. BS 3921: Specification for clay bricks.

Site Test ~ apart from the test outlined on page 93 site tests on materials which are to be combined to form another material such as concrete can also be tested to establish certain properties which if not known could affect the consistency and/or quality of the final material.

Typical Example ~ Testing Sand for Bulking ~

this data is required when batching concrete by volume — test made at commencement of mixing and if change in weather

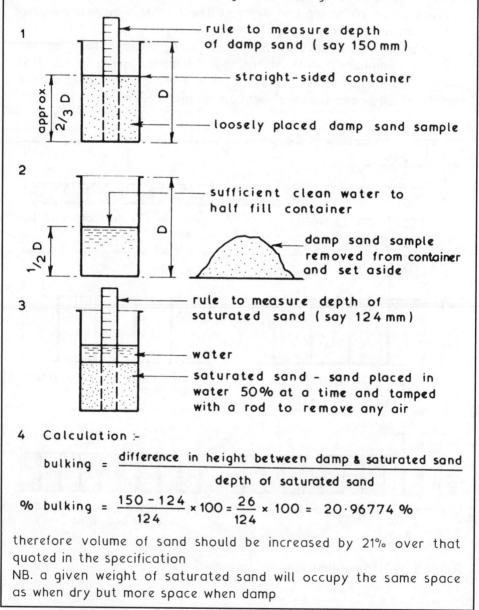

1 — rule to measure depth of damp sand (say 150 mm)

— straight-sided container

approx. 2/3 D

D

— loosely placed damp sand sample

2

1/2 D

D

— sufficient clean water to half fill container

— damp sand sample removed from container and set aside

3 — rule to measure depth of saturated sand (say 124 mm)

— water

— saturated sand - sand placed in water 50% at a time and tamped with a rod to remove any air

4   Calculation :-

$$\text{bulking} = \frac{\text{difference in height between damp \& saturated sand}}{\text{depth of saturated sand}}$$

$$\% \text{ bulking} = \frac{150 - 124}{124} \times 100 = \frac{26}{124} \times 100 = 20.96774\%$$

therefore volume of sand should be increased by 21% over that quoted in the specification

NB. a given weight of saturated sand will occupy the same space as when dry but more space when damp

Silt Test for Sand ~ the object of this test is to ascertain the cleanliness of sand by establishing the percentage of silt present in a natural sand since too much silt will weaken the concrete

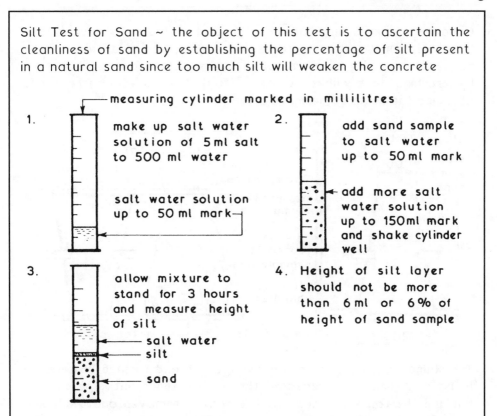

measuring cylinder marked in millilitres

1. make up salt water solution of 5 ml salt to 500 ml water

salt water solution up to 50 ml mark

2. add sand sample to salt water up to 50 ml mark

add more salt water solution up to 150 ml mark and shake cylinder well

3. allow mixture to stand for 3 hours and measure height of silt

salt water
silt
sand

4. Height of silt layer should not be more than 6 ml or 6% of height of sand sample

Obtaining Samples for Laboratory Testing ~ these tests may be required for checking aggregate grading by means of a sieve test, checking quality or checking for organic impurities but whatever the reason the sample must be truly representative of the whole:-

1.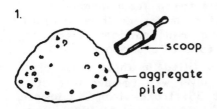

scoop

aggregate pile

samples extracted by means of a scoop from at least ten different positions in the pile

sample required :-
fine aggregate - 50 kg
coarse aggregate - 200 kg

2.

well mixed sample divided into four equal parts - opposite quarters are discarded - remainder of sample remixed and quartered - whole process is repeated until required size of sample is left.

samples required :-
fine aggregate ⇥ 6 mm - 3 kg
⇥ 10 mm - 6 kg
coarse aggregate ⇥ 20 mm - 25 kg
⇥ 32 mm - 50 kg

Concrete requires monitoring by means of tests to ensure that subsequent mixes are of the same consistency and this can be carried out on site by means of the slump test and in a laboratory by crushing test cubes to check that the cured concrete has obtained the required designed strength.

Slump Test ~

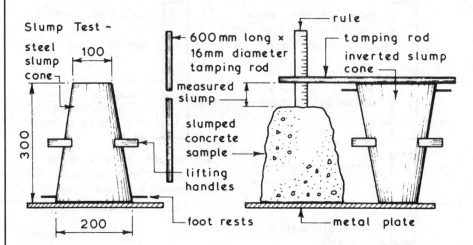

The slump cone is filled to a quarter depth and tamped 25 times — filling and tamping is repeated three more times until the cone is full and the top smoothed off. The cone is removed and the slump measured, for consistent mixes the slump should remain the same for all samples tested. Usual specification 50mm or 75mm slump.

Test Cubes - these are required for laboratory strength tests~

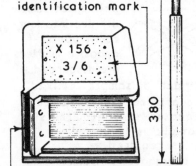

concrete sample with trowelled surface and identification mark

25 x 25mm square end tamping bar

1. Sample taken from discharge outlet of mixer or from point of placing using random selection by means of a scoop.

2. Mould filled in three equal layers each layer well tamped with at least 35 strokes from the tamping bar.

3. Sample left in mould for 24 hours and covered with a damp sack or similar at a temperature of 4·4 to 21°C

4. Remove sample from mould and store in water at temperature of 10 to 21°C until required for testing

150 x 150 x 150 standard steel test cube mould thinly coated inside with mould oil

Refs. BS 1881-102 (Slump) and BS 1881-108 (Cubes)

Non destructive testing of concrete. Also known as in-place or insitu tests.

Changes over time and in different exposures can be monitored.

References: BS 6089:1981 Guide to assessment of concrete strength in existing structures;
BS 1881:1970 on, Testing concrete.

Provides information on: strength insitu, voids, flaws, cracks and deterioration.

Rebound hammer test — attributed to Ernst Schmidt after he devised the impact hammer in 1948. It works on the principle of an elastic mass rebounding off a hard surface. Varying surface densities will affect impact and propagation of stress waves. These can be recorded on a numerical scale known as rebound numbers. It has limited application to smooth surfaces of concrete only. False results may occur where there are local variations in the concrete, such as a large piece of aggregate immediately below the impact surface. Rebound numbers can be graphically plotted to correspond with compressive strength.

Ref: BS 1881—202:1986.

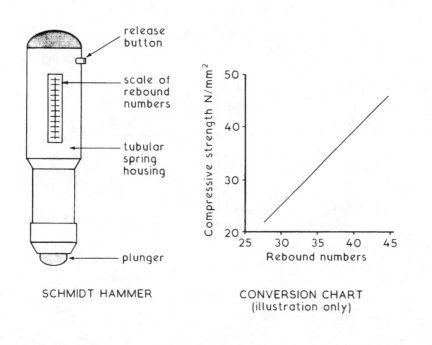

SCHMIDT HAMMER

CONVERSION CHART
(illustration only)

Penetration or Windsor probe test ~ there are various interpretations of this test. It is a measure of the penetration of a steel alloy rod, fired by a predetermined amount of energy into concrete. In principle, the depth of penetration is inversely proportional to the concrete compressive strength. Several recordings are necessary to obtain a fair assessment and some can be discarded particularly where the probe cannot penetrate some dense aggregates. The advantage over the rebound hammer is provision of test results at a greater depth (up to 50 mm).

Ref: BS1881–207:1992.

Pull out test ~ this is not entirely non destructive as there will be some surface damage, albeit easily repaired. A number of circular bars of steel with enlarged ends are cast into the concrete as work proceeds. This requires careful planning and location of bars with corresponding voids provided in the formwork. At the appropriate time, the bar and a piece of concrete are pulled out by tension jack. Although the concrete fails in tension and shear, the pull out force can be correlated to the compressive strength of the concrete.

Ref: BS 1881–207:1992.

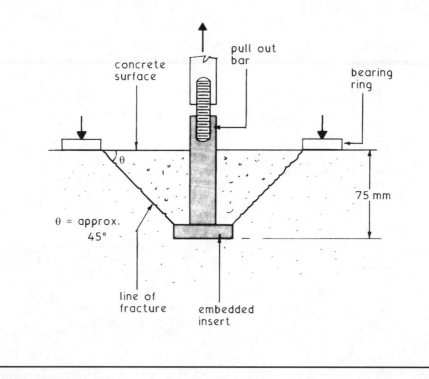

Vibration test ~ a number of electronic tests have been devised, which include measurement of ultrasonic pulse velocity through concrete. This applies the principle of recording a pulse at predetermined frequencies over a given distance. The apparatus includes transducers in contact with the concrete, pulse generator, amplifier, and time measurement to digital display circuit. Formulae for converting the data to concrete compressive strength are available in BS 1881–203:1986.

A variation, using resonant frequency, measures vibrations produced at one end of a concrete sample against a receiver or pick up at the other. The driving unit or exciter is activated by a variable frequency oscillator to generate vibrations varying in resonance, depending on the concrete quality. The calculation of compressive strength by conversion of amplified vibration data is by formulae found in BS 1881–209:1990.

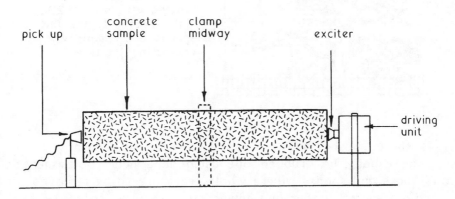

RESONANT FREQUENCY TEST

Other relevant standards:—
BS 1881–114: Testing concrete. Methods for determination of density of hardened concrete.
BS 1881–122: Testing concrete. Method for determination of water absorption.
BS 1881–124: Testing concrete. Methods for analysis of hardened concrete.

Trees ~ these are part of our national heritage and are also the source of timber — to maintain this source a control over tree felling has been established under the Forestry Act 1967 which places the control responsibility on the Forestry Commission. Local planning authorities also have powers under the Town and Country Planning Act 1990 and the Town and Country Amenities Act 1974 to protect trees by making tree preservation orders. Contravention of such an order can lead to a substantial fine and a compulsion to replace any protected tree which has been removed or destroyed. Trees on building sites which are covered by a tree preservation order should be protected by a suitable fence.

tree covered by a tree preservation order

cleft chestnut or similar fencing at least 1·200 high erected to the full spread and completely encircling the tree

Trees, shrubs, bushes and tree roots which are to be removed from site can usually be grubbed out using hand held tools such as saws, picks and spades. Where whole trees are to be removed for relocation special labour and equipment is required to ensure that the roots, root earth ball and bark are not damaged.

Structures ~ buildings which are considered to be of historic or architectural interest can be protected under the Town and Country Acts provisions. The Department of the Environment lists buildings according to age, architectural, historical and/or intrinsic value. It is an offence to demolish or alter a listed building without first obtaining 'listed building consent' from the local planning authority. Contravention is punishable by a fine and/or imprisonment. It is also an offence to demolish a listed building without giving notice to the Royal Commission on Historic Monuments, this is to enable them to note and record details of the building.

Services which may be encountered on construction sites and the authority responsible are:-

Water — Local Water Company

Electricity — transmission ~ National Power, PowerGen and Nuclear Electric

distribution ~ Area Electricity Companies in England and Wales. Scottish Power and Scottish Hydro-Electric.

Gas — Area Gas Board

Telephones — National Telecommunications Companys, eg. BT, C&W, etc.

Drainage — Local Authority unless a private drain or sewer when owner(s) is responsible.

All the above authorities must be notified of any proposed new services and alterations or terminations to existing services before any work is carried out.

Locating Existing Services on Site ~

Method 1 — By reference to maps and plans prepared and issued by the respective responsible authority.

Method 2 — Using visual indicators ~

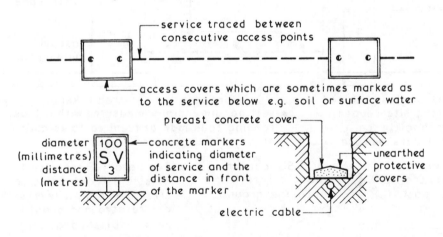

Method 3 — Detection specialist contractor employed to trace all forms of underground services using electronic sub-surface survey equipment.

Once located, position and type of service can be plotted on a map or plan, marked with special paint on hard surfaces and marked with wood pegs with indentification data on earth surfaces.

Setting Out the Building Outline ~ this task is usually undertaken once the site has been cleared of any debris or obstructions and any reduced level excavation work is finished. It is usually the responsibility of the contractor to set out the building(s) using the information provided by the designer or architect. Accurate setting out is of paramount importance and should therefore only be carried out by competent persons and all their work thoroughly checked, preferably by different personnel and by a different method.

The first task in setting out the building is to establish a base line to which all the setting out can be related. The base line very often coincides with the building line which is a line, whose position on site is given by the local authority in front of which no development is permitted.

Typical Setting Out Example ~

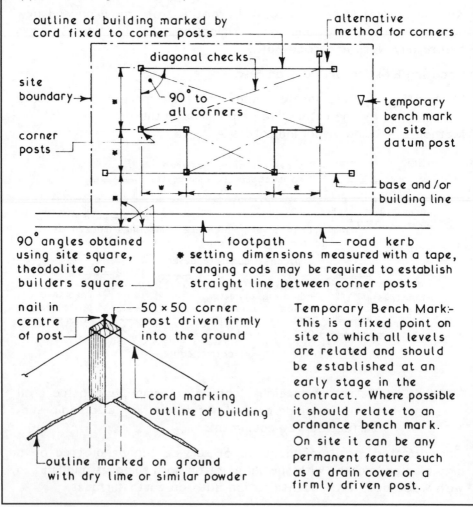

outline of building marked by cord fixed to corner posts

alternative method for corners

diagonal checks

site boundary →

90° to all corners

∇ ← temporary bench mark or site datum post

corner posts

base and/or building line

90° angles obtained using site square, theodolite or builders square

footpath   road kerb

∗ setting dimensions measured with a tape, ranging rods may be required to establish straight line between corner posts

nail in centre of post

50 × 50 corner post driven firmly into the ground

cord marking outline of building

outline marked on ground with dry lime or similar powder

Temporary Bench Mark:- this is a fixed point on site to which all levels are related and should be established at an early stage in the contract. Where possible it should relate to an ordnance bench mark. On site it can be any permanent feature such as a drain cover or a firmly driven post.

Setting Out Trenches ~ the objective of this task is twofold. Firstly it must establish the excavation size, shape and direction and secondly it must establish the width and position of the walls. The outline of building will have been set out and using this outline profile boards can be set up to control the position, width and possibly the depth of the proposed trenches. Profile boards should be set up at least 2·000 clear of trench positions so they do not obstruct the excavation work. The level of the profile crossboard should be related to the site datum and fixed at a convenient height above ground level if a traveller is to be used to control the depth of the trench. Alternatively the trench depth can be controlled using a level and staff related to site datum. The trench width can be marked on the profile with either nails or sawcuts and with a painted band if required for identification.

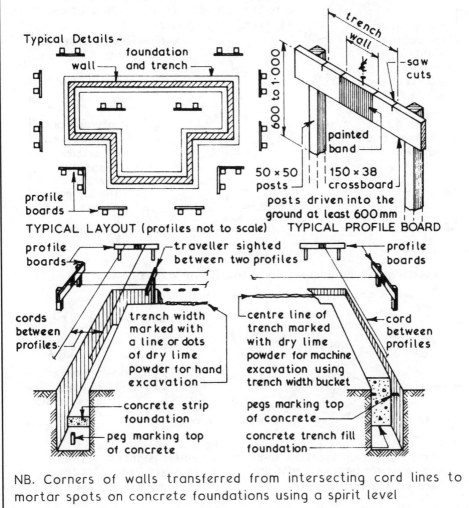

Typical Details ~

wall — foundation and trench

profile boards

TYPICAL LAYOUT (profiles not to scale)

trench
wall

600 to 1·000

saw cuts

painted band

50 × 50 posts

150 × 38 crossboard

posts driven into the ground at least 600 mm

TYPICAL PROFILE BOARD

profile boards

traveller sighted between two profiles

profile boards

cords between profiles

trench width marked with a line or dots of dry lime powder for hand excavation

concrete strip foundation

peg marking top of concrete

centre line of trench marked with dry lime powder for machine excavation using trench width bucket

pegs marking top of concrete

concrete trench fill foundation

cord between profiles

NB. Corners of walls transferred from intersecting cord lines to mortar spots on concrete foundations using a spirit level

Setting Out a Framed Building ~ framed buildings are usually related to a grid, the intersections of the grid lines being the centre point of an isolated or pad foundation. The grid is usually set out from a base line which does not always form part of the grid. Setting out dimensions for locating the grid can either be given on a drawing or they will have to be accurately scaled off a general layout plan. The grid is established using a theodolite and marking the grid line intersections with stout pegs. Once the grid has been set out offset pegs or profiles can be fixed clear of any subsequent excavation work. Control of excavation depth can be by means of a traveller sighted between sight rails or by level and staff related to site datum.

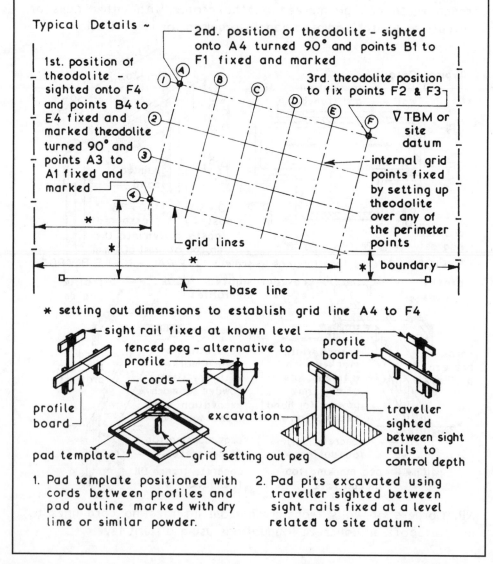

Typical Details ~

1st. position of theodolite - sighted onto F4 and points B4 to E4 fixed and marked theodolite turned 90° and points A3 to A1 fixed and marked

2nd. position of theodolite - sighted onto A4 turned 90° and points B1 to F1 fixed and marked

3rd. theodolite position to fix points F2 & F3

▽ TBM or site datum

internal grid points fixed by setting up theodolite over any of the perimeter points

grid lines

boundary

base line

* setting out dimensions to establish grid line A4 to F4

sight rail fixed at known level

fenced peg – alternative to profile

cords

profile board

profile board

pad template

grid setting out peg

excavation

traveller sighted between sight rails to control depth

1. Pad template positioned with cords between profiles and pad outline marked with dry lime or similar powder.

2. Pad pits excavated using traveller sighted between sight rails fixed at a level related to site datum.

Setting Out Reduced Level Excavations ~ the overall outline of the reduced level area can be set out using a theodolite, ranging rods, tape and pegs working from a base line. To control the depth of excavation, sight rails are set up at a convenient height and at positions which will enable a traveller to be used.

Typical Details ~

1. Setting up sight rails :-

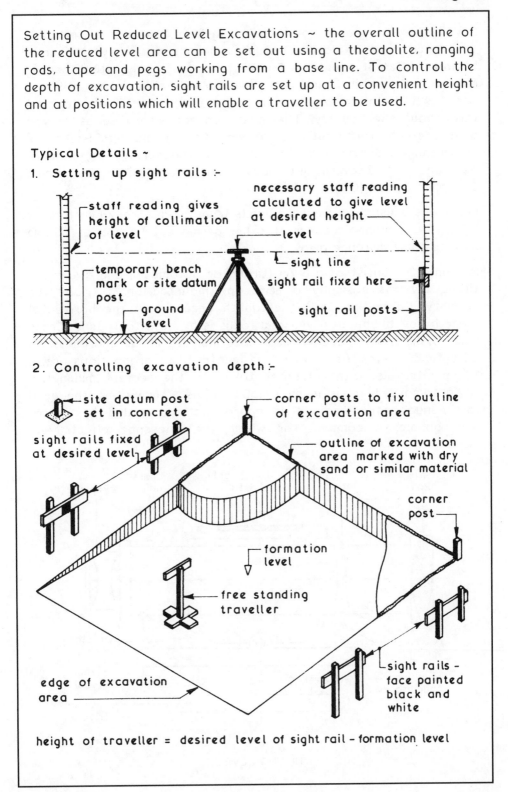

staff reading gives height of collimation of level

necessary staff reading calculated to give level at desired height

level

sight line

temporary bench mark or site datum post

sight rail fixed here

ground level

sight rail posts →

2. Controlling excavation depth :-

site datum post set in concrete

corner posts to fix outline of excavation area

sight rails fixed at desired level

outline of excavation area marked with dry sand or similar material

corner post

formation level

free standing traveller

edge of excavation area

sight rails - face painted black and white

height of traveller = desired level of sight rail - formation level

Levelling ~ the process of establishing height dimensions, relative to a fixed point or datum. Datum is mean sea level, which varies between different countries. For UK purposes this is established at Newlyn in Cornwall, from tide data recorded between May 1915 and April 1921. Relative levels defined by bench marks are located throughout the country. The most common, identified as carved arrows, can be found cut into walls of stable structures. Reference to Ordnance Survey maps of an area will indicate bench mark positions and their height above sea level, hence the name Ordnance Datum (OD).

On site it is usual to measure levels from a temporary bench mark (TBM), i.e. a manhole cover or other permanent fixture, as an OD may be some distance away.

Instruments consist of a level (tilting or automatic) and a staff. A tilting level is basically a telescope mounted on a tripod for stability. Correcting screws establish accuracy in the horizontal plane by air bubble in a vial and focus is by adjustable lens. Cross hairs of horizontal and vertical lines indicate image sharpness on an extending staff of 3, 4 or 5 m length. Staff graduations are in 10 mm intervals, with estimates taken to the nearest millimetre. An automatic level is much simpler to use, eliminating the need for manual adjustment. It is approximately levelled by centre bulb bubble. A compensator within the telescope effects fine adjustment.

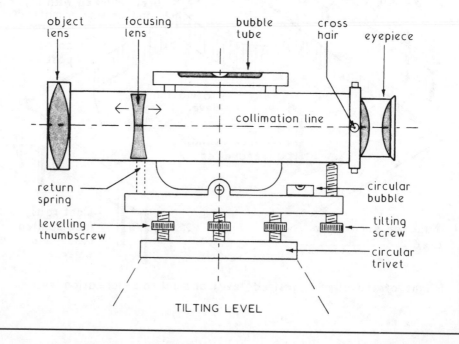

TILTING LEVEL

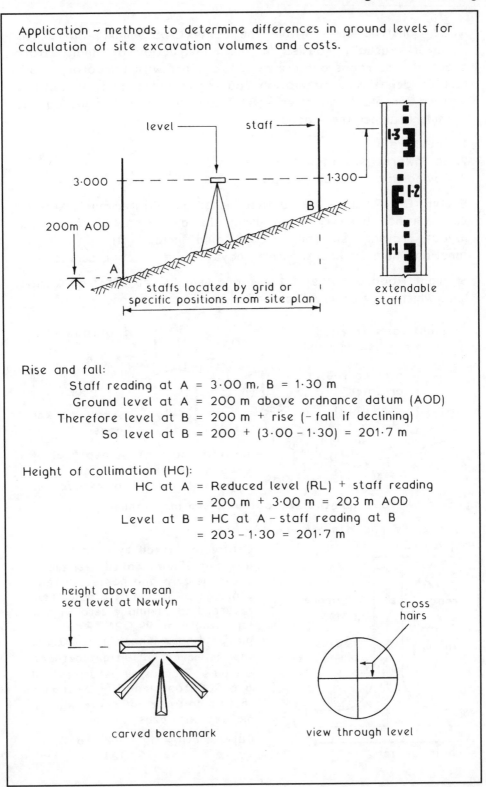

Application ~ methods to determine differences in ground levels for calculation of site excavation volumes and costs.

level   staff

3·000

1·300

200m AOD

B

A

staffs located by grid or specific positions from site plan

extendable staff

1·3

1·2

1·1

Rise and fall:
        Staff reading at A = 3·00 m, B = 1·30 m
        Ground level at A = 200 m above ordnance datum (AOD)
  Therefore level at B = 200 m + rise (− fall if declining)
        So level at B = 200 + (3·00 − 1·30) = 201·7 m

Height of collimation (HC):
               HC at A = Reduced level (RL) + staff reading
                        = 200 m + 3·00 m = 203 m AOD
        Level at B = HC at A − staff reading at B
                   = 203 − 1·30 = 201·7 m

height above mean sea level at Newlyn

cross hairs

carved benchmark

view through level

Road Construction ~ within the context of building operations roadworks usually consist of the construction of small estate roads, access roads and driveways together with temporary roads laid to define site circulation routes and/or provide a suitable surface for plant movements. The construction of roads can be considered under three headings :-

1. Setting out.
2. Earthworks (see page 109)
3. Paving Construction (see pages 110 & 111)

Setting Out Roads ~ this activity is usually carried out after the top soil has been removed using the dimensions given on the layout drawing (s). The layout could include straight lengths junctions, hammer heads, turning bays and intersecting curves.

Straight Road Lengths — these are usually set out from centre lines which have been established by traditional means

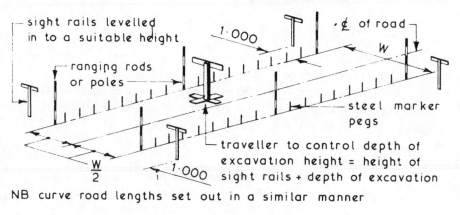

NB curve road lengths set out in a similar manner

Junctions and Hammer Heads -

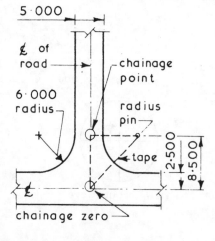

Centre lines fixed by traditional methods. Tape hooked over pin at chainage zero and passed around chainage point pin at 8·500 then returned to chainage zero with a tape length of 29·021. Radius pin held tape length 17·000 and tape is moved until tight between all pins. Radius pin is driven and a 6·000 tape length is swung from the pin to trace out curve which is marked with pegs or pins.

Tape length = 17 + √2 × 8·5
= 29·021

Earthworks ~ this will involve the removal of topsoil together with any vegetation, scraping and grading the required area down to formation level plus the formation of any cuttings or embankments. Suitable plant for these operations would be tractor shovels fitted with a 4 in 1 bucket (page 145); graders (page 144) and bulldozers (page 142). The soil immediately below the formation level is called the subgrade whose strength will generally decrease as its moisture content rises therefore if it is to be left exposed for any length of time protection may be required. Subgrade protection may take the form of a covering of medium gauge plastic sheeting with 300 mm laps or alternatively a covering of sprayed bituminous binder with a sand topping applied at a rate of 1 litre per m². To preserve the strength and durability of the subgrade it may be necessary to install cut off subsoil drains alongside the proposed road (see Road Drainage on page 600)

Paving Construction ~ once the subgrade has been prepared and any drainage or other buried services installed the construction of the paving can be undertaken. Paved surfaces can be either flexible or rigid in format. Flexible or bound surfaces are formed of materials applied in layers directly over the subgrade whereas rigid pavings consist of a concrete slab resting on a granular base (see pages 110 & 111)

Typical Flexible Paving Details ~

surfacing = base layer + wearing course

60 mm thick base course of dense bitumen macadam or asphalt laid to form the crossfalls and/or gradients

wearing course of coated macadam or asphalt having good non-skid properties; reasonable resistance to glare and an acceptable life — should be laid within 3 days of base layer

sub-base of crush stone or dry lean mix concrete (1:15) laid in 100 to 150 mm thick compacted layers - total thickness related to loading and subgrade strength

subgrade

NB. no road joints required

Rigid Pavings ~ these consist of a reinforced or unreinforced insitu concrete slab laid over a base course of crushed stone or similar material which has been blinded to receive a polythene sheet slip membrane. The primary objective of this membrane is to prevent grout loss from the insitu slab.

Typical Rigid Paving Details ~

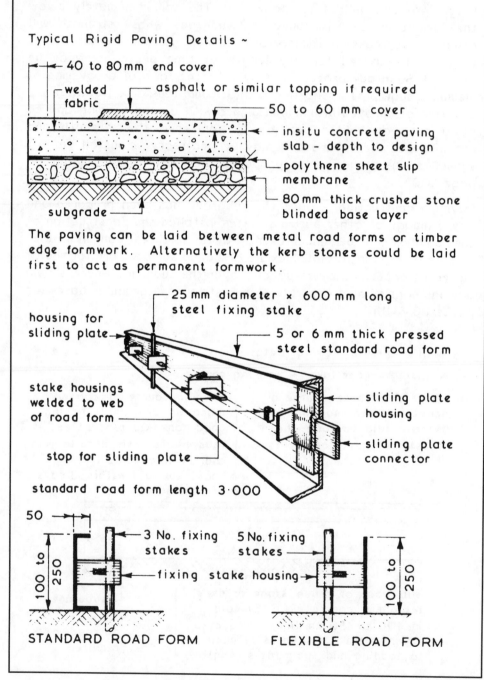

40 to 80 mm end cover

welded fabric

asphalt or similar topping if required

50 to 60 mm cover

insitu concrete paving slab – depth to design

polythene sheet slip membrane

80 mm thick crushed stone blinded base layer

subgrade

The paving can be laid between metal road forms or timber edge formwork. Alternatively the kerb stones could be laid first to act as permanent formwork.

25 mm diameter × 600 mm long steel fixing stake

housing for sliding plate

5 or 6 mm thick pressed steel standard road form

stake housings welded to web of road form

sliding plate housing

sliding plate connector

stop for sliding plate

standard road form length 3·000

50

3 No. fixing stakes

5 No. fixing stakes

100 to 250

fixing stake housing

100 to 250

STANDARD ROAD FORM

FLEXIBLE ROAD FORM

Joints in Rigid Pavings ~ longitudinal and transverse joints are required in rigid pavings to :-

1. Limit size of slab.
2. Limit stresses due to subgrade restraint.
3. Provide for expansion and contraction movements.

The main joints used are classified as expansion, contraction or longitudinal, the latter being the same in detail as the contraction joint differing only in direction. The spacing of road joints is determined by :-

1. Slab thickness.
2. Whether slab is reinforced or unreinforced.
3. Anticipated traffic load and flow rate.
4. Temperature at which concrete is laid.

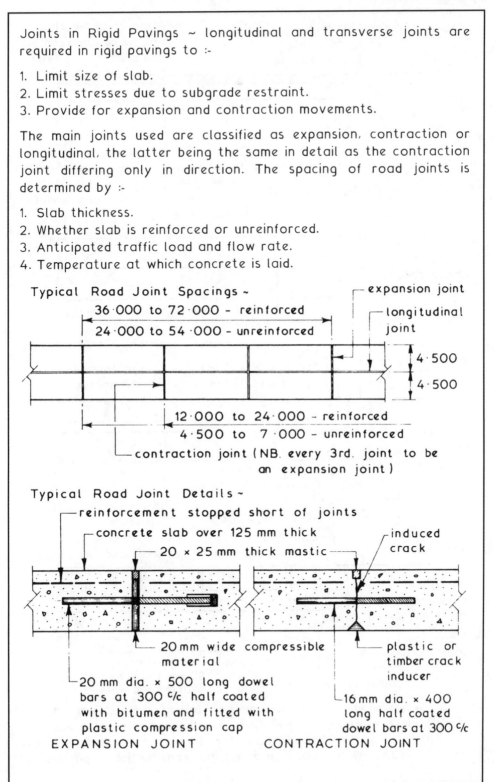

Typical Road Joint Spacings ~

- expansion joint
- longitudinal joint

36·000 to 72·000 – reinforced
24·000 to 54·000 – unreinforced

4·500
4·500

12·000 to 24·000 – reinforced
4·500 to 7·000 – unreinforced

contraction joint (NB. every 3rd. joint to be an expansion joint)

Typical Road Joint Details ~

reinforcement stopped short of joints
concrete slab over 125 mm thick
20 × 25 mm thick mastic
induced crack

20 mm wide compressible material

20 mm dia. × 500 long dowel bars at 300 c/c half coated with bitumen and fitted with plastic compression cap

plastic or timber crack inducer

16 mm dia. × 400 long half coated dowel bars at 300 c/c

EXPANSION JOINT   CONTRACTION JOINT

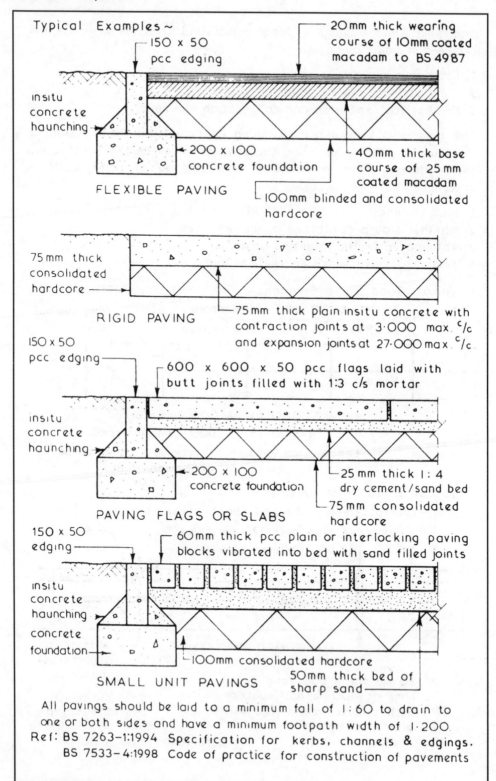

Typical Examples ~

150 × 50 pcc edging

20mm thick wearing course of 10mm coated macadam to BS 4987

insitu concrete haunching

200 × 100 concrete foundation

40mm thick base course of 25mm coated macadam

100mm blinded and consolidated hardcore

FLEXIBLE PAVING

75mm thick consolidated hardcore

75mm thick plain insitu concrete with contraction joints at 3·000 max. c/c and expansion joints at 27·000 max. c/c

RIGID PAVING

150 × 50 pcc edging

600 × 600 × 50 pcc flags laid with butt joints filled with 1:3 c/s mortar

insitu concrete haunching

200 × 100 concrete foundation

25mm thick 1 : 4 dry cement/sand bed

75mm consolidated hardcore

PAVING FLAGS OR SLABS

150 × 50 edging

60mm thick pcc plain or interlocking paving blocks vibrated into bed with sand filled joints

insitu concrete haunching

concrete foundation

100mm consolidated hardcore

50mm thick bed of sharp sand

SMALL UNIT PAVINGS

All pavings should be laid to a minimum fall of 1:60 to drain to one or both sides and have a minimum footpath width of 1·200.

Ref: BS 7263-1:1994 Specification for kerbs, channels & edgings.
BS 7533-4:1998 Code of practice for construction of pavements

Available sections ~ manufactured in 915mm lengths
from silver/grey aggregate concrete.

KERBS

splay – 12½° to 15°
r = 16 to 19mm radius

Half battered

size (mm)				
A	150	150	255	305
B	125	150	125	150
C	50	50	155	205
	*		*	*

Bullnose

size (mm)				
A	150	150	255	305
B	125	150	125	150
	*			

θ = 45°

Splayed

size (mm)			
A	150	255	305
B	125	125	150
C	75	180	230
D	50	50	75
		*	

CHANNELS

dish
bullnose } optional

	square				dished				bullnose	
A(mm)	125	150	150	125	150	90	125	75	150	125
B(mm)	255	230	150	150	305	305	255	230	305	255
	*				*				*	

EDGINGS

Round　　Flat　　Bullnose　r　　Chamfer

25mm

Round　　150/200/250 × 50mm*
Flat　　　150/200/250 × 50mm*
Bullnose 150/200 × 50mm*
Chamfer 178 × 63mm

*denotes BS sections

Further cpmponents such as drop/tapered kerbs are available for
vehicle accesses. Quadrants and angles provide for directional change.

Concrete paving flags — BS dimensions:

Type	Size (nominal)	Size (work)	Thickness (T)
A — plain	600×450	598×448	50 or 63
B — plain	600×600	598×598	50 or 63
C — plain	600×750	598×748	50 or 63
D — plain	600×900	598×898	50 or 63
E — plain	450×450	448×448	50 or 70
TA/E — tactile	450×450	448×448	50 or 70
TA/F — tactile	400×400	398×398	50 or 65
TA/G — tactile	300×300	298×298	50 or 60

Note: All dimensions in millimetres.

Tactile flags — manufactured with a blistered (shown) or ribbed surface. Used in walkways to provide warning of hazards or to enable recognition of locations for people whose visability is impaired. See also, Department of Transport Disability Circular DU 1/86[1], for uses and applications.

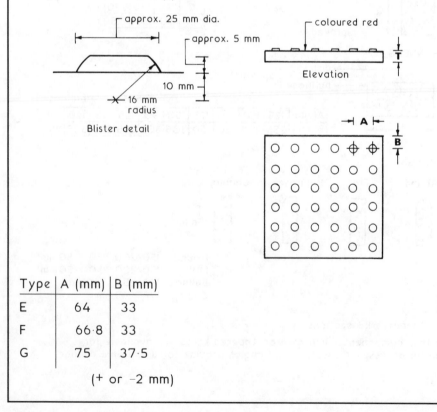

Blister detail

Elevation

Type	A (mm)	B (mm)
E	64	33
F	66·8	33
G	75	37·5

(+ or −2 mm)

Landscaping ~ in the context of building works this would involve reinstatement of the site as a preparation to the landscaping in the form of lawns, paths, pavings, flower and shrub beds and tree planting. The actual planning, lawn laying and planting activities are normally undertaken by a landscape subcontractor. The main contractor's work would involve clearing away all waste and unwanted materials, breaking up and levelling surface areas, removing all unwanted vegetation, preparing the subsoil for and spreading topsoil to a depth of at least 150 mm.

Services ~ the actual position and laying of services is the responsibility of the various service boards and undertakings. The best method is to use the common trench approach, avoid as far as practicable laying services under the highway.

Typical Common Trench Details ~
trench backfilled with selected granular material in 200mm thick compacted layers ─

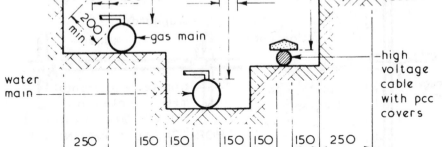

Road Signs ~ these can range from markings painted on roads to define traffic lanes, rights of way and warnings of hazards to signs mounted above the road level to give information, warning or directives, the latter being obligatory.

Typical Examples ~

blue ─► ONE WAY ─ white arrow and letters

red ─ black ─ white ─

INFORMATION   WARNING - ROAD WORKS   DIRECTIVE - NO LEFT TURN

Scaffolds ~ these are temporary working platforms erected around the perimeter of a building or structure to provide a safe working place at a convenient height. They are usually required when the working height or level is 1·500 or more above the ground level. All scaffolds must comply with the minimum requirements and objectives of the Construction (Health, Safety and Welfare) Regulations 1996.

Component Parts of a Tubular Scaffold ~

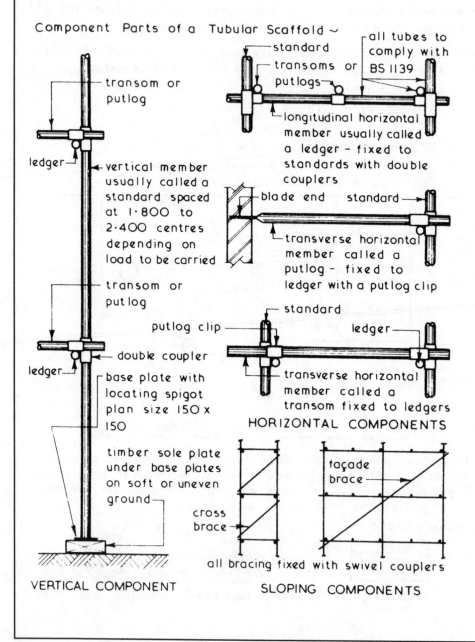

transom or putlog

ledger —

vertical member usually called a standard spaced at 1·800 to 2·400 centres depending on load to be carried

transom or putlog

putlog clip

double coupler

ledger —

base plate with locating spigot plan size 150 x 150

timber sole plate under base plates on soft or uneven ground

VERTICAL COMPONENT

standard

transoms or putlogs

all tubes to comply with BS 1139

longitudinal horizontal member usually called a ledger - fixed to standards with double couplers

blade end   standard

transverse horizontal member called a putlog - fixed to ledger with a putlog clip

standard

ledger

transverse horizontal member called a transom fixed to ledgers

HORIZONTAL COMPONENTS

façade brace

cross brace

all bracing fixed with swivel couplers

SLOPING COMPONENTS

Putlog Scaffolds ~ these are scaffolds which have an outer row of standards joined together by ledgers which in turn support the transverse putlogs which are built into the bed joints or perpends as the work proceeds, they are therefore only suitable for new work in bricks or blocks.

Typical Details ~

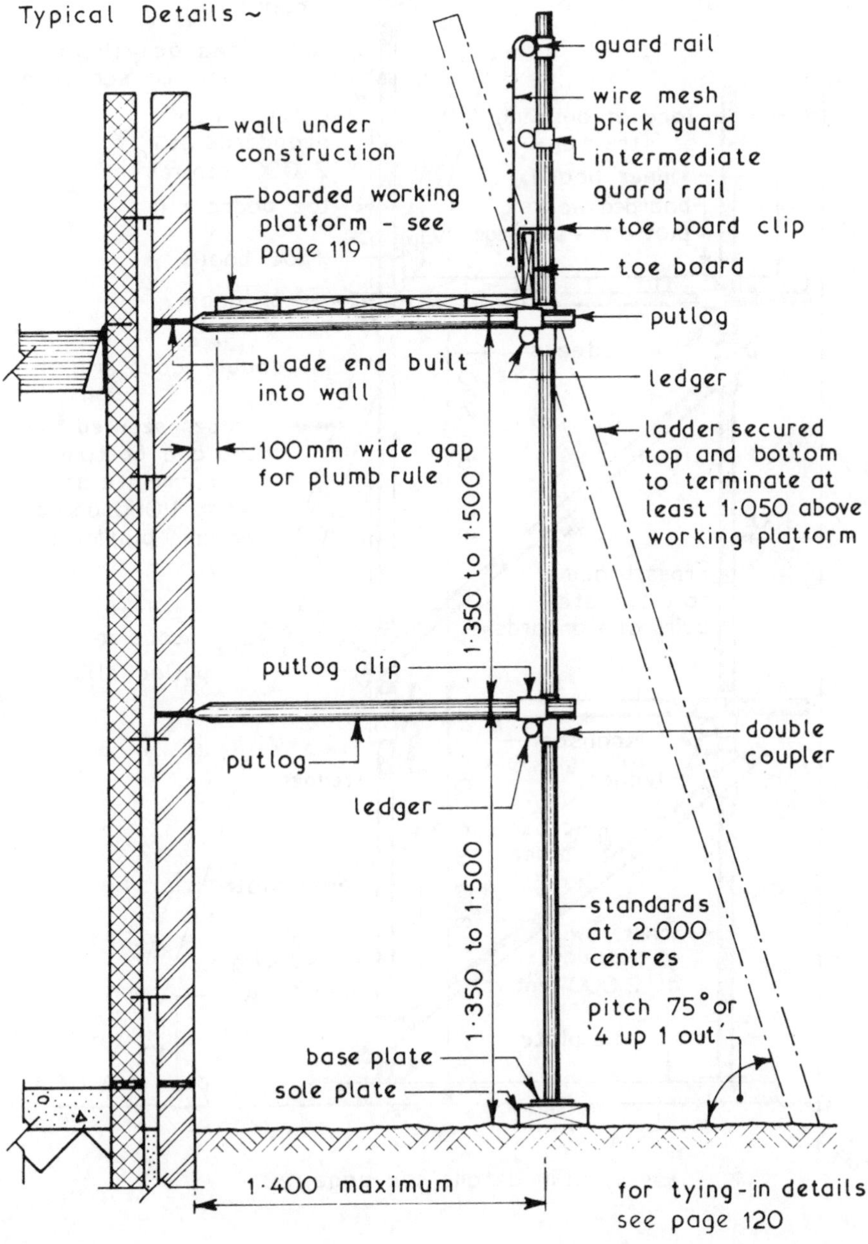

- guard rail
- wire mesh brick guard
- intermediate guard rail
- toe board clip
- toe board
- putlog
- ledger
- ladder secured top and bottom to terminate at least 1·050 above working platform

wall under construction

boarded working platform - see page 119

blade end built into wall

100 mm wide gap for plumb rule

1·350 to 1·500

putlog clip

putlog

ledger

double coupler

standards at 2·000 centres

pitch 75° or '4 up 1 out'

1·350 to 1·500

base plate
sole plate

1·400 maximum

for tying-in details see page 120

Independent Scaffolds ~ these are scaffolds which have two rows of standards each row joined together with ledgers which in turn support the transverse transoms. The scaffold is erected clear of the existing or proposed building but is tied to the building or structure at suitable intervals — see page 120

Typical Details ~

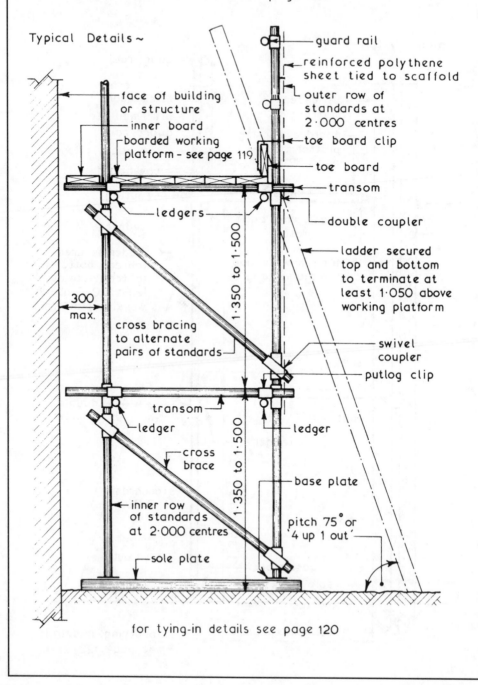

face of building or structure
inner board
boarded working platform - see page 119
ledgers
300 max.
cross bracing to alternate pairs of standards
transom
ledger
cross brace
inner row of standards at 2·000 centres
sole plate

guard rail
reinforced polythene sheet tied to scaffold
outer row of standards at 2·000 centres
toe board clip
toe board
transom
double coupler
ladder secured top and bottom to terminate at least 1·050 above working platform
swivel coupler
putlog clip
ledger
base plate
pitch 75° or '4 up 1 out'

1·350 to 1·500
1·350 to 1·500

for tying-in details see page 120

Working Platforms ~ these are close boarded or plated level surfaces at a height at which work is being carried out and they must provide a safe working place of sufficient strength to support the imposed loads of operatives and/or materials. All working platforms more than 2·000 above the ground level must be fitted with a toe board and a guard rail.

Typical Details ~

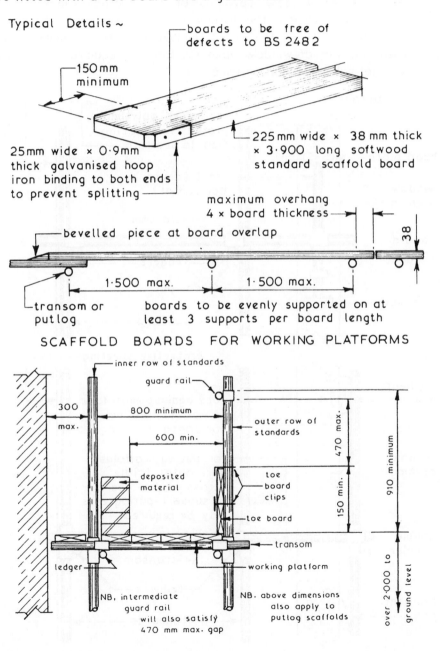

boards to be free of defects to BS 2482

150mm minimum

25mm wide × 0·9mm thick galvanised hoop iron binding to both ends to prevent splitting

225mm wide × 38 mm thick × 3·900 long softwood standard scaffold board

maximum overhang 4 × board thickness

bevelled piece at board overlap

38

transom or putlog

1·500 max.   1·500 max.

boards to be evenly supported on at least 3 supports per board length

SCAFFOLD BOARDS FOR WORKING PLATFORMS

inner row of standards

guard rail

300 max.

800 minimum

600 min.

outer row of standards

470 max.

910 minimum

deposited material

toe board clips

toe board

150 min.

transom

working platform

ledger

NB. intermediate guard rail will also satisfy 470 mm max. gap

NB. above dimensions also apply to putlog scaffolds

over 2·000 to ground level

Tying-in ~ all putlog and independent scaffolds should be tied securely to the building or structure at alternate lift heights vertically and at not more than 6·000 centres horizontally. Putlogs should not be classified as ties.

Suitable tying-in methods include connecting to tubes fitted between sides of window openings or to internal tubes fitted across window openings, the former method should not be used for more than 50% of the total number of ties. If there is an insufficient number of window openings for the required number of ties external rakers should be used.

Typical Details ~

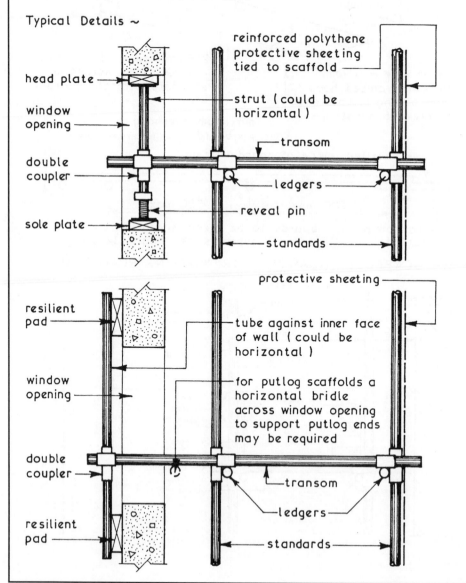

Mobile Scaffolds ~ sometimes called mobile tower scaffolds, are constructed to the basic principles as for independent tubular scaffolds and are used to provide access to restricted or small areas and/or where mobility is required.

Typical Details ~

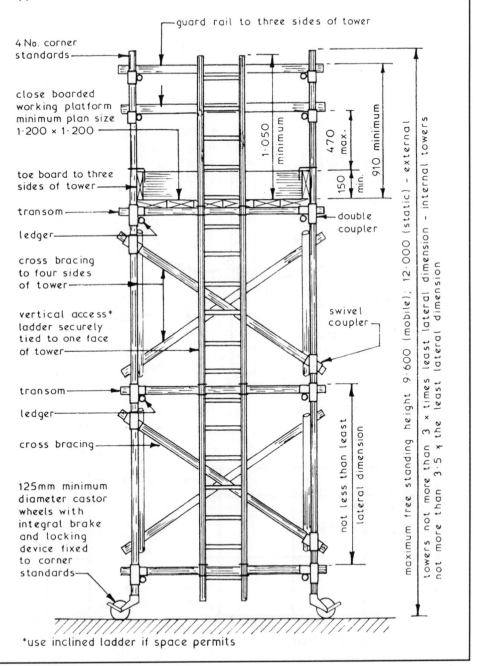

guard rail to three sides of tower

4 No. corner standards

close boarded working platform minimum plan size 1·200 × 1·200

toe board to three sides of tower

transom

ledger

cross bracing to four sides of tower

vertical access* ladder securely tied to one face of tower

transom

ledger

cross bracing

125mm minimum diameter castor wheels with integral brake and locking device fixed to corner standards

1·050 minimum

470 max.

150 min.

910 minimum

double coupler

swivel coupler

not less than least lateral dimension

maximum free standing height 9·600 (mobile); 12·000 (static) – external

towers not more than 3 × times least lateral dimension – internal towers

not more than 3·5 × the least lateral dimension

*use inclined ladder if space permits

Patent Scaffolding ~ these are systems based on an independent scaffold format in which the members are connected together using an integral locking device instead of conventional clips and couplers used with traditional tubular scaffolding. They have the advantages of being easy to assemble and take down using semi-skilled labour and will automatically comply with the majority of the requirements set out in the Construction (Health, Safety and Welfare) Regulations 1996. Generally cross bracing is not required with these systems but façade bracing can be fitted if necessary. Although simple in concept patent systems of scaffolding can lack the flexibility of traditional tubular scaffolds in complex layout situations.

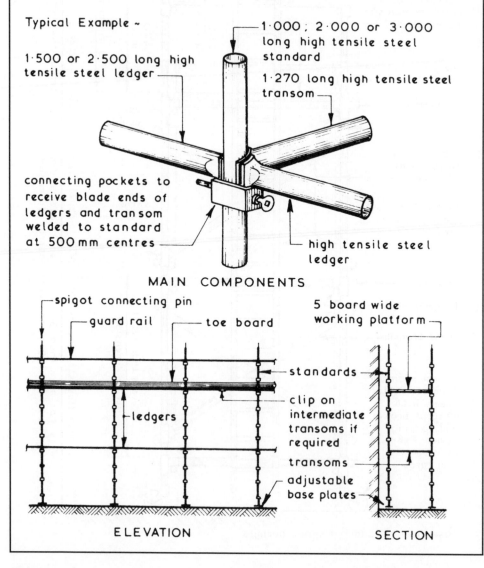

Typical Example ~

1·500 or 2·500 long high tensile steel ledger

1·000; 2·000 or 3·000 long high tensile steel standard

1·270 long high tensile steel transom

connecting pockets to receive blade ends of ledgers and transom welded to standard at 500 mm centres

high tensile steel ledger

MAIN COMPONENTS

spigot connecting pin
guard rail
toe board

5 board wide working platform

standards

clip on intermediate transoms if required

ledgers

transoms

adjustable base plates

ELEVATION

SECTION

Scaffolding Systems ~ these are temporary stagings to provide safe access to and egress from a working platform. The traditional putlog and independent scaffolds have been covered on pages 116 to 121 inclusive. The minimum legal requirements contained in the Construction (Health Safety and Welfare) Regulations 1996 applicable to traditional scaffolds apply equally to special scaffolds. Special scaffolds are designed to fulfil a specific function or to provide access to areas where it is not possible and or economic to use traditional formats. They can be constructed from standard tubes or patent systems, the latter complying with most regulation requirements are easy and quick to assemble but lack the complete flexibility of the traditional tubular scaffolds.

Birdcage Scaffolds ~ these are a form of independent scaffold normally used for internal work in large buildings such as public halls and churches to provide access to ceilings and soffits for light maintenance work like painting and cleaning. They consist of parallel rows of standards connected by ledgers in both directions, the whole arrangement being firmly braced in all directions. The whole birdcage scaffold assembly is designed to support a single working platform which should be double planked or underlined with polythene or similar sheeting as a means of restricting the amount of dust reaching the floor level.

Slung Scaffolds ~ these are a form of scaffold which is suspended from the main structure by means of wire ropes or steel chains and is not provided with a means of being raised or lowered. Each working platform of a slung scaffold consists of a supporting framework of ledgers and transoms which should not create a plan size in excess of 2·500×2·500 and be held in position by not less than six evenly spaced wire ropes or steel chains securely anchored at both ends. The working platform should be double planked or underlined with polythene or similar sheeting to restrict the amount of dust reaching the floor level. Slung scaffolds are an alternative to birdcage scaffolds and although more difficult to erect have the advantage of leaving a clear space beneath the working platform which makes them suitable for cinemas, theatres and high ceiling banking halls.

Suspended Scaffolds ~ these consist of a working platform in the form of a cradle which is suspended from cantilever beams or outriggers from the roof of a tall building to give access to the façade for carrying out light maintenance work and cleaning activities. The cradles can have manual or power control and be in single units or grouped together to form a continuous working platform. If grouped together they are connected to one another at their abutment ends with hinges to form a gap of not more than 25 mm wide. Many high rise buildings have a permanent cradle system installed at roof level and this is recommended for all buildings over 30·000 high.

Typical Example ~

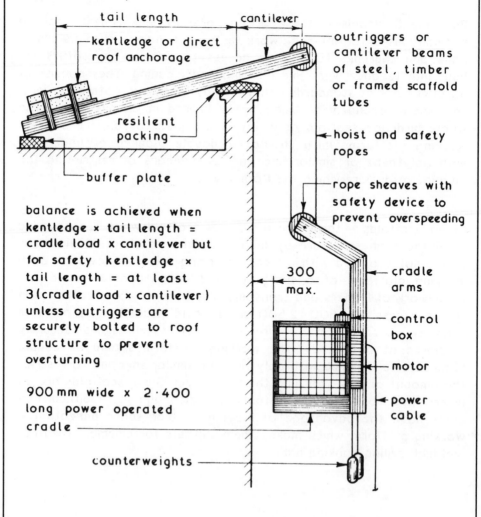

tail length   cantilever

kentledge or direct roof anchorage

outriggers or cantilever beams of steel, timber or framed scaffold tubes

resilient packing

hoist and safety ropes

buffer plate

rope sheaves with safety device to prevent overspeeding

balance is achieved when kentledge × tail length = cradle load × cantilever but for safety kentledge × tail length = at least 3(cradle load × cantilever) unless outriggers are securely bolted to roof structure to prevent overturning

300 max.

cradle arms

control box

motor

900 mm wide × 2·400 long power operated cradle

power cable

counterweights

124

Cantilever Scaffolds ~ these are a form of independent tied scaffold erected on cantilever beams and used where it is impracticable, undesirable or uneconomic to use a traditional scaffold raised from ground level. The assembly of a cantilever scaffold requires special skills and should therefore always be carried out by trained and experienced personnel

Typical Example ~

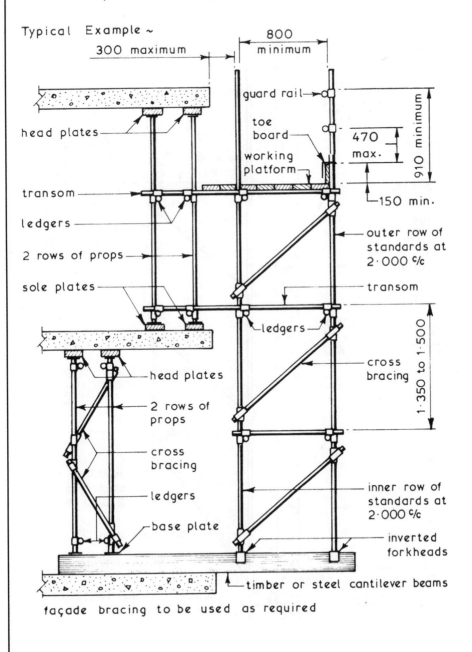

300 maximum

800 minimum

head plates

guard rail

toe board

working platform

470 max.

910 minimum

150 min.

transom

ledgers

2 rows of props

sole plates

outer row of standards at 2·000 c/c

transom

ledgers

cross bracing

1·350 to 1·500

head plates

2 rows of props

cross bracing

ledgers

base plate

inner row of standards at 2·000 c/c

inverted forkheads

timber or steel cantilever beams

façade bracing to be used as required

Truss-out Scaffold ~ this is a form of independent tied scaffold used where it is impracticable, undesirable or uneconomic to build a scaffold from ground level. The supporting scaffold structure is known as the truss-out. The assembly of this form of scaffold requires special skills and should therefore be carried out by trained and experienced personnel.

Typical Example ~

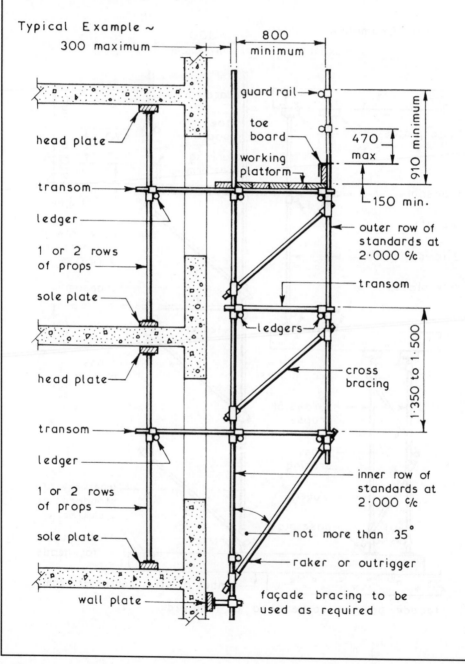

300 maximum

800 minimum

head plate

guard rail

toe board

working platform

470 max

910 minimum

150 min.

transom

ledger

outer row of standards at 2·000 c/c

1 or 2 rows of props

sole plate

transom

ledgers

head plate

cross bracing

1·350 to 1·500

transom

ledger

inner row of standards at 2·000 c/c

1 or 2 rows of props

sole plate

not more than 35°

raker or outrigger

wall plate

façade bracing to be used as required

Gantries ~ these are elevated platforms used when the building being maintained or under construction is adjacent to a public footpath. A gantry over a footpath can be used for storage of materials, housing units of accommodation and supporting an independent scaffold. Local authority permission will be required before a gantry can be erected and they have the power to set out the conditions regarding minimum sizes to be used for public walkways and lighting requirements. It may also be necessary to comply with police restrictions regarding the loading and unloading of vehicles at the gantry position. A gantry can be constructed of any suitable structural material and may need to be structurally designed to meet all the necessary safety requirements.

Typical Example ~

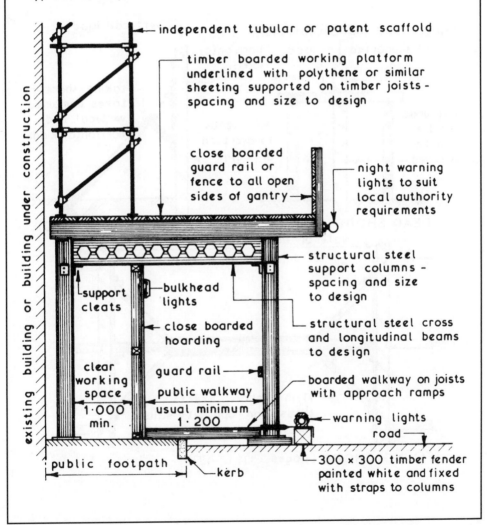

independent tubular or patent scaffold

timber boarded working platform underlined with polythene or similar sheeting supported on timber joists - spacing and size to design

close boarded guard rail or fence to all open sides of gantry

night warning lights to suit local authority requirements

existing building or building under construction

support cleats

bulkhead lights

close boarded hoarding

clear working space 1·000 min.

guard rail

public walkway

usual minimum 1·200

structural steel support columns - spacing and size to design

structural steel cross and longitudinal beams to design

boarded walkway on joists with approach ramps

warning lights

road

public footpath

kerb

300 × 300 timber fender painted white and fixed with straps to columns

Shoring ~ this is a form of temporary support which can be given to existing buildings with the primary function of providing the necessary precautions to avoid damage to any person from collapse of structure as required by the Construction (Health, Safety and Welfare) Regulations 1996.

Shoring Systems ~ there are three basic systems of shoring which can be used separately or in combination with one another to provide the support(s) and these are namely :-

1. Dead Shoring — used primarily to carry vertical loadings.
2. Raking Shoring — used to support a combination of vertical and horizontal loadings.
3. Flying Shoring — an alternative to raking shoring to give a clear working space at ground level.

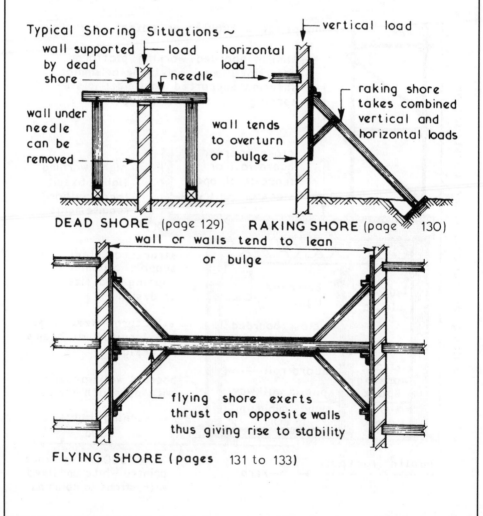

Typical Shoring Situations ~

wall supported by dead shore — load — needle / horizontal load — vertical load

wall under needle can be removed — wall tends to overturn or bulge → / raking shore takes combined vertical and horizontal loads

DEAD SHORE (page 129)   **RAKING SHORE** (page 130)

wall or walls tend to lean or bulge

flying shore exerts thrust on opposite walls thus giving rise to stability

FLYING SHORE (pages 131 to 133)

Dead Shores ~ these shores should be placed at approximately 2·000 c/c and positioned under the piers between the windows, any windows in the vicinity of the shores being strutted to prevent distortion of the openings. A survey should be carried out to establish the location of any underground services so that they can be protected as necessary. The sizes shown in the detail below are typical, actual sizes should be obtained from tables or calculated from first principles. Any suitable structural material such as steel can be substituted for the timber members shown.

Typical Detail ~

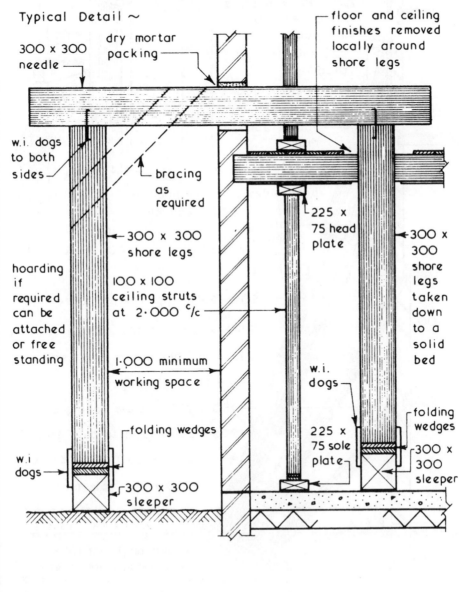

300 x 300 needle

dry mortar packing

floor and ceiling finishes removed locally around shore legs

w.i. dogs to both sides

bracing as required

300 x 300 shore legs

100 x 100 ceiling struts at 2·000 c/c

1·000 minimum working space

hoarding if required can be attached or free standing

folding wedges

w.i dogs

300 x 300 sleeper

225 x 75 head plate

w.i. dogs

225 x 75 sole plate

300 x 300 shore legs taken down to a solid bed

folding wedges

300 x 300 sleeper

Raking Shoring ~ these are placed at 3·000 to 4·500 c/c and can be of single, double, triple or multiple raker format. Suitable materials are timber structural steel and framed tubular scaffolding.

Typical Multiple Raking Shore Detail ~

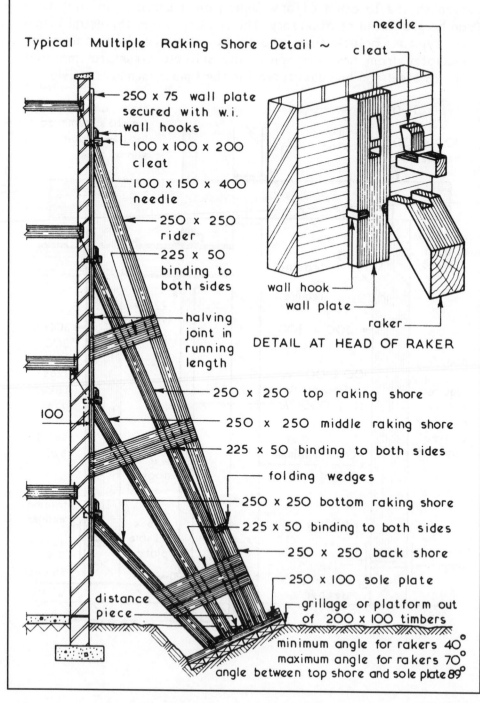

250 x 75 wall plate secured with w.i. wall hooks

100 x 100 x 200 cleat

100 x 150 x 400 needle

250 x 250 rider

225 x 50 binding to both sides

halving joint in running length

100

distance piece

needle

cleat

wall hook

wall plate

raker

DETAIL AT HEAD OF RAKER

250 x 250 top raking shore

250 x 250 middle raking shore

225 x 50 binding to both sides

folding wedges

250 x 250 bottom raking shore

225 x 50 binding to both sides

250 x 250 back shore

250 x 100 sole plate

grillage or platform out of 200 x 100 timbers

minimum angle for rakers 40°
maximum angle for rakers 70°
angle between top shore and sole plate 89°

Flying Shores ~ these are placed at 3·000 to 4·500 c/c and can be of a single or double format. They are designed, detailed and constructed to the same basic principles as that shown for raking shores on page 130. Unsymmetrical arrangements are possible providing the basic principles for flying shores are applied — see page 133.

Typical Single Flying Shore Detail ~

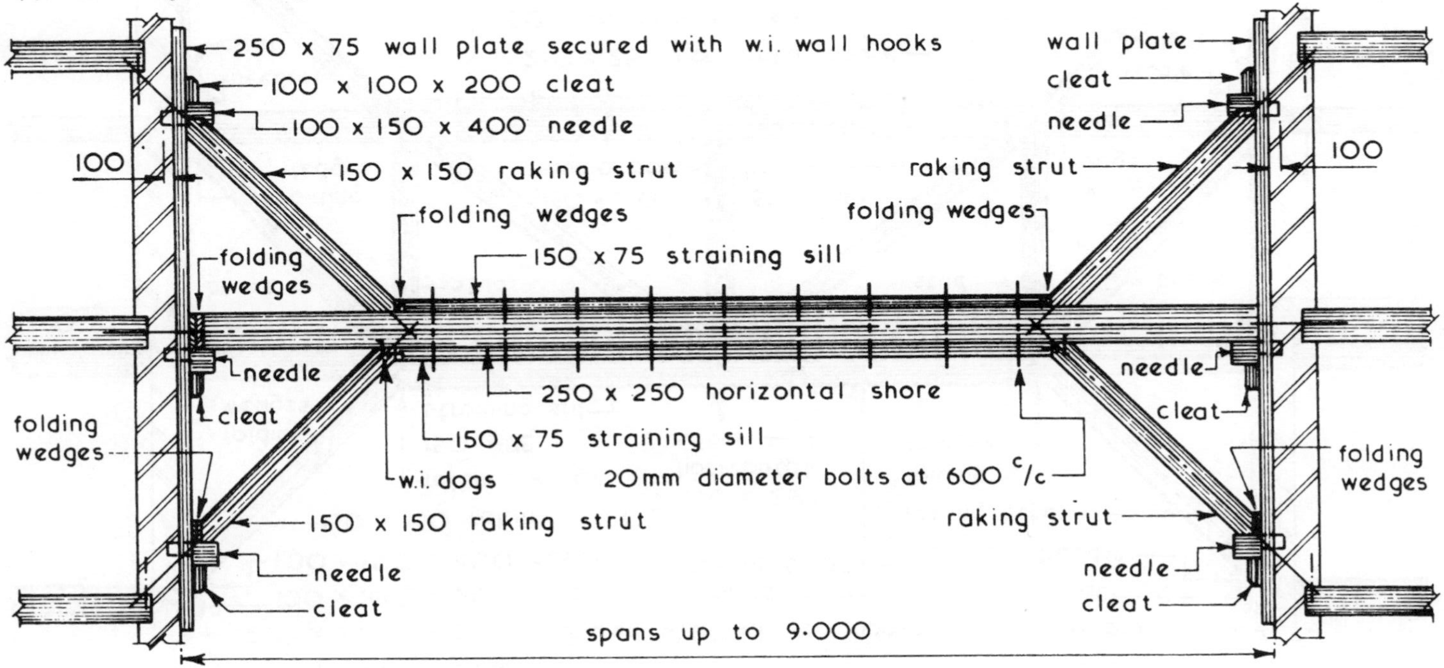

250 x 75 wall plate secured with w.i. wall hooks
100 x 100 x 200 cleat
100 x 150 x 400 needle
100
150 x 150 raking strut
folding wedges
150 x 75 straining sill
folding wedges
needle
cleat
folding wedges
250 x 250 horizontal shore
150 x 75 straining sill
w.i. dogs
20mm diameter bolts at 600 c/c
150 x 150 raking strut
needle
cleat
spans up to 9·000
wall plate
cleat
needle
100
raking strut
folding wedges
needle
cleat
folding wedges
raking strut
needle
cleat

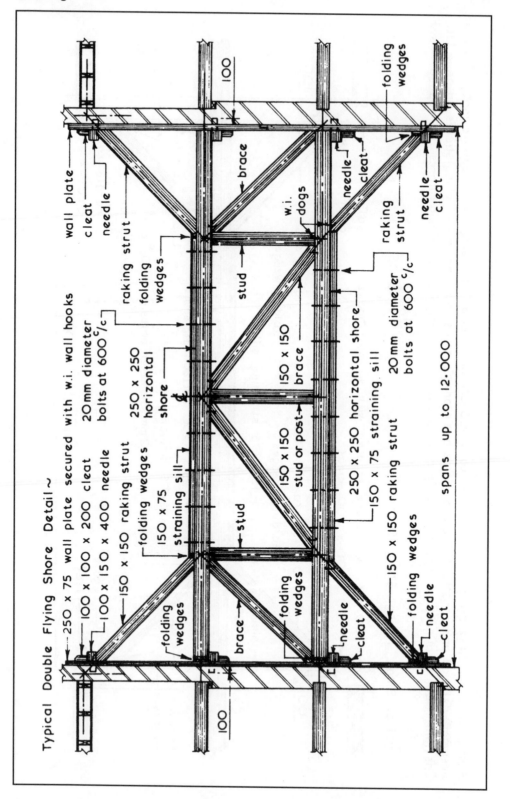

Typical Double Flying Shore Detail ~

250 x 75 wall plate secured with w.i. wall hooks
100 x 100 x 200 cleat
100 x 150 x 400 needle
150 x 150 raking strut
folding wedges
150 x 75 straining sill

wall plate
cleat
needle
100

20 mm diameter bolts at 600 c/c

250 x 250 horizontal shore
raking strut
folding wedges

brace

stud
w.i. dogs
150 x 150 brace
150 x 150 stud or post
250 x 250 horizontal shore
150 x 75 raking strut
20 mm diameter bolts at 600 c/c
250 x 250 straining sill

needle
cleat
raking strut
needle
cleat

folding wedges

stud
folding wedges

folding wedges
brace
folding wedges
needle
cleat
150 x 150 raking strut
folding wedges
needle
cleat

spans up to 12·000

100

Unsymmetrical Flying Shores ~ arrangements of flying shores for unsymmetrical situations can be devised if the basic principles for symmetrical shores is applied (see page 131). In some cases the arrangement will consist of a combination of both raking and flying shore principles.

Typical Examples~

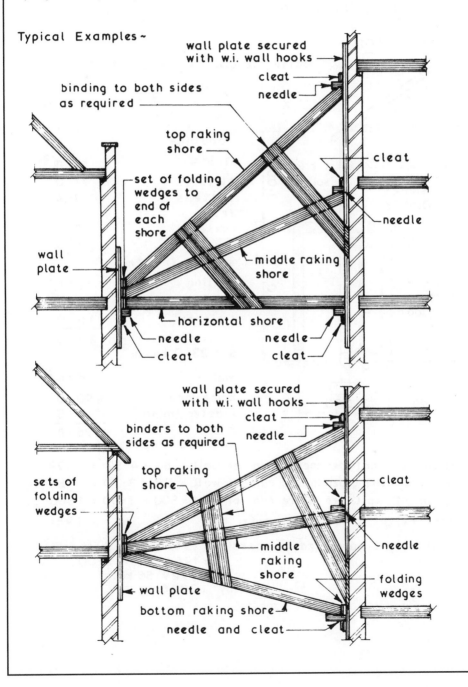

Temporary Support Determination ~ the basic sizing of most temporary supports follows the principles of elementary structural design. Readers with this basic knowledge should be able to calculate such support members which are required, particularly those used in the context of the maintenance and adaptation of buildings such as a dead shoring system.

Typical Example ~

Rafter length :- $\cos 35° = \dfrac{3 \cdot 7}{X}$

$\therefore X = \dfrac{3 \cdot 7}{\cos 35°} = \dfrac{3 \cdot 7}{0 \cdot 8192} = 4 \cdot 520$

Loadings :-

Roof ~	kg/m²
tiles	71·0
battens	3·4
felt	2·0
rafters	7·5
	83·9

say 84·0 kg/m²

Ceiling ~	kg/m²
joists	7·5
finishes	15·0
	22·5

say 23·0 kg/m²

Wall ~	kg/m²
brickwork	490·0
plaster finish	6·8
	496·8

say 500·0 kg/m²

Weight of roof per metre run of wall = 84 × 4·52 = 379·68
Weight of ceiling per metre run of wall = 23 × 3·70 = 85·10
Weight of wall per metre run of wall = 500 × 3·00 = 1500·00
Total weight of wall per meter run = 1964·78

Total weight supported by needle = 1964·78 × shore centres
= 1964·78 × 2·000
= 3929·56 kg
say 3930 kg

For design calculations see page 135

Design calculations reference page 134

Needle Design :-

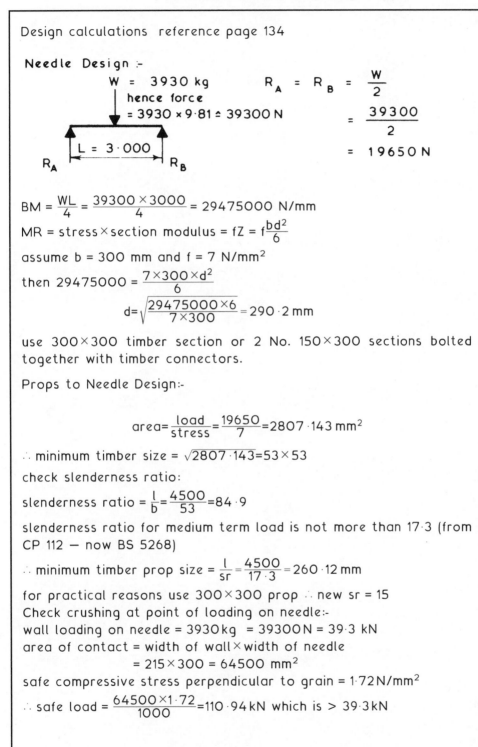

W = 3930 kg

hence force

= 3930 × 9·81 ≑ 39300 N

$R_A = R_B = \dfrac{W}{2}$

$= \dfrac{39300}{2}$

$= 19650\,N$

L = 3·000

R_A R_B

$BM = \dfrac{WL}{4} = \dfrac{39300 \times 3000}{4} = 29475000\ N/mm$

$MR = stress \times section\ modulus = fZ = f\dfrac{bd^2}{6}$

assume b = 300 mm and f = 7 N/mm²

then $29475000 = \dfrac{7 \times 300 \times d^2}{6}$

$d = \sqrt{\dfrac{29475000 \times 6}{7 \times 300}} = 290 \cdot 2\ mm$

use 300×300 timber section or 2 No. 150×300 sections bolted together with timber connectors.

Props to Needle Design:-

$$area = \frac{load}{stress} = \frac{19650}{7} = 2807 \cdot 143\ mm^2$$

∴ minimum timber size = $\sqrt{2807 \cdot 143} = 53 \times 53$

check slenderness ratio:

slenderness ratio = $\dfrac{l}{b} = \dfrac{4500}{53} = 84 \cdot 9$

slenderness ratio for medium term load is not more than 17·3 (from CP 112 — now BS 5268)

∴ minimum timber prop size = $\dfrac{l}{sr} = \dfrac{4500}{17 \cdot 3} = 260 \cdot 12\ mm$

for practical reasons use 300×300 prop ∴ new sr = 15
Check crushing at point of loading on needle:-
wall loading on needle = 3930 kg = 39300 N = 39·3 kN
area of contact = width of wall × width of needle
= 215×300 = 64500 mm²
safe compressive stress perpendicular to grain = 1·72 N/mm²

∴ safe load = $\dfrac{64500 \times 1 \cdot 72}{1000} = 110 \cdot 94\ kN$ which is > 39·3 kN

3 BUILDERS PLANT

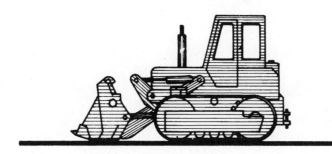

GENERAL CONSIDERATIONS
BULLDOZERS
SCRAPERS
GRADERS
TRACTOR SHOVELS
EXCAVATORS
TRANSPORT VEHICLES
HOISTS
RUBBLE CHUTES AND SKIPS
CRANES
CONCRETING PLANT

General Considerations ~ items of builders plant ranging from small hand held power tools to larger pieces of plant such as mechanical excavators and tower cranes can be considered for use for one or more of the following reasons :-

1. Increased production.
2. Reduction in overall construction costs.
3. Carry out activities which cannot be carried out by the traditional manual methods in the context of economics.
4. Eliminate heavy manual work thus reducing fatigue and as a consequence increasing productivity.
5. Replacing labour where there is a shortage of personnel with the necessary skills.
6. Maintain the high standards required particularly in the context of structural engineering works.

Economic Considerations ~ the introduction of plant does not always result in economic savings since extra temporary site works such as roadworks, hardstandings, foundations and anchorages may have to be provided at a cost which is in excess of the savings made by using the plant. The site layout and circulation may have to be planned around plant positions and movements rather than around personnel and material movements and accommodation. To be economic plant must be fully utilised and not left standing idle since plant, whether hired or owned, will have to be paid for even if it is non-productive. Full utilisation of plant is usually considered to be in the region of 85% of on site time, thus making an allowance for routine, daily and planned maintenance which needs to be carried out to avoid as far as practicable plant breakdowns which could disrupt the construction programme. Many pieces of plant work in conjunction with other items of plant such as excavators and their attendant haulage vehicles therefore a correct balance of such plant items must be obtained to achieve an economic result.

Maintenance Considerations ~ on large contracts where a number of plant items are to be used it may be advantageous to employ a skilled mechanic to be on site to carry out all the necessary daily, preventive and planned maintenance tasks together with any running repairs which could be carried out on site.

Plant Costing ~ with the exception of small pieces of plant, which are usually purchased, items of plant can be bought or hired or where there are a number of similar items a combination of buying and hiring could be considered. The choice will be governed by economic factors and the possibility of using the plant on future sites thus enabling the costs to be apportioned over several contracts.

Advantages of Hiring Plant:-

1. Plant can be hired for short periods.
2. Repairs and replacements are usually the responsibility of the hire company.
3. Plant is returned to the hire company after use thus relieving the building contractor of the problem of disposal or finding more work for the plant to justify its purchase or retention.
4. Plant can be hired with the operator, fuel and oil included in the hire rate.

Advantages of Buying Plant:-

1. Plant availability is totally within the control of the contractor.
2. Hourly cost of plant is generally less than hired plant.
3. Owner has choice of costing method used.

Typical Costing Methods ~

1. Straight Line — simple method

Captial Cost = £ 100 000
Anticipated life = 5 years
Year's working = 1500 hrs
Resale or scrap value = £ 9000
Annual depreciation ~

$$= \frac{100\,000 - 9000}{5} = £\,18\,200$$

Hourly depreciation ~

$$= \frac{18200}{1500} = 12 \cdot 13$$

Add 2% insurance =   0·27
10% maintenance =   1·33
          Hourly rate = £ 13 ·73

2. Interest on Capital Outlay- widely used more accurate method

Capital Cost	= £ 100 000
C.I. on capital (8% for 5 yrs)	= 46 930
	146 930
Deduct resale value	9 000
	137 930
+ Insurance at 2% =	2 000
+ Maintenance at 10% =	10 000
	149 930

Hourly rate ~

$$= \frac{149\,930}{5 \times 1500} = £\,20 \cdot 00$$

N.B. add to hourly rate running costs

Output and Cycle Times ~ all items of plant have optimum output and cycle times which can be used as a basis for estimating anticipated productivity taking into account the task involved, task efficiency of the machine, operator's efficiency and in the case of excavators the type of soil. Data for the factors to be taken into consideration can be obtained from timed observations, feedback information or published tables contained in manufacturer's literature or reliable textbooks.

Typical Example ~

Backacter with 1m³ capacity bucket engaged in normal trench excavation in a clayey soil and discharging directly into an attendant haulage vehicle.

Optimum output   = 60 bucket loads per hour
Task efficiency factor   = 0·8 (from tables)
Operator efficiency factor = 75% (typical figure)
∴ Anticipated output   = 60×0·8 × 0·75
                            = 36 bucket loads per hour
                            = 36 × 1 = 36 m³ per hour

An allowance should be made for the bulking or swell of the solid material due to the introduction of air or voids during the excavation process

∴ Net output allowing for a 30% swell = 36 −(36 × 0·3)
                                       = say 25 m³ per hr.

If the Bill of Quantities gives a total net excavation of 950 m³
time required = $\frac{950}{25}$ = 38 hours
or assuming an 8 hour day — 1/2 hour maintenance time in
days = $\frac{38}{7·5}$ = say 5 days
Haulage vehicles required = 1 + $\frac{\text{round trip time of vehicle}}{\text{loading time of vehicle}}$

If round trip time = 30 minutes and loading time = 10 mins.
number of haulage vehicles required = 1 + $\frac{30}{10}$ = 4
This gives a vehicle waiting overlap ensuring excavator is fully utilised which is economically desirable.

Bulldozers ~ these machines consist of a track or wheel mounted power unit with a mould blade at the front which is controlled by hydraulic rams. Many bulldozers have the capacity to adjust the mould blade to form an angledozer and the capacity to tilt the mould blade about a central swivel point. Some bulldozers can also be fitted with rear attachments such as rollers and scarifiers.

The main functions of a bulldozer are:-

1. Shallow excavations up to 300m deep either on level ground or sidehill cutting.
2. Clearance of shrubs and small trees.
3. Clearance of trees by using raised mould blade as a pusher arm.
4. Acting as a towing tractor.
5. Acting as a pusher to scraper machines (see page 143).

NB. Bulldozers push earth in front of the mould blade with some side spillage whereas angledozers push and cast the spoil to one side of the mould blade.

Typical Bulldozer Details ~

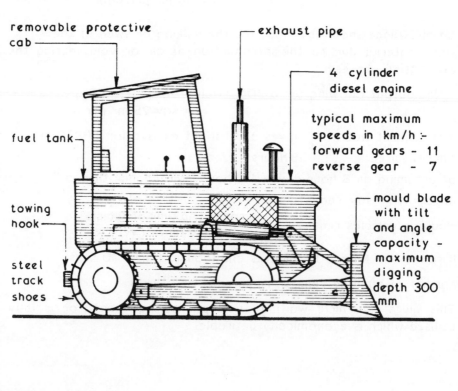

removable protective cab

exhaust pipe

4 cylinder diesel engine

typical maximum speeds in km/h :-
forward gears - 11
reverse gear - 7

fuel tank

mould blade with tilt and angle capacity - maximum digging depth 300 mm

towing hook

steel track shoes

Scrapers ~ these machines consist of a scraper bowl which is lowered to cut and collect soil where site stripping and levelling operations are required involving large volume of earth. When the scraper bowl is full the apron at the cutting edge is closed to retain the earth and the bowl is raised for travelling to the disposal area. On arrival the bowl is lowered, the apron opened and the spoil pushed out by the tailgate as the machine moves forwards. Scrapers are available in three basic formats:-

1. Towed Scrapers — these consist of a four wheeled scraper bowl which is towed behind a power unit such as a crawler tractor. They tend to be slower than other forms of scraper but are useful for small capacities with haul distances up to 300·00.

2. Two Axle Scrapers — these have a two wheeled scraper bowl with an attached two wheeled power unit. They are very manoeuvrable with a low rolling resistance and very good traction.

3. Three Axle Scrapers — these consist of a two wheeled scraper bowl which may have a rear engine to assist the four wheeled traction engine which makes up the complement. Generally these machines have a greater capacity potential than their counterparts, are easier to control and have a faster cycle time.

To obtain maximum efficiency scrapers should operate downhill if possible, have smooth haul roads, hard surfaces broken up before scraping and be assisted over the last few metres by a pushing vehicle such as a bulldozer.

Typical Scraper Details ~

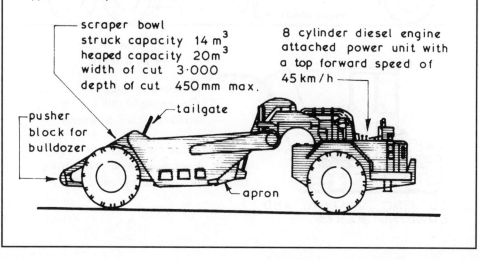

scraper bowl
struck capacity 14 m^3
heaped capacity 20m^3
width of cut 3·000
depth of cut 450mm max.

8 cylinder diesel engine attached power unit with a top forward speed of 45 km/h

pusher block for bulldozer

tailgate

apron

Graders

Graders ~ these machines are similar in concept to bulldozers in that they have a long slender adjustable mould blade, which is usually slung under the centre of the machine. A grader's main function is to finish or grade the upper surface of a large area usually as a follow up operation to scraping or bulldozing. They can produce a fine and accurate finish but do not have the power of a bulldozer therefore they are not suitable for oversite excavation work. The mould blade can be adjusted in both the horizontal and vertical planes through an angle of 300° the latter enabling it to be used for grading sloping banks.

Two basic formats of grader are available:-

1. Four Wheeled — all wheels are driven and steered which gives the machine the ability to offset and crab along its direction of travel.

2. Six Wheeled — this machine has 4 wheels in tandem drive at the rear and 2 front tilting idler wheels giving it the ability to counteract side thrust.

Typical Grader Details ~

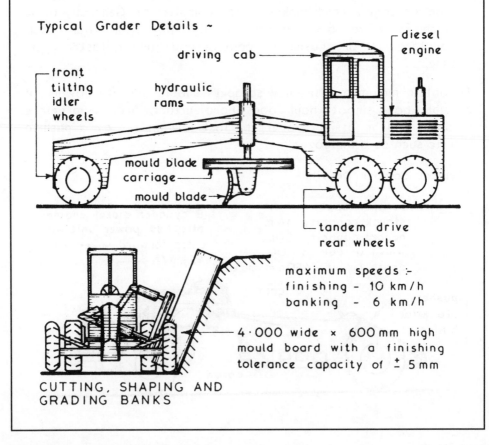

maximum speeds :-
finishing - 10 km/h
banking - 6 km/h

4·000 wide × 600 mm high mould board with a finishing tolerance capacity of ± 5mm

CUTTING, SHAPING AND GRADING BANKS

Tractor Shovels ~ these machines are sometimes called loaders or loader shovels and primary function is to scoop up loose materials in the front mounted bucket, elevate the bucket and manoeuvre into a position to deposit the loose material into an attendant transport vehicle. Tractor shovels are driven towards the pile of loose material with the bucket lowered, the speed and power of the machine will enable the bucket to be filled. Both tracked and wheeled versions are available, the tracked format being more suitable for wet and uneven ground conditions than the wheeled tractor shovel which has greater speed and manoeuvring capabilities. To increase their versatility tractor shovels can be fitted with a 4 in 1 bucket enabling them to carry out bulldozing, excavating, clam lifting and loading activities.

Typical Tractor Shovel Details ~

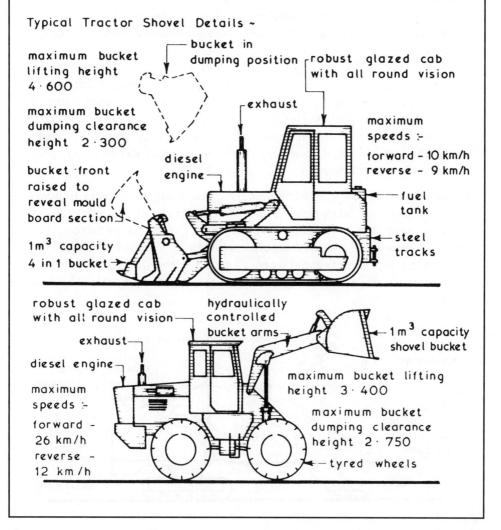

maximum bucket lifting height 4·600

maximum bucket dumping clearance height 2·300

bucket front raised to reveal mould board section

1m³ capacity 4 in 1 bucket

bucket in dumping position

exhaust

diesel engine

robust glazed cab with all round vision

maximum speeds :-

forward – 10 km/h
reverse – 9 km/h

fuel tank

steel tracks

robust glazed cab with all round vision

exhaust

diesel engine

maximum speeds :-

forward – 26 km/h

reverse – 12 km/h

hydraulically controlled bucket arms

1 m³ capacity shovel bucket

maximum bucket lifting height 3·400

maximum bucket dumping clearance height 2·750

tyred wheels

Excavating Machines ~ these are one of the major items of builders plant and are used primarily to excavate and load most types of soil. Excavating machines come in a wide variety of designs and sizes but all of them can be placed within one of three categories :-

1. Universal Excavators — this category covers most forms of excavators all of which have a common factor the power unit. The universal power unit is a tracked based machine with a slewing capacity of 360° and by altering the boom arrangement and bucket type different excavating functions can be obtained. These machines are selected for high output requirements and are rope controlled.

2. Purpose Designed Excavators — these are machines which have been designed specifically to carry out one mode of excavation and they usually have smaller bucket capacities than universal excavators; they are hydraulically controlled with a shorter cycle time.

3. Multi-purpose Excavators — these machines can perform several excavating functions having both front and rear attachments. They are designed to carry out small excavation operations of low output quickly and efficiently. Multi-purpose excavators can be obtained with a wheeled or tracked base and are ideally suited for a small building firm with low excavation plant utilisation requirements.

Skimmers ~ these excavators are rigged using a universal power unit for surface stripping and shallow excavation work up to 300mm deep where a high degree of accuracy is required. They usually require attendant haulage vehicles to remove the spoil and need to be transported between sites on a low-loader. Because of their limitations and the alternative machines available they are seldom used today.

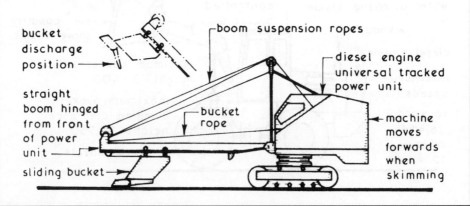

bucket discharge position →

boom suspension ropes

diesel engine universal tracked power unit

straight boom hinged from front of power unit

bucket rope

machine moves forwards when skimming

sliding bucket →

Face Shovels ~ the primary function of this piece of plant is to excavate above its own track or wheel level. They are available as a universal power unit based machine or as a hydraulic purpose designed unit. These machines can usually excavate any type of soil except rock which needs to be loosened, usually by blasting, prior to excavation. Face shovels generally require attendant haulage vehicles for the removal of spoil and a low loader transport lorry for travel between sites. Most of these machines have a limited capacity of between 300 and 400mm for excavation below their own track or wheel level.

Typical Face Shovel Details ~

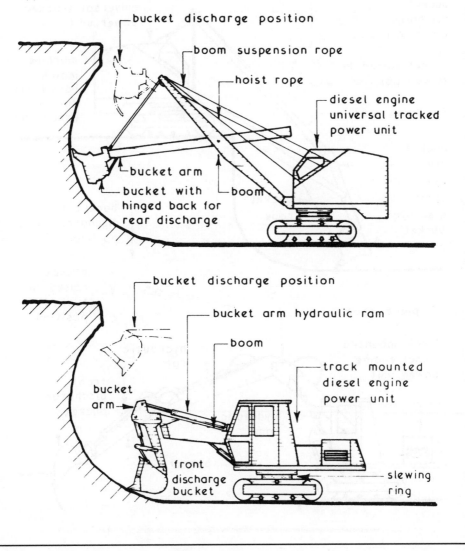

bucket discharge position

boom suspension rope

hoist rope

diesel engine universal tracked power unit

bucket arm

bucket with hinged back for rear discharge

boom

bucket discharge position

bucket arm hydraulic ram

boom

track mounted diesel engine power unit

bucket arm

front discharge bucket

slewing ring

Backacters ~ these machines are suitable for trench, foundation and basement excavations and are available as a universal power unit base machine or as a purpose designed hydraulic unit. They can be used with or without attendant haulage vehicles since the spoil can be placed alongside the excavation for use in backfilling. These machines will require a low loader transport vehicle for travel between sites. Backacters used in trenching operations with a bucket width equal to the trench width can be very accurate with a high output rating.

Typical Backacter Details ~

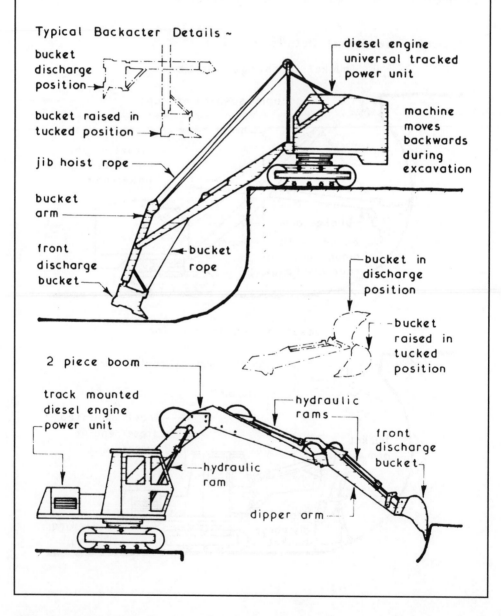

bucket discharge position

bucket raised in tucked position

jib hoist rope

bucket arm

front discharge bucket

bucket rope

diesel engine universal tracked power unit

machine moves backwards during excavation

bucket in discharge position

bucket raised in tucked position

2 piece boom

track mounted diesel engine power unit

hydraulic ram

hydraulic rams

front discharge bucket

dipper arm

Draglines ~ these machines are based on the universal power unit with basic crane rigging to which is attached a drag bucket. The machine is primarily designed for bulk excavation in loose soils up to 3·000 below its own track level by swinging the bucket out to the excavation position and hauling or dragging it back towards the power unit. Dragline machines can also be fitted with a grab or clamshell bucket for excavating in very loose soils.

Typical Dragline Details ~

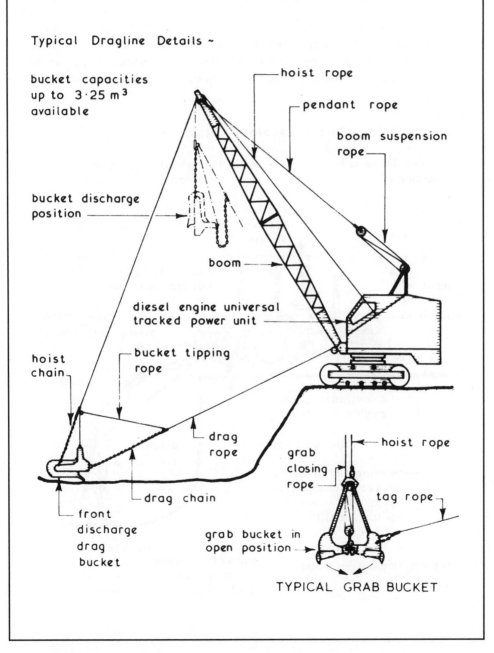

bucket capacities up to 3·25 m³ available

hoist rope

pendant rope

boom suspension rope

bucket discharge position

boom

diesel engine universal tracked power unit

hoist chain

bucket tipping rope

drag rope

drag chain

front discharge drag bucket

grab closing rope

hoist rope

tag rope

grab bucket in open position

TYPICAL GRAB BUCKET

149

Multi-purpose Excavators ~ these machines are usually based on the agricultural tractor with 2 or 4 wheel drive and are intended mainly for use in conjunction with small excavation works such as those encountered by the small to medium sized building contractor. Most multi-purpose excavators are fitted with a loading/excavating front bucket and a rear backacter bucket both being hydraulically controlled. When in operation using the backacter bucket the machine is raised off its axles by rear mounted hydraulic outriggers or jacks and in some models by placing the front bucket on the ground. Most machines can be fitted with a variety of bucket widths and various attachments such as bulldozer blades, scarifiers, grab buckets and post hole auger borers.

Typical Multi-purpose Excavator Details ~

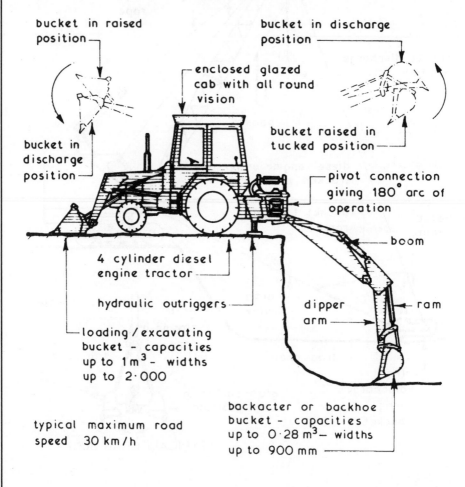

bucket in raised position

bucket in discharge position

enclosed glazed cab with all round vision

bucket raised in tucked position

bucket in discharge position

pivot connection giving 180° arc of operation

boom

4 cylinder diesel engine tractor

hydraulic outriggers

dipper arm

ram

loading/excavating bucket - capacities up to 1m³ - widths up to 2·000

backacter or backhoe bucket - capacities up to 0·28 m³ - widths up to 900 mm

typical maximum road speed 30 km/h

Transport Vehicles ~ these can be defined as vehicles whose primary function is to convey passengers and/or materials between and around building sites. The types available range from the conventional saloon car to the large low loader lorries designed to transport other items of builders plant between construction sites and the plant yard or depot.

Vans — these transport vehicles range from the small two person plus a limited amount of materials to the large vans with purpose designed bodies such as those built to carry large sheets of glass. Most small vans are usually fitted with a petrol engine and are based on the manufacturer's standard car range whereas the larger vans are purpose designed with either petrol or diesel engines. These basic designs can usually be supplied with an uncovered tipping or non-tipping container mounted behind the passenger cab for use as a `pick-up' truck.

Passenger Vehicles — these can range from a simple framed cabin which can be placed in the container of a small lorry or `pick-up' truck to a conventional bus or coach. Vans can also be designed to carry a limited number of seated passengers by having fixed or removable seating together with windows fitted in the van sides thus giving the vehicle a dual function. The number of passengers carried can be limited so that the driver does not have to hold a PSV (public service vehicle) licence.

Lorries — these are sometimes referred to as haul vehicles and are available as road or site only vehicles. Road haulage vehicles have to comply with all the requirements of the Road Traffic Acts which among other requirements limits size and axle loads. The off- highway or site only lorries are not so restricted and can be designed to carry two to three times the axle load allowed on the public highway. Site only lorries are usually specially designed to traverse and withstand the rough terrain encountered on many construction sites. Lorries are available as non-tipping, tipping and special purpose carriers such as those with removable skips and those equipped with self loading and unloading devices. Lorries specifically designed for the transportation of large items of plant are called low loaders and are usually fitted with integral or removable ramps to facilitate loading and some have a winching system to haul the plant onto the carrier platform.

Dumpers ~ these are used for the horizontal transportation of materials on and off construction sites generally by means of an integral tipping skip. Highway dumpers are of a similar but larger design and can be used to carry materials such as excavated spoil along the roads. A wide range of dumpers are available of various carrying capacities and options for gravity or hydraulic discharge control with front tipping, side tipping or elevated tipping facilities. Special format dumpers fitted with flat platforms, rigs to carry materials skips and rigs for concrete skips for crane hoisting are also obtainable. These machines are designed to traverse rough terrain but they are not designed to carry passengers and this misuse is the cause of many accidents involving dumpers.

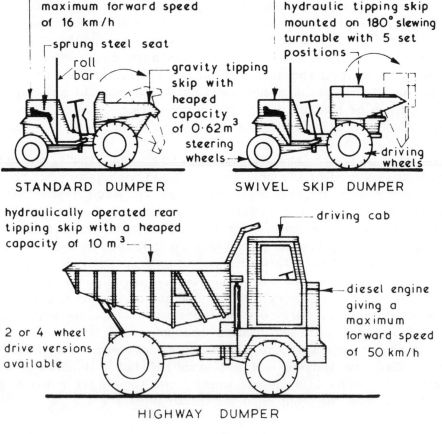

Typical Dumper Details ~

diesel engine giving a maximum forward speed of 16 km/h

sprung steel seat

roll bar

gravity tipping skip with heaped capacity of $0.62 \, m^3$

steering wheels

STANDARD DUMPER

diesel engine giving a maximum forward speed of 16 km/h

hydraulic tipping skip mounted on 180° slewing turntable with 5 set positions

driving wheels

SWIVEL SKIP DUMPER

hydraulically operated rear tipping skip with a heaped capacity of $10 \, m^3$

driving cab

diesel engine giving a maximum forward speed of 50 km/h

2 or 4 wheel drive versions available

HIGHWAY DUMPER

Fork Lift Trucks ~ these are used for the horizontal and limited vertical transportation of materials positioned on pallets or banded together such as brick packs. They are generally suitable for construction sites where the building height does not exceed three storeys. Although designed to negotiate rough terrain site fork lift trucks have a higher productivity on firm and level soils. Three basic fork lift truck formats are available namely straight mast, overhead and telescopic boom with various height, reach and lifting capacities. Scaffolds onto which the load(s) are to be placed should be strengthened locally or a specially constructed loading tower could be built as an attachment to or as an integral part of the main scaffold.

Typical Fork Lift Truck Details ~

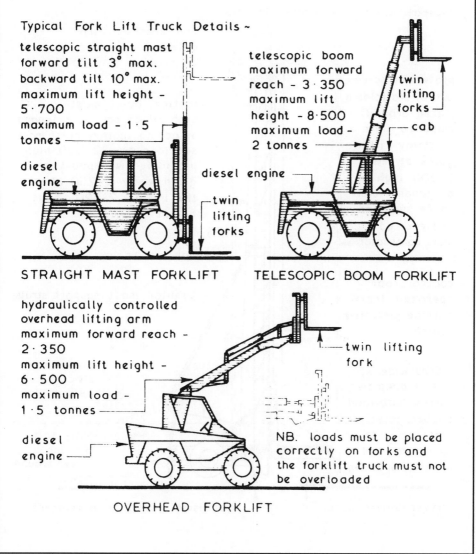

telescopic straight mast
forward tilt 3° max.
backward tilt 10° max.
maximum lift height –
5·700
maximum load – 1·5
tonnes

diesel
engine

telescopic boom
maximum forward
reach – 3·350
maximum lift
height – 8·500
maximum load –
2 tonnes

twin
lifting
forks

cab

diesel engine

twin
lifting
forks

STRAIGHT MAST FORKLIFT   TELESCOPIC BOOM FORKLIFT

hydraulically controlled
overhead lifting arm
maximum forward reach –
2·350
maximum lift height –
6·500
maximum load –
1·5 tonnes

diesel
engine

twin lifting
fork

NB. loads must be placed
correctly on forks and
the forklift truck must not
be overloaded

OVERHEAD FORKLIFT

Hoists ~ these are designed for the vertical transportation of materials, passengers or materials and passengers (see page 155). Materials hoists are designed for one specific use (i.e. the vertical transportation of materials) and under no circumstances should they be used to transport passengers. Most material hoists are of a mobile format which can be dismantled, folded onto the chassis and moved to another position or site under their own power or towed by a haulage vehicle. When in use material hoists need to be stabilised and/or tied to the structure and enclosed with a protective screen.

Typical Materials Hoist Details ~

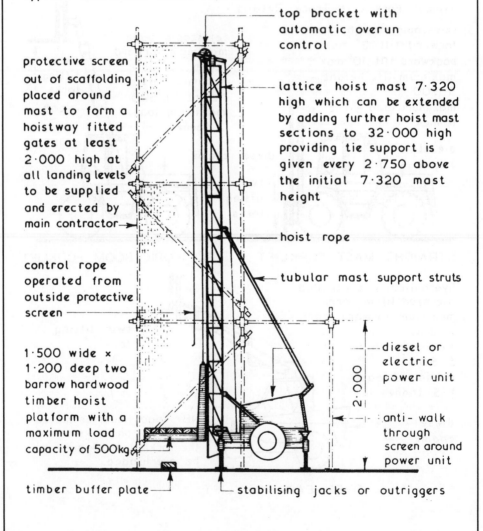

top bracket with automatic overrun control

protective screen out of scaffolding placed around mast to form a hoistway fitted gates at least 2·000 high at all landing levels to be supplied and erected by main contractor

lattice hoist mast 7·320 high which can be extended by adding further hoist mast sections to 32·000 high providing tie support is given every 2·750 above the initial 7·320 mast height

hoist rope

control rope operated from outside protective screen

tubular mast support struts

1·500 wide × 1·200 deep two barrow hardwood timber hoist platform with a maximum load capacity of 500kg

diesel or electric power unit

2·000

anti- walk through screen around power unit

timber buffer plate

stabilising jacks or outriggers

Passenger Hoists ~ these are designed to carry passengers although most are capable of transporting a combined load of materials and passengers within the lifting capacity of the hoist. A wide selection of hoists are available ranging from a single cage with rope suspension to twin cages with rack and pinion operation mounted on two sides of a static tower.

Typical Passenger Hoist Details ~

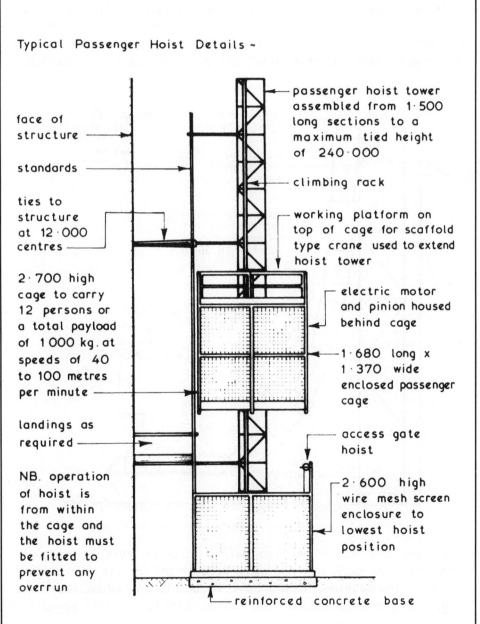

face of structure

standards

ties to structure at 12·000 centres

2·700 high cage to carry 12 persons or a total payload of 1000 kg. at speeds of 40 to 100 metres per minute

landings as required

NB. operation of hoist is from within the cage and the hoist must be fitted to prevent any overrun

passenger hoist tower assembled from 1·500 long sections to a maximum tied height of 240·000

climbing rack

working platform on top of cage for scaffold type crane used to extend hoist tower

electric motor and pinion housed behind cage

1·680 long x 1·370 wide enclosed passenger cage

access gate hoist

2·600 high wire mesh screen enclosure to lowest hoist position

reinforced concrete base

Rubble Chutes and Skips

Rubble Chutes ~ these apply to contracts involving demolition, repair, maintenance and refurbishment. The simple concept of connecting several perforated dustbins is reputed to have been conceived by an ingenious site operative for the expedient and safe conveyance of materials.

In purpose designed format, the tapered cylinders are produced from reinforced rubber with chain linkage for continuity. Overall unit lengths are generally 1100 mm, providing an effective length of 1m. Hoppers and side entry units are made for special applications.

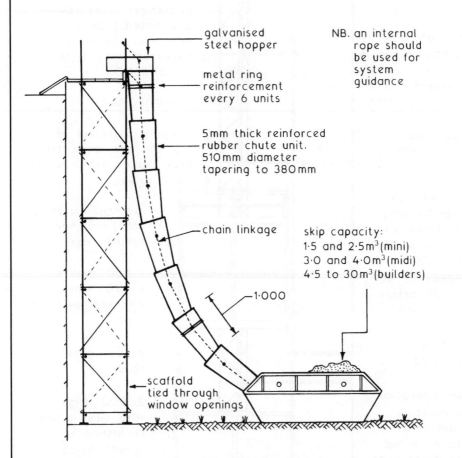

galvanised steel hopper

NB. an internal rope should be used for system guidance

metal ring reinforcement every 6 units

5mm thick reinforced rubber chute unit, 510mm diameter tapering to 380mm

chain linkage

skip capacity:
1·5 and 2·5m³(mini)
3·0 and 4·0m³(midi)
4·5 to 30m³(builders)

1·000

scaffold tied through window openings

Ref. Highways Act — written permit (license) must be obtained from the local authority highways department for use of a skip on a public thoroughfare. It will have to be illuminated at night and may require a temporary traffic light system to regulate vehicles.

Cranes ~ these are lifting devices designed to raise materials by means of rope operation and move the load horizontally within the limitations of any particular machine. The range of cranes available is very wide and therefore choice must be based on the loads to be lifted, height and horizontal distance to be covered, time period(s) of lifting operations, utilisation factors and degree of mobility required. Crane types can range from a simple rope and pulley or gin wheel to a complex tower crane but most can be placed within 1 of 3 groups namely mobile, static and tower cranes.

Typical Crane Classifications ~

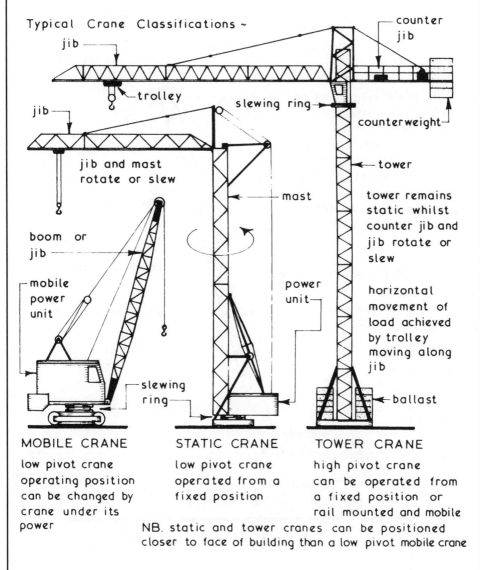

MOBILE CRANE
low pivot crane operating position can be changed by crane under its power

STATIC CRANE
low pivot crane operated from a fixed position

TOWER CRANE
high pivot crane can be operated from a fixed position or rail mounted and mobile

NB. static and tower cranes can be positioned closer to face of building than a low pivot mobile crane

Self Propelled Cranes ~ these are mobile cranes mounted on a wheeled chassis and have only one operator position from which the crane is controlled and the vehicle driven. The road speed of this type of crane is generally low usually not exceeding 30km p.h. A variety of self propelled crane formats are available ranging from short height lifting strut booms of fixed length to variable length lattice booms with a fly jib attachment.

Typical Self Propelled Crane Details ~

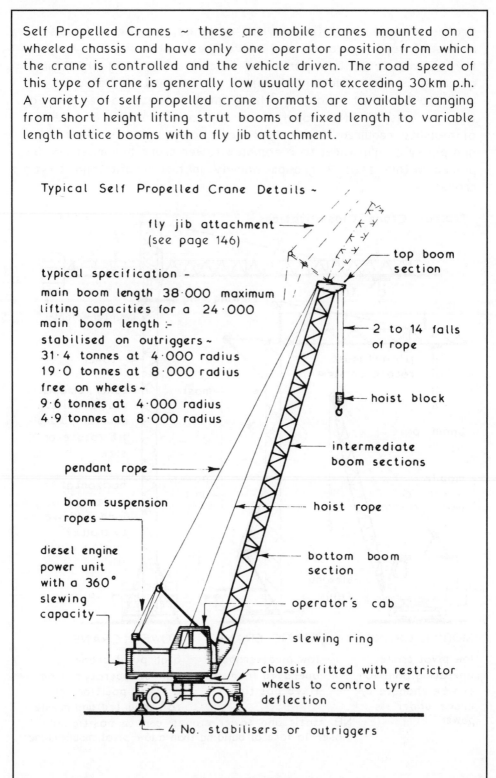

fly jib attachment ——
(see page 146)

typical specification -
main boom length 38·000 maximum
lifting capacities for a 24·000
main boom length :-
stabilised on outriggers ~
31·4 tonnes at 4·000 radius
19·0 tonnes at 8·000 radius
free on wheels ~
9·6 tonnes at 4·000 radius
4·9 tonnes at 8·000 radius

top boom section

2 to 14 falls of rope

hoist block

intermediate boom sections

pendant rope ——

boom suspension ropes ——

hoist rope

diesel engine power unit with a 360° slewing capacity ——

bottom boom section

operator's cab

slewing ring

chassis fitted with restrictor wheels to control tyre deflection

4 No. stabilisers or outriggers

Lorry Mounted Cranes ~ these mobile cranes consist of a lattice or telescopic boom mounted on a specially adapted truck or lorry. They have two operating positions: the lorry being driven from a conventional front cab and the crane being controlled from a different location. The lifting capacity of these cranes can be increased by using outrigger stabilising jacks and the approach distance to the face of building decreased by using a fly jib. Lorry mounted telescopic cranes require a firm surface from which to operate and because of their short site preparation time they are ideally suited for short hire periods.

Typical Lorry Mounted Telescopic Crane Details ~

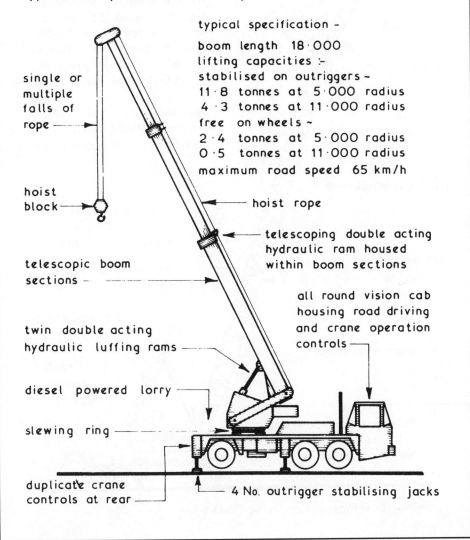

typical specification -
boom length 18·000
lifting capacities :-
stabilised on outriggers ~
11·8 tonnes at 5·000 radius
4·3 tonnes at 11·000 radius
free on wheels ~
2·4 tonnes at 5·000 radius
0·5 tonnes at 11·000 radius
maximum road speed 65 km/h

single or multiple falls of rope

hoist block

telescopic boom sections

twin double acting hydraulic luffing rams

diesel powered lorry

slewing ring

duplicate crane controls at rear

hoist rope

telescoping double acting hydraulic ram housed within boom sections

all round vision cab housing road driving and crane operation controls

4 No. outrigger stabilising jacks

Lorry Mounted Lattice Jib Cranes ~ these cranes follow the same basic principles as the lorry mounted telescopic cranes but they have a lattice boom and are designed as heavy duty cranes with lifting capacities in excess of 100 tonnes. These cranes will require a firm level surface from which to operate and can have a folding or sectional jib which will require the crane to be rigged on site before use.

Typical Lorry Mounted Lattice Jib Crane Details ~

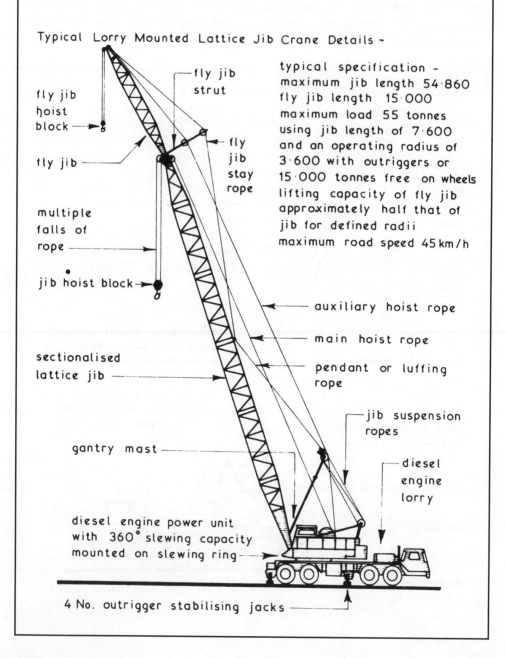

typical specification –
maximum jib length 54·860
fly jib length 15·000
maximum load 55 tonnes
using jib length of 7·600
and an operating radius of
3·600 with outriggers or
15·000 tonnes free on wheels
lifting capacity of fly jib
approximately half that of
jib for defined radii
maximum road speed 45 km/h

fly jib strut

fly jib hoist block

fly jib stay rope

fly jib

multiple falls of rope

jib hoist block

auxiliary hoist rope

main hoist rope

pendant or luffing rope

sectionalised lattice jib

jib suspension ropes

gantry mast

diesel engine lorry

diesel engine power unit with 360° slewing capacity mounted on slewing ring

4 No. outrigger stabilising jacks

Track Mounted Cranes ~ these machines can be a universal power unit rigged as a crane (see page 149) or a purpose designed track mounted crane with or without a fly jib attachment. The latter type are usually more powerful with lifting capacities up to 45 tonnes. Track mounted cranes can travel and carry out lifting operations on most sites without the need for special road and hardstand provisions but they have to be rigged on arrival after being transported to site on a low loader lorry.

Typical Track Mounted or Crawler Crane Details ~

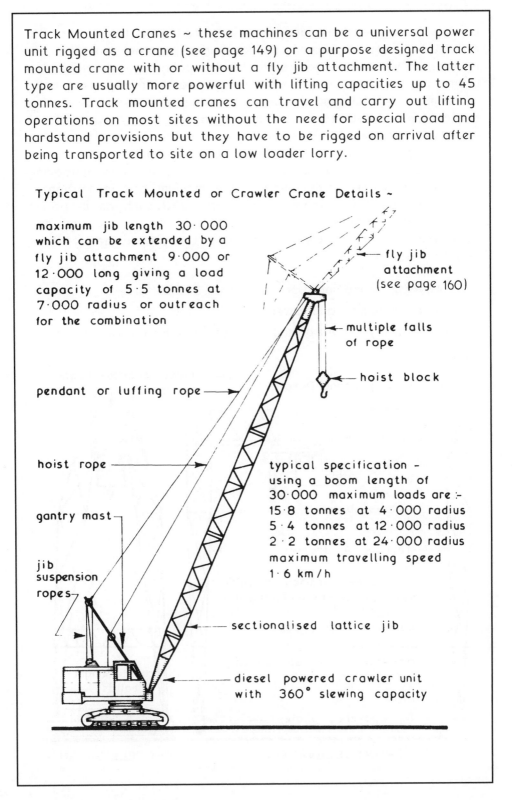

maximum jib length 30·000 which can be extended by a fly jib attachment 9·000 or 12·000 long giving a load capacity of 5·5 tonnes at 7·000 radius or outreach for the combination

fly jib attachment (see page 160)

multiple falls of rope

hoist block

pendant or luffing rope ——→

hoist rope ——————→

gantry mast

jib suspension ropes

typical specification – using a boom length of 30·000 maximum loads are :-
15·8 tonnes at 4·000 radius
5·4 tonnes at 12·000 radius
2·2 tonnes at 24·000 radius
maximum travelling speed 1·6 km/h

sectionalised lattice jib

diesel powered crawler unit with 360° slewing capacity

Cranes

Gantry Cranes ~ these are sometimes called portal cranes and consist basically of two 'A' frames joined together with a cross member on which transverses the lifting appliance. In small gantry cranes (up to 10 tonnes lifting capacity) the 'A' frames are usually wheel mounted and manually propelled whereas in the large gantry cranes (up to 100 tonnes lifting capacity) the 'A' frames are mounted on powered bogies running on rail tracks with the driving cab and lifting gear mounted on the cross beam or gantry. Small gantry cranes are used primarily for loading and unloading activities in stock yards whereas the medium and large gantry cranes are used to straddle the work area such as in power station construction or in repetitive low to medium rise developments. All gantry cranes have the advantage of three direction movement —

1. Transverse by moving along the cross beam.

2. Vertical by raising and lowering the hoist block.

3. Horizontal by forward and reverse movements of the whole gantry crane.

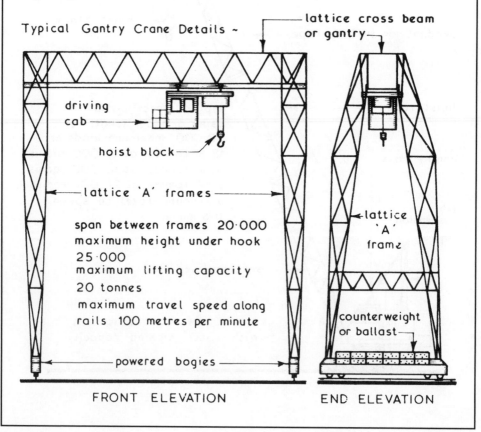

Typical Gantry Crane Details ~

lattice cross beam or gantry

driving cab

hoist block

lattice 'A' frames

span between frames 20·000
maximum height under hook 25·000
maximum lifting capacity 20 tonnes
maximum travel speed along rails 100 metres per minute

lattice 'A' frame

counterweight or ballast

powered bogies

FRONT ELEVATION

END ELEVATION

Mast Cranes ~ these are similar in appearance to the familiar tower cranes but they have one major difference in that the mast or tower is mounted on the slewing ring and thus rotates whereas a tower crane has the slewing ring at the top of the tower and therefore only the jib portion rotates. Mast cranes are often mobile, self erecting, of relatively low lifting capacity and are usually fitted with a luffing jib. A wide variety of models are available and have the advantage over most mobile low pivot cranes of a closer approach to the face of the building.

Typical Mast Crane Details ~

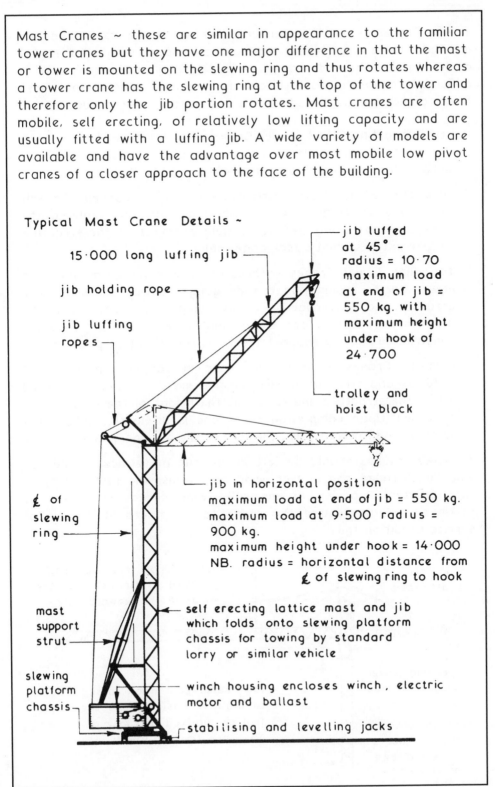

15·000 long luffing jib

jib holding rope

jib luffing ropes

jib luffed at 45° - radius = 10·70 maximum load at end of jib = 550 kg. with maximum height under hook of 24·700

trolley and hoist block

jib in horizontal position
maximum load at end of jib = 550 kg.
maximum load at 9·500 radius = 900 kg.
maximum height under hook = 14·000
NB. radius = horizontal distance from ₵ of slewing ring to hook

₵ of slewing ring

mast support strut

self erecting lattice mast and jib which folds onto slewing platform chassis for towing by standard lorry or similar vehicle

slewing platform chassis

winch housing encloses winch, electric motor and ballast

stabilising and levelling jacks

Tower Cranes ~ most tower cranes have to be assembled and erected on site prior to use and can be equipped with a horizontal or luffing jib. The wide range of models available often make it difficult to choose a crane suitable for any particular site but most tower cranes can be classified into one of four basic groups thus:-

1. Self Supporting Static Tower Cranes — high lifting capacity with the mast or tower fixed to a foundation base — they are suitable for confined and open sites. (see page 165)

2. Supported Static Tower Cranes — similar in concept to self supporting cranes and are used where high lifts are required, the mast or tower being tied at suitable intervals to the structure to give extra stability. (see page 166)

3. Travelling Tower Cranes — these are tower cranes mounted on power bogies running on a wide gauge railway track to give greater site coverage — only slight gradients can be accommodated therefore a reasonably level site or specially constructed railway support trestle is required. (see page 167)

4. Climbing Cranes — these are used in conjunction with tall buildings and structures. The climbing mast or tower is housed within the structure and raised as the height of the structure is increased. Upon completion the crane is dismantled into small sections and lowered down the face of the building. (see page 168)

All tower cranes should be left in an `out of service` condition when unattended and in high wind conditions, the latter varying with different models but generally wind speeds in excess of 60 km p.h. would require the crane to be placed in an out of service condition thus:-

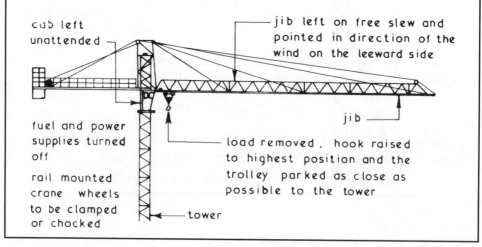

cab left unattended

jib left on free slew and pointed in direction of the wind on the leeward side

fuel and power supplies turned off

rail mounted crane wheels to be clamped or chocked

load removed, hook raised to highest position and the trolley parked as close as possible to the tower

jib

tower

Typical Self Supporting Static Tower Crane Details ~

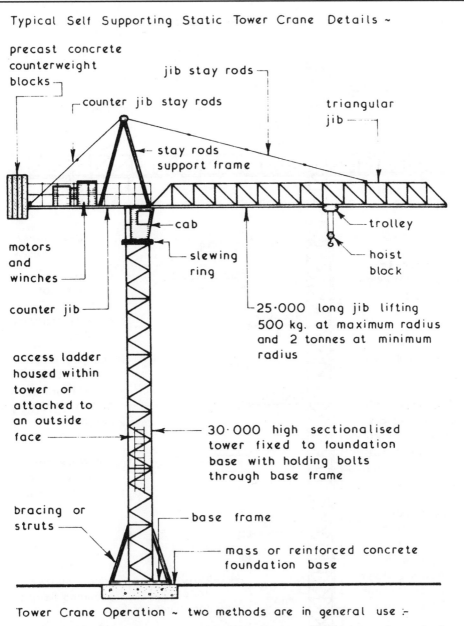

precast concrete counterweight blocks

counter jib stay rods

jib stay rods

triangular jib

stay rods support frame

cab

slewing ring

trolley

hoist block

motors and winches

counter jib

25·000 long jib lifting 500 kg. at maximum radius and 2 tonnes at minimum radius

access ladder housed within tower or attached to an outside face

30·000 high sectionalised tower fixed to foundation base with holding bolts through base frame

bracing or struts

base frame

mass or reinforced concrete foundation base

Tower Crane Operation ~ two methods are in general use :-

1 Cab Control - the crane operator has a good view of most of the lifting operations from the cab mounted at top of the tower but a second person or banksman is required to give clear signals to the crane operator and to load the crane

2 Remote Control - the crane operator carries a control box linked by a wandering lead to the crane controls.

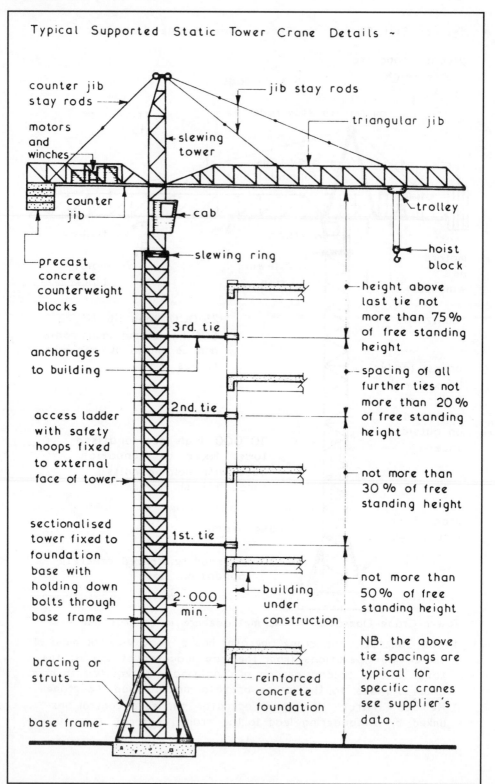

Typical Supported Static Tower Crane Details ~

counter jib stay rods

jib stay rods

motors and winches

triangular jib

slewing tower

counter jib

cab

trolley

hoist block

precast concrete counterweight blocks

slewing ring

anchorages to building

3rd. tie

height above last tie not more than 75% of free standing height

2nd. tie

spacing of all further ties not more than 20% of free standing height

access ladder with safety hoops fixed to external face of tower

not more than 30% of free standing height

1st. tie

not more than 50% of free standing height

sectionalised tower fixed to foundation base with holding down bolts through base frame

2·000 min.

building under construction

bracing or struts

reinforced concrete foundation

base frame

NB. the above tie spacings are typical for specific cranes see supplier's data.

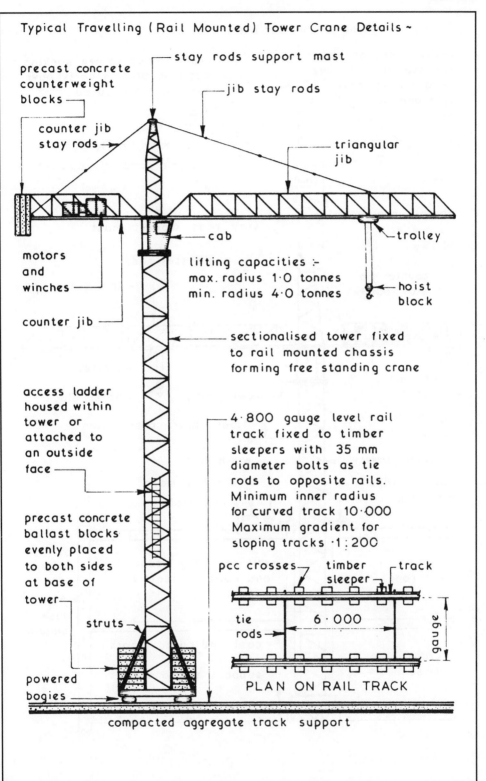

Typical Travelling (Rail Mounted) Tower Crane Details ~

stay rods support mast

precast concrete counterweight blocks

jib stay rods

counter jib stay rods →

triangular jib

cab

trolley

motors and winches

lifting capacities :-
max. radius 1·0 tonnes
min. radius 4·0 tonnes

hoist block

counter jib

sectionalised tower fixed to rail mounted chassis forming free standing crane

access ladder housed within tower or attached to an outside face

4·800 gauge level rail track fixed to timber sleepers with 35 mm diameter bolts as tie rods to opposite rails. Minimum inner radius for curved track 10·000 Maximum gradient for sloping tracks ·1:200

precast concrete ballast blocks evenly placed to both sides at base of tower

pcc crosses

timber sleeper

track

struts

tie rods

6·000

gauge

powered bogies

PLAN ON RAIL TRACK

compacted aggregate track support

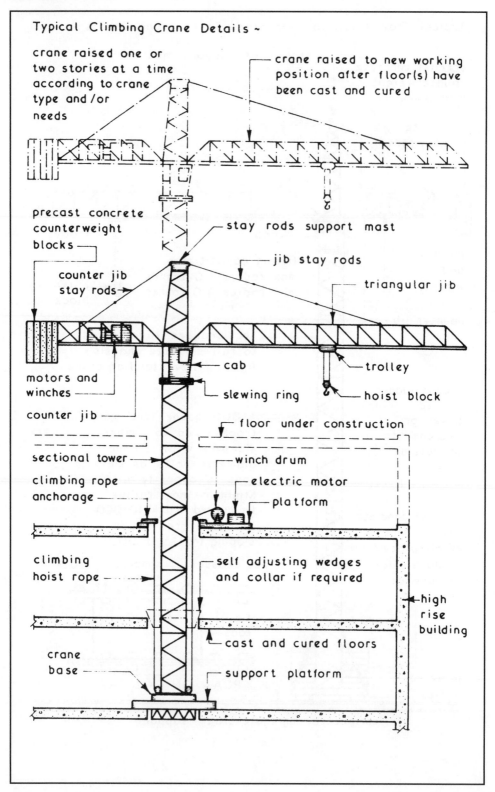

Typical Climbing Crane Details ~

crane raised one or two stories at a time according to crane type and/or needs

crane raised to new working position after floor(s) have been cast and cured

precast concrete counterweight blocks

stay rods support mast

counter jib stay rods

jib stay rods

triangular jib

motors and winches

cab

trolley

counter jib

slewing ring

hoist block

floor under construction

sectional tower

winch drum

climbing rope anchorage

electric motor

platform

climbing hoist rope

self adjusting wedges and collar if required

high rise building

cast and cured floors

crane base

support platform

Concreting ~ this site activity consists of four basic procedures —

1. Material Supply and Storage — this is the receiving on site of the basic materials namely cement, fine aggregate and coarse aggregate and storing them under satisfactory conditions. (see Concrete Production — Materials on pages 247 & 248)

2. Mixing — carried out in small batches this requires only simple hand held tools whereas when demand for increased output is required mixers or ready mixed supplies could be used. (see Concrete Production on pages 249 to 252 and Concreting Plant on pages 170 to 172)

3. Transporting — this can range from a simple bucket to barrows and dumpers for small amounts. For larger loads, especially those required at high level, crane skips could be used:-

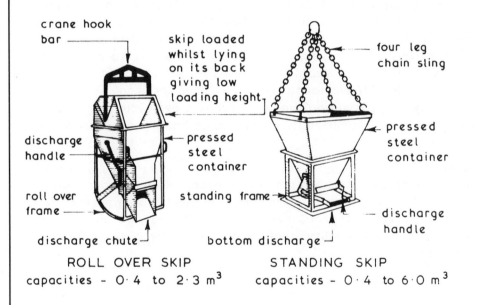

crane hook bar
skip loaded whilst lying on its back giving low loading height
four leg chain sling
discharge handle
pressed steel container
pressed steel container
roll over frame
standing frame
discharge chute
bottom discharge
discharge handle

ROLL OVER SKIP
capacities - 0.4 to 2.3 m³

STANDING SKIP
capacities - 0.4 to 6.0 m³

For the transportation of large volumes of concrete over a limited distance concrete pumps could be used. (see page 173)

4. Placing Concrete — this activity involves placing the wet concrete in the excavation, formwork or mould; working the concrete between and around any reinforcement; vibrating and/ or tamping and curing in accordance with the recommendations of BS 8110 which also covers the striking or removal of the formwork. (see Concreting Plant on page 174 and Formwork on page 350)

Concrete Mixers ~ apart from the very large output mixers most concrete mixers in general use have a rotating drum designed to produce a concrete without segregation of the mix

Concreting Plant ~ the selection of concreting plant can be considered under three activity headings —
1. Mixing.   2. Transportng.   3. Placing.

Choice of Mixer ~ the factors to be taken into consideration when selecting the type of concrete mixer required are —

1. Maximum output required (m³/ hour).
2. Total output required (m³)
3. Type or method of transporting the mixed concrete.
4. Discharge height of mixer (compatibility with transporting method).

Concrete mixer types are generally related to their designed output performance, therefore when the answer to the question 'How much concrete can be placed in a given time period ?' or alternatively 'What mixing and placing methods are to be employed to mix and place a certain amount of concrete in a given time period ?' has been found the actual mixer can be selected. Generally a batch mixing time of 5 minutes per cycle or 12 batches per hour can be assumed as a reasonable basis for assessing mixer output.

Small Batch Mixers ~ these mixers have outputs of up to 200 litres per batch with wheelbarrow transportation an hourly placing rate of 2 to 3m³ can be achieved. Most small batch mixers are of the tilting drum type. Generally these mixers are hand loaded which makes the quality control of successive mixes difficult to regulate.

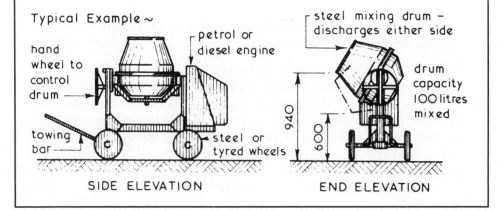

Typical Example ~

hand wheel to control drum

petrol or diesel engine

towing bar

steel or tyred wheels

SIDE ELEVATION

steel mixing drum – discharges either side

drum capacity 100 litres mixed

940

600

END ELEVATION

Medium Batch Mixers ~ outputs of these mixers range from 200 to 750 litres and can be obtained at the lower end of the range as a tilting drum mixer or over the complete range as a non-tilting drum mixer with either reversing drum or chute discharge. The latter usually having a lower discharge height. These mixers usually have integral weight batching loading hoppers, scraper shovels and water tanks thus giving better quality control than the small batch mixers. Generally they are unsuitable for wheelbarrow transportation because of their high output.

Typical Examples ~

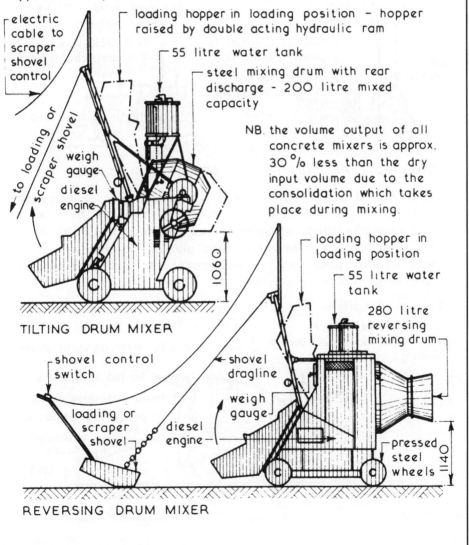

electric cable to scraper shovel control

to loading or scraper shovel

weigh gauge

diesel engine

1060

TILTING DRUM MIXER

loading hopper in loading position – hopper raised by double acting hydraulic ram

55 litre water tank

steel mixing drum with rear discharge – 200 litre mixed capacity

NB. the volume output of all concrete mixers is approx. 30 % less than the dry input volume due to the consolidation which takes place during mixing.

loading hopper in loading position

55 litre water tank

280 litre reversing mixing drum

shovel control switch

shovel dragline

loading or scraper shovel

weigh gauge

diesel engine

pressed steel wheels

1140

REVERSING DRUM MIXER

Transporting Concrete ~ the usual means of transporting mixed concrete produced in a small capacity mixer is by wheelbarrow. The run between the mixing and placing positions should be kept to a minimum and as smooth as possible by using planks or similar materials to prevent segregation of the mix within the wheelbarrow.

Dumpers ~ these can be used for transporting mixed concrete from mixers up to 600 litre capacity when fitted with an integral skip and for lower capacities when designed to take a crane skip.

Typical Examples ~

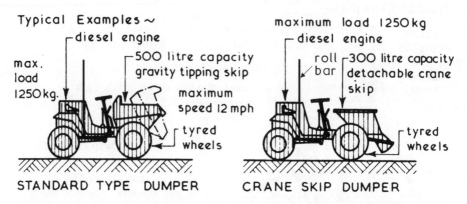

STANDARD TYPE DUMPER   CRANE SKIP DUMPER

Ready Mixed Concrete Trucks ~ these are used to transport mixed concrete from a mixing plant or depot to the site. Usual capacity range of ready mixed concrete trucks is 4 to 6 m³. Discharge can be direct into placing position via a chute or into some form of site transport such as a dumper, crane skip or concrete pump.

Typical Details ~

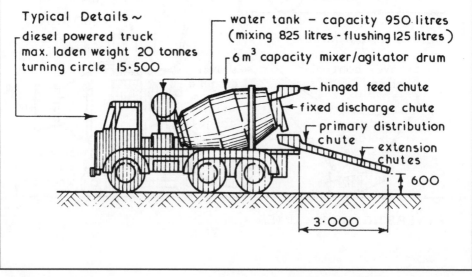

Concrete Pumps ~ these are used to transport large volumes of concrete in a short time period (up to 100 m³ per hour) in both the vertical and horizontal directions from the pump position to the point of placing. Concrete pumps can be trailer or lorry mounted and are usually of a twin cylinder hydraulically driven format with a small bore pipeline (100 mm diameter) with pumping ranges of up to 85·000 vertically and 200·000 horizontally depending on the pump model and the combination of vertical and horizontal distances. It generally requires about 45 minutes to set up a concrete pump on site including coating the bore of the pipeline with a cement grout prior to pumping the special concrete mix. The pump is supplied with pumpable concrete by means of a constant flow of ready mixed concrete lorries throughout the pumping period after which the pipeline is cleared and cleaned. Usually a concrete pump and its operator(s) are hired for the period required.

Typical Concrete Pump Details ~

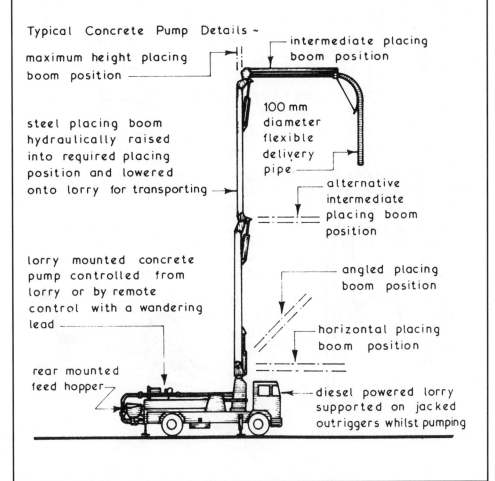

maximum height placing boom position

intermediate placing boom position

steel placing boom hydraulically raised into required placing position and lowered onto lorry for transporting

100 mm diameter flexible delivery pipe

alternative intermediate placing boom position

lorry mounted concrete pump controlled from lorry or by remote control with a wandering lead

angled placing boom position

horizontal placing boom position

rear mounted feed hopper

diesel powered lorry supported on jacked outriggers whilst pumping

Placing Concrete ~ this activity is usually carried out by hand with the objectives of filling the mould, formwork or excavated area to the correct depth, working the concrete around any inserts or reinforcement and finally compacting the concrete to the required consolidation. The compaction of concrete can be carried out using simple tamping rods or boards or alternatively it can be carried out with the aid of plant such as vibrators.

Poker Vibrators ~ these consist of a hollow steel tube casing in which is a rotating impellor which generates vibrations as its head comes into contact with the casing —

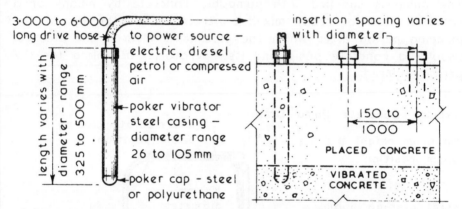

Poker vibrators should be inserted vertically and allowed to penetrate 75mm into any previously vibrated concrete.

Clamp or Tamping Board Vibrators ~ clamp vibrators are powered either by compressed air or electricity whereas tamping board vibrators are usually petrol driven —

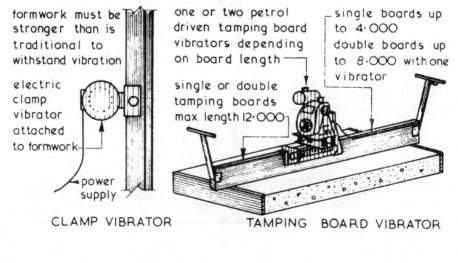

CLAMP VIBRATOR   TAMPING BOARD VIBRATOR

4 SUBSTRUCTURE

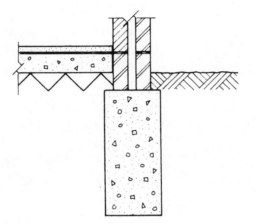

FOUNDATIONS — FUNCTION, MATERIALS AND SIZING
FOUNDATION BEDS
SHORT BORED PILE FOUNDATIONS
FOUNDATION TYPES AND SELECTION
PILED FOUNDATIONS
RETAINING WALLS
BASEMENT CONSTRUCTION
WATERPROOFING BASEMENTS
EXCAVATIONS
CONCRETE PRODUCTION
COFFERDAMS
CAISSONS
UNDERPINNING
GROUND WATER CONTROL
SOIL STABILISATION AND IMPROVEMENT

Foundations ~ the function of any foundation is to safely sustain and transmit to the ground on which it rests the combined dead, imposed and wind loads in such a manner as not to cause any settlement or other movement which would impair the stability or cause damage to any part of the building.

Example ~

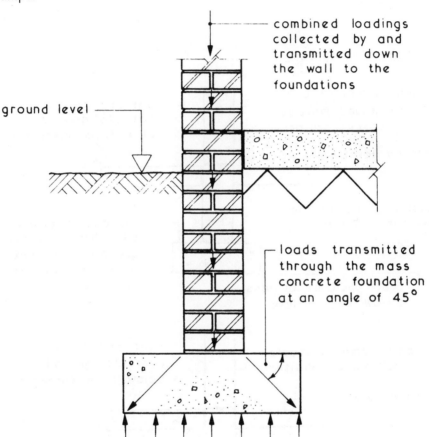

combined loadings collected by and transmitted down the wall to the foundations

ground level

loads transmitted through the mass concrete foundation at an angle of 45°

Subsoil beneath foundation is compressed and reacts by exerting an upward pressure to resist foundation loading. If foundation load exceeds maximum passive pressure of ground (i.e. bearing capacity) a downward movement of the foundation could occur. Remedy is to increase plan size of foundation to reduce the load per unit area or alternatively reduce the loadings being carried by the foundations.

Subsoil Movements ~ these are due primarily to changes in volume when the subsoil becomes wet or dry and occurs near the upper surface of the soil. Compact granular soils such as gravel suffer very little movement whereas cohesive soils such as clay do suffer volume changes near the upper surface. Similar volume changes can occur due to water held in the subsoil freezing and expanding — this is called Frost Heave.

Typical Examples ~

wall tends to tilt with ground movement and cracks can occur →

roof and floors protects ground below from elements

ground level

soil exposed to the weather elements

foundation tends to tilt thus since the ground movement is greater at the outer edge

depth 600mm or less below ground level

wall remains stable under most conditions →

roof and floors protects ground below from elements

ground level

soil exposed to the weather elements

no ground movement should occur at depths below 1·800

depth at least 1·000 below ground level will produce only very slight ground movement which should not affect stability of foundations

Trees ~ damage to foundations. Substructural damage to buildings can occur with direct physical contact by tree roots. More common is the indirect effect of moisture shrinkage or heave, particularly apparent in clay subsoils.

Shrinkage is most evident in long periods of dry weather, compounded by moisture abstraction from vegetation. Notably broad leaved trees such as oak, elm and poplar in addition to the thirsty willow species. Heave is the opposite. It occurs during wet weather and is compounded by previous removal of moisture-dependent trees that would otherwise effect some drainage and balance to subsoil conditions.

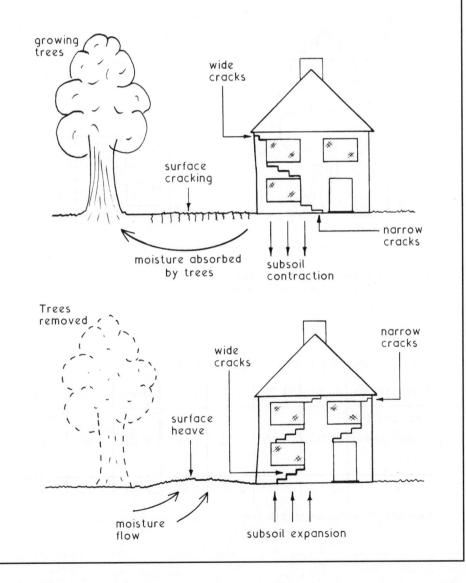

Trees ~ effect on foundations. Trees up to 30 m distance may have an effect on foundations, therefore reference to local authority building control policy should be undertaken before specifying construction techniques.

Traditional strip foundations are practically unsuited, but at excavation depths up to 2·5 or 3·0 m, deep strip or trench fill (preferably reinforced) may be appropriate. Short bored pile foundations are likely to be more economical and particularly suited to depths exceeding 3·0 m.

For guidance only, the illustration and table provide an indication of foundation depths in shrinkable subsoils.

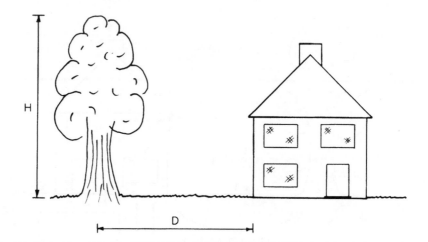

H = Mature height of tree
D = Distance to centre of tree

D/H – Distance from tree/Height of tree

Tree species	0·10	0·25	0·33	0·50	0·66	0·75	1·00
Oak, elm, poplar and willow	3·00	2·80	2·60	2·30	2·10	1·90	1·50
All others	2·80	2·40	2·10	1·80	1·50	1·20	1·00

Minimum foundation depth (m)

Trees ~ preservation orders (see page 100) may be waived by the local planning authority. Permission for tree felling is by formal application and will be considered if the proposed development is in the economic and business interests of the community. However, tree removal is only likely to be acceptable if there is an agreement for replacement stock being provided elsewhere on the site.

In these circumstances, there is potential for ground heave within the 'footprint' of felled trees. To resist this movement, foundations must incorporate an absorbing layer or compressible filler with ground floor suspended above the soil.

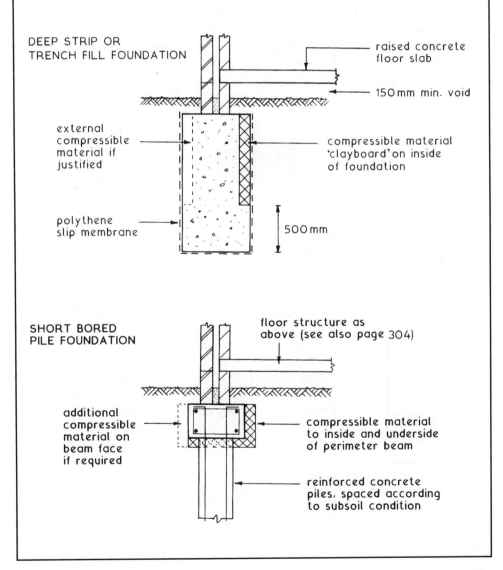

DEEP STRIP OR
TRENCH FILL FOUNDATION

raised concrete
floor slab

150 mm min. void

external
compressible
material if
justified

compressible material
'clayboard' on inside
of foundation

polythene
slip membrane

500 mm

SHORT BORED
PILE FOUNDATION

floor structure as
above (see also page 304)

additional
compressible
material on
beam face
if required

compressible material
to inside and underside
of perimeter beam

reinforced concrete
piles, spaced according
to subsoil condition

Foundation Materials ~ from page 177 one of the functions of a foundation can be seen to be the ability to spread its load evenly over the ground on which it rests. It must of course be constructed of a durable material of adequate strength. Experience has shown that the most suitable material is concrete.

Concrete is a mixture of cement + aggregates + water in controlled proportions.

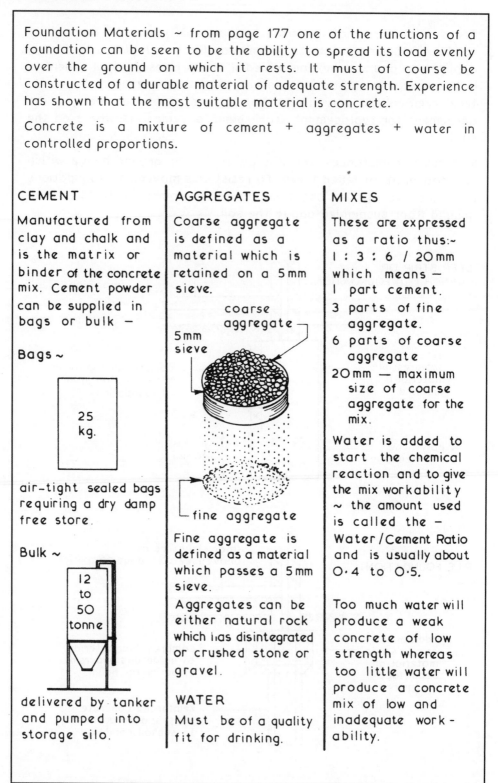

CEMENT

Manufactured from clay and chalk and is the matrix or binder of the concrete mix. Cement powder can be supplied in bags or bulk —

Bags ~

25 kg.

air-tight sealed bags requiring a dry damp free store.

Bulk ~

12 to 50 tonne

delivered by tanker and pumped into storage silo.

AGGREGATES

Coarse aggregate is defined as a material which is retained on a 5mm sieve.

coarse aggregate

5mm sieve

fine aggregate

Fine aggregate is defined as a material which passes a 5mm sieve.

Aggregates can be either natural rock which has disintegrated or crushed stone or gravel.

WATER

Must be of a quality fit for drinking.

MIXES

These are expressed as a ratio thus:-
1 : 3 : 6 / 20 mm
which means —
1 part cement.
3 parts of fine aggregate.
6 parts of coarse aggregate
20 mm — maximum size of coarse aggregate for the mix.

Water is added to start the chemical reaction and to give the mix workability ~ the amount used is called the — Water / Cement Ratio and is usually about 0.4 to 0.5.

Too much water will produce a weak concrete of low strength whereas too little water will produce a concrete mix of low and inadequate work-ability.

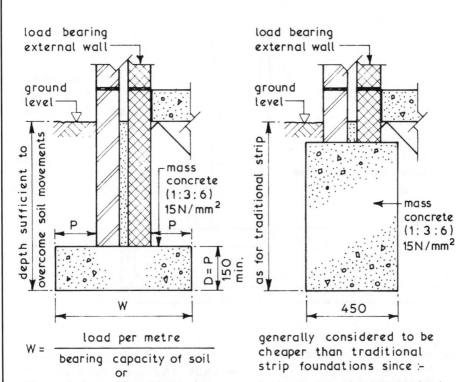

W = $\dfrac{\text{load per metre}}{\text{bearing capacity of soil}}$

or

W = not less than that given in Table 12 of AD 'A'

NB. In all cases W must give adequate working space which is usually 450 to 600 mm minimum depending on depth of excavation.

TRADITIONAL STRIP

generally considered to be cheaper than traditional strip foundations since :-

1. fewer man hours required.
2. requires less skilled trades.
3. uses ready mix concrete therefore less material is stored on site making it clearer and easier to manage.

DEEP STRIP OR TRENCH FILL

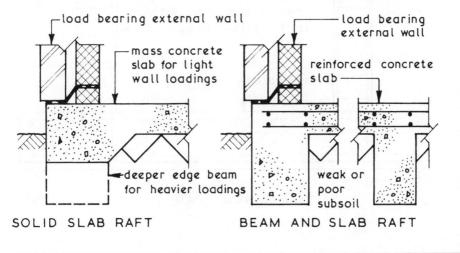

SOLID SLAB RAFT

BEAM AND SLAB RAFT

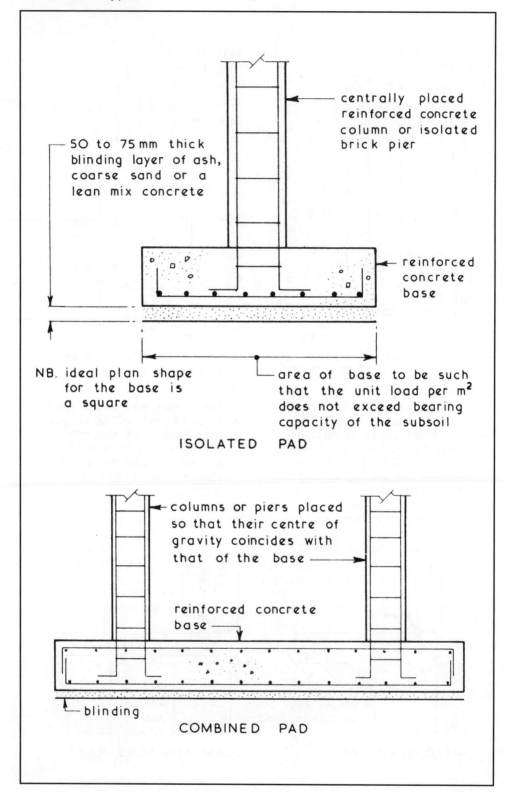

centrally placed
reinforced concrete
column or isolated
brick pier

50 to 75 mm thick
blinding layer of ash,
coarse sand or a
lean mix concrete

reinforced
concrete
base

NB. ideal plan shape
for the base is
a square

area of base to be such
that the unit load per m²
does not exceed bearing
capacity of the subsoil

ISOLATED PAD

columns or piers placed
so that their centre of
gravity coincides with
that of the base

reinforced concrete
base

blinding

COMBINED PAD

Bed ~ a concrete slab resting on and supported by the subsoil, usually forming the ground floor surface. Beds (sometimes called oversite concrete) are usually cast on a layer of hardcore which is used to make up the reduced level excavation and thus raise the level of the concrete bed to a position above ground level.

Typical Example ~

mass concrete bed (1 : 3 : 6 / 20 mm mix 15 N / mm^2). Thickness for domestic work is usually 100 to 150 mm and the bed is constructed so as to prevent the passage of moisture from the ground to the upper surface of the floor - this is usually achieved by incorporating into the design a damp-proof membrane ~ for details see page 532

100 to 150 mm thick layer of hardcore ~ material used should be inert and not affected by water. Suitable materials are gravel; crushed rock; quarry waste; concrete rubble; brick or tile rubble; blast furnace slag and pulverised fuel ash. The hardcore material should be laid evenly and well compacted with the upper surface blinded with fine grade material as required.

Foundations—Basic Sizing

Basic Sizing ~ the size of a foundation is basically dependent on two factors —

1. Load being transmitted.
2. Bearing capacity of subsoil under proposed foundation.

Bearing capacities for different types of subsoils may be obtained from tables such as those in BS 8004: Code of practice for foundations and BS 8103: Structural design of low rise buildings, or from soil investigation results.

Typical Examples ~

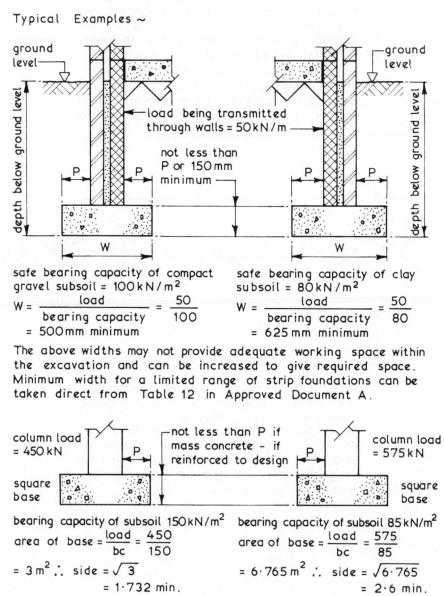

safe bearing capacity of compact gravel subsoil = 100 kN/m²

$W = \dfrac{\text{load}}{\text{bearing capacity}} = \dfrac{50}{100}$
= 500 mm minimum

safe bearing capacity of clay subsoil = 80 kN/m²

$W = \dfrac{\text{load}}{\text{bearing capacity}} = \dfrac{50}{80}$
= 625 mm minimum

The above widths may not provide adequate working space within the excavation and can be increased to give required space. Minimum width for a limited range of strip foundations can be taken direct from Table 12 in Approved Document A.

column load = 450 kN

square base

column load = 575 kN

square base

bearing capacity of subsoil 150 kN/m²

$\text{area of base} = \dfrac{\text{load}}{bc} = \dfrac{450}{150}$
= 3 m² ∴ side = $\sqrt{3}$
= 1·732 min.

bearing capacity of subsoil 85 kN/m²

$\text{area of base} = \dfrac{\text{load}}{bc} = \dfrac{575}{85}$
= 6·765 m² ∴ side = $\sqrt{6·765}$
= 2·6 min.

Typical procedure (for guidance only) —

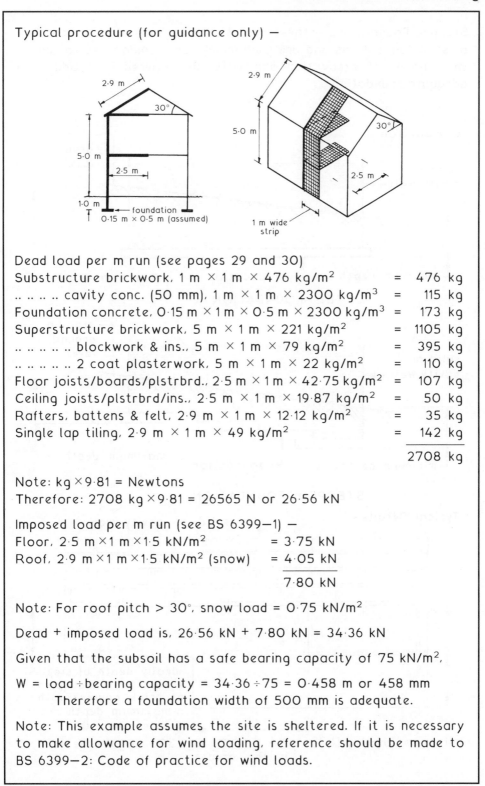

Dead load per m run (see pages 29 and 30)

Substructure brickwork, 1 m × 1 m × 476 kg/m²	=	476	kg
.. .. cavity conc. (50 mm), 1 m × 1 m × 2300 kg/m³	=	115	kg
Foundation concrete, 0·15 m × 1 m × 0·5 m × 2300 kg/m³	=	173	kg
Superstructure brickwork, 5 m × 1 m × 221 kg/m²	=	1105	kg
.. .. blockwork & ins., 5 m × 1 m × 79 kg/m²	=	395	kg
.. .. 2 coat plasterwork, 5 m × 1 m × 22 kg/m²	=	110	kg
Floor joists/boards/plstrbrd., 2·5 m × 1 m × 42·75 kg/m²	=	107	kg
Ceiling joists/plstrbrd/ins., 2·5 m × 1 m × 19·87 kg/m²	=	50	kg
Rafters, battens & felt, 2·9 m × 1 m × 12·12 kg/m²	=	35	kg
Single lap tiling, 2·9 m × 1 m × 49 kg/m²	=	142	kg
		2708	kg

Note: kg × 9·81 = Newtons
Therefore: 2708 kg × 9·81 = 26565 N or 26·56 kN

Imposed load per m run (see BS 6399—1) —

Floor, 2·5 m × 1 m × 1·5 kN/m²	= 3·75 kN
Roof, 2·9 m × 1 m × 1·5 kN/m² (snow)	= 4·05 kN
	7·80 kN

Note: For roof pitch > 30°, snow load = 0·75 kN/m²

Dead + imposed load is, 26·56 kN + 7·80 kN = 34·36 kN

Given that the subsoil has a safe bearing capacity of 75 kN/m²,

W = load ÷ bearing capacity = 34·36 ÷ 75 = 0·458 m or 458 mm
    Therefore a foundation width of 500 mm is adequate.

Note: This example assumes the site is sheltered. If it is necessary to make allowance for wind loading, reference should be made to BS 6399—2: Code of practice for wind loads.

Stepped Foundations

Stepped Foundations ~ these are usually considered in the context of strip foundations and are used mainly on sloping sites to reduce the amount of excavation and materials required to produce an adequate foundation.

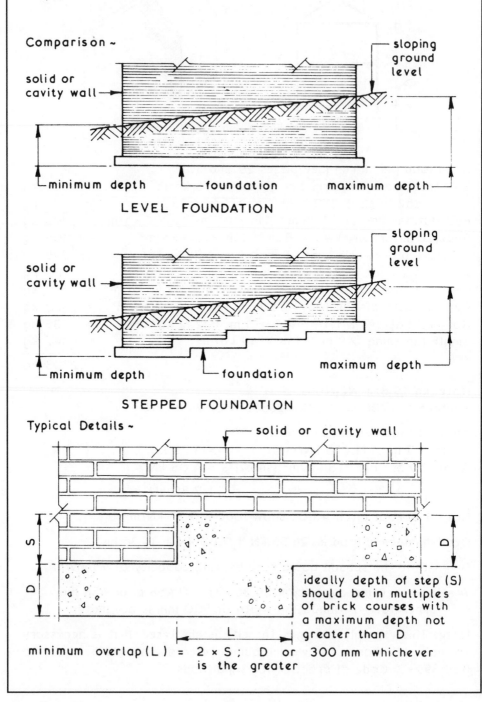

Comparison ~

sloping ground level

solid or cavity wall →

minimum depth   foundation   maximum depth

LEVEL FOUNDATION

solid or cavity wall →

sloping ground level

minimum depth   foundation   maximum depth

STEPPED FOUNDATION

Typical Details ~   solid or cavity wall

ideally depth of step (S) should be in multiples of brick courses with a maximum depth not greater than D

minimum overlap (L) = 2 × S; D or 300 mm whichever is the greater

Concrete Foundations ~ concrete is a material which is strong in compression but weak in tension. If its tensile strength is exceeded cracks will occur resulting in a weak and unsuitable foundation. One method of providing tensile resistance is to include in the concrete foundation bars of steel as a form of reinforcement to resist all the tensile forces induced into the foundation. Steel is a material which is readily available and has high tensile strength.

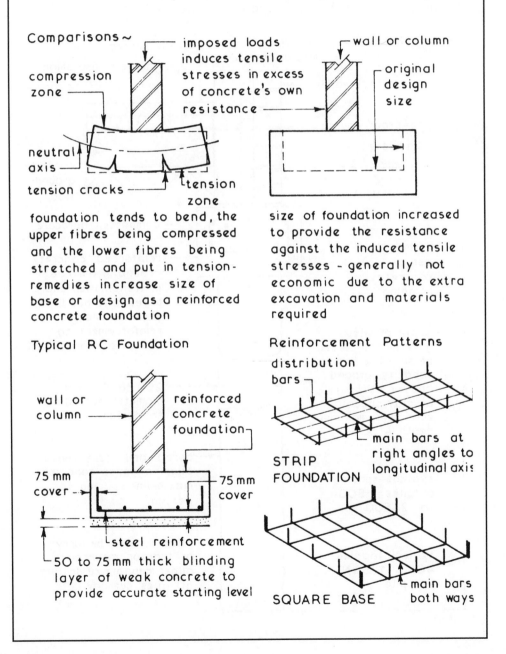

Comparisons ~

compression zone

imposed loads induces tensile stresses in excess of concrete's own resistance

neutral axis

tension cracks — tension zone

wall or column

original design size

foundation tends to bend, the upper fibres being compressed and the lower fibres being stretched and put in tension-remedies increase size of base or design as a reinforced concrete foundation

size of foundation increased to provide the resistance against the induced tensile stresses - generally not economic due to the extra excavation and materials required

Typical RC Foundation

wall or column →

reinforced concrete foundation

75 mm cover →

75 mm cover

steel reinforcement

50 to 75 mm thick blinding layer of weak concrete to provide accurate starting level

Reinforcement Patterns

distribution bars

main bars at right angles to longitudinal axis

STRIP FOUNDATION

main bars both ways

SQUARE BASE

Short Bored Pile Foundations

Short Bored Piles ~ these are a form of foundation which are suitable for domestic loadings and clay subsoils where ground movements can occur below the 1·000 depth associated with traditional strip and trench fill foundations. They can be used where trees are planted close to a new building since the trees may eventually cause damaging ground movements due to extracting water from the subsoil and root growth. Conversely where trees have been removed this may lead to ground swelling.

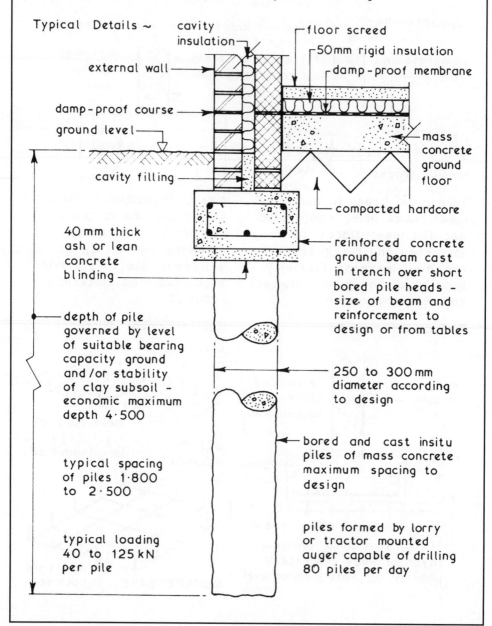

Typical Details ~

cavity insulation

floor screed

50mm rigid insulation

external wall

damp-proof membrane

damp-proof course

ground level

mass concrete ground floor

cavity filling

compacted hardcore

40 mm thick ash or lean concrete blinding

reinforced concrete ground beam cast in trench over short bored pile heads - size of beam and reinforcement to design or from tables

depth of pile governed by level of suitable bearing capacity ground and /or stability of clay subsoil - economic maximum depth 4·500

250 to 300 mm diameter according to design

bored and cast insitu piles of mass concrete maximum spacing to design

typical spacing of piles 1·800 to 2·500

typical loading 40 to 125 kN per pile

piles formed by lorry or tractor mounted auger capable of drilling 80 piles per day

Simple Raft Foundations ~ these can be used for lightly loaded buildings on poor soils or where the top 450 to 600 mm of soil is overlaying a poor quality substrata.

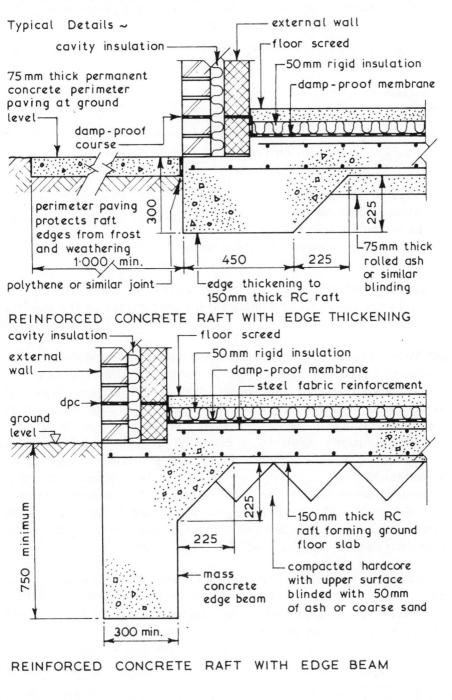

Typical Details ~

external wall
cavity insulation
floor screed
50 mm rigid insulation
damp-proof membrane

75 mm thick permanent concrete perimeter paving at ground level

damp-proof course

perimeter paving protects raft edges from frost and weathering

1·000 min.

300

225

75 mm thick rolled ash or similar blinding

450   225

polythene or similar joint

edge thickening to 150 mm thick RC raft

REINFORCED CONCRETE RAFT WITH EDGE THICKENING

cavity insulation
external wall
floor screed
50 mm rigid insulation
damp-proof membrane
steel fabric reinforcement

dpc

ground level

750 minimum

225

225

150 mm thick RC raft forming ground floor slab

mass concrete edge beam

compacted hardcore with upper surface blinded with 50 mm of ash or coarse sand

300 min.

REINFORCED CONCRETE RAFT WITH EDGE BEAM

Foundation Design Principles ~ the main objectives of foundation design are to ensure that the structural loads are transmitted to the subsoil(s) safely, economically and without any unacceptable movement during the construction period and throughout the anticipated life of the building or structure.

Basic Design Procedure ~ this can be considered as a series of steps or stages —

1. Assessment of site conditions in the context of the site and soil investigation report.

2. Calculation of anticipated structural loading(s).

3. Choosing the foundation type taking into consideration —

   a. Soil conditions;
   b. Type of structure;
   c. Structural loading(s);
   d. Economic factors;
   e. Time factors relative to the proposed contract period;
   f. Construction problems.

4. Sizing the chosen foundation in the context of loading(s), ground bearing capacity and any likely future movements of the building or structure.

Foundation Types ~ apart from simple domestic foundations most foundation types are constructed in reinforced concrete and may be considered as being shallow or deep. Most shallow types of foundation are constructed within 2·000 of the ground level but in some circumstances it may be necessary to take the whole or part of the foundations down to a depth of 2·000 to 5·000 as in the case of a deep basement where the structural elements of the basement are to carry the superstructure loads. Generally foundations which need to be taken below 5·000 deep are cheaper when designed and constructed as piled foundations and such foundations are classified as deep foundations. (For piled foundation details see pages 197 to 214)

Foundations are usually classified by their type such as strips, pads, rafts and piles. It is also possible to combine foundation types such as strip foundations connected by beams to and working in conjunction with pad foundations.

Strip Foundations ~ these are suitable for most subsoils and light structural loadings such as those encountered in low to medium rise domestic dwellings where mass concrete can be used. Reinforced concrete is usually required for all other situations.

Typical Strip Foundation Types ~

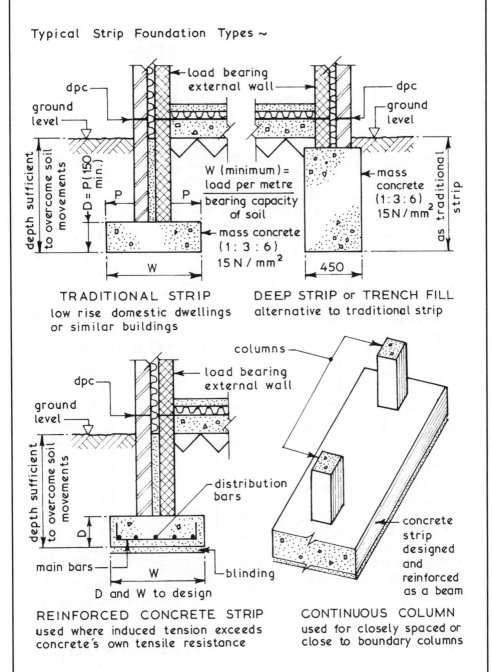

TRADITIONAL STRIP
low rise domestic dwellings or similar buildings

DEEP STRIP or TRENCH FILL
alternative to traditional strip

REINFORCED CONCRETE STRIP
used where induced tension exceeds concrete's own tensile resistance

CONTINUOUS COLUMN
used for closely spaced or close to boundary columns

Pad Foundations ~ suitable for most subsoils except loose sands, loose gravels and filled areas. Pad foundations are usually constructed of reinforced concrete and where possible are square in plan.

Typical Pad Foundation Types ~

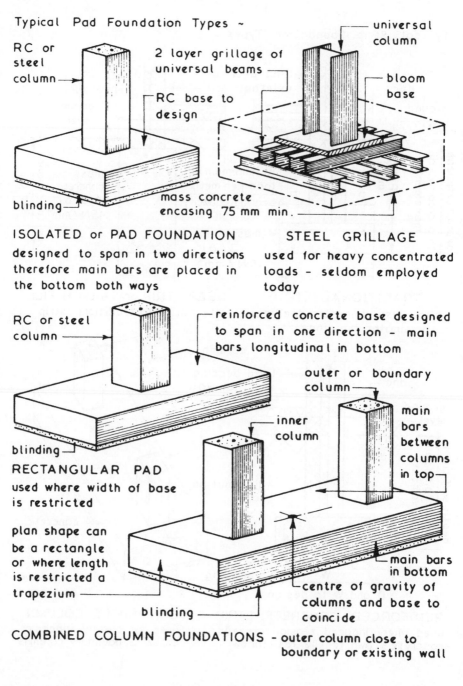

RC or steel column

2 layer grillage of universal beams

RC base to design

universal column

bloom base

blinding

mass concrete encasing 75 mm min.

ISOLATED or PAD FOUNDATION
designed to span in two directions therefore main bars are placed in the bottom both ways

STEEL GRILLAGE
used for heavy concentrated loads - seldom employed today

RC or steel column

reinforced concrete base designed to span in one direction - main bars longitudinal in bottom

blinding

RECTANGULAR PAD
used where width of base is restricted

plan shape can be a rectangle or where length is restricted a trapezium

outer or boundary column

inner column

main bars between columns in top

blinding

main bars in bottom

centre of gravity of columns and base to coincide

COMBINED COLUMN FOUNDATIONS - outer column close to boundary or existing wall

Raft Foundations ~ these are used to spread the load of the superstructure over a large base to reduce the load per unit area being imposed on the ground and this is particularly useful where low bearing capacity soils are encountered and where individual column loads are heavy.

Typical Raft Foundation Types ~

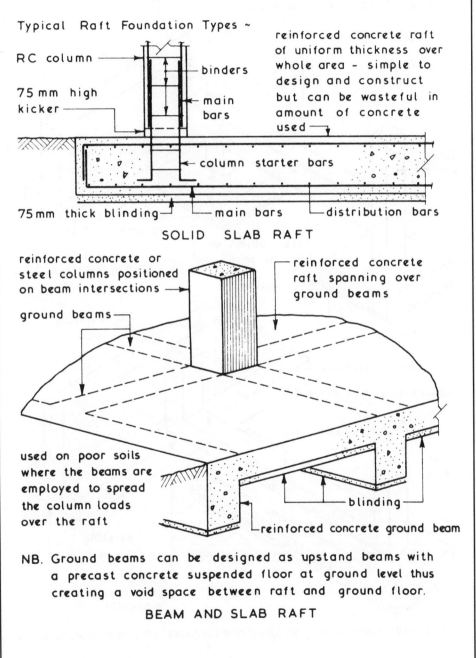

RC column

75 mm high kicker

binders

main bars

reinforced concrete raft of uniform thickness over whole area - simple to design and construct but can be wasteful in amount of concrete used

column starter bars

75 mm thick blinding

main bars

distribution bars

SOLID SLAB RAFT

reinforced concrete or steel columns positioned on beam intersections

ground beams

reinforced concrete raft spanning over ground beams

used on poor soils where the beams are employed to spread the column loads over the raft

blinding

reinforced concrete ground beam

NB. Ground beams can be designed as upstand beams with a precast concrete suspended floor at ground level thus creating a void space between raft and ground floor.

BEAM AND SLAB RAFT

Cantilever Foundations ~ these can be used where it is necessary to avoid imposing any pressure on an adjacent foundation or underground service.

Typical Cantilever Foundation Types ~

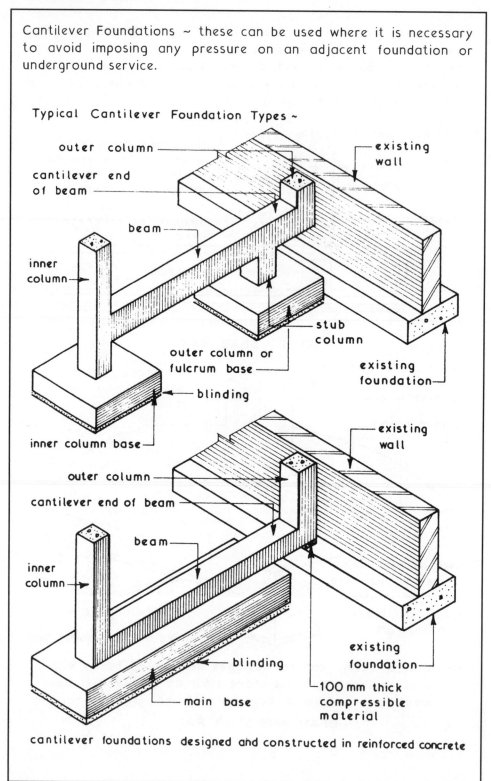

cantilever foundations designed and constructed in reinforced concrete

Piled Foundations ~ these can be defined as a series of columns constructed or inserted into the ground to transmit the load(s) of a structure to a lower level of subsoil. Piled foundations can be used when suitable foundation conditions are not present at or near ground level making the use of deep traditional foundations uneconomic. The lack of suitable foundation conditions may be caused by :-

1. Natural low bearing capacity of subsoil.
2. High water table — giving rise to high permanent dewatering costs.
3. Presence of layers of highly compressible subsoils such as peat and recently placed filling materials which have not sufficiently consolidated.
4. Subsoils which may be subject to moisture movement or plastic failure.

Classification of Piles ~ piles may be classified by their basic design function or by their method of construction :-

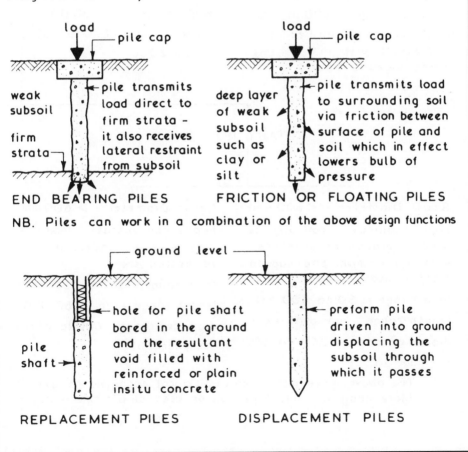

END BEARING PILES

FRICTION OR FLOATING PILES

NB. Piles can work in a combination of the above design functions

REPLACEMENT PILES

DISPLACEMENT PILES

Replacement Piles ~ these are often called bored piles since the removal of the spoil to form the hole for the pile is always carried out by a boring technique. They are used primarily in cohesive subsoils for the formation of friction piles and when forming pile foundations close to existing buildings where the allowable amount of noise and/or vibration is limited.

Replacement Pile Types ~

PERCUSSION BORED

small or medium size contracts with up to 300 piles

load range - 300 to 1300 kN

length range - up to 24·000

diameter range - 300 to 900

may have to be formed as a pressure pile in waterlogged subsoils -

see page 199

FLUSH BORED

large projects - these are basically a rotary bored pile using bentonite as a drilling fluid

load range - 1000 to 5000 kN

length range - up to 30·000

diameter range - 600 to 1500

see page 200

ROTARY BORED

Small Diameter - < 600 mm

light loadings - can also be used in groups or clusters with a common pile cap to receive heavy loads

load range - 50 to 400 kN

length range - up to 15·000

diameter range - 240 to 600

see page 201

Large Diameter - > 600 mm

heavy concentrated loadings - may have an underreamed or belled toe

load range - 800 to 15000 kN

length range - up to 60·000

diameter range - 600 to 2400

see page 202

NB. The above given data depicts typical economic ranges. More than one pile type can be used on a single contract.

Percussion Bored Piles ~ reference page 198

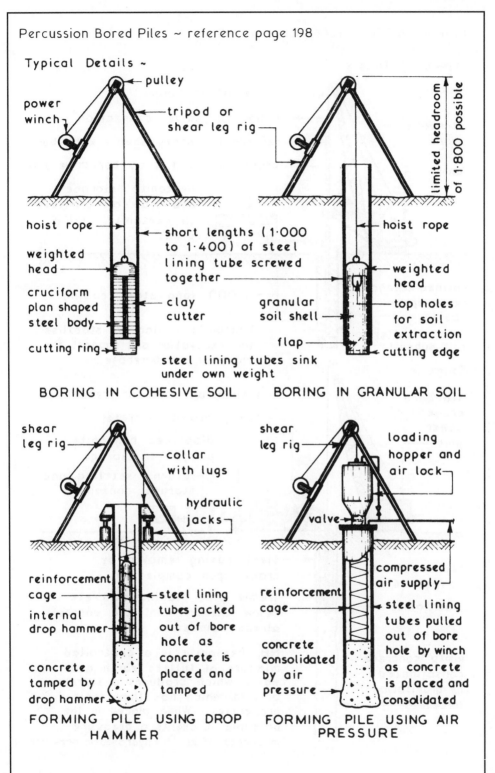

Typical Details ~

pulley

power winch

tripod or shear leg rig

limited headroom of 1·800 possible

hoist rope

weighted head

cruciform plan shaped steel body

cutting ring

short lengths (1·000 to 1·400) of steel lining tube screwed together

clay cutter

steel lining tubes sink under own weight

hoist rope

weighted head

granular soil shell

flap

top holes for soil extraction

cutting edge

BORING IN COHESIVE SOIL   BORING IN GRANULAR SOIL

shear leg rig

collar with lugs

hydraulic jacks

reinforcement cage

internal drop hammer

concrete tamped by drop hammer

steel lining tubes jacked out of bore hole as concrete is placed and tamped

FORMING PILE USING DROP HAMMER

shear leg rig

loading hopper and air lock

valve

compressed air supply

reinforcement cage

concrete consolidated by air pressure

steel lining tubes pulled out of bore hole by winch as concrete is placed and consolidated

FORMING PILE USING AIR PRESSURE

Flush Bored Piles ~ reference page 198

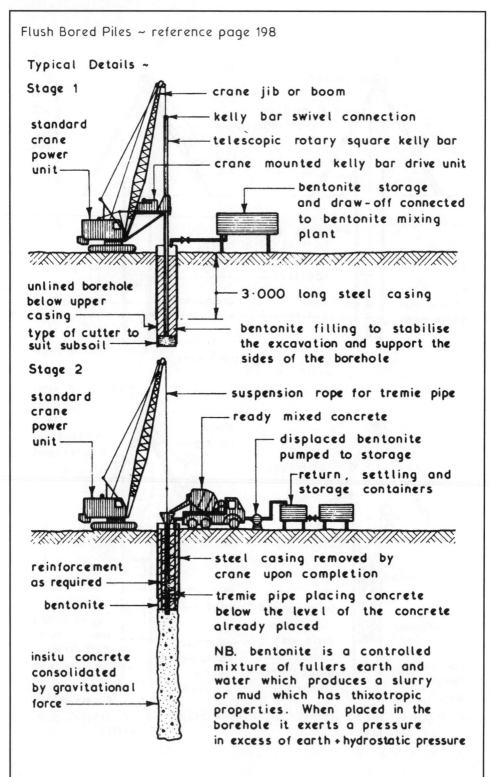

Typical Details ~

Stage 1

standard crane power unit

crane jib or boom

kelly bar swivel connection

telescopic rotary square kelly bar

crane mounted kelly bar drive unit

bentonite storage and draw-off connected to bentonite mixing plant

unlined borehole below upper casing

3·000 long steel casing

type of cutter to suit subsoil

bentonite filling to stabilise the excavation and support the sides of the borehole

Stage 2

standard crane power unit

suspension rope for tremie pipe

ready mixed concrete

displaced bentonite pumped to storage

return, settling and storage containers

reinforcement as required

steel casing removed by crane upon completion

bentonite

tremie pipe placing concrete below the level of the concrete already placed

insitu concrete consolidated by gravitational force

NB. bentonite is a controlled mixture of fullers earth and water which produces a slurry or mud which has thixotropic properties. When placed in the borehole it exerts a pressure in excess of earth + hydrostatic pressure

Small Diameter Rotary Bored Piles ~ reference page 198

Typical Details ~

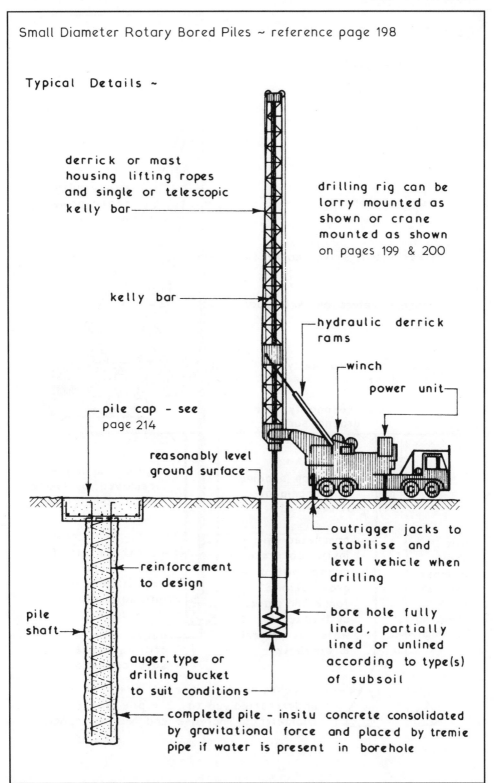

derrick or mast housing lifting ropes and single or telescopic kelly bar

drilling rig can be lorry mounted as shown or crane mounted as shown on pages 199 & 200

kelly bar

hydraulic derrick rams

winch

power unit

pile cap - see page 214

reasonably level ground surface

outrigger jacks to stabilise and level vehicle when drilling

reinforcement to design

pile shaft

bore hole fully lined, partially lined or unlined according to type(s) of subsoil

auger.type or drilling bucket to suit conditions

completed pile - insitu concrete consolidated by gravitational force and placed by tremie pipe if water is present in borehole

Large Diameter Rotary Bored Piles ~ reference page 198

Typical Details ~

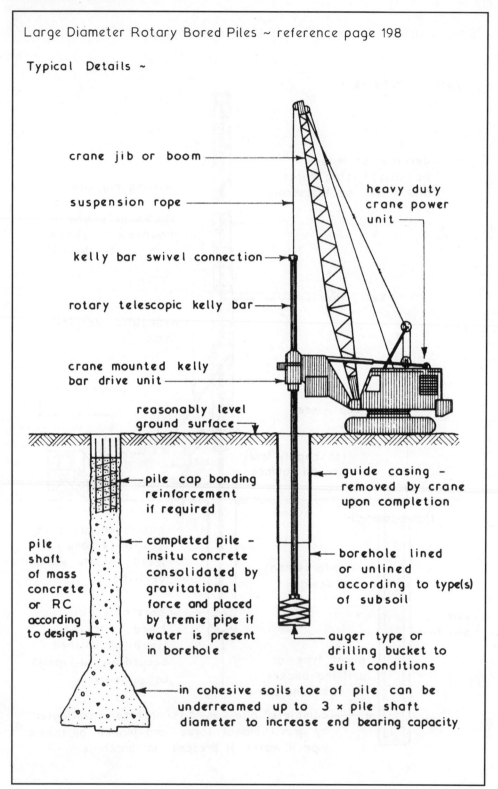

crane jib or boom

suspension rope

kelly bar swivel connection

rotary telescopic kelly bar

crane mounted kelly bar drive unit

reasonably level ground surface

heavy duty crane power unit

pile cap bonding reinforcement if required

completed pile - insitu concrete consolidated by gravitational force and placed by tremie pipe if water is present in borehole

pile shaft of mass concrete or RC according to design

guide casing - removed by crane upon completion

borehole lined or unlined according to type(s) of subsoil

auger type or drilling bucket to suit conditions

in cohesive soils toe of pile can be underreamed up to 3 × pile shaft diameter to increase end bearing capacity

Displacement Piles ~ these are often called driven piles since they are usually driven into the ground displacing the earth around the pile shaft. These piles can be either preformed or partially preformed if they are not cast insitu and are available in a wide variety of types and materials. The pile or forming tube is driven into the required position to a predetermined depth or to the required 'set' which is a measure of the subsoils resistance to the penetration of the pile and hence its bearing capacity by noting the amount of penetration obtained by a fixed number of hammer blows.

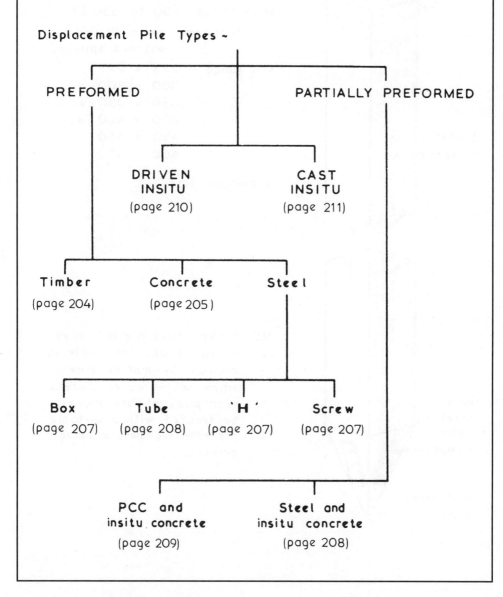

Displacement Pile Types ~

PREFORMED   PARTIALLY PREFORMED

DRIVEN   CAST
INSITU   INSITU
(page 210)   (page 211)

Timber   Concrete   Steel
(page 204)   (page 205)

Box   Tube   'H'   Screw
(page 207)   (page 208)   (page 207)   (page 207)

PCC and   Steel and
insitu concrete   insitu concrete
(page 209)   (page 208)

Timber Piles ~ these are usually square sawn and can be used for small contracts on sites with shallow alluvial deposits overlying a suitable bearing strata (e.g. river banks and estuaries.) Timber piles are percussion driven.

Typical Example ~

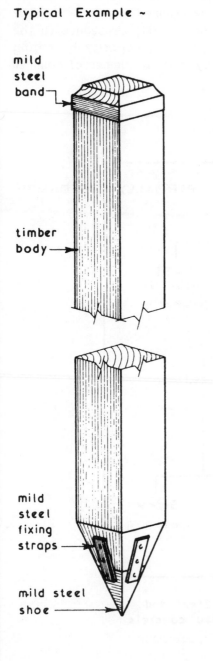

mild
steel
band

timber
body

mild
steel
fixing
straps

mild steel
shoe

Typical Data :-

load range - 50 to 350 kN

length range - up to 12·000
                without splicing

size range -   225 × 225
               300 × 300 *
               350 × 350 *
               400 × 400 *
               450 × 450
               600 × 600

* common sizes

NB. timber piles are not easy
    to splice and are liable to
    attack by marine borers
    when set in water therefore
    such piles should always
    be treated with a suitable
    preservative before being
    driven.

Preformed Concrete Piles ~ variety of types available which are generally used on medium to large contracts of not less than one hundred piles where soft soil deposits overlie a firmer strata. These piles are percussion driven using a drop or single acting hammer.

Typical Example [West's Hardrive Precast Modular Pile] ~

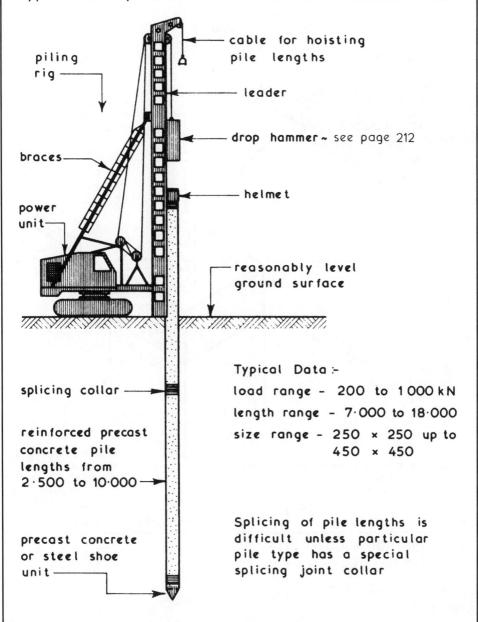

piling rig

braces

power unit

cable for hoisting pile lengths

leader

drop hammer ~ see page 212

helmet

reasonably level ground surface

splicing collar

reinforced precast concrete pile lengths from 2·500 to 10·000

precast concrete or steel shoe unit

Typical Data :-
load range - 200 to 1000 kN
length range - 7·000 to 18·000
size range - 250 × 250 up to
450 × 450

Splicing of pile lengths is difficult unless particular pile type has a special splicing joint collar

Preformed Concrete Piles - jointing with a peripheral steel splicing collar as shown on the preceding page is adequate for most concentrically or directly loaded situations. Where very long piles are to be used and/or high stresses due to compression, tension and bending from the superstructure or the ground conditions are anticipated, the 4 or 8 lock pile joint [AARSLEFF PILING] may be considered.

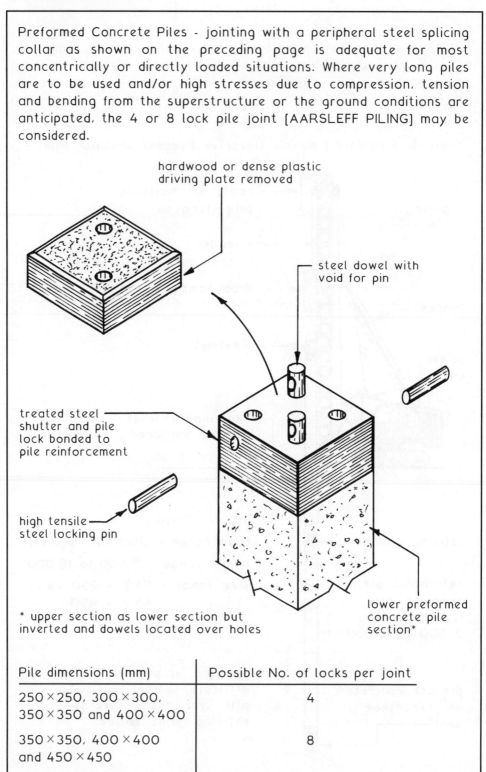

hardwood or dense plastic
driving plate removed

steel dowel with
void for pin

treated steel
shutter and pile
lock bonded to
pile reinforcement

high tensile
steel locking pin

lower preformed
concrete pile
section*

* upper section as lower section but
inverted and dowels located over holes

Pile dimensions (mm)	Possible No. of locks per joint
250×250, 300×300, 350×350 and 400×400	4
350×350, 400×400 and 450×450	8

Steel Box and 'H' Sections ~ standard steel sheet pile sections can be used to form box section piles whereas the 'H' section piles are cut from standard rolled sections. These piles are percussion driven and are used mainly in connection with marine structures.

Typical Examples ~

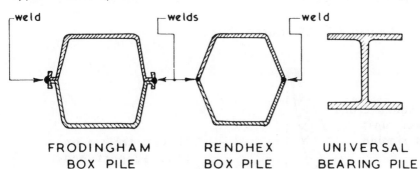

FRODINGHAM
BOX PILE

RENDHEX
BOX PILE

UNIVERSAL
BEARING PILE

Typical Data :-

load range - box piles 300 to 1500 kN
             bearing piles 300 to 1700 kN

length range - all types up to 36·000

size range - various sizes and profiles available

Steel Screw Piles ~ rotary driven and used for dock and jetty works where support at shallow depths in soft silts and sands is required.

Typical Example ~

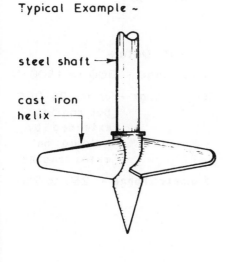

steel shaft

cast iron
helix

Typical Data :-

load range - 400 to 3000 kN

length range - up to 24·000

size range -   shafts 150 to
               350 mm dia.
               overall blades
               600 to 1200

Steel Tube Piles ~ used on small to medium size contracts for marine structures and foundations in soft subsoils over a suitable bearing strata. Tube piles are usually bottom driven with an internal drop hammer. The loading can be carried by the tube alone but it is usual to fill the tube with mass concrete to form a composite pile. Reinforcement, except for pile cap bonding bars, is not normally required.

Typical Example [BSP Cased Pile] ~

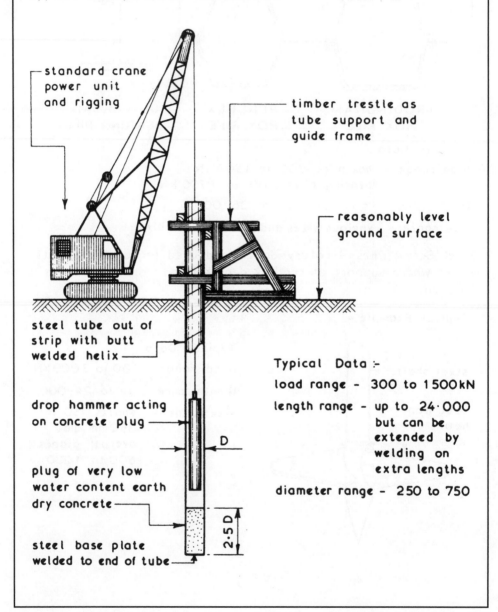

standard crane power unit and rigging

timber trestle as tube support and guide frame

reasonably level ground surface

steel tube out of strip with butt welded helix

drop hammer acting on concrete plug

D

plug of very low water content earth dry concrete

2·5 D

steel base plate welded to end of tube

Typical Data :-

load range - 300 to 1500kN

length range - up to 24·000 but can be extended by welding on extra lengths

diameter range - 250 to 750

Partially Preformed Piles ~ these are composite piles of precast concrete and insitu concrete or steel and insitu concrete (see page 208). These percussion driven piles are used on medium to large contracts where bored piles would not be suitable owing to running water or very loose soils.

Typical Example [West's Shell Pile] ~

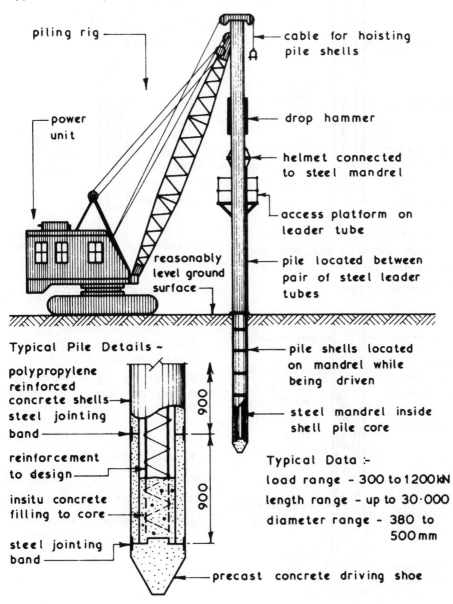

piling rig

cable for hoisting pile shells

power unit

drop hammer

helmet connected to steel mandrel

access platform on leader tube

reasonably level ground surface

pile located between pair of steel leader tubes

Typical Pile Details ~

polypropylene reinforced concrete shells

steel jointing band

reinforcement to design

insitu concrete filling to core

steel jointing band

900

900

pile shells located on mandrel while being driven

steel mandrel inside shell pile core

Typical Data :-
load range – 300 to 1200 kN
length range – up to 30·000
diameter range – 380 to
500 mm

precast concrete driving shoe

Driven Insitu Piles ~ used on medium to large contracts as an alternative to preformed piles particularly where final length of pile is a variable to be determined on site.

Typical Example [Franki Driven Insitu Pile] ~

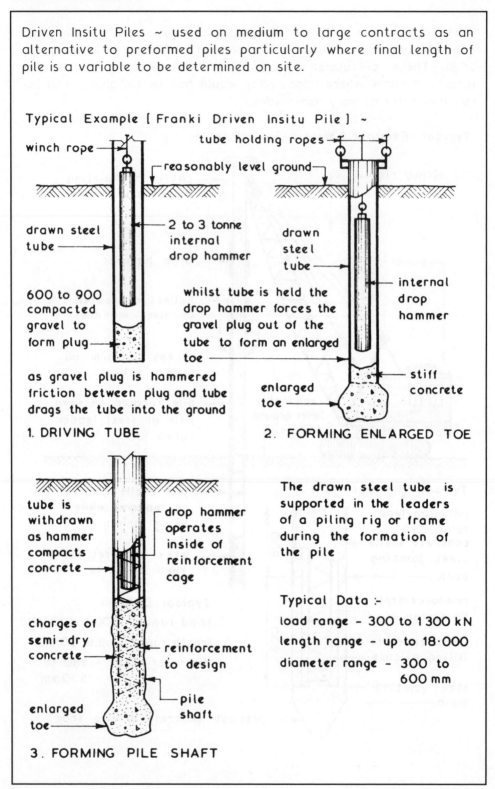

winch rope

tube holding ropes

reasonably level ground

drawn steel tube

2 to 3 tonne internal drop hammer

drawn steel tube

internal drop hammer

600 to 900 compacted gravel to form plug

whilst tube is held the drop hammer forces the gravel plug out of the tube to form an enlarged toe

stiff concrete

as gravel plug is hammered friction between plug and tube drags the tube into the ground

enlarged toe

1. DRIVING TUBE

2. FORMING ENLARGED TOE

tube is withdrawn as hammer compacts concrete

drop hammer operates inside of reinforcement cage

charges of semi-dry concrete

reinforcement to design

enlarged toe

pile shaft

3. FORMING PILE SHAFT

The drawn steel tube is supported in the leaders of a piling rig or frame during the formation of the pile

Typical Data :-

load range - 300 to 1300 kN

length range - up to 18.000

diameter range - 300 to 600 mm

Cast Insitu Piles ~ an alternative to the driven insitu piles (see page 210)

Typical Example [Vibro Cast Insitu Pile] ~

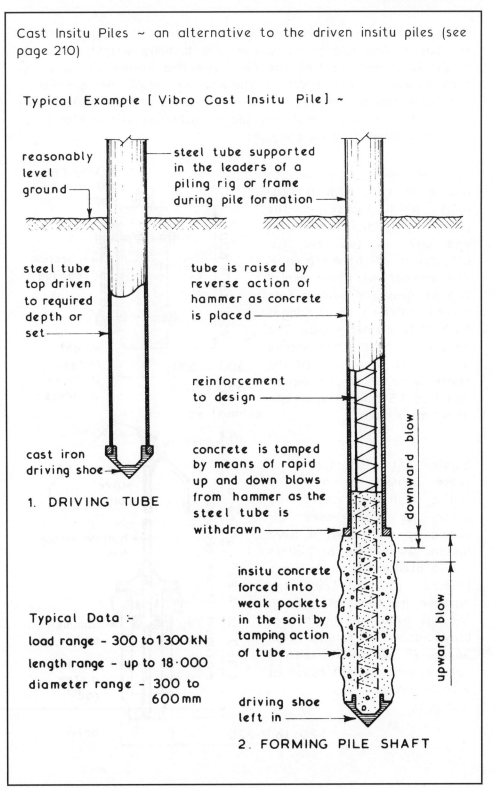

reasonably level ground

steel tube supported in the leaders of a piling rig or frame during pile formation

steel tube top driven to required depth or set

tube is raised by reverse action of hammer as concrete is placed

reinforcement to design

cast iron driving shoe

concrete is tamped by means of rapid up and down blows from hammer as the steel tube is withdrawn

1. DRIVING TUBE

insitu concrete forced into weak pockets in the soil by tamping action of tube

downward blow

upward blow

Typical Data :-

load range - 300 to 1300 kN

length range - up to 18·000

diameter range - 300 to 600 mm

driving shoe left in

2. FORMING PILE SHAFT

Piling Hammers ~ these are designed to deliver an impact blow to the top of the pile to be driven. The hammer weight and drop height is chosen to suit the pile type and nature of subsoil(s) through which it will be driven. The head of the pile being driven is protected against damage with a steel helmet which is padded with a sand bed or similar material and is cushioned with a plastic or hardwood block called a dolly.

Drop Hammers ~ these are blocks of iron with a rear lug(s) which locate in the piling rig guides or leaders and have a top eye for attachment of the winch rope. The number of blows which can be delivered with a free fall of 1·200 to 1·500 ranges from 10 to 20 per minute. The weight of the hammer should be not less than 50% of the concrete or steel pile weight and 1 to 1·5 times the weight of a timber pile.

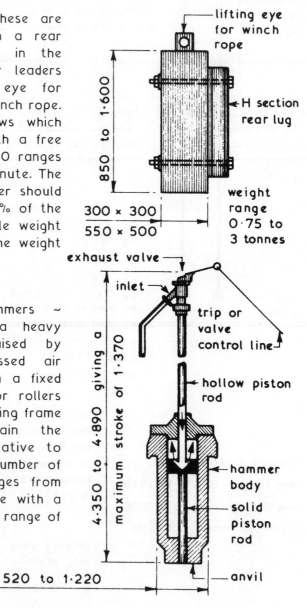

Single Acting Hammers ~ these consist of a heavy falling cylinder raised by steam or compressed air sliding up and down a fixed piston. Guide lugs or rollers are located in the piling frame leaders to maintain the hammer position relative to the pile head. The number of blows delivered ranges from 36 to 75 per minute with a total hammer weight range of 2 to 15 tonnes.

Double Acting Hammers ~ these consist of a cast iron cylinder which remains stationary on the pile head whilst a ram powered by steam or compressed air for both up and down strokes delivers a series of rapid blows which tends to keep the pile on the move during driving. The blow delivered is a smaller force than that from a drop or single acting hammer. The number of blows delivered ranges from 95 to 300 per minute with a total hammer weight range of 0·7 to 6·5 tonnes. Diesel powered double acting hammers are also available.

Diesel Hammers ~ these are self contained hammers which are located in the leaders of a piling rig and rest on the head of the pile. The driving action is started by raising the ram within the cylinder which activates the injection of a measured amount of fuel. The free falling ram compresses the fuel above the anvil causing the fuel to explode and expand resulting in a downward force on the anvil and upward force which raises the ram to recommence the cycle which is repeated until the fuel is cut off. The number of blows delivered ranges from 40 to 60 per minute with a total hammer weight range of 1·0 to 4·5 tonnes.

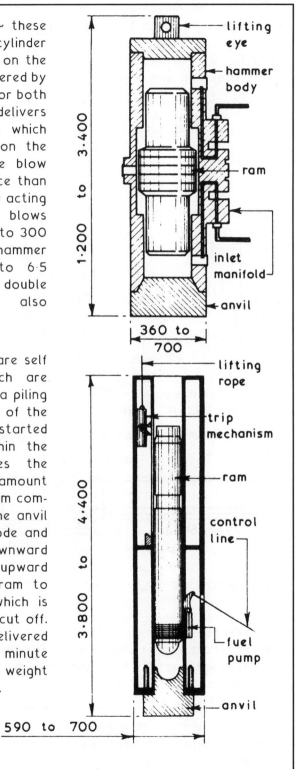

Pile Caps ~ piles can be used singly to support the load but often it is more economical to use piles in groups or clusters linked together with a reinforced concrete cap. The pile caps can also be linked together with reinforced concrete ground beams.

The usual minimum spacing for piles is :-

1. Friction Piles — 1·100 or not less than 3×pile diameter, whichever is the greater.
2. Bearing Piles — 750 mm or not less than 2×pile diameter, whichever is the greater.

Typical Examples ~

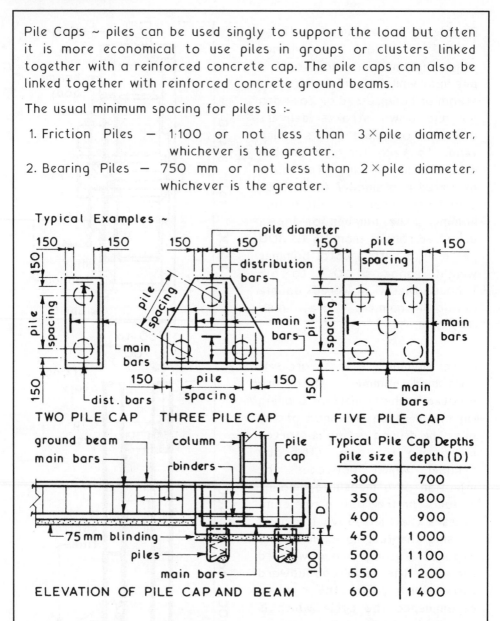

TWO PILE CAP   THREE PILE CAP   FIVE PILE CAP

ELEVATION OF PILE CAP AND BEAM

Typical Pile Cap Depths	
pile size	depth (D)
300	700
350	800
400	900
450	1000
500	1100
550	1200
600	1400

Pile Testing ~ it is advisable to test load at least one pile per scheme. The test pile should be overloaded by at least 50% of its working load and this load should be held for 24 hours. The test pile should not form part of the actual foundations. Suitable testing methods are :-

1. Jacking against kentledge placed over test pile.
2. Jacking against a beam fixed to anchor piles driven in on two sides of the test pile.

Retaining Walls ~ the major function of any retaining wall is to act as on earth retaining structure for the whole or part of its height on one face, the other being exposed to the elements. Most small height retaining walls are built entirely of brickwork or a combination of brick facing and blockwork or mass concrete backing. To reduce hydrostatic pressure on the wall from ground water an adequate drainage system in the form of weep holes should be used, alternatively subsoil drainage behind the wall could be employed.

Typical Example of Combination Retaining Wall ~

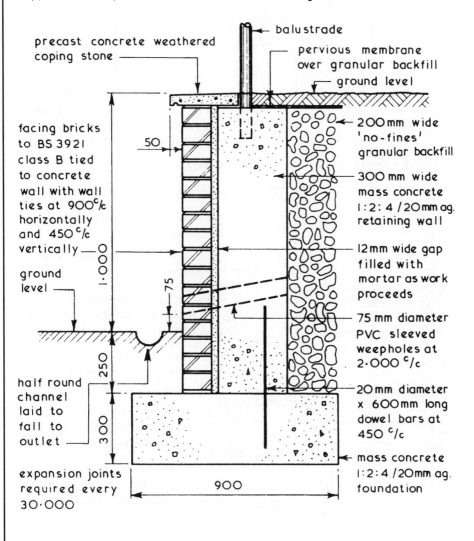

precast concrete weathered coping stone

balustrade

pervious membrane over granular backfill

ground level

facing bricks to BS 3921 class B tied to concrete wall with wall ties at 900 c/c horizontally and 450 c/c vertically

50

200 mm wide 'no-fines' granular backfill

300 mm wide mass concrete 1:2:4 /20mm ag. retaining wall

12mm wide gap filled with mortar as work proceeds

1·000

ground level

75

75 mm diameter PVC sleeved weepholes at 2·000 c/c

20 mm diameter x 600mm long dowel bars at 450 c/c

half round channel laid to fall to outlet

250

300

mass concrete 1:2:4 /20mm ag. foundation

expansion joints required every 30·000

900

Small Height Retaining Walls ~ retaining walls must be stable and the usual rule of thumb for small height brick retaining walls is for the height to lie between 2 and 4 times the wall thickness. Stability can be checked by applying the middle third rule —

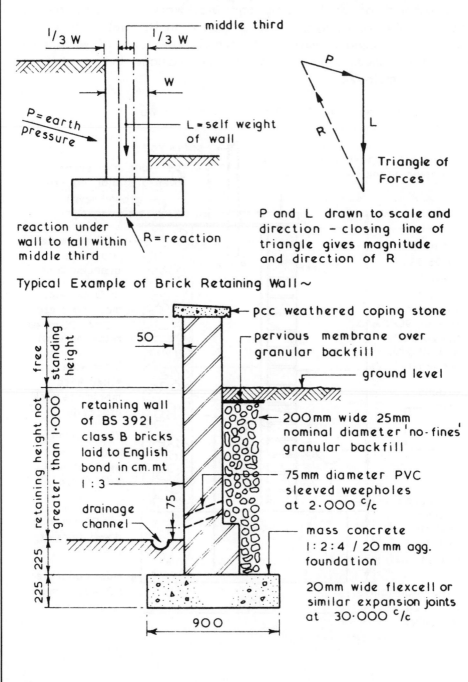

P and L drawn to scale and direction – closing line of triangle gives magnitude and direction of R

Typical Example of Brick Retaining Wall~

Retaining Walls up to 6·000 High ~ these can be classified as medium height retaining walls and have the primary function of retaining soils at an angle in excess of the soil's natural angle of repose. Walls within this height range are designed to provide the necessary resistance by either their own mass or by the principles of leverage.

Design ~ the actual design calculations are usually carried out by a structural engineer who endeavours to ensure that :-

1. Overturning of the wall does not occur.
2. Forward sliding of the wall does not occur.
3. Materials used are suitable and not overstressed .
4. The subsoil is not overloaded.
5. In clay subsoils slip circle failure does not occur.

The factors which the designer will have to take into account:-

1. Nature and characteristics of the subsoil(s).
2. Height of water table - the presence of water can create hydrostatic pressure on the rear face of the wall, it can also affect the bearing capacity of the subsoil together with its shear strength, reduce the frictional resistance between the underside of the foundation and the subsoil and reduce the passive pressure in front of the toe of the wall.
3. Type of wall.
4. Material(s) to be used in the construction of the wall.

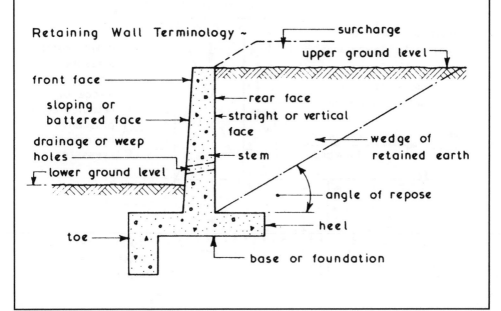

Retaining Wall Terminology ~

Earth Pressures ~ these can take one of two forms namely :-

1. Active Earth Pressures — these are those pressures which tend to move the wall at all times and consist of the wedge of earth retained plus any hydrostatic pressure. The latter can be reduced by including a subsoil drainage system behind and/or through the wall.

2. Passive Earth Pressures ~ these are a reaction of an equal and opposite force to any imposed pressure thus giving stability by resisting movement.

Typical Examples ~

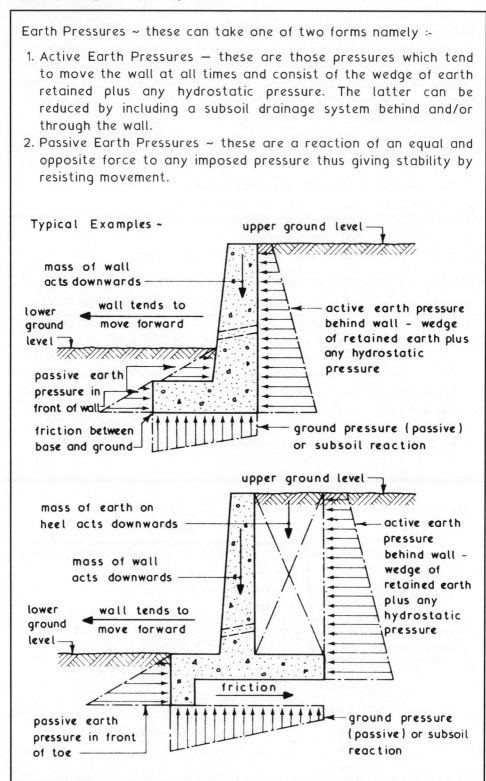

Mass Retaining Walls ~ these walls rely mainly on their own mass to overcome the tendency to slide forwards. Mass retaining walls are not generally considered to be economic over a height of 1·800 when constructed of brick or concrete and 1·000 high in the case of natural stonework. Any mass retaining wall can be faced with another material but generally any applied facing will not increase the strength of the wall and is therefore only used for aesthetic reasons.

Typical Brick Mass Retaining Wall Details ~

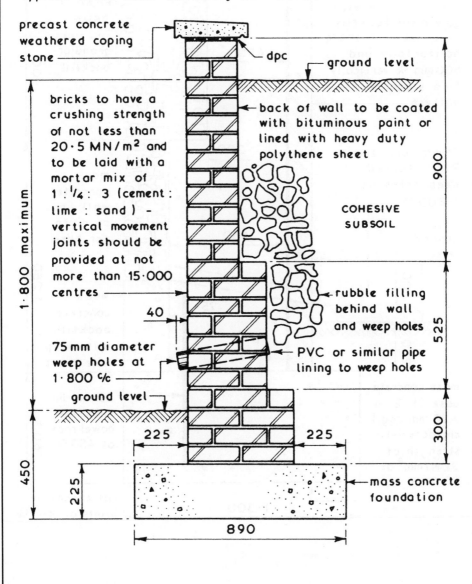

precast concrete weathered coping stone

dpc

ground level

bricks to have a crushing strength of not less than 20·5 MN/m² and to be laid with a mortar mix of 1 : ¹/₄ : 3 (cement : lime : sand) – vertical movement joints should be provided at not more than 15·000 centres

back of wall to be coated with bituminous paint or lined with heavy duty polythene sheet

COHESIVE SUBSOIL

900

1·800 maximum

40

75 mm diameter weep holes at 1·800 c/c

ground level

rubble filling behind wall and weep holes

PVC or similar pipe lining to weep holes

525

225   225

300

450

225

mass concrete foundation

890

Typical Brick Faced Mass Concrete Retaining Wall Detail ~

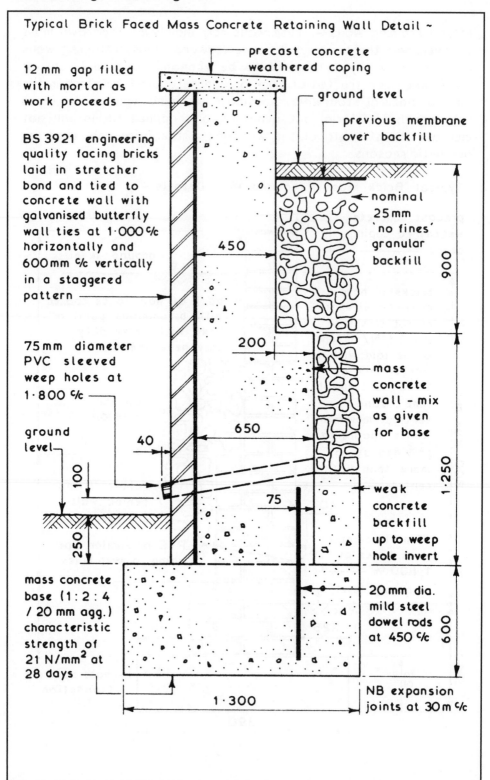

12 mm gap filled with mortar as work proceeds

precast concrete weathered coping

ground level

previous membrane over backfill

BS 3921 engineering quality facing bricks laid in stretcher bond and tied to concrete wall with galvanised butterfly wall ties at 1·000 ℅ horizontally and 600 mm ℅ vertically in a staggered pattern

nominal 25 mm 'no fines' granular backfill

450

900

75 mm diameter PVC sleeved weep holes at 1·800 ℅

200

mass concrete wall – mix as given for base

ground level

40

650

weak concrete backfill up to weep hole invert

100

75

1·250

250

mass concrete base (1 : 2 : 4 / 20 mm agg.) characteristic strength of 21 N/mm² at 28 days

20 mm dia. mild steel dowel rods at 450 ℅

600

1·300

NB expansion joints at 30 m ℅

Cantilever Retaining Walls ~ these are constructed of reinforced concrete with an economic height range of 1·200 to 6·000. They work on the principles of leverage where the stem is designed as a cantilever fixed at the base and base is designed as a cantilever fixed at the stem. Several formats are possible and in most cases a beam is placed below the base to increase the total passive resistance to sliding. Facing materials can be used in a similar manner to that shown on page 220.

Typical Formats ~

stem
base
beam

stem
base
beam

stem
base
beam

Typical Details ~

concrete to be 1 : 2 : 4/ 20 mm aggregate with minimum cube crushing strength of 21 N/mm² at 28 days

welded fabric to control shrinkage cracks

75 mm diameter weep holes at 1·800 ℅

ground level

450 mm wide beam

300

ground level

previous membrane over nominal 25 mm 'no fines' granular backfill

main bars at 300 ℅

nominal diameter distribution bars

main bars at 150 ℅

distribution bars

4·200

300

ground level

100

450

900

450

450

75 mm blinding

3·000

welded fabric

221

Formwork ~ concrete retaining walls can be cast in one of three ways - full height; climbing (page 223) or against earth face (page 224).

Full Height Casting ~ this can be carried out if the wall is to be cast as a freestanding wall and allowed to cure and gain strength before the earth to be retained is backfilled behind the wall. Considerations are the height of the wall, anticipated pressure of wet concrete, any strutting requirements and the availability of suitable materials to fabricate the formwork. As with all types of formwork a traditional timber format or a patent system using steel forms could be used.

Typical Details ~

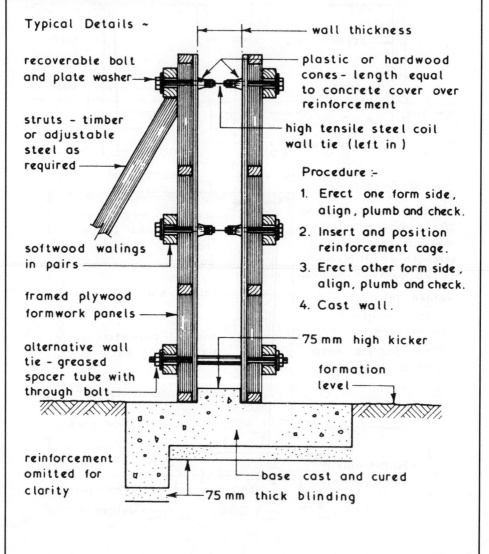

recoverable bolt and plate washer

plastic or hardwood cones - length equal to concrete cover over reinforcement

struts - timber or adjustable steel as required

high tensile steel coil wall tie (left in)

wall thickness

softwood walings in pairs

framed plywood formwork panels

alternative wall tie - greased spacer tube with through bolt

reinforcement omitted for clarity

Procedure :-

1. Erect one form side, align, plumb and check.

2. Insert and position reinforcement cage.

3. Erect other form side, align, plumb and check.

4. Cast wall.

75 mm high kicker

formation level

base cast and cured

75 mm thick blinding

Climbing Formwork or Lift Casting ~ this method can be employed on long walls, high walls or where the amount of concrete which can be placed in a shift is limited.

Typical Details ~

100 × 50 softwood studs in pairs fixed to back of wall forms at 900 c/c

1·200 high × 2·400 long plywood faced framed wall forms

formation level

reinforcement omitted for clarity

spacer if required

raking struts to be used as required

1·125

wall thickness

1·125

bolts and spacer tubes or steel coil wall ties

75mm high kicker

base cast and cured

75mm thick blinding

STAGE ONE OR FIRST LIFT

bolts and spacer tubes or steel coil wall ties

bolt holes from first lift no longer required to be made good

hardwood folding wedges

through bolt fixings

NB. all subsequent lifts as for second lift

first lift forms reversed

1·200

raking struts not required after first lift

75

first lift of wall cast and cured sufficiently to support second lift formwork

1·125

wall can be constructed using climbing shoes instead of studs

STAGE TWO OR SECOND LIFT

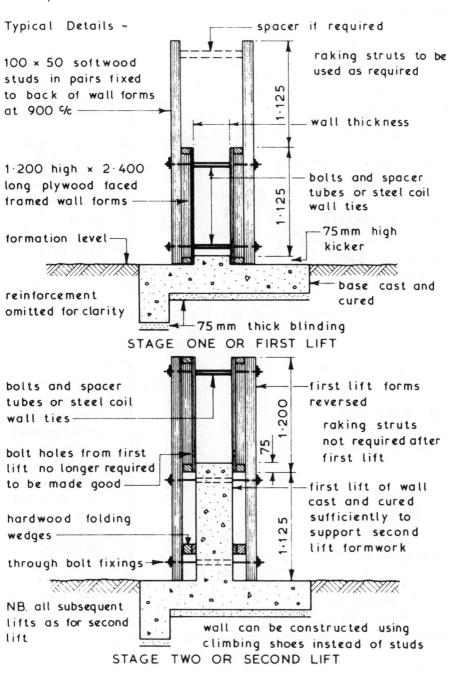

Casting Against Earth Face ~ this method can be an adaptation of the full height or climbing formwork systems. The latter uses a steel wire loop tie fixing to provide the support for the second and subsequent lifts.

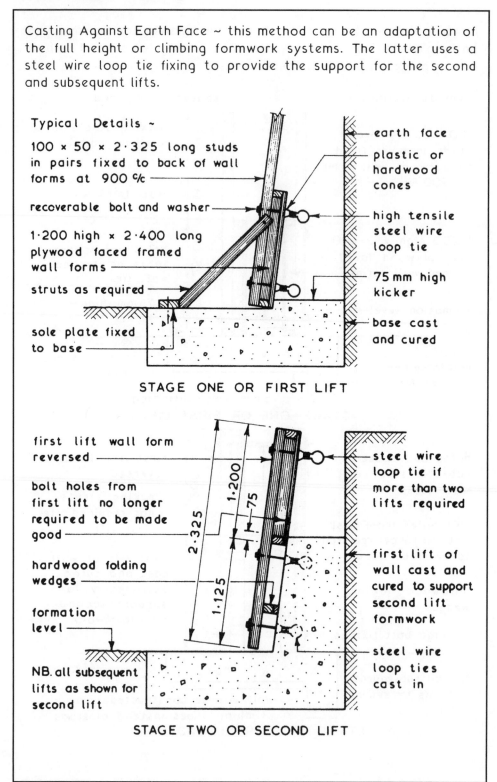

Typical Details ~

100 × 50 × 2·325 long studs in pairs fixed to back of wall forms at 900 ℃ —

recoverable bolt and washer —

1·200 high × 2·400 long plywood faced framed wall forms —

struts as required —

sole plate fixed to base —

earth face

plastic or hardwood cones

high tensile steel wire loop tie

75 mm high kicker

base cast and cured

STAGE ONE OR FIRST LIFT

first lift wall form reversed —

bolt holes from first lift no longer required to be made good —

hardwood folding wedges —

formation level —

NB. all subsequent lifts as shown for second lift

1·200
75
2·325
1·125

steel wire loop tie if more than two lifts required

first lift of wall cast and cured to support second lift formwork

steel wire loop ties cast in

STAGE TWO OR SECOND LIFT

Crib Retaining Walls — a system of precast concrete or treated timber components comprising headers and stretchers which interlock to form a 3 dimensional framework. During assembly the framework is filled with graded stone to create sufficient mass to withstand ground pressures.

Principle —

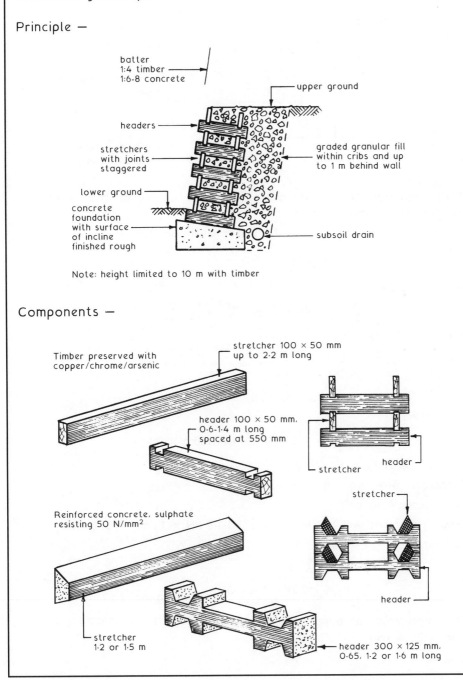

batter
1:4 timber
1:6-8 concrete

upper ground

headers

stretchers with joints staggered

graded granular fill within cribs and up to 1 m behind wall

lower ground

concrete foundation with surface of incline finished rough

subsoil drain

Note: height limited to 10 m with timber

Components —

Timber preserved with copper/chrome/arsenic

stretcher 100 × 50 mm up to 2·2 m long

header 100 × 50 mm, 0·6-1·4 m long spaced at 550 mm

stretcher

header

stretcher

Reinforced concrete, sulphate resisting 50 N/mm²

header

stretcher 1·2 or 1·5 m

header 300 × 125 mm, 0·65, 1·2 or 1·6 m long

Design of Retaining Walls ~ this should allow for the effect of hydrostatics or water pressure behind the wall and the pressure created by the retained earth (see page 218). Calculations are based on a 1m unit length of wall, from which it is possible to ascertain:

1. The resultant thrust

2. The overturning or bending moment

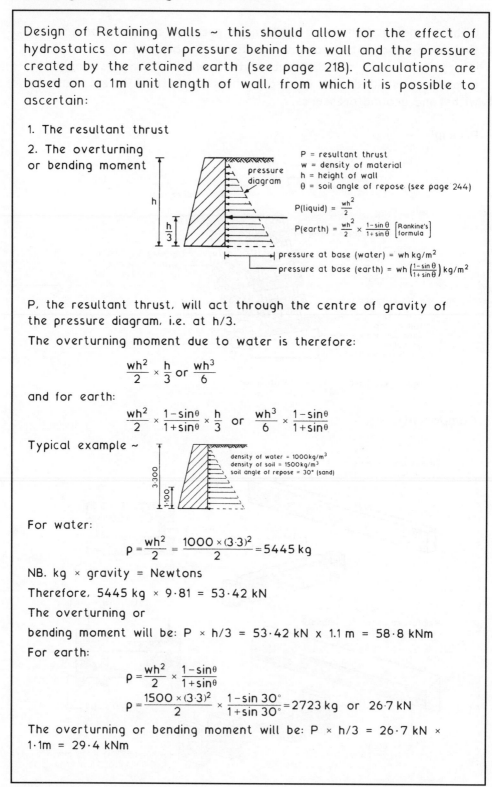

P = resultant thrust
w = density of material
h = height of wall
θ = soil angle of repose (see page 244)

$$P(\text{liquid}) = \frac{wh^2}{2}$$

$$P(\text{earth}) = \frac{wh^2}{2} \times \frac{1-\sin\theta}{1+\sin\theta} \quad \left[\begin{array}{l}\text{Rankine's} \\ \text{formula}\end{array}\right]$$

pressure at base (water) = wh kg/m^2

pressure at base (earth) = $wh\left(\frac{1-\sin\theta}{1+\sin\theta}\right)$ kg/m^2

P, the resultant thrust, will act through the centre of gravity of the pressure diagram, i.e. at h/3.

The overturning moment due to water is therefore:

$$\frac{wh^2}{2} \times \frac{h}{3} \text{ or } \frac{wh^3}{6}$$

and for earth:

$$\frac{wh^2}{2} \times \frac{1-\sin\theta}{1+\sin\theta} \times \frac{h}{3} \text{ or } \frac{wh^3}{6} \times \frac{1-\sin\theta}{1+\sin\theta}$$

Typical example ~

density of water = 1000kg/m^3
density of soil = 1500kg/m^3
soil angle of repose = 30° (sand)

For water:

$$P = \frac{wh^2}{2} = \frac{1000 \times (3 \cdot 3)^2}{2} = 5445 \text{ kg}$$

NB. kg × gravity = Newtons

Therefore, 5445 kg × 9·81 = 53·42 kN

The overturning or bending moment will be: P × h/3 = 53·42 kN x 1.1 m = 58·8 kNm

For earth:

$$P = \frac{wh^2}{2} \times \frac{1-\sin\theta}{1+\sin\theta}$$

$$P = \frac{1500 \times (3 \cdot 3)^2}{2} \times \frac{1-\sin 30°}{1+\sin 30°} = 2723 \text{ kg or } 26 \cdot 7 \text{ kN}$$

The overturning or bending moment will be: P × h/3 = 26·7 kN × 1·1m = 29·4 kNm

A graphical design solution, to determine the earth thrust (P) behind a retaining wall. Data from previous page:

h = 3·300 m
θ = 30°
w = 1500 kg/m³

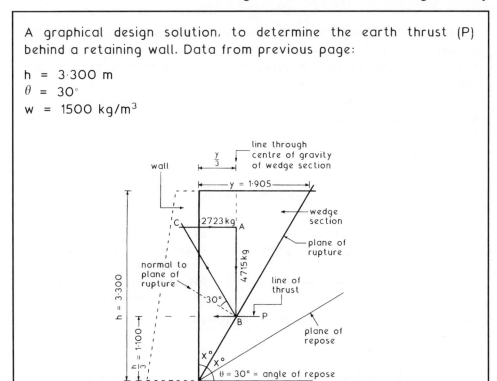

Wall height is drawn to scale and plane of repose plotted. The wedge section is obtained by drawing the plane of rupture through an angle bisecting the plane of repose and vertical back of the wall. Dimension 'y' can be scaled or calculated:

Tangent $x = \dfrac{y}{3 \cdot 3}$ x = 30°, and tan 30° = 0·5774

therefore, y = 3·3 × 0·5774 = 1·905 m

Area of wedge section = $\dfrac{3 \cdot 3}{2}$ × 1·905 m

Volume of wedge per metre run of wall = 3·143 × 1 = 3·143 m³

Weight .. .. = 3·143 × 1500 = 4715 kg

Vector line A – B is drawn to a scale through centre of gravity of wedge section, line of thrust and plane of rupture to represent 4715 kg.

Vector line B – C is drawn at the angle of earth friction (usually same as angle of repose, i.e. 30° in this case), to the normal to the plane of rupture until it meets the horizontal line C – A.

Triangle ABC represents the triangle of forces for the wedge section of earth, so C – A can be scaled at 2723 kg to represent (P), the earth thrust behind the retaining wall.

227

Open Excavations ~ one of the main problems which can be encountered with basement excavations is the need to provide temporary support or timbering to the sides of the excavation. This can be intrusive when the actual construction of the basement floor and walls is being carried out. One method is to use battered excavation sides cut back to a safe angle of repose thus eliminating the need for temporary support works to the sides of the excavation.

Typical Example of Open Basement Excavations ~

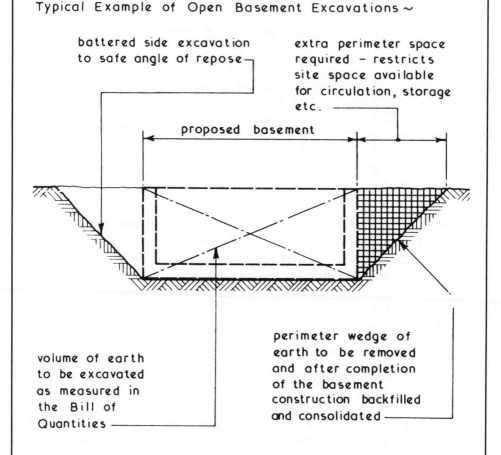

battered side excavation to safe angle of repose —

extra perimeter space required – restricts site space available for circulation, storage etc. —

proposed basement

volume of earth to be excavated as measured in the Bill of Quantities —

perimeter wedge of earth to be removed and after completion of the basement construction backfilled and consolidated —

In economic terms the costs of plant and manpower to cover the extra excavation, backfilling and consolidating must be offset by the savings made by omitting the temporary support works to the sides of the excavation. The main disadvantage of this method is the large amount of free site space required.

Perimeter Trench Excavations ~ in this method a trench wide enough for the basement walls to be constructed is excavated and supported with timbering as required. It may be necessary for runners or steel sheet piling to be driven ahead of the excavation work. This method can be used where weak subsoils are encountered so that the basement walls act as permanent timbering whilst the mound or dumpling is excavated and the base slab cast. Perimeter trench excavations can also be employed in firm subsoils when the mechanical plant required for excavating the dumpling is not available at the right time.

Typical Details ~

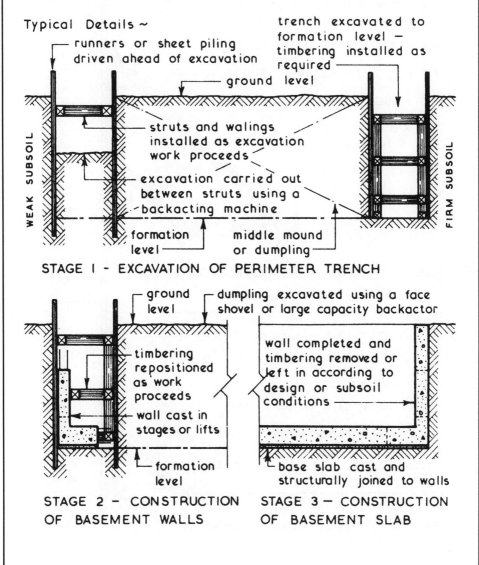

STAGE 1 - EXCAVATION OF PERIMETER TRENCH

STAGE 2 - CONSTRUCTION OF BASEMENT WALLS

STAGE 3 - CONSTRUCTION OF BASEMENT SLAB

Complete Excavation ~ this method can be used in firm subsoils where the centre of the proposed basement can be excavated first to enable the basement slab to be cast thus giving protection to the subsoil at formation level. The sides of excavation to the perimeter of the basement can be supported from the formation level using raking struts or by using raking struts pitched from the edge of the basement slab.

Typical Details ~

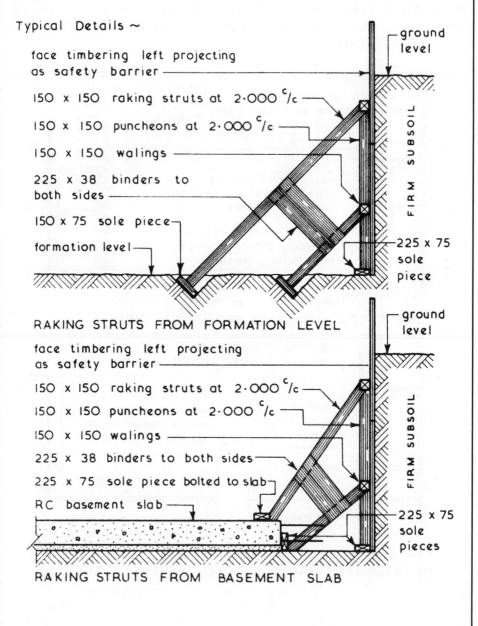

RAKING STRUTS FROM FORMATION LEVEL

RAKING STRUTS FROM BASEMENT SLAB

Excavating Plant ~ the choice of actual pieces of plant to be used in any construction activity is a complex matter taking into account many factors. Specific details of various types of excavators are given on pages 146 to 150. At this stage it is only necessary to consider basic types for particular operations. In the context of basement excavation two forms of excavator could be considered.

1. Backactors – these machines are available as cable rigged or hydraulic excavators suitable for trench and bulk excavating. Cable rigged backactors are usually available with larger bucket sizes and deeper digging capacities than the hydraulic machines but these have a more positive control and digging operation and are also easier to operate.

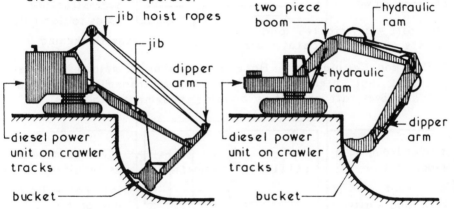

2. Face Shovels – these are robust machines designed to excavate above their own wheel or track level and are suitable for bulk excavation work. In basement work they will require a ramp approach unless they are to be lifted out of the excavation area by means of a crane. Like backactors face shovels are available as cable rigged or hydraulic machines.

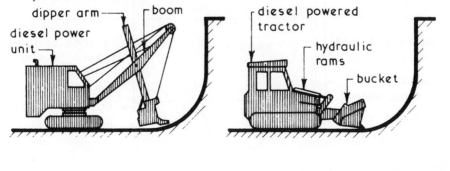

Basement Construction ~ in the general context of buildings a basement can be defined as a storey which is below the ground storey and is therefore constructed below ground level. Most basements can be classified into one of three groups:-

1. Retaining Wall and Raft Basements - this is the general format for basement construction and consists of a slab raft foundation which forms the basement floor and helps to distribute the structural loads transmitted down the retaining walls.

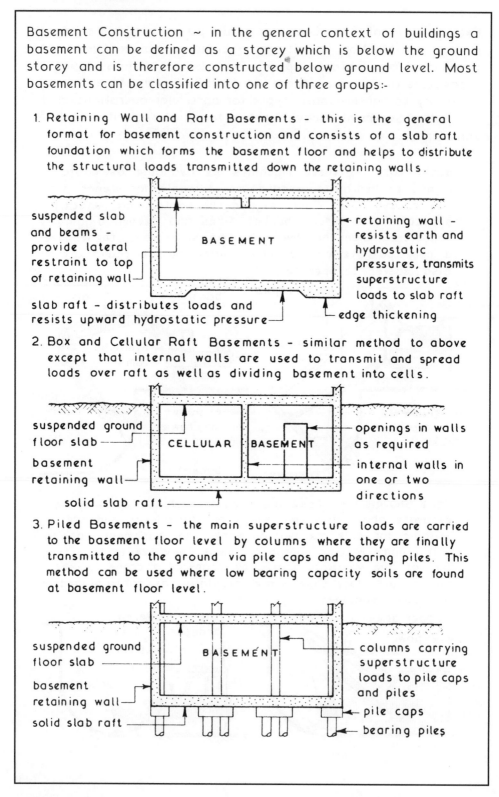

suspended slab and beams – provide lateral restraint to top of retaining wall

BASEMENT

retaining wall – resists earth and hydrostatic pressures, transmits superstructure loads to slab raft

slab raft - distributes loads and resists upward hydrostatic pressure

edge thickening

2. Box and Cellular Raft Basements - similar method to above except that internal walls are used to transmit and spread loads over raft as well as dividing basement into cells.

suspended ground floor slab

basement retaining wall

solid slab raft

CELLULAR BASEMENT

openings in walls as required

internal walls in one or two directions

3. Piled Basements - the main superstructure loads are carried to the basement floor level by columns where they are finally transmitted to the ground via pile caps and bearing piles. This method can be used where low bearing capacity soils are found at basement floor level.

suspended ground floor slab

basement retaining wall

solid slab raft

BASEMENT

columns carrying superstructure loads to pile caps and piles

pile caps

bearing piles

232

Deep Basement Construction ~ basements can be constructed within a cofferdam or other temporary supported excavation (see Basement Excavations on pages 228 to 230) up to the point when these methods become uneconomic, unacceptable or both due to the amount of necessary temporary support work. Deep basements can be constructed by installing diaphragm walls within a trench and providing permanent support with ground anchors or by using the permanent lateral support given by the internal floor during the excavation period (see page 234). Temporary lateral support during the excavation period can be provided by lattice beams spanning between the diaphragm walls (see page 234).

Typical Ground Anchor Support Details ~

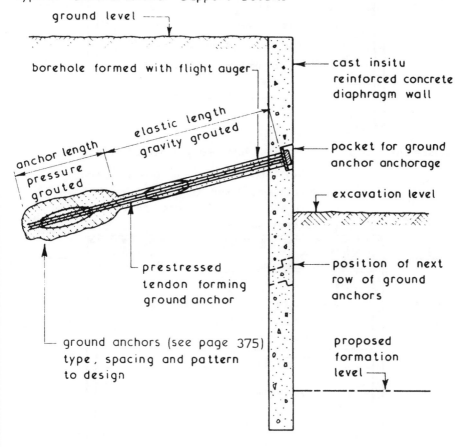

NB vertical ground anchors installed through the lowest floor can be used to overcome any tendency to floatation during the construction period

Basement Construction

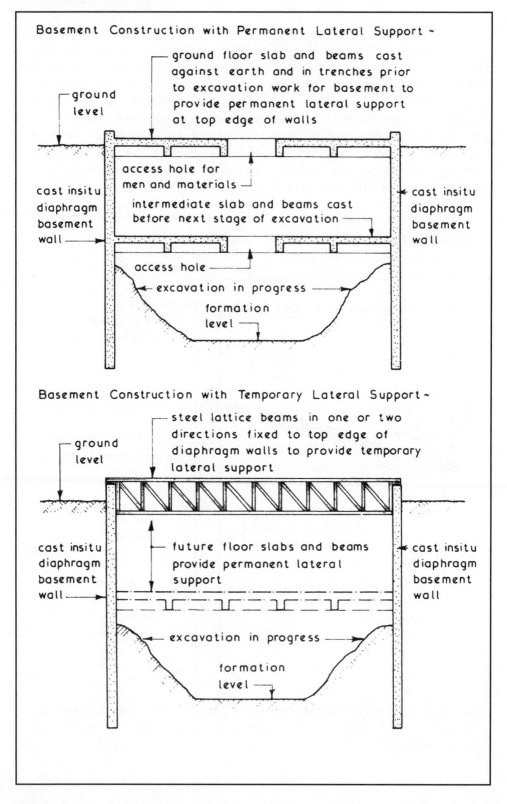

Basement Construction with Permanent Lateral Support ~

ground floor slab and beams cast against earth and in trenches prior to excavation work for basement to provide permanent lateral support at top edge of walls

ground level

access hole for men and materials

cast insitu diaphragm basement wall

intermediate slab and beams cast before next stage of excavation

cast insitu diaphragm basement wall

access hole

excavation in progress

formation level

Basement Construction with Temporary Lateral Support ~

steel lattice beams in one or two directions fixed to top edge of diaphragm walls to provide temporary lateral support

ground level

cast insitu diaphragm basement wall

future floor slabs and beams provide permanent lateral support

cast insitu diaphragm basement wall

excavation in progress

formation level

Waterproofing Basements ~ basements can be waterproofed by one of three basic methods namely:-

1. Use of dense monolithic concrete walls and floor
2. Tanking techniques (see pages 237 & 238)
3. Drained cavity system (see page 239)

Dense Monolithic Concrete — the main objective is to form a watertight basement using dense high quality reinforced or prestressed concrete by a combination of good materials, good workmanship, attention to design detail and on site construction methods. If strict control of all aspects is employed a sound watertight structure can be produced but it should be noted that such structures are not always water vapourproof. If the latter is desirable some waterproof coating, lining or tanking should be used. The watertightness of dense concrete mixes depends primarily upon two factors:-

1. Water/cement ratio.
2. Degree of compaction.

The hydration of cement during the hardening process produces heat therefore to prevent early stage cracking the temperature changes within the hardening concrete should be kept to a minimum. The greater the cement content the more is the evolution of heat therefore the mix should contain no more cement than is necessary to fulfil design requirements. Concrete with a free water/cement ratio of 0·5 is watertight and although the permeability is three time more at a ratio of 0·6 it is for practical purposes still watertight but above this ratio the concrete becomes progressively less watertight. For lower water/cement ratios the workability of the mix would have to be increased, usually by adding more cement, to enable the concrete to be fully compacted.

Admixtures — if the ingredients of good design, materials and workmanship are present watertight concrete can be produced without the use of admixtures. If admixtures are used they should be carefully chosen and used to obtain a specific objective:-

1. Water-reducing admixtures — used to improve workability
2. Retarding admixtures — slow down rate of hardening
3. Accelerating admixtures — increase rate of hardening — useful for low temperatures — calcium chloride not suitable for reinforced concrete.
4. Water-repelling admixtures — effective only with low water head, will not improve poor quality or porous mixes.
5. Air-entraining admixtures — increases workability — lowers water content.

Joints ~ in general these are formed in basement constructions to provide for movement accommodation (expansion joints) or to create a convenient stopping point in the construction process (construction joints). Joints are lines of weakness which will leak unless carefully designed and constructed therefore they should be simple in concept and easy to construct.

Basement slabs ~ these are usually designed to span in two directions and as a consequence have relatively heavy top and bottom reinforcement. To enable them to fulfil their basic functions they usually have a depth in excess of 250mm. The joints, preferably of the construction type, should be kept to a minimum and if waterbars are specified they must be placed to ensure that complete compaction of the concrete is achieved.

Typical Basement Slab Joint Details ~

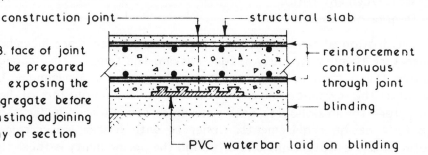

NB. face of joint to be prepared by exposing the aggregate before casting adjoining bay or section

Basement Walls ~ joints can be horizontal and/or vertical according to design requirements. A suitable waterbar should be incorporated in the joint to prevent the ingress of water. The top surface of a kicker used in conjunction with single lift pouring if adequately prepared by exposing the aggregate should not require a waterbar but if one is specified it should be either placed on the rear face or consist of a centrally placed mild steel strip inserted into the kicker whilst the concrete is still in a plastic state.

Typical Basement Wall Joint Details ~

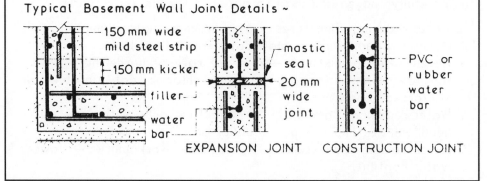

EXPANSION JOINT   CONSTRUCTION JOINT

Mastic Asphalt Tanking ~ the objective of tanking is to provide a continuous waterproof membrane which is applied to the base slab and walls with complete continuity between the two applications. The tanking can be applied externally or internally according to the circumstances prevailing on site. Alternatives to mastic asphalt are polythene sheeting: bituminous compounds: epoxy resin compounds and bitumen laminates.

External Mastic Asphalt Tanking ~ this is the preferred method since it not only prevents the ingress of water it also protects the main structure of the basement from aggressive sulphates which may be present in the surrounding soil or ground water.

Typical External Tanking Details ~

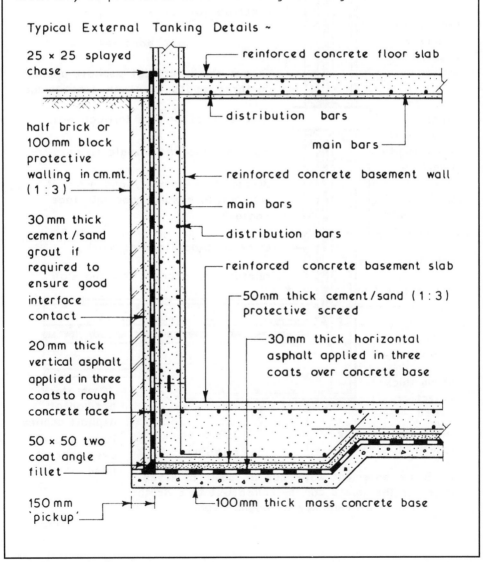

25 × 25 splayed chase

reinforced concrete floor slab

half brick or 100 mm block protective walling in cm.mt. (1 : 3)

distribution bars

main bars

reinforced concrete basement wall

30 mm thick cement / sand grout if required to ensure good interface contact

main bars

distribution bars

reinforced concrete basement slab

50 mm thick cement/sand (1 : 3) protective screed

20 mm thick vertical asphalt applied in three coats to rough concrete face

30 mm thick horizontal asphalt applied in three coats over concrete base

50 × 50 two coat angle fillet

150 mm 'pickup'

100 mm thick mass concrete base

Waterproofing Basements

Internal Mastic Asphalt Tanking ~ this method should only be adopted if external tanking is not possible since it will not give protection to the main structure and unless adequately loaded may be forced away from the walls and/or floor by hydrostatic pressure. To be effective the horizontal and vertical coats of mastic asphalt must be continuous.

Typical Internal Tanking Details ~

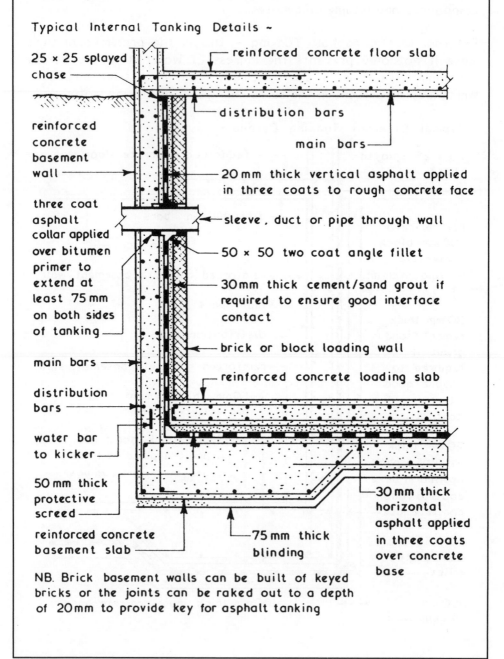

25 × 25 splayed chase

reinforced concrete floor slab

reinforced concrete basement wall

distribution bars

main bars

20 mm thick vertical asphalt applied in three coats to rough concrete face

three coat asphalt collar applied over bitumen primer to extend at least 75 mm on both sides of tanking

sleeve, duct or pipe through wall

50 × 50 two coat angle fillet

30 mm thick cement/sand grout if required to ensure good interface contact

brick or block loading wall

reinforced concrete loading slab

main bars

distribution bars

water bar to kicker

50 mm thick protective screed

reinforced concrete basement slab

75 mm thick blinding

30 mm thick horizontal asphalt applied in three coats over concrete base

NB. Brick basement walls can be built of keyed bricks or the joints can be raked out to a depth of 20mm to provide key for asphalt tanking

Drained Cavity System ~ this method of waterproofing basements can be used for both new and refurbishment work. The basic concept is very simple in that it accepts that a small amount of water seepage is possible through a monolithic concrete wall and the best method of dealing with such moisture is to collect it and drain it away. This is achieved by building an inner non-load bearing wall to form a cavity which is joined to a floor composed of special triangular tiles laid to falls which enables the moisture to drain away to a sump from which it is either discharged direct or pumped into the surface water drainage system. The inner wall should be relatively vapour tight or alternatively the cavity should be ventilated.

Typical Details ~

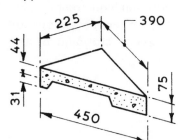

225 390
44
31 75
450

TOP VIEW - HALF TILE

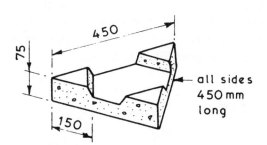

450
75
150
all sides 450 mm long

BOTTOM VIEW - STANDARD TILE

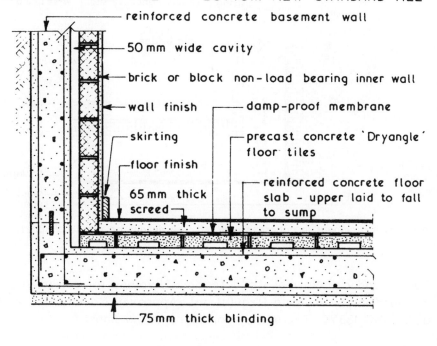

reinforced concrete basement wall

50 mm wide cavity

brick or block non-load bearing inner wall

wall finish damp-proof membrane

skirting precast concrete 'Dryangle' floor tiles

floor finish

65 mm thick screed

reinforced concrete floor slab - upper laid to fall to sump

75mm thick blinding

Basements benefit considerably from the insulating properties of the surrounding soil. However, that alone is insufficient to satisfy the typical requirements for wall and floor U-values of 0·35 and 0·30 W/m²K, respectively.

Refurbishment of existing basements may include insulation within dry lined walls and under the floor screed or particle board overlay. This should incorporate an integral vapour control layer to minimise risk of condensation.

External insulation of closed cell rigid polystyrene slabs is generally applied to new construction. These slabs combine low thermal conductivity with low water absorption and high compressive strength. The external face of insulation is grooved to encourage moisture run off. It is also filter faced to prevent clogging of the grooves. Backfill is granular.

Typical application -

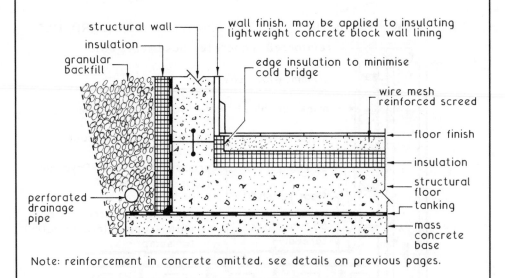

Note: reinforcement in concrete omitted, see details on previous pages.

Tables and calculations to determine U-values for basements are provided in the Building Regulations, Approved Document L and in BS EN ISO 13370: Thermal performance of buildings.

Excavation ~ to hollow out — in building terms to remove earth to form a cavity in the ground.

Types of Excavation ~

Oversite – the removal of top soil (Building Regulations requirement.)

depth varies from site to site but is usually in a 150 to 300 mm range. Top soil contains plant life animal life and decaying matter which makes the soil compressible and therefore unsuitable for supporting buildings.

s u b s o i l

Reduce Level – carried out below oversite level to form a level surface on which to build and can consist of both cutting and filling operations. The level to which the ground is reduced is called the formation level.

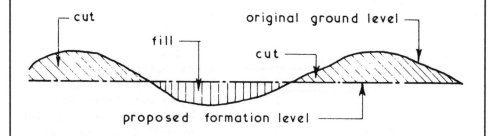

cut

original ground level

fill

cut

proposed formation level

NB. Water in Excavations — this should be removed since it can:~

1. Undermine sides of excavation.
2. Make it impossible to adequately compact bottom of excavation to receive foundations.
3. Cause puddling which can reduce the bearing capacity of the subsoil.

Trench Excavations ~ narrow excavations primarily for strip foundations and buried services — excavation can be carried out by hand or machine.

Typical Examples ~

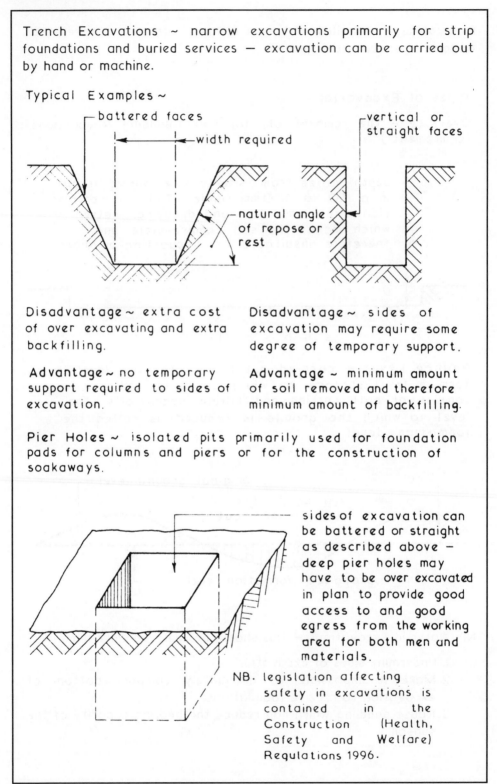

┌ battered faces

←— width required —→

vertical or straight faces

natural angle of repose or rest

Disadvantage ~ extra cost of over excavating and extra backfilling.

Advantage ~ no temporary support required to sides of excavation.

Disadvantage ~ sides of excavation may require some degree of temporary support.

Advantage ~ minimum amount of soil removed and therefore minimum amount of backfilling.

Pier Holes ~ isolated pits primarily used for foundation pads for columns and piers or for the construction of soakaways.

sides of excavation can be battered or straight as described above — deep pier holes may have to be over excavated in plan to provide good access to and good egress from the working area for both men and materials.

NB. legislation affecting safety in excavations is contained in the Construction (Health, Safety and Welfare) Regulations 1996.

Site Clearance and Removal of Top Soil ~

On small sites this could be carried out by manual means using hand held tools such as picks, shovels and wheelbarrows.

On all sites mechanical methods could be used the actual plant employed being dependent on factors such as volume of soil involved, nature of site and time elements.

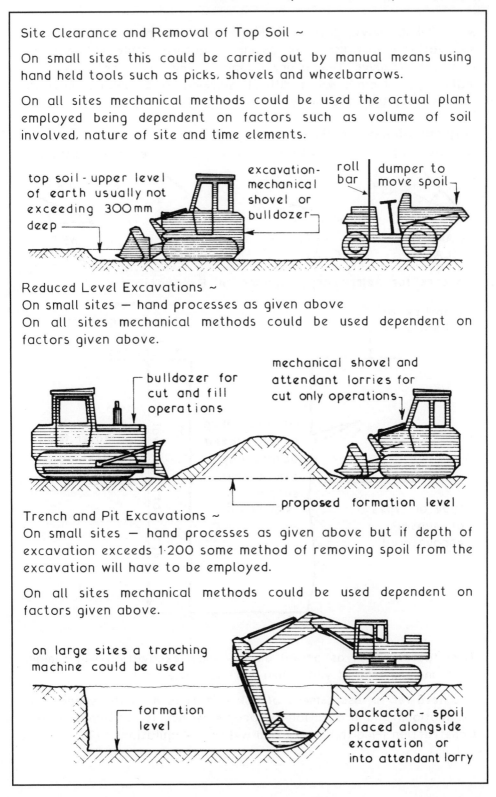

top soil - upper level of earth usually not exceeding 300 mm deep

excavation-mechanical shovel or bulldozer

roll bar

dumper to move spoil

Reduced Level Excavations ~
On small sites — hand processes as given above
On all sites mechanical methods could be used dependent on factors given above.

bulldozer for cut and fill operations

mechanical shovel and attendant lorries for cut only operations

proposed formation level

Trench and Pit Excavations ~
On small sites — hand processes as given above but if depth of excavation exceeds 1·200 some method of removing spoil from the excavation will have to be employed.

On all sites mechanical methods could be used dependent on factors given above.

on large sites a trenching machine could be used

formation level

backactor - spoil placed alongside excavation or into attendant lorry

All subsoils have different abilities in remaining stable during excavation works. Most will assume a natural angle of repose or rest unless given temporary support. The presence of ground water apart from creating difficult working conditions can have an adverse effect on the subsoil's natural angle of repose.

Typical Angles of Repose ~
Excavations cut to a natural angle of repose are called battered.

DRAINED CLAY — 45°
WET CLAY — 16°
GRAVEL & DRY SAND — 40°
WET SAND — 22°

Factors for Temporary Support of Excavations ~

stability of subsoil

nearness of surcharges such as buildings and vehicles

can excavation be completed before any temporary support is required?

depth of excavation

water table level

type or types of subsoil encountered

Time factors such as period during which excavation will remain open and the time of year when work is carried out.

The need for an assessment of risk with regard to the support of excavations and protection of people within, is contained in the Construction (Health, Safety and welfare) Regulations 1996.

Temporary Support ~ in the context of excavations this is called timbering irrespective of the actual materials used. If the sides of the excavation are completely covered with timbering it is known as close timbering whereas any form of partial covering is called open timbering.

An adequate supply of timber or other suitable material must be available and used to prevent danger to any person employed in an excavation from a fall or dislodgement of materials forming the sides of an excavation.

A suitable barrier or fence must be provided to the sides of all excavations or alternatively they must be securely covered

Materials must not be placed near to the edge of any excavation, nor must plant be placed or moved near to any excavation so that persons employed in the excavation are endangered.

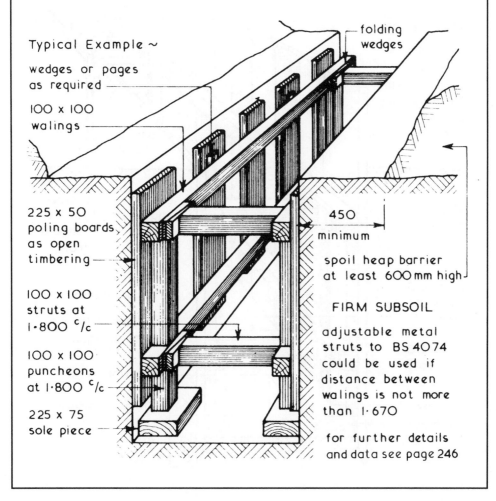

Typical Example ~

wedges or pages as required

100 x 100 walings

folding wedges

225 x 50 poling boards as open timbering

100 x 100 struts at 1·800 $^c/_c$

100 x 100 puncheons at 1·800 $^c/_c$

225 x 75 sole piece

450 minimum

spoil heap barrier at least 600 mm high

FIRM SUBSOIL

adjustable metal struts to BS 4074 could be used if distance between walings is not more than 1·670

for further details and data see page 246

Poling Boards ~ a form of temporary support which is placed in position against the sides of excavation after the excavation work has been carried out. Poling boards are placed at centres according to the stability of the subsoils encountered.

Runners ~ a form of temporary support which is driven into position ahead of the excavation work either to the full depth or by a drive and dig technique where the depth of the runner is always lower than that of the excavation.

Trench Sheeting ~ form of runner made from sheet steel with a trough profile — can be obtained with a lapped joint or an interlocking joint.

Water ~ if present or enters an excavation a pit or sump should be excavated below the formation level to act as collection point from which the water can be pumped away

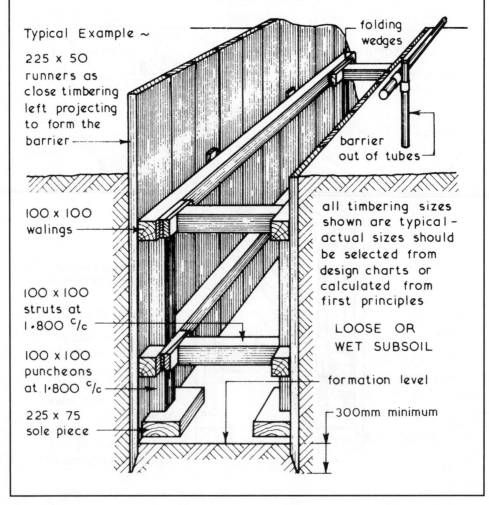

Typical Example ~

225 x 50 runners as close timbering left projecting to form the barrier

folding wedges

barrier out of tubes

100 x 100 walings

100 x 100 struts at 1·800 ᶜ/c

100 x 100 puncheons at 1·800 ᶜ/c

225 x 75 sole piece

all timbering sizes shown are typical - actual sizes should be selected from design charts or calculated from first principles

LOOSE OR WET SUBSOIL

formation level

300mm minimum

Concrete ~ a mixture of cement + fine aggregate + coarse aggregate + water in controlled proportions and of a suitable quality.

Cement ~ powder produced from clay and chalk or limestone. In general most concrete is made with ordinary or rapid hardening Portland cement, both types being manufactured to the recommendations of BS 12. Ordinary Portland cement is adequate for most purposes but has a low resistance to attack by acids and sulphates. Rapid hardening Portland cement does not set faster than ordinary Portland cement but it does develop its working strength at a faster rate. For a concrete which must have an acceptable degree of resistance to sulphate attack sulphate resisting Portland cement made to the recommendations of BS 4027 could be specified.

25 kg

BAGS

12 t to 50 t

SILOS

Aggregates ~ shape, surface texture and grading (distribution of particle sizes) are factors which influence the workability and strength of a concrete mix. Fine aggregates are those materials which pass through a 5mm sieve whereas coarse aggregates are those materials which are retained on a 5mm sieve. Dense aggregates are those with a density of more than 1200kg/m^3 for coarse aggregates and more than 1250kg/m^3 for fine aggregates and are covered by BS 882 - Coarse and Fine Aggregates from Natural Sources and BS 1047 - Air-cooled Blastfurnace Slag Coarse Aggregates for Concrete. Lightweight aggregates include clinker; foamed or expanded blastfurnace slag and exfoliated and expanded materials such as vermiculite, perlite, clay and sintered pulverized-fuel ash to BS 3797.

coarse aggregate

5mm sieve

fine aggregate

Water ~ must be clean and free from impurities which are likely to affect the quality or strength of the resultant concrete. Pond, river, canal and sea water should not be used and only water which is fit for drinking should be specified.

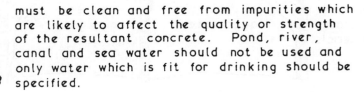

drinking water quality

Cement ~ whichever type of cement is being used it must be properly stored on site to keep it in good condition. The cement must be kept dry since contact with any moisture whether direct or airborne could cause it to set. A rotational use system should be introduced to ensure that the first batch of cement delivered is the first to be used.

Typical Storage Methods ~

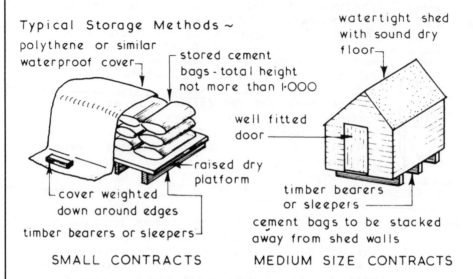

polythene or similar waterproof cover

stored cement bags - total height not more than 1·000

well fitted door

raised dry platform

cover weighted down around edges

timber bearers or sleepers

watertight shed with sound dry floor

timber bearers or sleepers

cement bags to be stacked away from shed walls

SMALL CONTRACTS   MEDIUM SIZE CONTRACTS

LARGE CONTRACTS — for bagged cement watertight shed as above for bulk delivery loose cement a cement storage silo.

Aggregates ~ essentials of storage are to keep different aggregate types and/or sizes separate, store on a clean, hard, free draining surface and to keep the stored aggregates clean and free of leaves and rubbish.

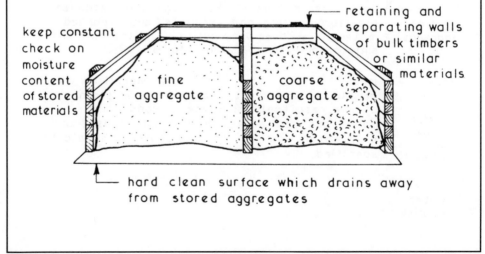

keep constant check on moisture content of stored materials

retaining and separating walls of bulk timbers or similar materials

fine aggregate

coarse aggregate

hard clean surface which drains away from stored aggregates

Concrete Batching ~ a batch is one mixing of concrete and can be carried out by measuring the quantities of materials required by volume or weight. The main aim of both methods is to ensure that all consecutive batches are of the same standard and quality.

Volume Batching ~ concrete mixes are often quoted by ratio such as 1 : 2 : 4 (cement : fine aggregate or sand : coarse aggregate). Cement weighing 50 kg has a volume of $0.033\,m^3$ therefore for the above mix 2×0.033 ($0.066\,m^3$) of sand and 4×0.033 ($0.132\,m^3$) of coarse aggregate is required. To ensure accurate amounts of materials are used for each batch a gauge box should be employed its size being based on convenient handling. Ideally a batch of concrete should be equated to using 50 kg of cement per batch. Assuming a gauge box 300 mm deep and 300 mm wide with a volume of half the required sand the gauge box size would be —

$$\text{volume} = \text{length} \times \text{width} \times \text{depth} = \text{length} \times 300 \times 300$$

$$\therefore \text{length} = \frac{\text{volume}}{\text{width} \times \text{depth}} = \frac{0.033}{0.3 \times 0.3} = 0.366\,m$$

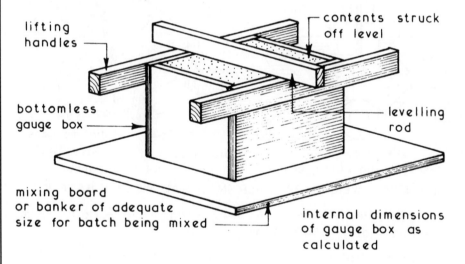

lifting handles

contents struck off level

bottomless gauge box

levelling rod

mixing board or banker of adequate size for batch being mixed

internal dimensions of gauge box as calculated

For the above given mix fill gauge box once with cement, twice with sand and four times with coarse aggregate.

An allowance must be made for the bulking of damp sand which can be as much as $33^1/_3$ %. General rule of thumb unless using dry sand allow for 25% bulking.

Materials should be well mixed dry before adding water.

Weight Batching ~ this is a more accurate method of measuring materials for concrete than volume batching since it reduces considerably the risk of variation between different batches. The weight of sand is affected very little by its dampness which in turn leads to greater accuracy in proportioning materials. When loading a weighing hopper the materials should be loaded in a specific order —

1. Coarse aggregates — tends to push other materials out and leaves the hopper clean.
2. Cement — this is sandwiched between the other materials since some of the fine cement particles could be blown away if cement is put in last.
3. Sand or fine Aggregates — put in last to stabilise the fine lightweight particles of cement powder.

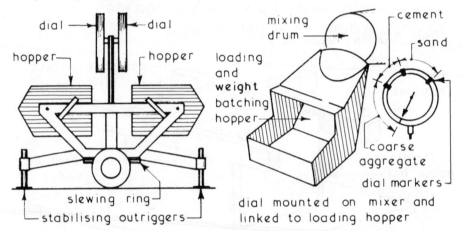

INDEPENDENT WEIGHT BATCHER INTEGRAL WEIGHT BATCHER

Typical Densities ~ cement — 1440 kg/m^3 sand — 1600 kg/m^3 coarse aggregate — 1440 kg/m^3

Water/Cement Ratio ~ water in concrete has two functions -

1. Start the chemical reaction which causes the mixture to set into a solid mass.
2. Give the mix workability so that it can be placed, tamped or vibrated into the required position.

Very little water is required to set concrete (approximately 0·2 w/c ratio) the surplus evaporates leaving minute voids therefore the more water added to the mix to increase its workability the weaker is the resultant concrete. Generally w/c ratios of 0·4 to 0·5 are adequate for most purposes.

Concrete ~ a composite with many variables, represented by numerous gradings which indicate components, quality and manufacturing control.

Grade mixes: C7·5, C10, C15, C20, C25, C30, C35, C40, C45, C50, C55, and C60; F3, F4 and F5; IT2, IT2·5, and IT3.

C = Characteristic compressive ⎫
F = Flexural   ⎬ strengths at 28 days (N/mm²)
IT = Indirect tensile   ⎭

NB. If the grade is followed by a 'P', e.g. C30P, this indicates a prescribed mix (see below).

Grades C7·5 and C10 — Unreinforced plain concrete.
Grades C15 and C20 — Plain concrete or if reinforced containing lightweight aggregate.
Grades C25 — Reinforced concrete containing dense aggregate.
Grades C30 and C35 — Post-tensioned reinforced concrete.
Grades C40 to C60 — Pre-tensioned reinforced concrete.

Categories of mix: 1. Standard; 2. Prescribed; 3. Designed; 4. Designated.

1. Standard Mix — BS guidelines provide this for minor works or in situations limited by available material and manufacturing data. Volume or weight batching is appropriate, but no grade over C30 is recognised.

2. Prescribed Mix — components are predetermined (to a recipe) to ensure strength requirements. Variations exist to allow the purchaser to specify particular aggregates, admixtures and colours. All grades permitted.

3. Designed Mix — concrete is specified to an expected performance. Criteria can include characteristic strength, durability and workability, to which a concrete manufacturer will design and supply an appropriate mix. All grades permitted.

4. Designated Mix — selected for specific applications. General (GEN) graded 0-4, 7·5—25 N/mm² for foundations, floors and external works. Foundations (FND) graded 2, 3, 4A and 4B, 35 N/mm² mainly for sulphate resisting foundations.

Paving (PAV) graded 1 or 2, 35 or 45 N/mm² for roads and drives.

Reinforced (RC) graded 30, 35, 40, 45 and 50 N/mm² mainly for prestressing.

ref BS 5328—2: Methods for specifying concrete.

Concrete Supply ~ this is usually geared to the demand or the rate at which the mixed concrete can be placed. Fresh concrete should always be used or placed within 30 minutes of mixing to prevent any undue drying out. Under no circumstances should more water be added after the initial mixing.

Small Batches ~ small easily transported mixers with output capacities of up to 100 litres can be used for small and intermittent batches. These mixers are versatile and robust machines which can be used for mixing mortars and plasters as well as concrete.

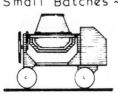

Medium to Large Batches ~ mixers with output capacities from 100 litres to $10 \, m^3$ with either diesel or electric motors. Many models are available with tilting or reversing drum discharge, integral weigh batching and loading hopper and a controlled water supply

Ready Mixed Concrete ~ used mainly for large concrete batches of up to $6 \, m^3$. This method of concrete supply has the advantages of eliminating the need for site space to accommodate storage of materials, mixing plant and the need to employ adequately trained site staff who can constantly produce reliable and consistent concrete mixes. Ready mixed concrete supply depot also have better facilities and arrangements for producing and supplying mixed concrete in winter or inclement weather conditions. In many situations it is possible to place the ready mixed concrete into the required position direct from the delivery lorry via the delivery chute or by feeding it into a concrete pump. The site must be capable of accepting the 20 tonnes laden weight of a typical ready mixed concrete lorry with a turning circle of about 15·000. The supplier will want full details of mix required and the proposed delivery schedule.

Ref. BS 5328—3: Specification for the procedures to be used in producing and transporting concrete.

Cofferdams ~ these are temporary enclosures installed in soil or water to prevent the ingress of soil and/or water into the working area with the cofferdam. They are usually constructed from inter-locking steel sheet piles which are suitably braced or tied back with ground anchors. Alternatively a cofferdam can be installed using any structural material which will fulfil the required function.

Typical Cofferdam Details ~

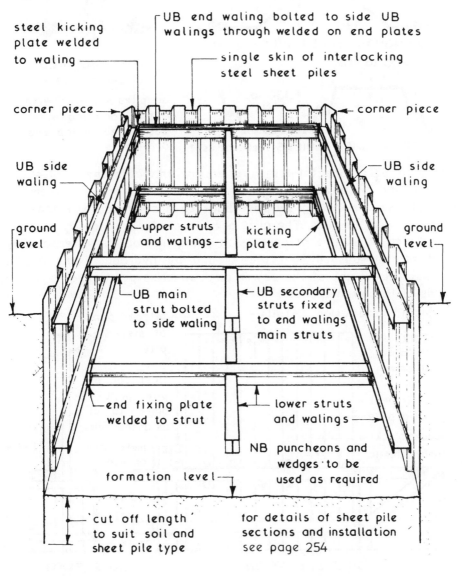

steel kicking plate welded to waling

UB end waling bolted to side UB walings through welded on end plates

single skin of interlocking steel sheet piles

corner piece

corner piece

UB side waling

UB side waling

ground level

upper struts and walings

kicking plate

ground level

UB main strut bolted to side waling

UB secondary struts fixed to end walings main struts

end fixing plate welded to strut

lower struts and walings

NB puncheons and wedges to be used as required

formation level

cut off length to suit soil and sheet pile type

for details of sheet pile sections and installation see page 254

Steel Sheet Piling ~ apart from cofferdam work steel sheet can be used as a conventional timbering material in excavations and to form permanent retaining walls. Three common formats of steel sheet piles with interlocking joints are available with a range of section sizes and strengths up to a usual maximum length of 18·000:-

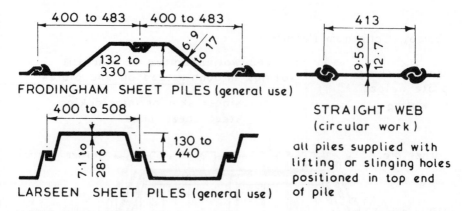

FRODINGHAM SHEET PILES (general use)

LARSEEN SHEET PILES (general use)

STRAIGHT WEB (circular work)

all piles supplied with lifting or slinging holes positioned in top end of pile

Installing Steel Sheet Piles ~ to ensure that the sheet piles are pitched and installed vertically a driving trestle or guide frame is used. These are usually purpose built to accommodate a panel of 10 to 12 pairs of piles. The piles are lifted into position by a crane and driven by means of percussion piling hammer or alternatively they can be pushed into the ground by hydraulic rams acting against the weight of the power pack which is positioned over the heads of the pitched piles.

Typical Installation Details ~

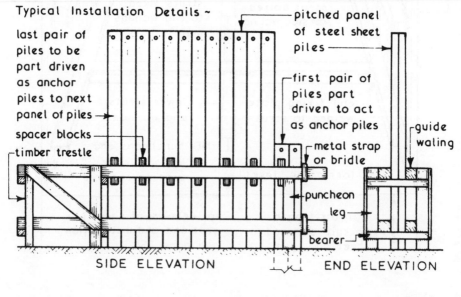

last pair of piles to be part driven as anchor piles to next panel of piles

spacer blocks

timber trestle

pitched panel of steel sheet piles

first pair of piles part driven to act as anchor piles

metal strap or bridle

guide waling

puncheon

leg

bearer

SIDE ELEVATION   END ELEVATION

Caissons ~ these are box-like structures which are similar in concept to cofferdams but they usually form an integral part of the finished structure. They can be economically constructed and installed in water or soil where the depth exceeds 18·000. There are 4 basic types of caisson namely:-

1. Box Caissons
2. Open Caissons
3. Monolithic Caissons
}
usually of precast concrete and used in water being towed or floated into position and sunk — land caissons are of the open type and constructed insitu.

4. Pneumatic Caissons — used in water — see page 256

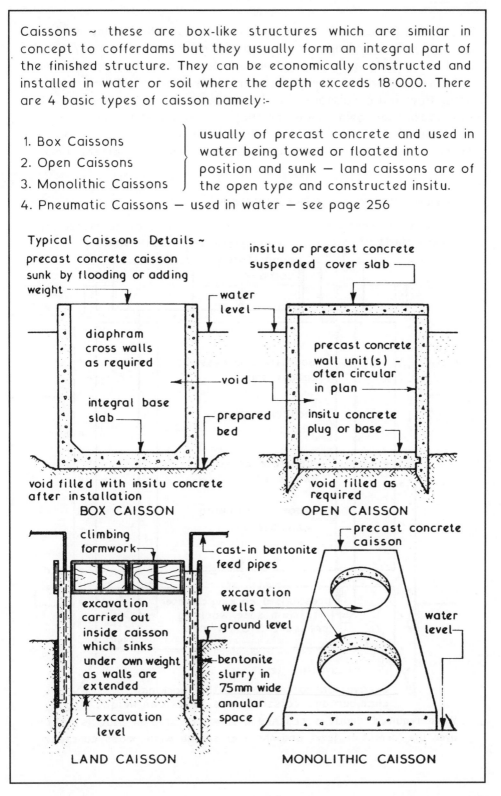

Typical Caissons Details ~
precast concrete caisson sunk by flooding or adding weight

insitu or precast concrete suspended cover slab

water level

diaphram cross walls as required

void

integral base slab

prepared bed

void filled with insitu concrete after installation

BOX CAISSON

precast concrete wall unit(s) - often circular in plan

insitu concrete plug or base

void filled as required

OPEN CAISSON

climbing formwork

cast-in bentonite feed pipes

excavation carried out inside caisson which sinks under own weight as walls are extended

excavation wells

ground level

bentonite slurry in 75mm wide annular space

excavation level

LAND CAISSON

precast concrete caisson

water level

MONOLITHIC CAISSON

Pneumatic Caissons ~ these are sometimes called compressed air caissons and are similar in concept to open caissons. They can be used in difficult subsoil conditions below water level and have a pressurised lower working chamber to provide a safe dry working area. Pneumatic caissons can be made of concrete whereby they sink under their own weight or they can be constructed from steel with hollow walls which can be filled with water to act as ballast. These caissons are usually designed to form part of the finished structure.

Typical Pneumatic Caisson Details ~

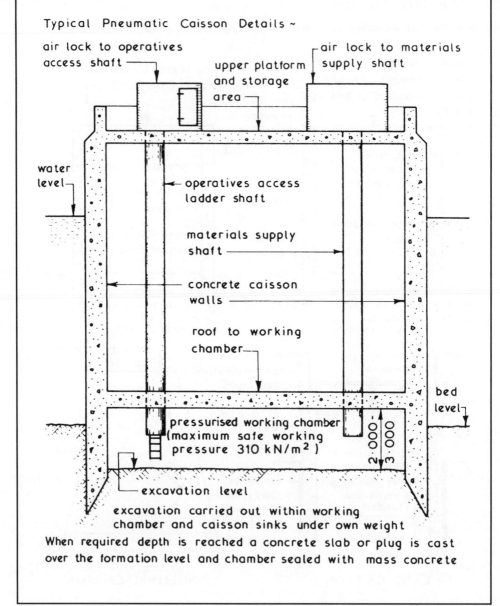

air lock to operatives access shaft

upper platform and storage area

air lock to materials supply shaft

water level

operatives access ladder shaft

materials supply shaft

concrete caisson walls

roof to working chamber

bed level

pressurised working chamber (maximum safe working pressure 310 kN/m²)

2·000 - 3·000

excavation level

excavation carried out within working chamber and caisson sinks under own weight

When required depth is reached a concrete slab or plug is cast over the formation level and chamber sealed with mass concrete

Underpinning ~ the main objective of most underpinning work is to transfer the load carried by a foundation from its existing bearing level to a new level at a lower depth. Underpinning techniques can also be used to replace an existing weak foundation. An underpinning operation may be necessary for one or more of the following reasons:-

1. Uneven Settlement — this could be caused by uneven loading of the building, unequal resistance of the soil action of tree roots or cohesive soil settlement.

2. Increase in Loading — this could be due to the addition of an extra storey or an increase in imposed loadings such as that which may occur with a change of use.

3. Lowering of Adjacent Ground — usually required when constructing a basement adjacent to existing foundations.

General Precautions ~ before any form of underpinning work is commenced the following precautions should be taken :-

1. Notify adjoining owners of proposed works giving full details and temporary shoring or tying.

2. Carry out a detailed survey of the site, the building to be underpinned and of any other adjoining or adjacent building or structures. A careful record of any defects found should be made and where possible agreed with the adjoining owner(s) before being lodged in a safe place.

3. Indicators or 'tell tales' should be fixed over existing cracks so that any subsequent movements can be noted and monitored.

4. If settlement is the reason for the underpinning works a thorough investigation should be carried out to establish the cause and any necessary remedial work put in hand before any underpinning works are started.

5. Before any underpinning work is started the loads on the building to be underpinned should be reduced as much as possible by removing the imposed loads from the floors and installing any props and/or shoring which is required.

6. Any services which are in the vicinity of the proposed underpinning works should be identified, traced, carefully exposed, supported and protected as necessary.

Underpinning to Walls ~ to prevent fracture, damage or settlement of the wall(s) being underpinned the work should always be carried out in short lengths called legs or bays. The length of these bays will depend upon the following factors:-

1. Total length of wall to be underpinned.

2. Wall loading.

3. General state of repair and stability of wall and foundation to be underpinned.

4. Nature of subsoil beneath existing foundation.

5. Estimated spanning ability of existing foundation.

Generally suitable bay lengths are:-

1·000 to 1·500 for mass concrete strip foundations supporting walls of traditional construction.

1·500 to 3·000 for reinforced concrete strip foundations supporting walls of moderate loading.

In all the cases the total sum of the unsupported lengths of wall should not exceed 25% of the total wall length.

The sequence of bays should be arranged so that working in adjoining bays is avoided until one leg of underpinning has been completed, pinned and cured sufficiently to support the wall above.

Typical Underpinning Schedule ~

schedule shows 2 bay working - once all the underpinning legs have been completed the working bays will have combined to form a trench enabling a complete and final check to be made before backfilling

bay width to give adequate working space - minimum 1·000

wall and foundation to be underpinned

1·500 long working bays

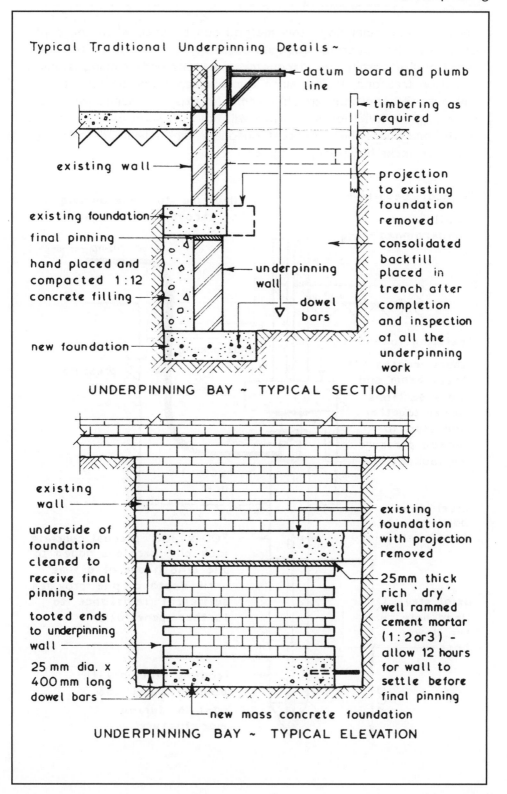

Typical Traditional Underpinning Details ~

datum board and plumb line

timbering as required

existing wall

existing foundation

final pinning

hand placed and compacted 1:12 concrete filling

underpinning wall

dowel bars

projection to existing foundation removed

consolidated backfill placed in trench after completion and inspection of all the underpinning work

new foundation

UNDERPINNING BAY ~ TYPICAL SECTION

existing wall

underside of foundation cleaned to receive final pinning

tooted ends to underpinning wall

25 mm dia. x 400 mm long dowel bars

existing foundation with projection removed

25mm thick rich `dry' well rammed cement mortar (1 : 2 or 3) - allow 12 hours for wall to settle before final pinning

new mass concrete foundation

UNDERPINNING BAY ~ TYPICAL ELEVATION

259

Jack Pile Underpinning ~ this method can be used when the depth of a suitable bearing capacity subsoil is too deep to make traditional underpinning uneconomic. Jack pile underpinning is quiet, vibration free and flexible since the pile depth can be adjusted to suit subsoil conditions encountered. The existing foundations must be in a good condition since they will have to span over the heads of the pile caps which are cast onto the jack pile heads after the hydraulic jacks have been removed.

Typical Details ~

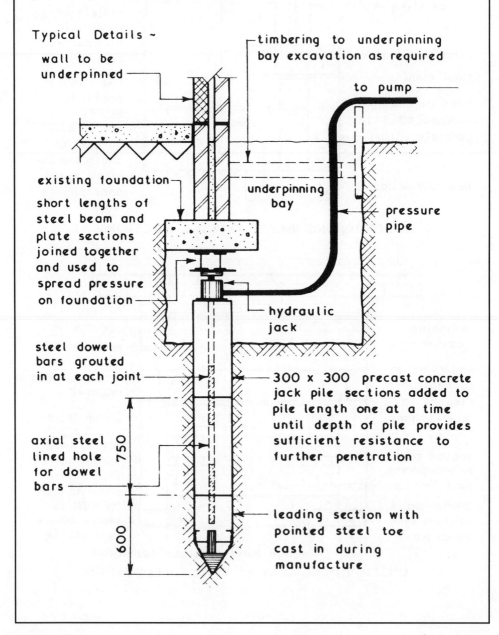

wall to be underpinned

timbering to underpinning bay excavation as required

to pump

existing foundation

short lengths of steel beam and plate sections joined together and used to spread pressure on foundation

underpinning bay

pressure pipe

hydraulic jack

steel dowel bars grouted in at each joint

axial steel lined hole for dowel bars

750

600

300 x 300 precast concrete jack pile sections added to pile length one at a time until depth of pile provides sufficient resistance to further penetration

leading section with pointed steel toe cast in during manufacture

Needle and Pile Underpinning ~ this method of underpinning can be used where the condition of the existing foundation is unsuitable for traditional or jack pile underpinning techniques. The brickwork above the existing foundation must be in a sound condition since this method relies on the 'arching effect' of the brick bonding to transmit the wall loads onto the needles and ultimately to the piles. The piles used with this method are usually small diameter bored piles — see page 199.

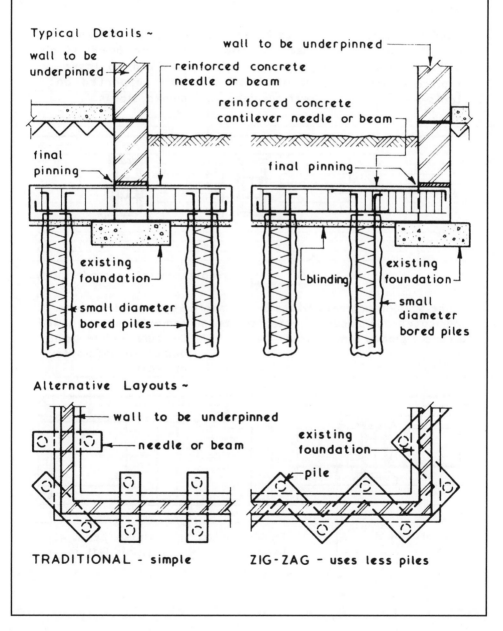

Typical Details ~

wall to be underpinned

reinforced concrete needle or beam

wall to be underpinned

reinforced concrete cantilever needle or beam

final pinning

final pinning

existing foundation

small diameter bored piles

blinding

existing foundation

small diameter bored piles

Alternative Layouts ~

wall to be underpinned

needle or beam

existing foundation

pile

TRADITIONAL - simple

ZIG-ZAG - uses less piles

'Pynford' Stool Method of Underpinning ~ this method can be used where the existing foundations are in a poor condition and it enables the wall to be underpinned in a continuous run without the need for needles or shoring. The reinforced concrete beam formed by this method may well be adequate to spread the load of the existing wall or it may be used in conjunction with other forms of underpinning such as traditional and jack pile.

Typical Details ~

Stage 1 - holes formed in wall to recieve steel or precast concrete stools

Stage 2 - stools inserted and pinned to soffit of brickwork over opening

Stage 3 - brickwork between pinned stools removed to leave wall supported on pinned stools

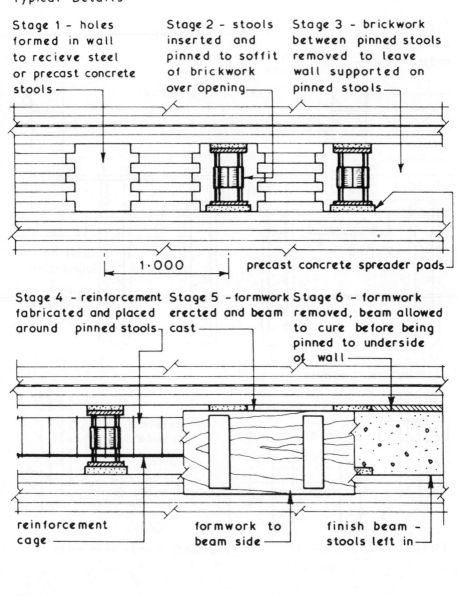

1·000   precast concrete spreader pads

Stage 4 - reinforcement fabricated and placed around pinned stools

Stage 5 - formwork erected and beam cast

Stage 6 - formwork removed, beam allowed to cure before being pinned to underside of wall

reinforcement cage

formwork to beam side

finish beam - stools left in

'Bullivant' Patent Angle Piling ~ this is a much simpler alternative to traditional underpinning techniques, applying modern concrete drilling equipment to achieve cost benefits through time saving. The process is also considerably less disruptive, as large volumes of excavation are avoided. Where sound bearing strata can be located within a few metres of the surface, wall stability is achieved through lined reinforced concrete piles installed in pairs, at opposing angles. The existing floor, wall and foundation are pre-drilled with air flushed percussion auger, giving access for a steel lining to be driven through the low grade/clay subsoil until it impacts with firm strata. The lining is cut to terminate at the underside of the foundation and the void steel reinforced prior to concreting.

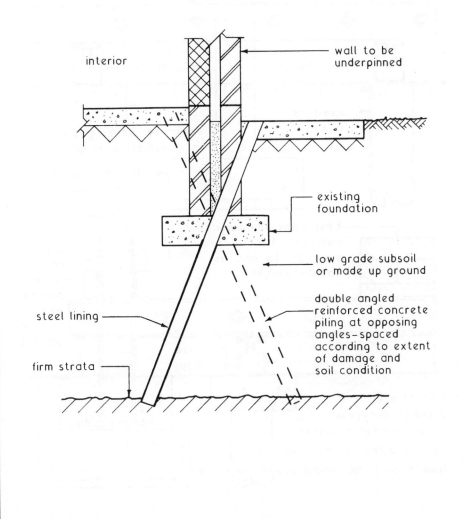

interior

wall to be underpinned

existing foundation

low grade subsoil or made up ground

double angled reinforced concrete piling at opposing angles-spaced according to extent of damage and soil condition

steel lining

firm strata

Underpinning Columns ~ columns can be underpinned in the some manner as walls using traditional or jack pile methods after the columns have been relieved of their loadings. The beam loads can usually be transferred from the columns by means of dead shores and the actual load of the column can be transferred by means of a pair of beams acting against a collar attached to the base of the column shaft.

Typical Details ~

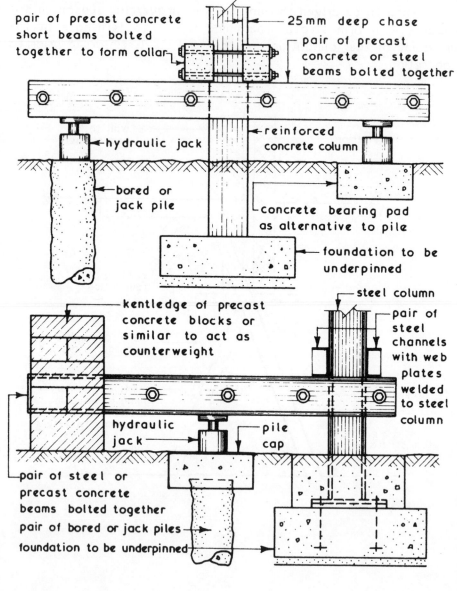

Classification of Water ~ water can be classified by its relative position to or within the ground thus —

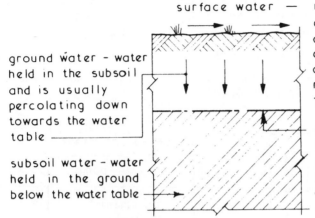

surface water — run off from an earth surface ≈ 10 % of water falling onto surface but on hard paved areas run off is usually 75 to 90 %

ground water - water held in the subsoil and is usually percolating down towards the water table

water table - upper level of water held in the soil which varies with wet and dry periods

subsoil water - water held in the ground below the water table

Problems of Water in the Subsoil ~

1. A high water table could cause flooding during wet periods.
2. Subsoil water can cause problems during excavation works by its natural tendency to flow into the voids created by the excavation activities.
3. It can cause an unacceptable humidity level around finished buildings and structures.

Control of Ground Water ~ this can take one of two forms which are usually referred to as temporary and permanent exclusion —

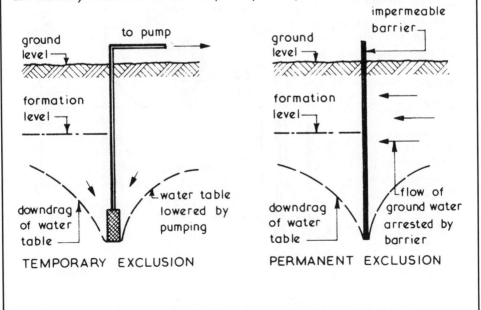

ground level

to pump

formation level

downdrag of water table

water table lowered by pumping

TEMPORARY EXCLUSION

ground level

impermeable barrier

formation level

downdrag of water table

flow of ground water arrested by barrier

PERMANENT EXCLUSION

Permanent Exclusion ~ this can be defined as the insertion of an impermeable barrier to stop the flow of water within the ground.

Temporary Exclusion ~ this can be defined as the lowering of the water table and within the economic depth range of 1·500 can be achieved by subsoil drainage methods, for deeper treatment a pump or pumps are usually involved.

Simple Sump Pumping ~ suitable for trench work and/or where small volumes of water are involved.

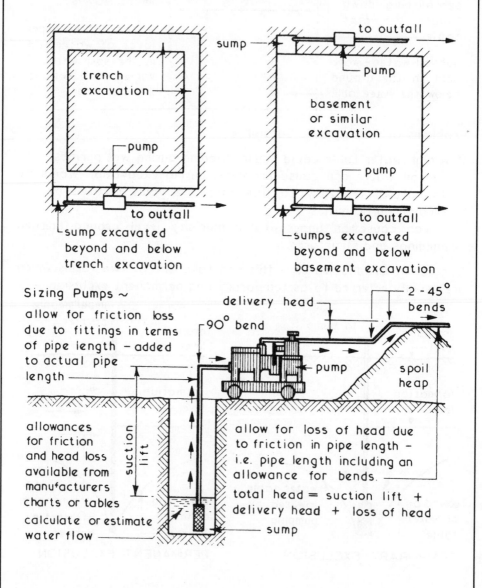

total head = suction lift + delivery head + loss of head

Jetted Sumps ~ this method achieves the same objectives as the simple sump methods of dewatering (page 266) but it will prevent the soil movement associated with this and other open sump methods. A borehole is formed in the subsoil by jetting a metal tube into the ground by means of pressurised water, to a depth within the maximum suction lift of the extract pump. The metal tube is withdrawn to leave a void for placing a disposable wellpoint and plastic suction pipe. The area surrounding the pipe is filled with coarse sand to function as a filtering media.

Typical Example ~

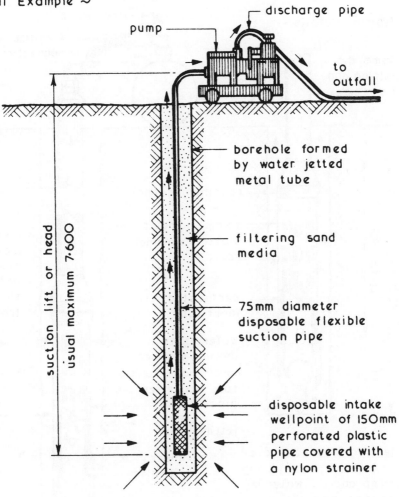

discharge pipe

pump

to outfall

suction lift or head
usual maximum 7·600

borehole formed by water jetted metal tube

filtering sand media

75mm diameter disposable flexible suction pipe

disposable intake wellpoint of 150mm perforated plastic pipe covered with a nylon strainer

Wellpoint Systems ~ method of lowering the water table to a position below the formation level to give a dry working area. The basic principle is to jet into the subsoil a series of wellpoints which are connected to a common header pipe which is connected to a vacuum pump. Wellpoint systems are suitable for most subsoils and can encircle an excavation or be laid progressively alongside as in the case of a trench excavation. If the proposed formation level is below the suction lift capacity of the pump a multi-stage system can be employed — see page 269.

Typical Details ~

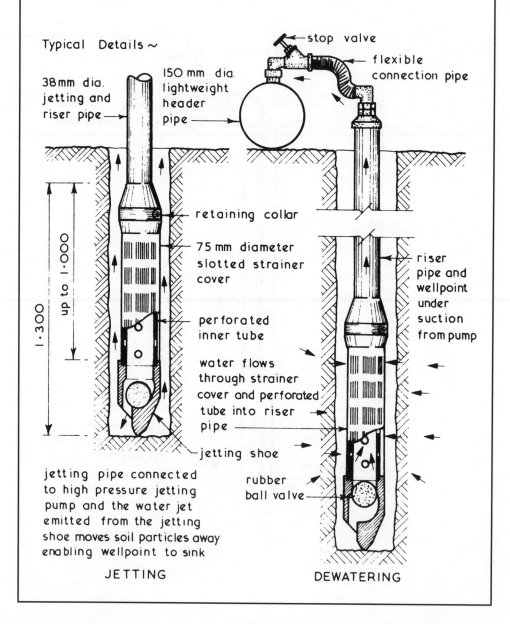

38mm dia. jetting and riser pipe

150 mm dia. lightweight header pipe

stop valve

flexible connection pipe

retaining collar

75 mm diameter slotted strainer cover

perforated inner tube

water flows through strainer cover and perforated tube into riser pipe

jetting shoe

riser pipe and wellpoint under suction from pump

rubber ball valve

up to 1·000

1·300

jetting pipe connected to high pressure jetting pump and the water jet emitted from the jetting shoe moves soil particles away enabling wellpoint to sink

JETTING

DEWATERING

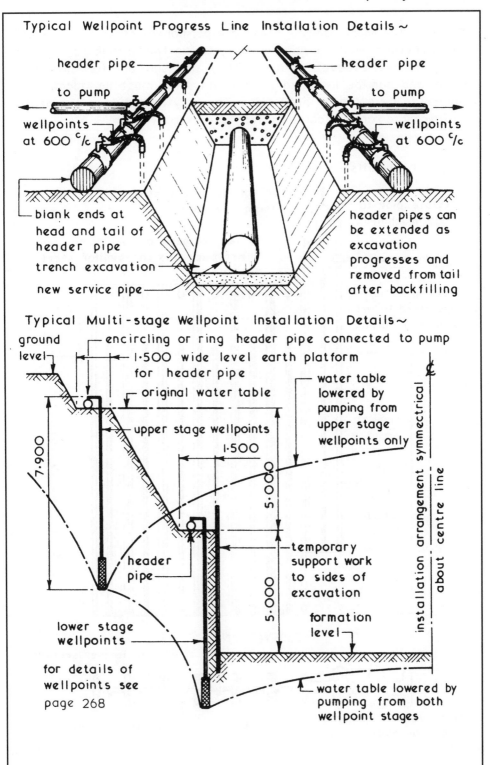

Typical Wellpoint Progress Line Installation Details ~

header pipe

header pipe

to pump

to pump

wellpoints at 600 c/c

wellpoints at 600 c/c

blank ends at head and tail of header pipe

trench excavation

new service pipe

header pipes can be extended as excavation progresses and removed from tail after backfilling

Typical Multi-stage Wellpoint Installation Details~

ground level

encircling or ring header pipe connected to pump

1·500 wide level earth platform for header pipe

original water table

upper stage wellpoints

water table lowered by pumping from upper stage wellpoints only

1·500

7·900

5·000

installation arrangement symmetrical about centre line

header pipe

temporary support work to sides of excavation

5·000

formation level

lower stage wellpoints

for details of wellpoints see page 268

water table lowered by pumping from both wellpoint stages

Thin Grouted Membranes ~ these are permanent curtain or cut-off non structural walls or barriers inserted in the ground to enclose the proposed excavation area. They are suitable for silts and sands and can be installed rapidly but they must be adequately supported by earth on both sides. The only limitation is the depth to which the formers can be driven and extracted.

Typical Details ~

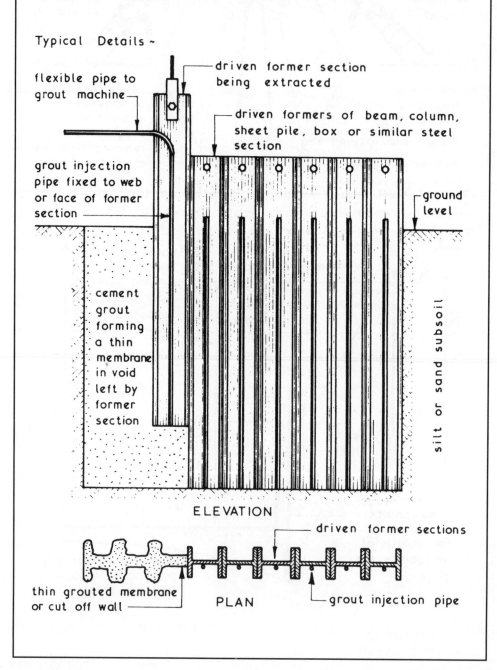

flexible pipe to grout machine

driven former section being extracted

driven formers of beam, column, sheet pile, box or similar steel section

grout injection pipe fixed to web or face of former section

ground level

cement grout forming a thin membrane in void left by former section

silt or sand subsoil

ELEVATION

driven former sections

thin grouted membrane or cut off wall

PLAN

grout injection pipe

Contiguous Piling ~ this forms a permanent structural wall of interlocking bored piles. Alternate piles are bored and cast by traditional methods after which the interlocking piles are bored using a special auger or cutter. This system is suitable for most types of subsoil and has the main advantages of being economical on small and confined sites; capable of being formed close to existing foundations and can be installed with the minimum of vibration and noise. Ensuring a complete interlock of all piles over the entire length may be difficult to achieve in practice therefore the exposed face of the piles is usually covered with a mesh or similar fabric and face with rendering or sprayed concrete. Alternatively a reinforced concrete wall could be cast in front of the contiguous piling. This method of ground water control is suitable for structures such as basements, road underpasses and underground car parks.

Typical Details ~

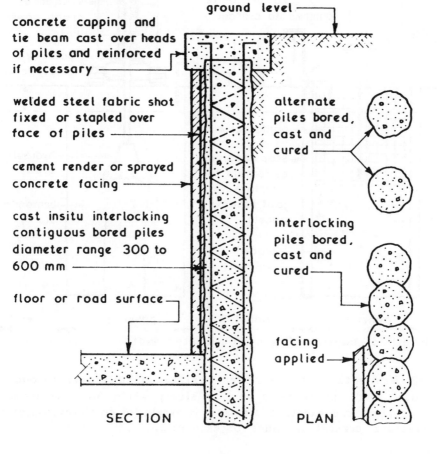

concrete capping and tie beam cast over heads of piles and reinforced if necessary

welded steel fabric shot fixed or stapled over face of piles

cement render or sprayed concrete facing

cast insitu interlocking contiguous bored piles diameter range 300 to 600 mm

floor or road surface

ground level

alternate piles bored, cast and cured

interlocking piles bored, cast and cured

facing applied

SECTION   PLAN

271

Diaphragm Walls ~ these are structural concrete walls which can be cast insitu (usually by the bentonite slurry method) or constructed using precast concrete components (see page 273). They are suitable for most subsoils and their installation generates only a small amount of vibration and noise making them suitable for works close to existing buildings. The high cost of these walls makes them uneconomic unless they can be incorporated into the finished structure. Diaphragm walls are suitable for basements, underground car parks and similar structures.

Typical Cast Insitu Concrete Diaphragm Wall Details ~

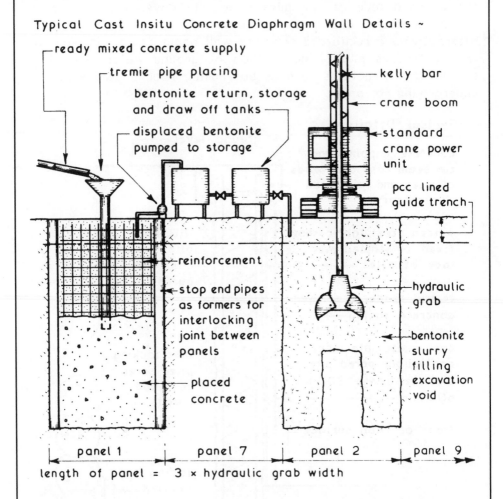

length of panel = 3 × hydraulic grab width

NB. Bentonite is a controlled mixture of fullers earth and water which produces a mud or slurry which has thixotropic properties and exerts a pressure in excess of earth + hydrostatic pressure present on sides of excavation.

Precast Concrete Diaphragm Walls ~ these walls have the some applications as their insitu counterparts and have the advantages of factory produced components but lack the design flexibility of cast insitu walls. The panel or post and panel units are installed in a trench filled with a special mixture of bentonite and cement with a retarder to control the setting time. This mixture ensures that the joints between the wall components are effectively sealed. To provide stability the panels or posts are tied to the retained earth with ground anchors.

Typical Precast Concrete Diaphragm Wall Details ~

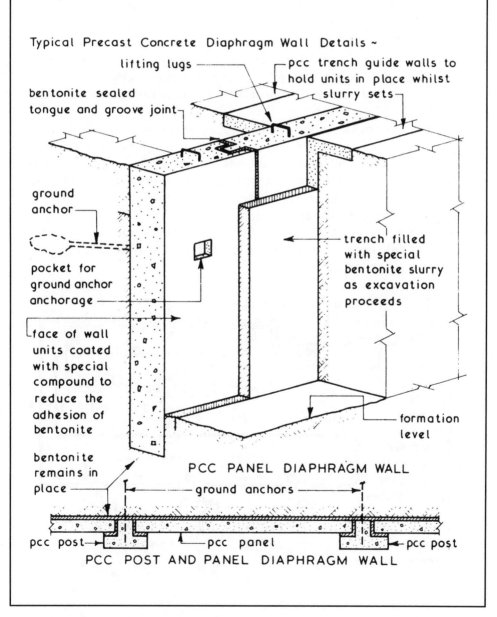

lifting lugs

pcc trench guide walls to hold units in place whilst slurry sets

bentonite sealed tongue and groove joint

ground anchor

pocket for ground anchor anchorage

trench filled with special bentonite slurry as excavation proceeds

face of wall units coated with special compound to reduce the adhesion of bentonite

formation level

bentonite remains in place

PCC PANEL DIAPHRAGM WALL

ground anchors

pcc post

pcc panel

pcc post

PCC POST AND PANEL DIAPHRAGM WALL

Grouting Methods ~ these techniques are used to form a curtain or cut off wall in high permeability soils where pumping methods could be uneconomic. The curtain walls formed by grouting methods are non-structural therefore adequate earth support will be required and in some cases this will be a distance of at least 4·000 from the face of the proposed excavation. Grout mixtures are injected into the soil by pumping the grout at high pressure through special injection pipes inserted in the ground. The pattern and spacing of the injection pipes will depend on the grout type and soil conditions.

Grout Types ~

1. Cement Grouts — mixture of neat cement and water cement sand up to 1 : 4 or PFA (pulverized fuel ash) cement to a 1 : 1 ratio. Suitable for coarse grained soils and fissured and jointed rock strata.
2. Chemical Grouts — one shot (premixed) of two shot (first chemical is injected followed immediately by second chemical resulting in an immediate reaction) methods can be employed to form a permanent gel in the soil to reduce its permeability and at the same time increase the soil's strength. Suitable for medium to coarse sands and gravels.
3. Resin Grouts — these are similar in application to chemical grouts but have a low viscosity and can therefore penetrate into silty fine sands.

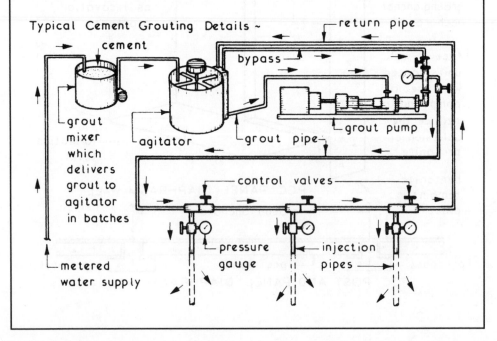

Typical Cement Grouting Details ~

Ground Freezing Techniques ~ this method is suitable for all types of saturated soils and rock and for soils with a moisture content in excess of 8% of the voids. The basic principle is to insert into the ground a series of freezing tubes to form an ice wall thus creating an impermeable barrier. The treatment takes time to develop and the initial costs are high therefore it is only suitable for large contracts of reasonable duration. The freezing tubes can be installed vertically for conventional excavations and horizontally for tunneling works. The usual circulating brines employed are magnesium chloride and calcium chloride with a temperature of −15° to −25°C which would take 10 to 17 days to form an ice wall 1·000 thick. Liquid nitrogen could be used as the freezing medium to reduce the initial freezing period if the extra cost can be justified.

Typical Ground Freezing Details ~

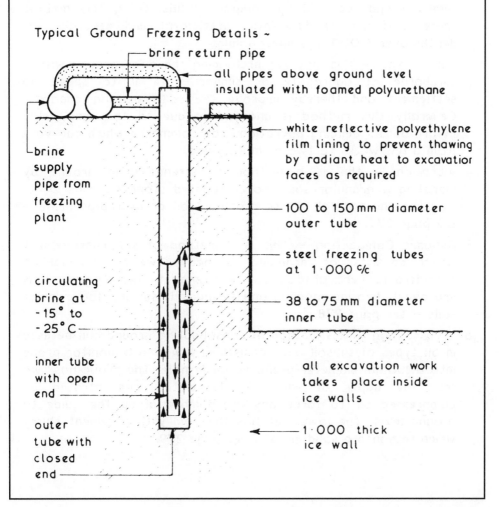

- brine return pipe
- all pipes above ground level insulated with foamed polyurethane
- brine supply pipe from freezing plant
- white reflective polyethylene film lining to prevent thawing by radiant heat to excavation faces as required
- 100 to 150 mm diameter outer tube
- steel freezing tubes at 1·000 %c
- circulating brine at -15° to -25°C
- 38 to 75 mm diameter inner tube
- inner tube with open end
- all excavation work takes place inside ice walls
- outer tube with closed end
- 1·000 thick ice wall

Soil Stabilization and Improvement

Soil Investigation ~ before a decision is made as to the type of foundation which should be used on any particular site a soil investigation should be carried out to establish existing ground conditions and soil properties. The methods which can be employed together with other sources of information such as local knowledge, ordnance survey and geological maps, mining records and aerial photography should be familiar to students at this level. If such an investigation reveals a naturally poor subsoil or extensive filling the designer has several options:-

1. Not to Build — unless a new and suitable site can be found building is only possible if the poor ground is localised and the proposed foundations can be designed around these areas with the remainder of the structure bridging over these positions.

2. Remove and Replace — the poor ground can be excavated, removed and replaced by compacted fills. Using this method there is a risk of differential settlement and generally for depths over 4·000 it is uneconomic.

3. Surcharging — this involves preloading the poor ground with a surcharge of aggregate or similar material to speed up settlement and thereby improve the soil's bearing capacity. Generally this method is uneconomic due to the time delay before actual building operations can commence which can vary from a few weeks to two or more years.

4. Vibration — this is a method of strengthening ground by vibrating a granular soil into compacted stone columns either by using the natural coarse granular soil or by replacement — see page 277.

5. Dynamic Compaction — this is a method of soil improvement which consists of dropping a heavy weight through a considerable vertical distance to compact the soil and thus improve its bearing capacity and is especially suitable for granular soils — see page 278.

6. Jet Grouting — this method of consolidating ground can be used in all types of subsoil and consists of lowering a monitor probe into a 150mm diameter prebored guide hole. The probe has two jets the upper of which blasts water, concentrated by compressed air to force any loose material up the guide to ground level. The lower jet fills the void with a cement slurry which sets into a solid mass — see page 279.

Ground Vibration ~ the objective of this method is to strengthen the existing soil by rearranging and compacting coarse granular particles to form stone columns with the ground. This is carried out by means of a large poker vibrator which has an effective compacting radius of 1·500 to 2·700. On large sites the vibrator is inserted on a regular triangulated grid pattern with centres ranging from 1·500 to 3·000. In coarse grained soils extra coarse aggregate is tipped into the insertion positions to make up levels as required whereas in clay and other fine particle soils the vibrator is surged up and down enabling the water jetting action to remove the surrounding soft material thus forming a borehole which is backfilled with a coarse granular material compacted insitu by the vibrator. The backfill material is usually of 20 to 70mm size of uniform grading within the chosen range. Ground vibration is not a piling system but a means of strengthening ground to increase the bearing capacity within a range of 200 to 500 kN/m².

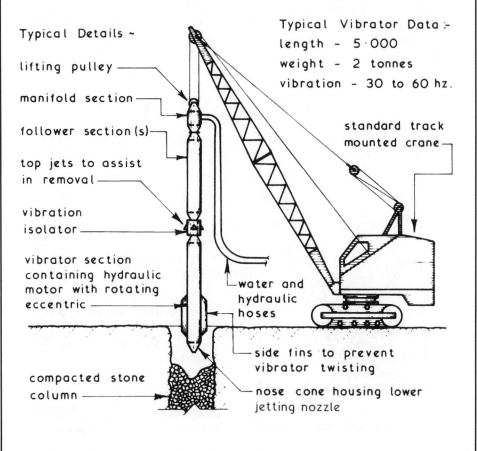

Typical Details ~

lifting pulley

manifold section

follower section (s)

top jets to assist in removal

vibration isolator

vibrator section containing hydraulic motor with rotating eccentric

compacted stone column

Typical Vibrator Data :-
length - 5·000
weight - 2 tonnes
vibration - 30 to 60 hz.

standard track mounted crane

water and hydraulic hoses

side fins to prevent vibrator twisting

nose cone housing lower jetting nozzle

Soil Stabilization and Improvement

Dynamic Compaction ~ this method of ground improvement consists of dropping a heavy weight from a considerable height and is particularly effective in granular soils. Where water is present in the subsoil, trenches should be excavated to allow the water to escape and not collect in the craters formed by the dropped weight. The drop pattern, size of weight and height of drop are selected to suit each individual site but generally 3 or 4 drops are made in each position forming a crater up to 2·500 deep and 5·000 in diameter. Vibration through the subsoil can be a problem with dynamic compaction operations therefore the proximity and condition of nearby buildings must be considered together with the depth position and condition of existing services on site.

Typical Details ~

NB. Final ground level after compaction treatment and final levelling could be up to 1·500 lower than original ground level

heavy duty track mounted crane

weight range 10 to 20 tonnes

depth up to 2· 500 after 4 blows

crater up to 5·000 in diameter

free fall distance range 15·000 to 25·000

compacted soil

20° to 40° spread

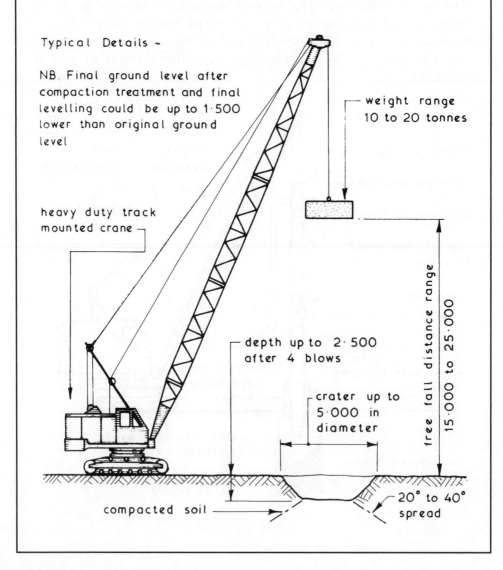

Jet Grouting ~ this is a means of consolidating ground by lowering into preformed bore holes a monitor probe. The probe is rotated and the sides of the bore hole are subjected to a jet of pressurised water and air from a single outlet which enlarges and compacts the bore hole sides. At the same time a cement grout is being introduced under pressure to fill the void being created. The water used by the probe and any combined earth is forced up to the surface in the form of a sludge. If the monitor probe is not rotated grouted panels can be formed. The spacing, depth and layout of the bore holes is subject to specialist design.

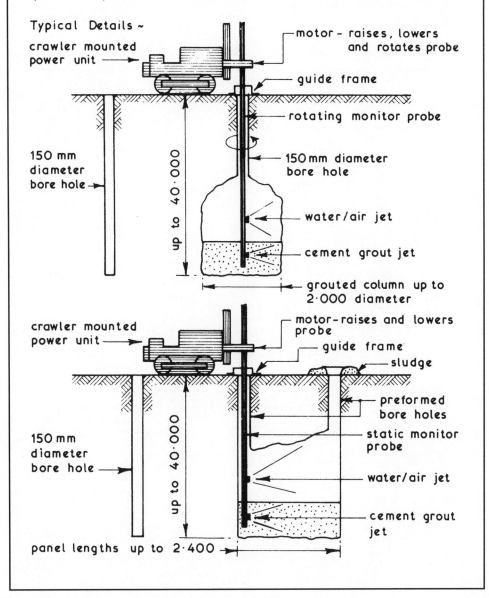

Typical Details ~

crawler mounted power unit →

motor - raises, lowers and rotates probe

guide frame

rotating monitor probe

150 mm diameter bore hole →

up to 40·000

150 mm diameter bore hole

water/air jet

cement grout jet

grouted column up to 2·000 diameter

crawler mounted power unit →

motor-raises and lowers probe

guide frame

sludge

preformed bore holes

150 mm diameter bore hole →

up to 40·000

static monitor probe

water/air jet

cement grout jet

panel lengths up to 2·400 →

5 SUPERSTRUCTURE

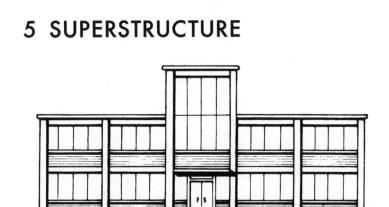

CHOICE OF MATERIALS
BRICK AND BLOCK WALLS
GAS RESISTANT MEMBRANES
ARCHES AND OPENINGS
WINDOWS, GLASS AND GLAZING
DOMESTIC AND INDUSTRIAL DOORS
TIMBER FRAME CONSTRUCTION
REINFORCED CONCRETE FRAMED STRUCTURES
FORMWORK
PRECAST CONCRETE FRAMES
STRUCTURAL STEELWORK
COMPOSITE TIMBER BEAMS
TIMBER PITCHED AND FLAT ROOFS
LONG SPAN ROOFS
SHELL ROOF CONSTRUCTION
RAINSCREEN CLADDING
PANEL WALLS AND CURTAIN WALLING
CONCRETE CLADDINGS
PRESTRESSED CONCRETE
THERMAL INSULATION
THERMAL BRIDGING
SOUND INSULATION
ACCESS FOR THE DISABLED

STAGE 1

Consideration to be given to the following :~

1. Building type and usage.
2. Building owner's requirements and preferences.
3. Local planning restrictions.
4. Legal restrictions and requirements.
5. Site restrictions.
6. Capital resources.
7. Future policy in terms of maintenance and adaptation.

STAGE 2

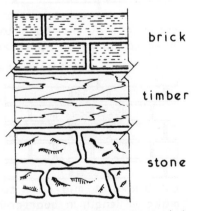

Decide on positions, sizes and shapes of openings.

STAGE 3

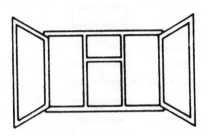

Decide on style, character and materials for openings

STAGE 4

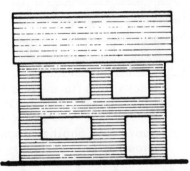

brick

timber

stone

Decide on basic materials for fabric of roof and walls

STAGE 5

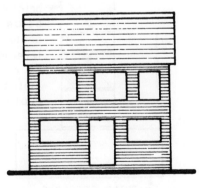

Review all decisions and make changes if required

Bricks ~ these are walling units within a length of 337·5 mm, a width of 225 mm and a height of 112·5 mm. The usual size of bricks in common use is length 215 mm, width 112·5 mm and height 65 mm and like blocks they must be laid in a definite pattern or bond if they are to form a structural wall. Bricks are usually made from clay (BS 3921) or from sand and lime (BS 187) and are available in a wide variety of strengths, types, textures, colours and special shaped bricks to BS 4729.

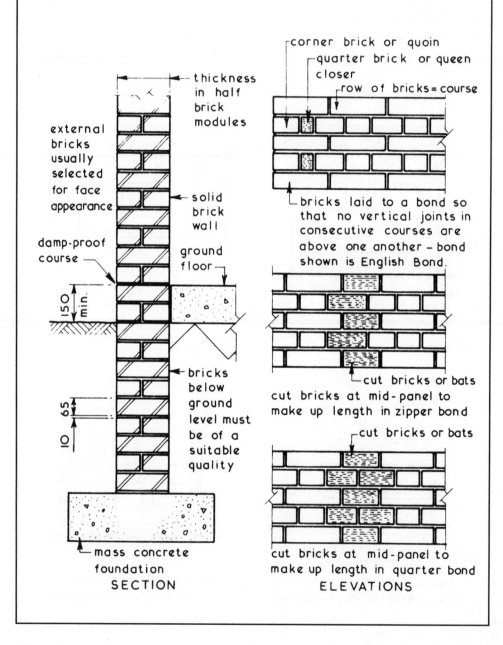

thickness in half brick modules

external bricks usually selected for face appearance

solid brick wall

damp-proof course

ground floor

150 min.

bricks below ground level must be of a suitable quality

65

10

mass concrete foundation
SECTION

corner brick or quoin

quarter brick or queen closer

row of bricks = course

bricks laid to a bond so that no vertical joints in consecutive courses are above one another – bond shown is English Bond.

cut bricks or bats

cut bricks at mid-panel to make up length in zipper bond

cut bricks or bats

cut bricks at mid-panel to make up length in quarter bond
ELEVATIONS

Typical Details ~
Bonding ~ an arrangement of bricks in a wall, column or pier laid to a set pattern to maintain an adequate lap.

Purposes of Brick Bonding ~

1. Obtain maximum strength whilst distributing the loads to be carried throughout the wall, column or pier.
2. Ensure lateral stability and resistance to side thrusts.
3. Create an acceptable appearance.

Lap Forms ~

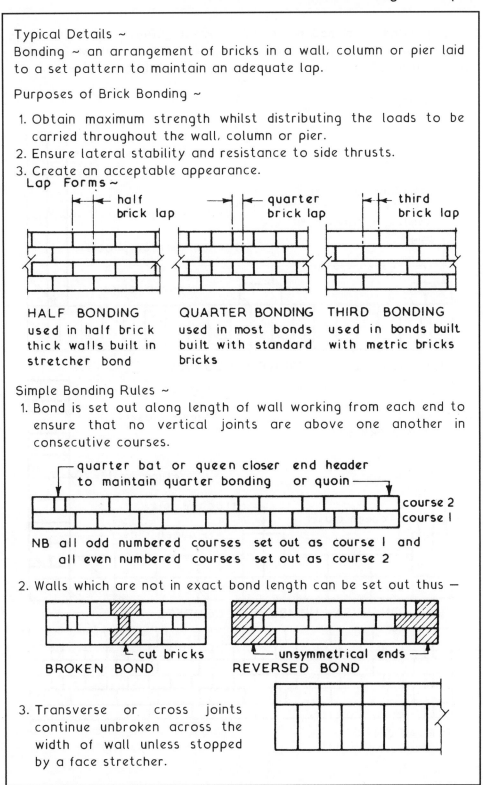

HALF BONDING
used in half brick thick walls built in stretcher bond

QUARTER BONDING
used in most bonds built with standard bricks

THIRD BONDING
used in bonds built with metric bricks

Simple Bonding Rules ~

1. Bond is set out along length of wall working from each end to ensure that no vertical joints are above one another in consecutive courses.

NB all odd numbered courses set out as course 1 and all even numbered courses set out as course 2

2. Walls which are not in exact bond length can be set out thus —

BROKEN BOND   REVERSED BOND

3. Transverse or cross joints continue unbroken across the width of wall unless stopped by a face stretcher.

English Bond ~ formed by laying alternate courses of stretchers and headers it is one of the strongest bonds but it will require more facing bricks than other bonds (89 facing bricks per m²)

Typical Example ~

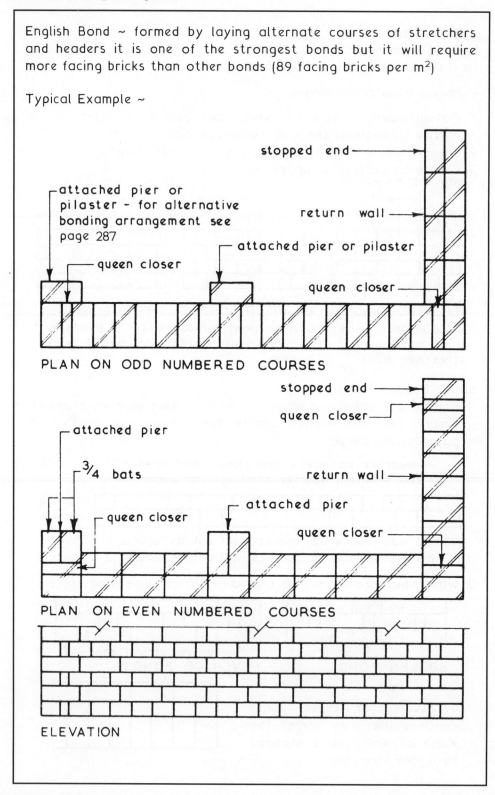

stopped end

attached pier or pilaster - for alternative bonding arrangement see page 287

return wall

attached pier or pilaster

queen closer

queen closer

PLAN ON ODD NUMBERED COURSES

stopped end

queen closer

attached pier

return wall

³/₄ bats

attached pier

queen closer

queen closer

PLAN ON EVEN NUMBERED COURSES

ELEVATION

Flemish Bond ~ formed by laying headers and stretchers alternately in each course. Not as strong as English bond but is considered to be aesthetically superior uses less facing bricks. (79 facing brick per m²)

Typical Example

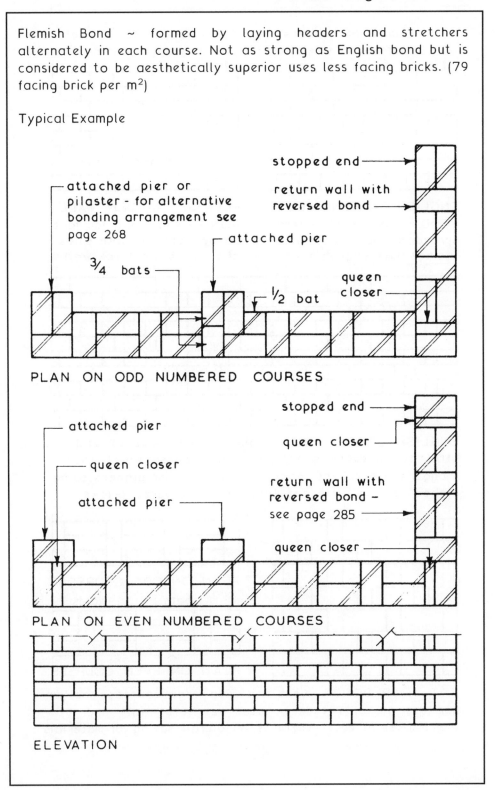

PLAN ON ODD NUMBERED COURSES

PLAN ON EVEN NUMBERED COURSES

ELEVATION

I course of headers to
3 courses of stretchers

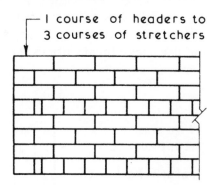

I header to 3 stretchers
in each course

ENGLISH GARDEN WALL
BOND – gives quick lateral
spread of load – uses less
facings than English bond.

FLEMISH GARDEN WALL
BOND – enables a fair face
to be kept on both sides
of a one brick thick wall.

ENGLISH CROSS BOND – header placed next to end
stretcher in every other stretcher course which thus
staggers stretchers enabling patterns or diapers to be
picked out in different texture or coloured bricks.

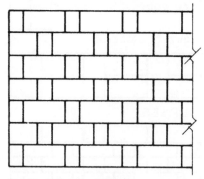

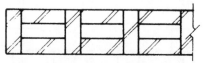

PLAN ON ODD COURSES

$^2/_3$ bats ⌐ voids

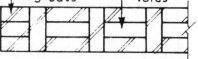

PLAN ON EVEN COURSES

RAT TRAP BOND – uses brick on edge courses – hollow
pockets or voids reduce total weight of wall and by
the bricks on edge there is an overall saving of materials.

Attached Piers ~ the main function of an attached pier is to give lateral support to the wall of which it forms part from the base to the top of the wall. It also has the subsidiary function of dividing a wall into distinct lengths whereby each length can be considered as a wall. Generally walls must be tied at end to an attached pier, buttressing or return wall.

Typical Examples ~

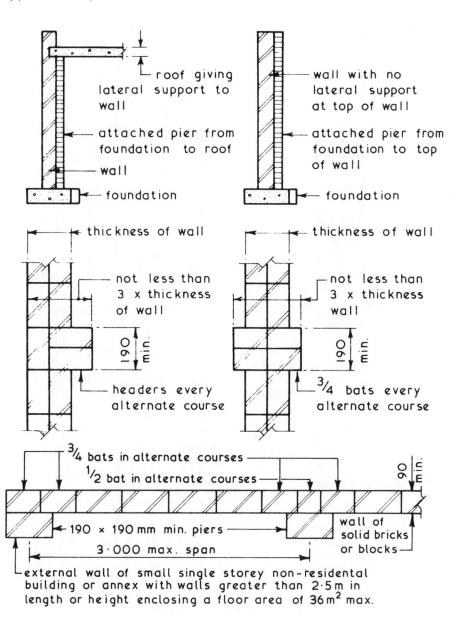

thickness of wall

roof giving lateral support to wall

wall with no lateral support at top of wall

attached pier from foundation to roof

attached pier from foundation to top of wall

wall

foundation

foundation

thickness of wall

not less than 3 x thickness of wall

not less than 3 x thickness wall

190 min.

190 min.

headers every alternate course

³⁄₄ bats every alternate course

³⁄₄ bats in alternate courses

¹⁄₂ bat in alternate courses

90 min.

190 x 190 mm min. piers

wall of solid bricks or blocks

3·000 max. span

external wall of small single storey non-residental building or annex with walls greater than 2·5m in length or height enclosing a floor area of 36m² max.

Solid Block Walls

Blocks ~ these are walling units exceeding in length, width or height the dimensions specified for bricks in BS 3921. Precast concrete blocks should comply with the recommendations set out in BS 6073. Blocks suitable for external solid walls are classified as loadbearing and are required to have a minimum average crushing strength of $2 \cdot 8 \text{N/mm}^2$.

Typical Details ~

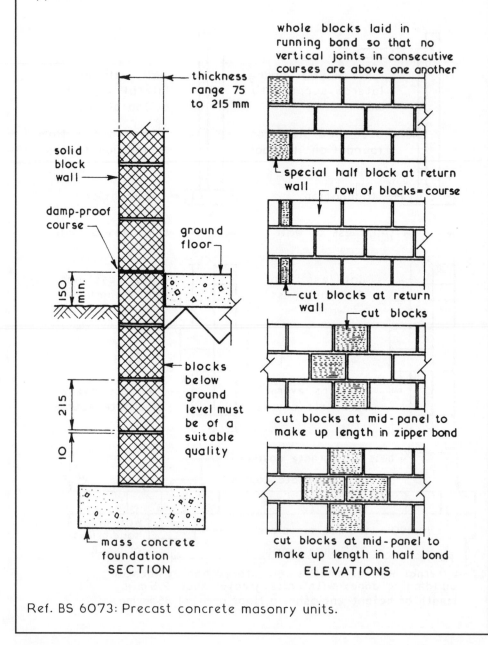

whole blocks laid in running bond so that no vertical joints in consecutive courses are above one another

thickness range 75 to 215 mm

solid block wall

damp-proof course

ground floor

150 min.

215

10

blocks below ground level must be of a suitable quality

mass concrete foundation
SECTION

special half block at return wall

row of blocks = course

cut blocks at return wall

cut blocks

cut blocks at mid-panel to make up length in zipper bond

cut blocks at mid-panel to make up length in half bond
ELEVATIONS

Ref. BS 6073: Precast concrete masonry units.

290

Cavity Walls ~ these consist of an outer brick or block leaf or skin separated from an inner brick or block leaf or skin by an air space called a cavity. These walls have better thermal insulation and weather resistance properties than a comparable solid brick or block wall and therefore are in general use for the enclosing walls of domestic buildings. The two leaves of a cavity wall are tied together with wall ties at not less than the spacings given in Table 5 in Approved Document A — Building Regulations (see below).

The width of the cavity should be between 50 and 75mm unless vertical twist type ties are used at not more than the centres given in Table 5 when the cavity width can be between 75 and 300mm. Cavities are not normally ventilated and should be sealed at eaves level.

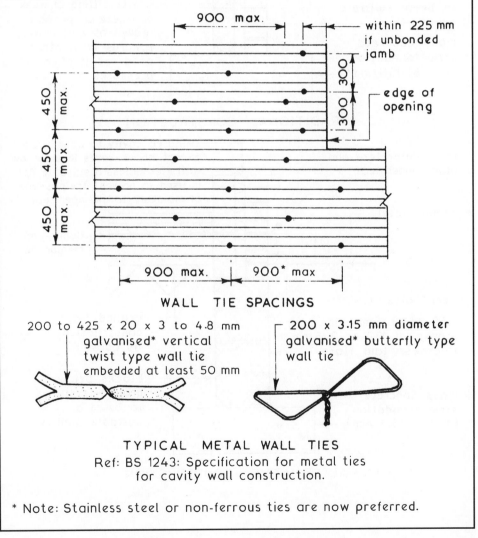

WALL TIE SPACINGS

200 to 425 x 20 x 3 to 4.8 mm galvanised* vertical twist type wall tie embedded at least 50 mm

200 x 3.15 mm diameter galvanised* butterfly type wall tie

TYPICAL METAL WALL TIES
Ref: BS 1243: Specification for metal ties for cavity wall construction.

* Note: Stainless steel or non-ferrous ties are now preferred.

Cavity Walls

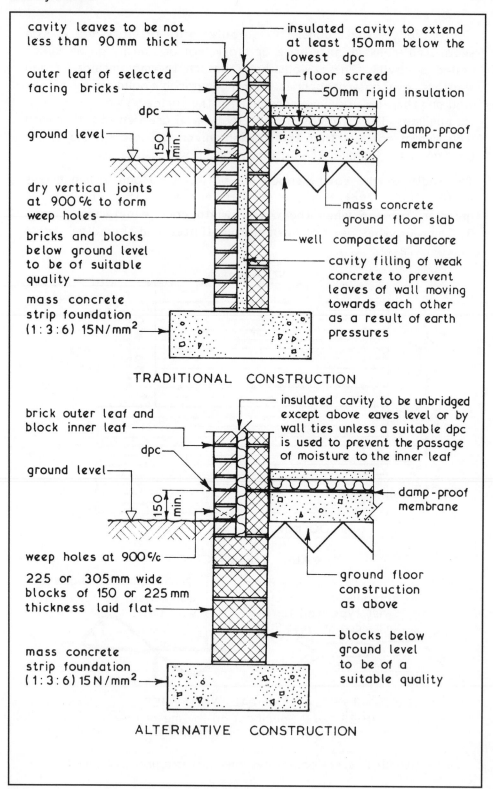

cavity leaves to be not less than 90mm thick

insulated cavity to extend at least 150mm below the lowest dpc

outer leaf of selected facing bricks

floor screed

50mm rigid insulation

dpc

ground level

150 min.

damp-proof membrane

dry vertical joints at 900% to form weep holes

mass concrete ground floor slab

bricks and blocks below ground level to be of suitable quality

well compacted hardcore

cavity filling of weak concrete to prevent leaves of wall moving towards each other as a result of earth pressures

mass concrete strip foundation (1:3:6) 15N/mm²

TRADITIONAL CONSTRUCTION

brick outer leaf and block inner leaf

insulated cavity to be unbridged except above eaves level or by wall ties unless a suitable dpc is used to prevent the passage of moisture to the inner leaf

dpc

ground level

150 min.

damp-proof membrane

weep holes at 900%

ground floor construction as above

225 or 305mm wide blocks of 150 or 225mm thickness laid flat

blocks below ground level to be of a suitable quality

mass concrete strip foundation (1:3:6) 15N/mm²

ALTERNATIVE CONSTRUCTION

Parapet ~ a low wall projecting above the level of a roof, bridge or balcony forming a guard or barrier at the edge. Parapets are exposed to the elements on three faces namely front, rear and top and will therefore need careful design and construction if they are to be durable and reliable.

Typical Details ~

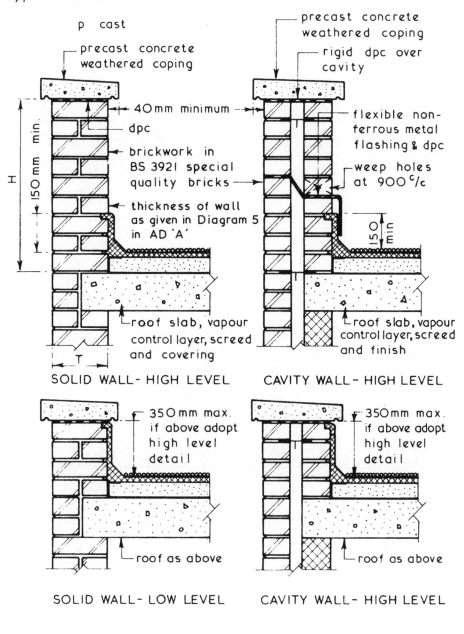

p cast

precast concrete weathered coping

precast concrete weathered coping

rigid dpc over cavity

40mm minimum

dpc

brickwork in BS 3921 special quality bricks

thickness of wall as given in Diagram 5 in AD 'A'

150 mm min.

H

flexible non-ferrous metal flashing & dpc

weep holes at 900 c/c

150 min

roof slab, vapour control layer, screed and covering

roof slab, vapour control layer, screed and finish

T

SOLID WALL - HIGH LEVEL

CAVITY WALL - HIGH LEVEL

350 mm max. if above adopt high level detail

350 mm max. if above adopt high level detail

roof as above

roof as above

SOLID WALL - LOW LEVEL

CAVITY WALL - HIGH LEVEL

Historically, finned or buttressed walls have been used to provide lateral support to tall single storey masonry structures such as churches and cathedrals. Modern applications are similar in principle and include theatres, gymnasiums, warehouses, etc. Where space permits, they are an economic alternative to masonry cladding of steel or reinforced concrete framed buildings. The fin or pier is preferably brick bonded to the main wall. It may also be connected with horizontally bedded wall ties, sufficient to resist vertical shear stresses between fin and wall.

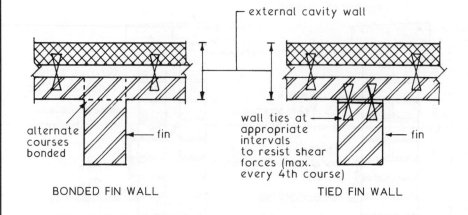

BONDED FIN WALL   TIED FIN WALL

Structurally, the fins are deep piers which reinforce solid or cavity masonry walls. For design purposes the wall may be considered as a series of 'T' sections composed of a flange and a pier. If the wall is of cavity construction, the inner leaf is not considered for bending moment calculations, although it does provide stiffening to the outer leaf or flange.

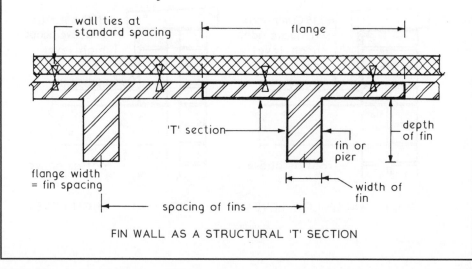

FIN WALL AS A STRUCTURAL 'T' SECTION

Masonry diaphragm walls are an alternative means of constructing tall, single storey buildings such as warehouses, sports centres, churches, assembly halls, etc. They can also be used as retaining and boundary walls with planting potential within the voids. These voids may also be steel reinforced and concrete filled to resist the lateral stresses in high retaining walls.

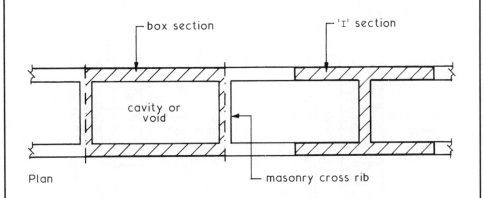

A diaphragm wall is effectively a cavity wall where the two leaves of masonry are bonded together with cross ribs and not wall ties. It is stronger than a conventionally tied cavity wall and for structural purposes may be considered as a series of bonded 'I' sections or box sections. The voids may be useful for housing services, but any access holes in the construction must not disturb the integrity of the wall. The voids may also be filled with insulation to reduce heat energy losses from the building, and to prevent air circulatory heat losses within the voids. Where thermal insulation standards apply, this type of wall will have limitations as the cross ribs will provide a route for cold bridging. U values will increase by about 10% compared with conventional cavity wall construction of the same materials.

Ref. BS 5628: Code of practice for use of masonry.
    BS 5628-3: Materials and components, design and workmanship.

Damp-proof Courses and Membranes

Function ~ the primary function of any damp-proof course (dpc) or damp-proof membrane (dpm) is to provide an impermeable barrier to the passage of moisture. The three basic ways in which damp-proof courses are used is to:-

1. Resist moisture penetration from below (rising damp).
2. Resist moisture penetration from above.
3. Resist moisture penetration from horizontal entry.

Typical Examples ~

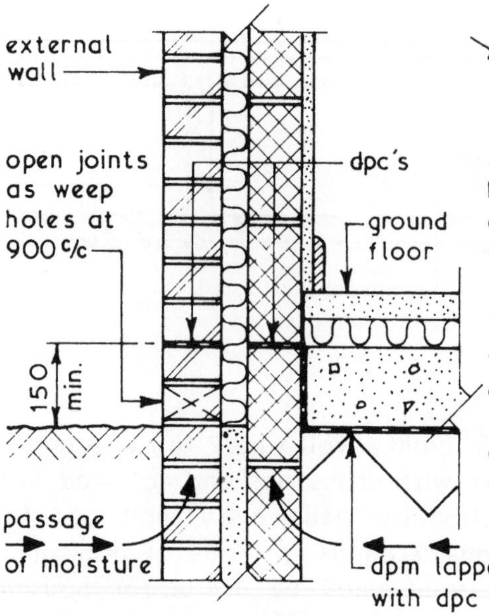

PENETRATION FROM BELOW
(Ground Floor / External Wall)

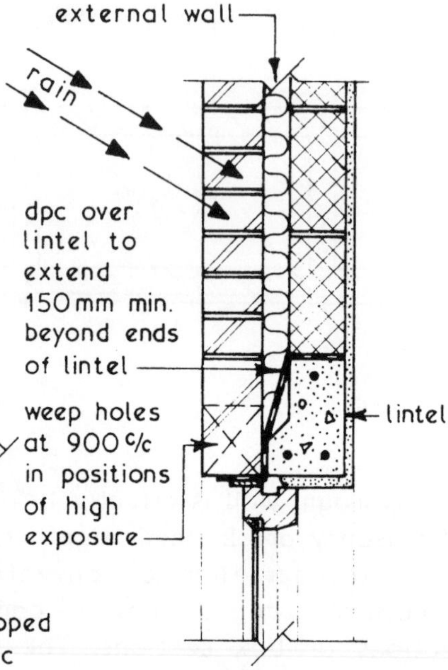

PENETRATION FROM ABOVE
(Traditional Window / Door Head)

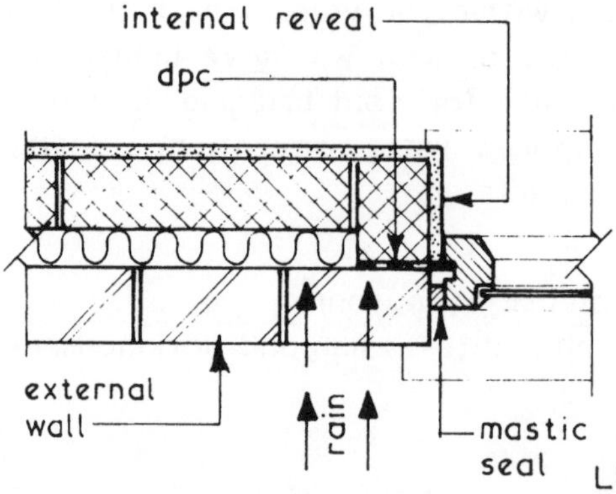

HORIZONTAL ENTRY
(Window / Door Jamb)

SUITABLE DPC MATERIALS
Engineering bricks - BS 3921
Slates - BS 680
Lead - BS 1178
Copper - BS 2870
Bitumen BS 6398 Products
Propriety emulsions
Polythene BS 6515
Pitch polymers
Mastic Asphalt a BS 6925
and BS 6577

SUITABLE DPM MATERIALS
LDPE— 1200 gauge (0.3 mm)
Bitumen— 3 coats applied cold
Refs: BS s 743 & 8102

296

Building Regulations, Approved Document C4, Section 4.4:
A wall may be built with a 'damp-proof course of bituminous material, engineering bricks or slates in cement mortar, or any other material that will prevent the passage of moisture.'

Material	Remarks
Lead BS 1178 Code 4 (1·8 mm)	May corrode in the presence of mortar. Both surfaces to be coated with bituminous paint. Workable for application to cavity trays, etc.
Copper BS 2870 0·25 mm	Can cause staining to adjacent masonry. Resistant to corrosion.
Bitumen BS 6398 in various bases: Hessian 3·8 kg/m² Fibre 3·3 .. .. Asbestos 3·8 .. .. Hessian & lead 4·4 .. .. Fibre & lead 4·4 .. .. Asbestos & lead 4·9	Hessian or fibre may decay with age, but this will not affect efficiency. Tearable if not protected. Lead bases are suited where there may be a high degree of movement in the wall.
LDPE BS 6515 0·46 mm (polyethylene)	No deterioration likely, but may be difficult to bond, hence the profiled surface finish. Not suited under light loads.
Bitumen polymer and pitch polymer 1·10 mm	Absorbs movement well. Pitch polymers have been associated with skin cancer!

Note: All the above dpc's to be lapped at least 100 mm at joints and adhesive sealed.

Material			Remarks
Mastic asphalt	BS 6925	12kg/m²	Does not deteriorate. Requires surface treatment with sand or scoring to effect a mortar key.
Engineering bricks	BS 3921	<4.5% absorption	Min. 2 courses laid breaking joint in cement mortar 1:3. No deterioration, but may not blend with adjacent facings.
Slate	BS 680	4mm	Min. 2 courses laid as above. Will not deteriorate, but brittle so may fracture if building settles.

Refs:

BS 743: Specification for materials for damp proof courses.

BS 5628: Code of practice for use of masonry.

BS 5628–3: Materials and components, design and workmanship.

BS 8215: Code of practice for design and installation of damp proof courses in masonry construction.

BRE Digest 380: Damp proof courses.

Note: It was not until the Public Health Act of 1875, that it became mandatory to instal damp proof courses in new buildings. Structures constructed before that time, and those since, which have suffered dpc failure due to deterioration or incorrect installation, will require remedial treatment. This could involve cutting out the mortar bed joint two brick courses above ground level in stages of about 1m in length. A new dpc can then be inserted with mortar packing, before proceeding to the next length. No two adjacent sections should be worked consecutively. This process is very time consuming and may lead to some structural settlement. Therefore, the measures explained on the following two pages are usually preferred.

Materials — Silicon solutions in organic solvent.
Aluminium stearate solutions.
Water soluble silicon formulations (siliconates).

Methods — High pressure injection (0·70 — 0·90 MPa) solvent based.
Low pressure injection (0·15 — 0·30 MPa) water based.
Gravity feed, water based.
Insertion/injection, mortar based.

Pressure injection — 12 mm diameter holes are bored to about two-thirds the depth of masonry, at approximately 150 mm horizontal intervals at the appropriate depth above ground (normally 2—3 brick courses). These holes can incline slightly downwards. With high (low) pressure injection, walls in excess of 120 mm (460 mm) thickness should be drilled from both sides. The chemical solution is injected by pressure pump until it exudes from the masonry. Cavity walls are treated as each leaf being a solid wall.

Gravity feed — 25 mm diameter holes are bored as above. Dilute chemical is transfused from containers which feed tubes inserted in the holes. This process can take from a few hours to several days to effect. An alternative application is insertion of frozen pellets placed in the bore holes. On melting, the solution disperses into the masonry to be replaced with further pellets until the wall is saturated.

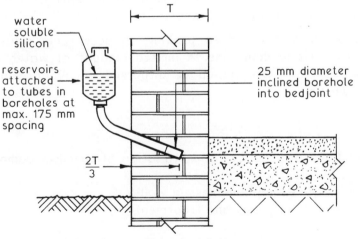

water soluble silicon

reservoirs attached to tubes in boreholes at max. 175 mm spacing

T

25 mm diameter inclined borehole into bedjoint

$\dfrac{2T}{3}$

Gravity infusion of silicon in solution

Injection mortars — 19mm diameter holes are bored from both sides of a wall, at the appropriate level and no more than 230mm apart horizontally, to a depth equating to three-fifths of the wall thickness. They should be inclined downwards at an angle of 20 to 30°. The drill holes are flushed out with water, before injecting mortar from the base of the hole and outwards. This can be undertaken with a hand operated caulking gun. Special cement mortars contain styrene butadiene resin (SDR) or epoxy resin and must be mixed in accordance with the manufacturer's guidance.

Notes relating to all applications of chemical dpc's:

* Before commencing work, old plasterwork and rendered undercoats are removed to expose the masonry. This should be to a height of at least 300mm above the last detectable (moisture meter reading) signs of rising dampness (1 metre min.).

* If the wall is only accessible from one side and both sides need treatment, a second deeper series of holes may be bored from one side, to penetrate the inaccessible side.

* On completion of work, all boreholes are made good with cement mortar. Where dilute chemicals are used for the dpc, the mortar is rammed the full length of the hole with a piece of timber dowelling.

* The chemicals are effective by bonding to, and lining the masonry pores by curing and solvent evaporation.

* The process is intended to provide an acceptable measure of control over rising dampness. A limited amount of water vapour may still rise, but this should be dispersed by evaporation in a heated building.

Refs.
BS 6576: Code of practice for installation of chemical damp proof courses.
BRE Digest 245: Rising dampness in walls: diagnosis and treatment.

In addition to damp proof courses failing due to deterioration or damage, they may be bridged as a result of:

* Faults occurring during construction.
* Work undertaken after construction, with disregard for the damp proof course.

Typical examples —

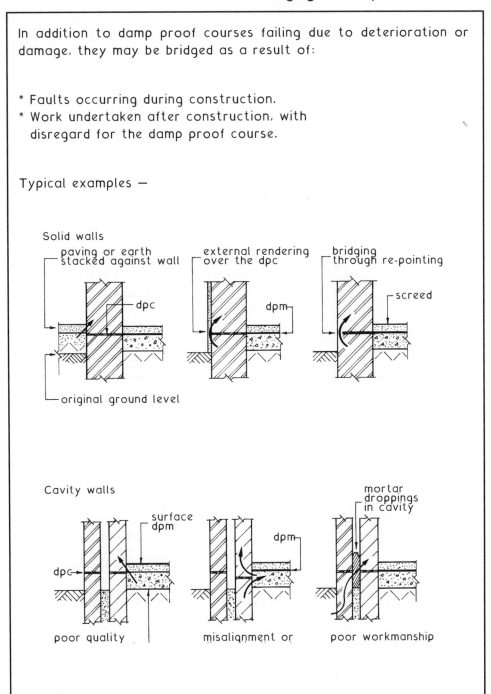

Solid walls

paving or earth stacked against wall

external rendering over the dpc

bridging through re-pointing

dpc

dpm

screed

original ground level

Cavity walls

surface dpm

mortar droppings in cavity

dpc

dpm

poor quality

misalignment or

poor workmanship

Thermal insulation regulations may require insulating dpc's to prevent cold bridging around window and door openings in cavity wall construction (see pages 497 and 498). By locating a vertical dpc with a bonded insulant at the cavity closure, the dpc prevents penetration of dampness from the outside, and the insulation retains the structural temperature of the internal reveal. This will reduce heat losses by maintaining the temperature above dewpoint, preventing condensation, wall staining and mould growth.

Application —

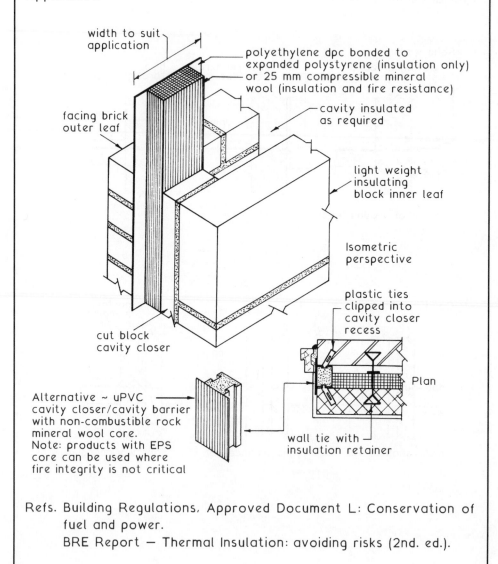

width to suit application

polyethylene dpc bonded to expanded polystyrene (insulation only) or 25 mm compressible mineral wool (insulation and fire resistance)

cavity insulated as required

facing brick outer leaf

light weight insulating block inner leaf

Isometric perspective

plastic ties clipped into cavity closer recess

cut block cavity closer

Plan

Alternative ~ uPVC cavity closer/cavity barrier with non-combustible rock mineral wool core.
Note: products with EPS core can be used where fire integrity is not critical

wall tie with insulation retainer

Refs. Building Regulations, Approved Document L: Conservation of fuel and power.
BRE Report — Thermal Insulation: avoiding risks (2nd. ed.).

Penetrating Gases ~ Methane and Radon

Methane — methane is produced by deposited organic material decaying in the ground. It often occurs with carbon dioxide and traces of other gases to form a cocktail known as landfill gas. It has become an acute problem in recent years, as planning restrictions on 'green-field' sites have forced development of derelict and reclaimed 'brown-field' land.

The gas would normally escape to the atmosphere, but under a building it pressurizes until percolating through cracks, cavities and junctions with services. Being odourless, it is not easily detected until contacting a naked flame, then the result is devastating!

Radon ~ a naturally occurring colour/odourless gas produced by radioactive decay of radium. It originates in uranium deposits of granite subsoils as far apart as the south-west and north of England and the Grampian region of Scotland. Concentrations of radon are considerably increased if the building is constructed of granite masonry. The combination of radon gas and the tiny radioactive particles known as radon daughters are inhaled. In some people with several years' exposure, research indicates a high correlation with cancer related illness and death.

Protection of buildings and the occupants from subterranean gases can be achieved by passive or active measures incorporated within the structure.

1. Passive protection consists of a complete airtight seal integrated within the ground floor and walls. A standard LDPE damp proof membrane of 0.3 mm thickness should be adequate if carefully sealed at joints, but thicknesses up to 1mm are preferred, combined with foil and/or wire reinforcement.

2. Active protection requires installation of a permanently running extract fan connected to a gas sump below the ground floor. It is an integral part of the building services system and will incur operating and maintenance costs throughout the building's life.

(See next page for construction details)

Gas Resistant Construction

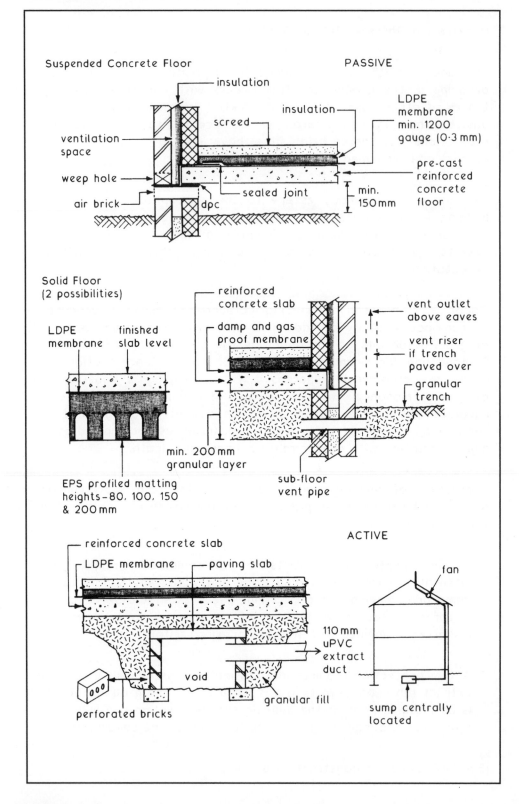

Suspended Concrete Floor   PASSIVE

insulation

insulation

ventilation space

screed

LDPE membrane min. 1200 gauge (0·3 mm)

weep hole

air brick

dpc

sealed joint

pre-cast reinforced concrete floor

min. 150mm

Solid Floor (2 possibilities)

LDPE membrane   finished slab level

reinforced concrete slab

damp and gas proof membrane

vent outlet above eaves

vent riser if trench paved over

granular trench

min. 200 mm granular layer

sub-floor vent pipe

EPS profiled matting heights – 80, 100, 150 & 200 mm

reinforced concrete slab

LDPE membrane   paving slab

ACTIVE

fan

110 mm uPVC extract duct

void

perforated bricks

granular fill

sump centrally located

Calculated Brickwork ~ for small and residential buildings up to three storeys high the sizing of load bearing brick walls can be taken from data given in Part C of Approved Document A. The alternative methods for these and other load bearing brick walls are given in BS 5628 — Code of practice for use of masonry.

The main factors governing the loadbearing capacity of brick walls and columns are:-

1. Thickness of wall.
2. Strength of bricks used.
3. Type of mortar used.
4. Slenderness ratio of wall or column.
5. Eccentricity of applied load.

Thickness of wall ~ this must always be sufficient throughout its entire body to carry the design loads and induced stresses. Other design requirements such as thermal and sound insulation properties must also be taken into account when determining the actual wall thickness to be used.

Effective Thickness ~ this is the assumed thickness of the wall or column used for the purpose of calculating its slenderness ratio — see page 307

Typical Examples ~

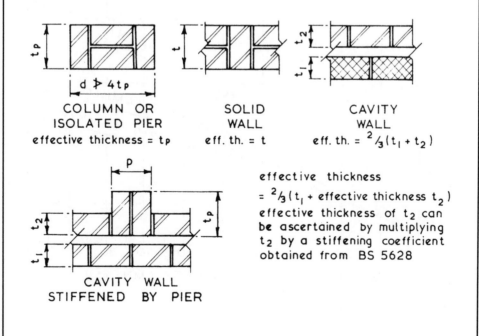

COLUMN OR
ISOLATED PIER
effective thickness = t_p

SOLID
WALL
eff. th. = t

CAVITY
WALL
eff. th. = $\frac{2}{3}(t_1 + t_2)$

CAVITY WALL
STIFFENED BY PIER

effective thickness

= $\frac{2}{3}(t_1$ + effective thickness t_2)

effective thickness of t_2 can be ascertained by multiplying t_2 by a stiffening coefficient obtained from BS 5628

Strength of Bricks ~ due to the wide variation of the raw materials and methods of manufacture bricks can vary greatly in their compressive strength. The compressive strength of a particular type of brick or batch of bricks is taken as the arithmetic mean of a sample of ten bricks tested in accordance with the appropriate British Standard. A typical range for clay bricks would be from 20 to 170 MN/m^2 the majority of which would be in the 20 to 90 MN/m^2 band. Generally calcium silicate bricks have a lower compressive strength than clay bricks with a typical strength range of 10 to 65 MN/m^2.

Strength of Mortars ~ mortars consist of an aggregate (sand) and a binder which is usually cement; cement plus additives to improve workability; or cement and lime. The factors controlling the strength of any particular mix are the ratio of binder to aggregate plus the water:cement ratio. The strength of any particular mix can be ascertained by taking the arithmetic mean of a series of test cubes.

Wall Design Strength ~ the basic stress of any brickwork depends on the crushing strength of the bricks and the type of mortar used to form the wall unit. This relationship can be plotted on a graph using data given in BS 5628 as shown below:-

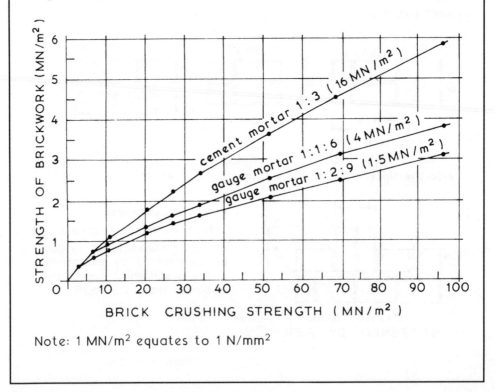

Note: 1 MN/m^2 equates to 1 N/mm^2

Slenderness Ratio ~ this is the relationship of the effective height to the effective thickness thus:-

$$\text{Slenderness ratio} = \frac{\text{effective height}}{\text{effective thickness}} = \frac{h}{t}$$

Effective Height ~ this is the dimension taken to calculate the slenderness ratio as opposed to the actual height.

Typical Examples– actual height = H   effective height = h

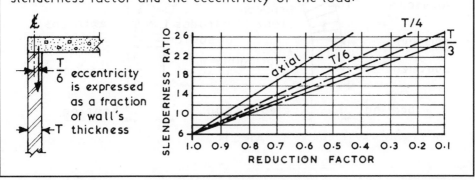

NO LATERAL SUPPORT   CONCRETE FLOOR   FLOOR EFFECTIVELY
AT TOP   BEARING ON WALL   TIED TO WALL
(see pages 530 & 532)

Effective Thickness ~ this is the dimension taken to calculate the slenderness ratio as opposed to the actual thickness.

Typical Examples — actual thickness = T effective thickness = t

$$t = T \qquad t = 0.66(T_1 + T_2) \qquad t = T \times \text{stiffening coefficient (see BS 5628)}$$

stiffening pier or wall

SOLID WALLS   CAVITY WALLS   WALLS STIFFENED BY PIERS

Stress Reduction ~ the permissible stress for a wall is based on the basic stress multiplied by a reduction factor related to the slenderness factor and the eccentricity of the load:-

$\frac{T}{6}$ eccentricity is expressed as a fraction of wall's thickness

SLENDERNESS RATIO — axial, T/6, T/4, T/3
REDUCTION FACTOR: 1.0 0.9 0.8 0.7 0.6 0.5 0.4 0.3 0.2 0.1

307

Supports Over Openings ~ the primary function of any support over an opening is to carry the loads above the opening and transmit them safely to the abutments, jambs or piers on both sides. A support over an opening is usually required since the opening infilling such as a door or window frame will not have sufficient strength to carry the load through its own members.

Type of Support ~

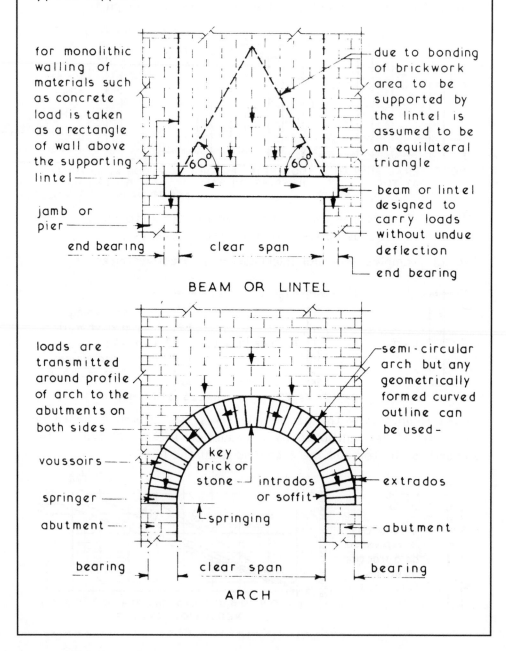

for monolithic walling of materials such as concrete load is taken as a rectangle of wall above the supporting lintel —

jamb or pier ———

due to bonding of brickwork area to be supported by the lintel is assumed to be an equilateral triangle

beam or lintel designed to carry loads without undue deflection

end bearing

BEAM OR LINTEL

loads are transmitted around profile of arch to the abutments on both sides ——

voussoirs ———

springer ———

abutment ———

bearing

semi-circular arch but any geometrically formed curved outline can be used –

key brick or stone — intrados or soffit

extrados

abutment

bearing

clear span

ARCH

Arch Construction ~ by the arrangement of the bricks or stones in an arch over an opening it will be self supporting once the jointing material has set and gained adequate strength. The arch must therefore be constructed over a temporary support until the arch becomes self supporting. The traditional method is to use a framed timber support called a centre. Permanent arch centres are also available for small spans and simple formats.

Typical Arch Formats ~

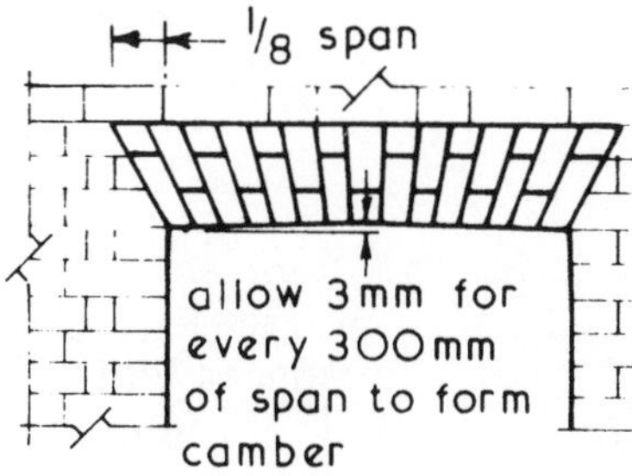

1/8 span

allow 3mm for every 300mm of span to form camber

CAMBER ARCH

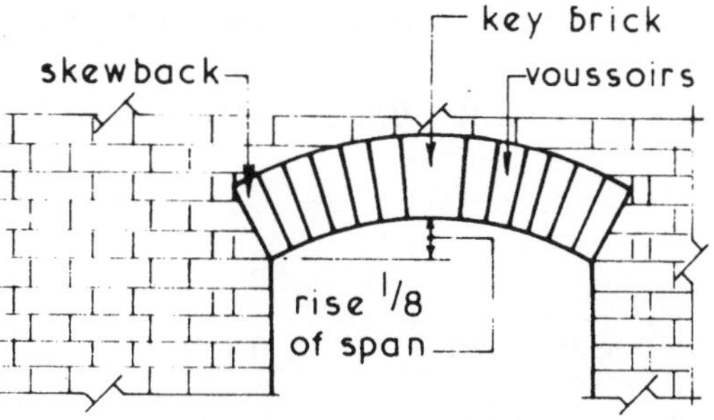

skewback

key brick

voussoirs

rise 1/8 of span

SEGMENTAL ARCH

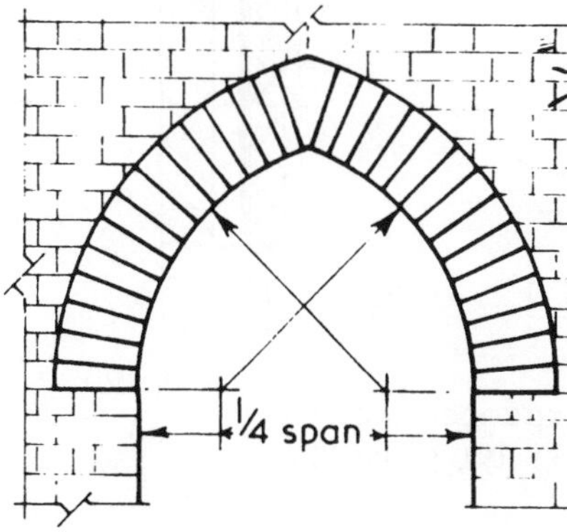

1/4 span

DROP ARCH

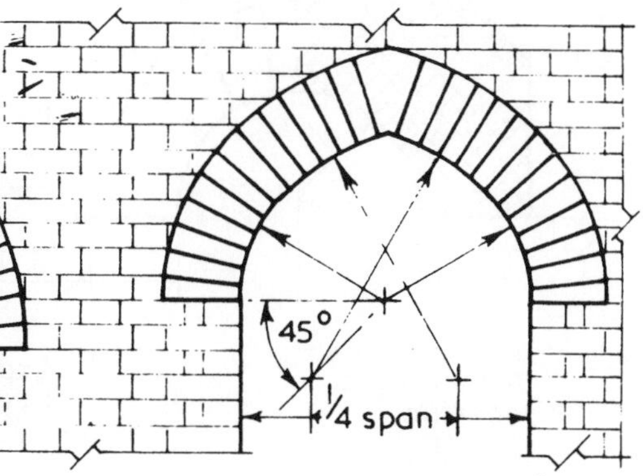

45°

1/4 span

3 CENTRE ARCH

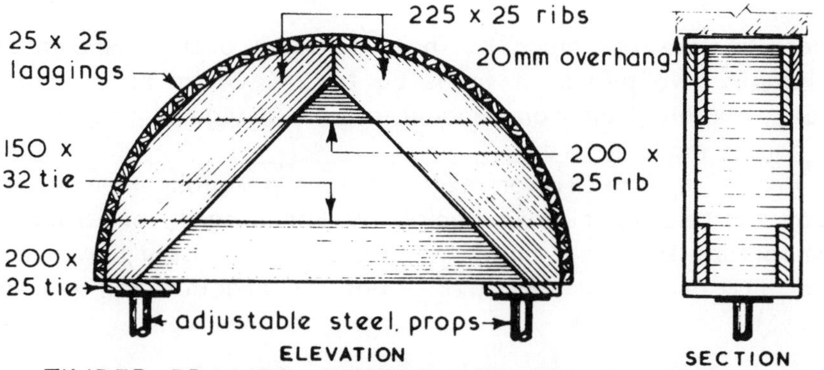

25 x 25 laggings

225 x 25 ribs

20mm overhang

150 x 32 tie

200 x 25 rib

200 x 25 tie

adjustable steel props

ELEVATION

SECTION

TIMBER FRAMED CENTRE FOR SPANS UP TO 1·500

The profile of an arch does not lend itself to simple positioning of a damp proof course. At best, it can be located horizontally at upper extrados level. This leaves the depth of the arch and masonry below the dpc vulnerable to dampness. Proprietary galvanised or stainless steel cavity trays resolve this problem by providing:

* Continuity of dpc around the extrados.

* Arch support/centring during construction.

* Arch and wall support after construction.

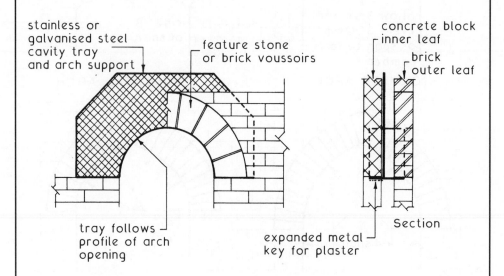

Standard profiles are made to the traditional outlines shown on the previous two pages, in spans up to 2m. Other options may also be available from some manufacturers. Irregular shapes and spans can be made to order.

Note: Arches in semi-circular, segmental or parabolic form up to 2m span can be proportioned empirically. For integrity of structure it is important to ensure sufficient provision of masonry over and around any arch, see BS 5628: Code of practice for use of masonry.

Openings ~ these consist of a head, jambs and sill and the different methods and treatments which can be used in their formation is very wide but they are all based on the same concepts. Application limited — see pages 497—8

Typical Head Details ~

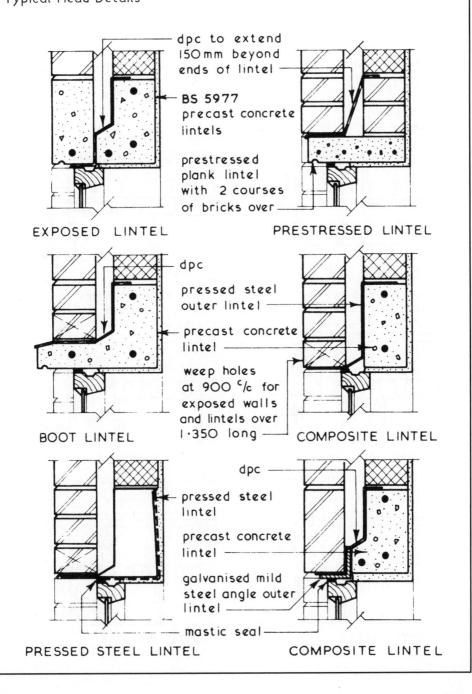

dpc to extend 150 mm beyond ends of lintel

BS 5977 precast concrete lintels

prestressed plank lintel with 2 courses of bricks over

EXPOSED LINTEL

PRESTRESSED LINTEL

dpc

pressed steel outer lintel

precast concrete lintel

weep holes at 900 $^c/c$ for exposed walls and lintels over 1·350 long

BOOT LINTEL

COMPOSITE LINTEL

dpc

pressed steel lintel

precast concrete lintel

galvanised mild steel angle outer lintel

mastic seal

PRESSED STEEL LINTEL

COMPOSITE LINTEL

Jambs ~ these may be bonded as in solid walls or unbonded as in cavity walls. The latter must have some means of preventing the ingress of moisture from the outer leaf to the inner leaf and hence the interior of the building.

Application limited — see pages 497—8

Typical Jamb Details ~

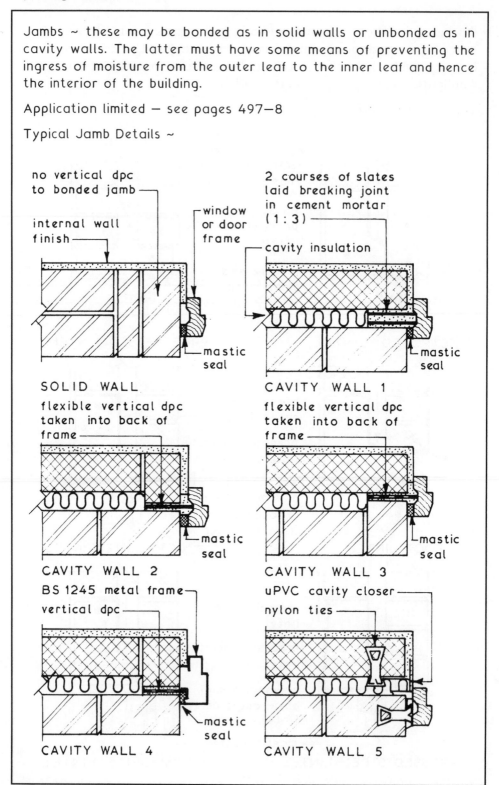

no vertical dpc
to bonded jamb

internal wall
finish

window
or door
frame

mastic
seal

SOLID WALL

2 courses of slates
laid breaking joint
in cement mortar
(1:3)

cavity insulation

mastic
seal

CAVITY WALL 1

flexible vertical dpc
taken into back of
frame

mastic
seal

CAVITY WALL 2

flexible vertical dpc
taken into back of
frame

mastic
seal

CAVITY WALL 3

BS 1245 metal frame
vertical dpc

mastic
seal

CAVITY WALL 4

uPVC cavity closer
nylon ties

CAVITY WALL 5

Sills ~ the primary function of any sill is to collect the rainwater which has run down the face of the window or door and shed it clear of the wall below.

Application limited — see page 497—8

Typical Sill Details ~

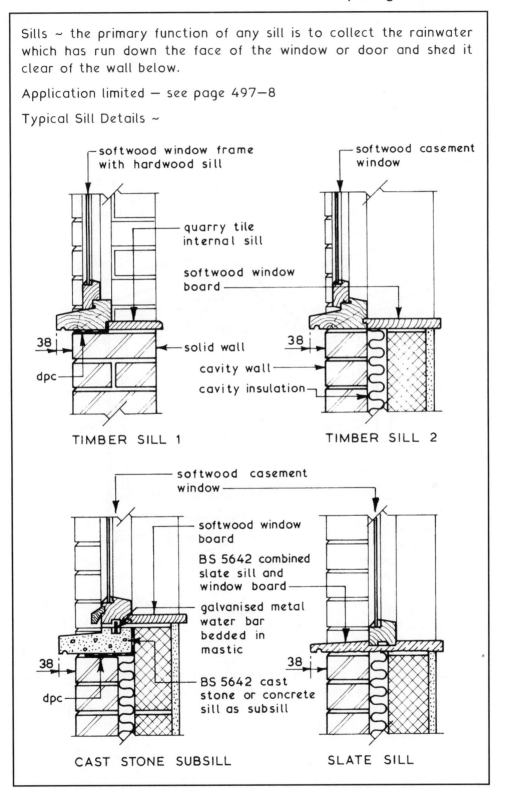

softwood window frame with hardwood sill

quarry tile internal sill

softwood window board

solid wall

cavity wall

cavity insulation

38

dpc

TIMBER SILL 1

softwood casement window

TIMBER SILL 2

softwood casement window

softwood window board

BS 5642 combined slate sill and window board

galvanised metal water bar bedded in mastic

BS 5642 cast stone or concrete sill as subsill

38

dpc

CAST STONE SUBSILL

SLATE SILL

A window must be aesthetically acceptable in the context of building design and surrounding environment

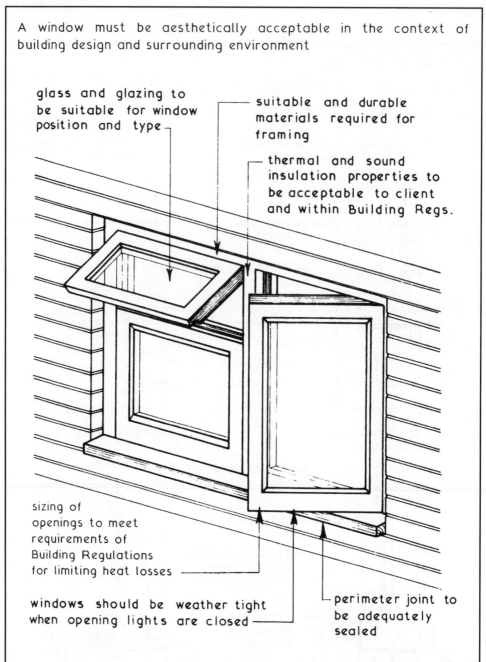

glass and glazing to be suitable for window position and type

suitable and durable materials required for framing

thermal and sound insulation properties to be acceptable to client and within Building Regs.

sizing of openings to meet requirements of Building Regulations for limiting heat losses

windows should be weather tight when opening lights are closed

perimeter joint to be adequately sealed

Windows should be selected or designed to resist wind loadings, be easy to clean and provide for safety and security. They should be sited to provide a vision out and therefore visual contact with the world outside the building

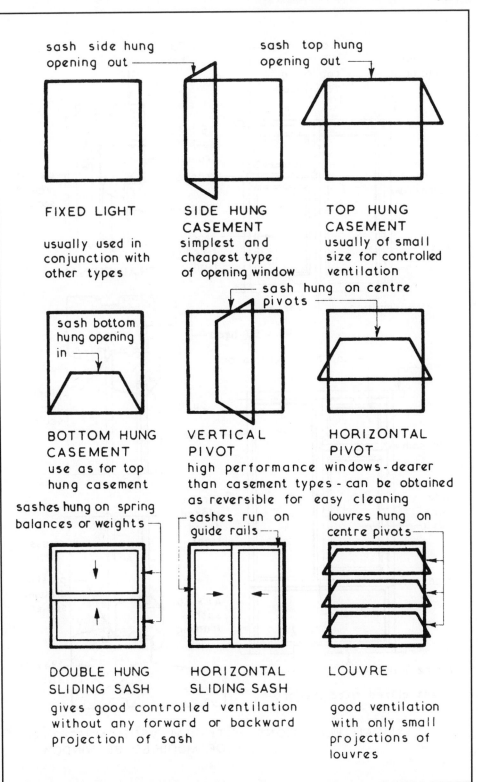

sash side hung
opening out

sash top hung
opening out

FIXED LIGHT

usually used in
conjunction with
other types

**SIDE HUNG
CASEMENT**
simplest and
cheapest type
of opening window

**TOP HUNG
CASEMENT**
usually of small
size for controlled
ventilation

sash hung on centre
pivots

sash bottom
hung opening
in

**BOTTOM HUNG
CASEMENT**
use as for top
hung casement

**VERTICAL
PIVOT**

**HORIZONTAL
PIVOT**

high performance windows - dearer
than casement types - can be obtained
as reversible for easy cleaning

sashes hung on spring
balances or weights

sashes run on
guide rails

louvres hung on
centre pivots

**DOUBLE HUNG
SLIDING SASH**

**HORIZONTAL
SLIDING SASH**

LOUVRE

gives good controlled ventilation
without any forward or backward
projection of sash

good ventilation
with only small
projections of
louvres

Timber Casement Windows

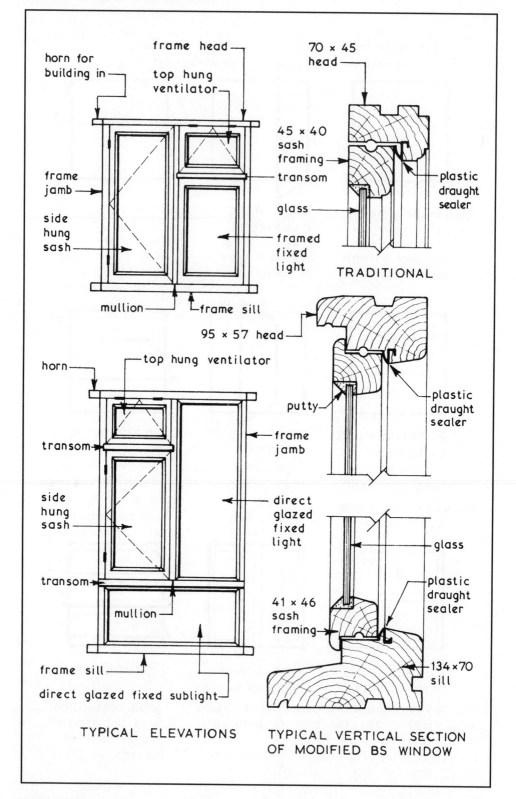

horn for building in

frame head

top hung ventilator

70 × 45 head

45 × 40 sash framing

transom

glass

plastic draught sealer

frame jamb

side hung sash

framed fixed light

TRADITIONAL

mullion

frame sill

95 × 57 head

putty

plastic draught sealer

horn

top hung ventilator

transom

frame jamb

side hung sash

direct glazed fixed light

glass

transom

mullion

41 × 46 sash framing

plastic draught sealer

frame sill

134 × 70 sill

direct glazed fixed sublight

TYPICAL ELEVATIONS

TYPICAL VERTICAL SECTION OF MODIFIED BS WINDOW

The standard range of casement windows used in the UK was derived from the English Joinery Manufacturer's Association (EJMA) designs of some 50 years ago. These became adopted in BS 644–1: Wood windows, Specification for factory assembled windows of various types. A modified type is shown on the preceding page. Contemporary building standards require higher levels of performance in terms of thermal and sound insulation (Bldg. Regs. Pt. L and E), air permeability, water tightness and wind resistance (BS 5368: Methods of testing windows). This has been achieved by adapting Scandinavian designs with double and triple glazing to attain U values as low as 1·2 W/m²K and a sound reduction of 50 dB.

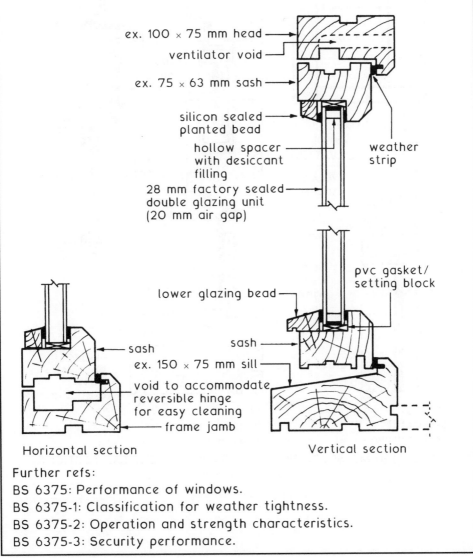

ex. 100 × 75 mm head
ventilator void
ex. 75 × 63 mm sash
silicon sealed planted bead
hollow spacer with desiccant filling
28 mm factory sealed double glazing unit (20 mm air gap)
weather strip

pvc gasket/ setting block
lower glazing bead
sash
sash
ex. 150 × 75 mm sill
void to accommodate reversible hinge for easy cleaning
frame jamb

Horizontal section

Vertical section

Further refs:
BS 6375: Performance of windows.
BS 6375-1: Classification for weather tightness.
BS 6375-2: Operation and strength characteristics.
BS 6375-3: Security performance.

Metal Windows ~ these can be obtained in steel (BS 6510) or in aluminium alloy (BS 4873). Steel windows are cheaper in initial cost than aluminium alloy but have higher maintenance costs over their anticipated life, both can be obtained fitted into timber subframes. Generally they give a larger glass area for any given opening size than similar timber windows but they can give rise to condensation on the metal components.

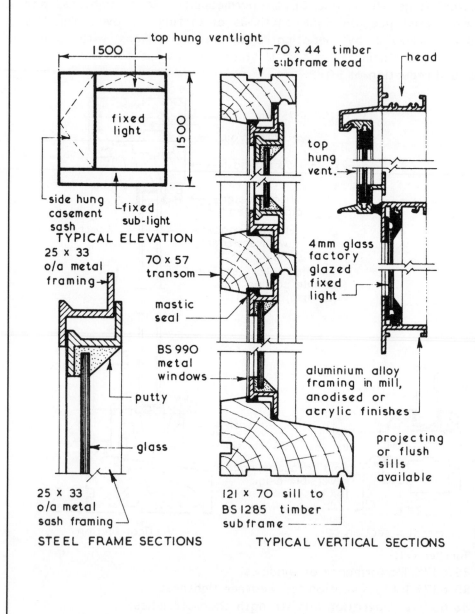

TYPICAL ELEVATION

STEEL FRAME SECTIONS

TYPICAL VERTICAL SECTIONS

Timber Windows ~ wide range of ironmongery available which can be factory fitted or supplied and fixed on site.

Metal Windows ~ ironmongery usually supplied with and factory fitted to the windows.

Typical Examples ~

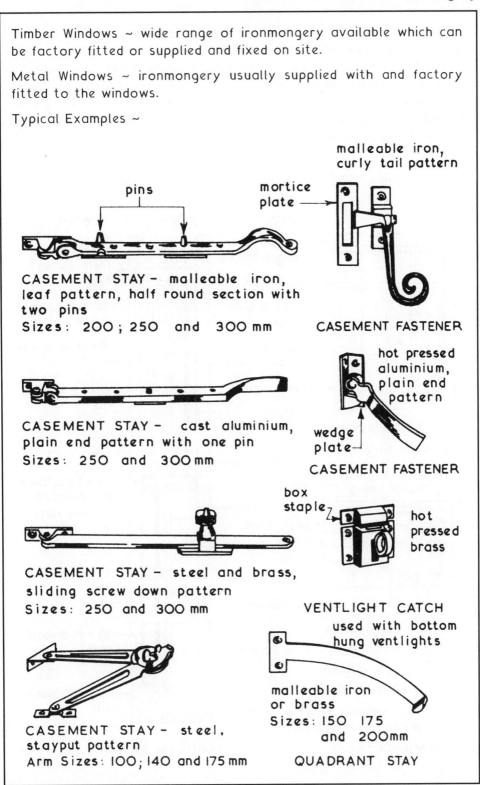

pins

mortice plate

malleable iron, curly tail pattern

CASEMENT STAY – malleable iron, leaf pattern, half round section with two pins
Sizes: 200; 250 and 300 mm

CASEMENT FASTENER

CASEMENT STAY – cast aluminium, plain end pattern with one pin
Sizes: 250 and 300 mm

hot pressed aluminium, plain end pattern

wedge plate

CASEMENT FASTENER

box staple

hot pressed brass

CASEMENT STAY – steel and brass, sliding screw down pattern
Sizes: 250 and 300 mm

VENTLIGHT CATCH
used with bottom hung ventlights

malleable iron or brass
Sizes: 150 175 and 200mm

CASEMENT STAY – steel, stayput pattern
Arm Sizes: 100; 140 and 175 mm

QUADRANT STAY

Sliding Sash Windows

Sliding Sash Windows ~ these are an alternative format to the conventional side hung casement windows and can be constructed as a vertical or double hung sash window or as a horizontal sliding window in timber, metal, plastic or in any combination of these materials. The performance and design functions of providing daylight, ventilation, vision out, etc., are the same as those given for traditional windows in Windows — Performance Requirements on page 314

Typical Double Hung Weight Balanced Window Details ~

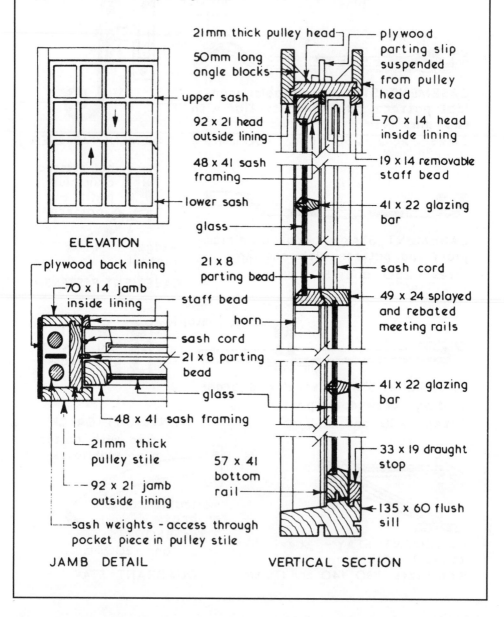

21mm thick pulley head

plywood parting slip suspended from pulley head

50mm long angle blocks

upper sash

92 x 21 head outside lining

70 x 14 head inside lining

48 x 41 sash framing

19 x 14 removable staff bead

lower sash

41 x 22 glazing bar

glass

ELEVATION

21 x 8 parting bead

sash cord

plywood back lining

staff bead

49 x 24 splayed and rebated meeting rails

70 x 14 jamb inside lining

horn

sash cord

21 x 8 parting bead

41 x 22 glazing bar

glass

48 x 41 sash framing

21mm thick pulley stile

57 x 41 bottom rail

33 x 19 draught stop

92 x 21 jamb outside lining

135 x 60 flush sill

sash weights - access through pocket piece in pulley stile

JAMB DETAIL

VERTICAL SECTION

Double Hung Sash Windows ~ these vertical sliding sash windows come in two formats when constructed in timber. The weight balanced format is shown on page 320 the alternative spring balanced type is illustrated below. Both formats are usually designed and constructed to the recommendations set out in BS 644-2: Wood double hung sash windows.

Typical Double Hung Spring Balanced Window Details ~

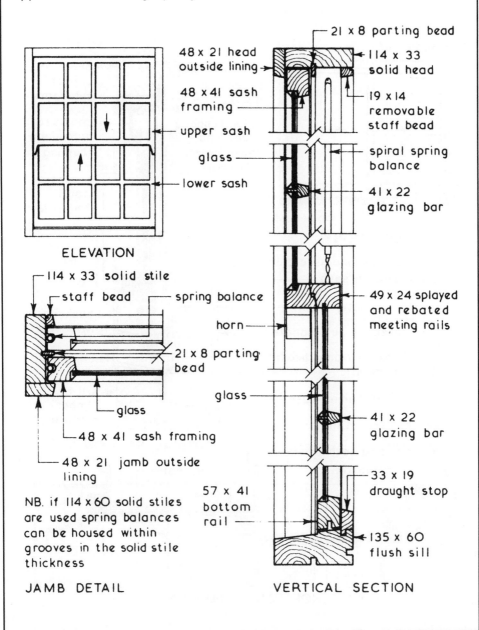

ELEVATION

JAMB DETAIL

VERTICAL SECTION

NB. if 114 x 60 solid stiles are used spring balances can be housed within grooves in the solid stile thickness

Sliding Sash Windows

Horizontally Sliding Sash Windows ~ these are an alternative format to the vertically sliding or double hung sash windows shown on pages 320 & 321 and can be constructed in timber, metal, plastic or combinations of these materials with single or double glazing. A wide range of arrangements are available with two or more sliding sashes which can have a ventlight incorporated in the outer sliding sash.

Typical Horizontally Sliding Sash Window Details ~

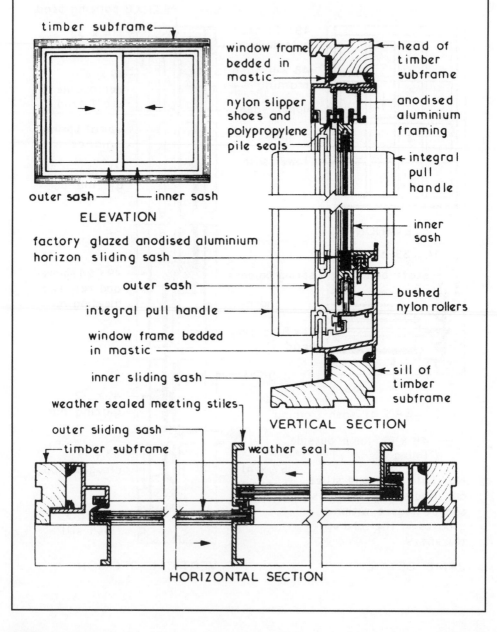

timber subframe

outer sash — — inner sash

ELEVATION

window frame bedded in mastic

nylon slipper shoes and polypropylene pile seals

factory glazed anodised aluminium horizon sliding sash

outer sash

integral pull handle

window frame bedded in mastic

inner sliding sash

weather sealed meeting stiles

outer sliding sash

timber subframe

head of timber subframe

anodised aluminium framing

integral pull handle

inner sash

bushed nylon rollers

sill of timber subframe

VERTICAL SECTION

weather seal

HORIZONTAL SECTION

Pivot Windows ~ like other windows these are available in timber, metal, plastic or in combinations of these materials.

They can be constructed with centre jamb pivots enabling the sash to pivot or rotate in the horizontal plane or alternatively the pivots can be fixed in the head and sill of the frame so that the sash rotates in the vertical plane.

Typical Example ~

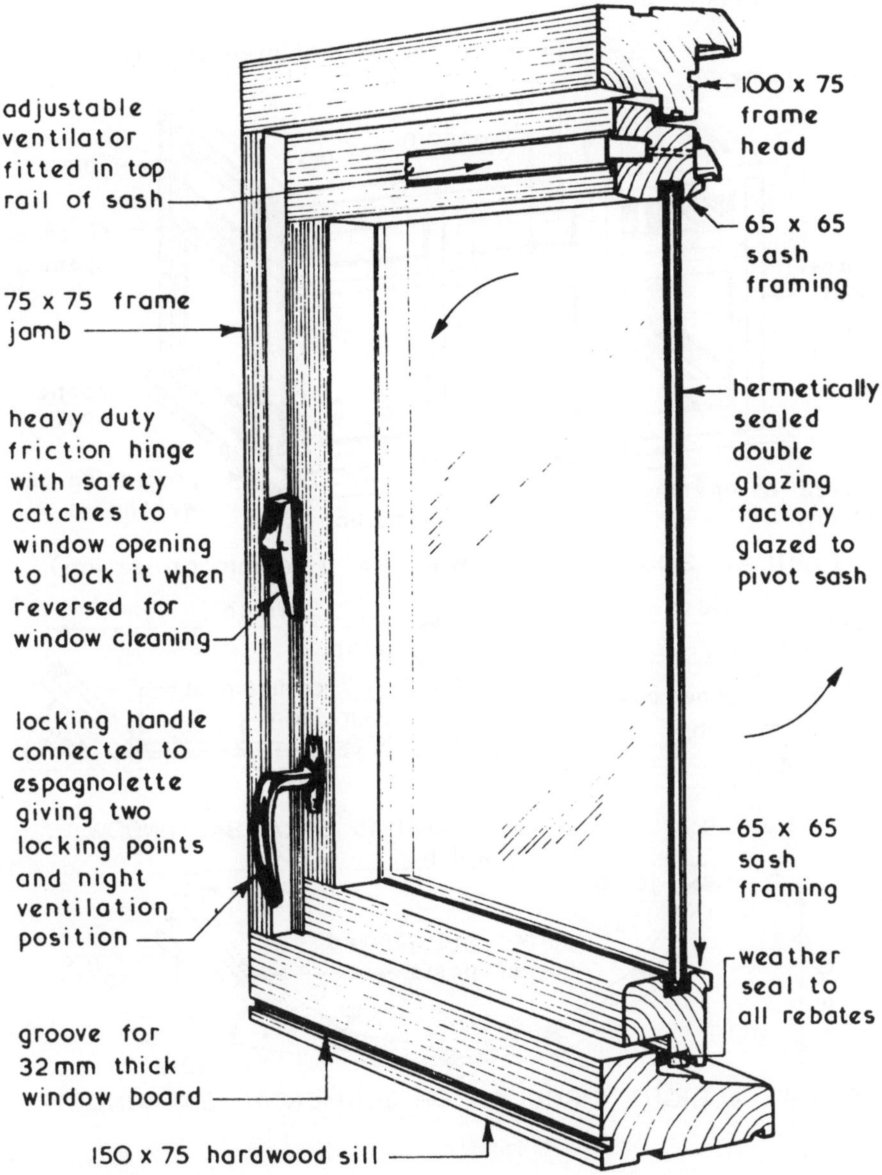

adjustable ventilator fitted in top rail of sash

75 x 75 frame jamb

heavy duty friction hinge with safety catches to window opening to lock it when reversed for window cleaning

locking handle connected to espagnolette giving two locking points and night ventilation position

groove for 32 mm thick window board

150 x 75 hardwood sill

100 x 75 frame head

65 x 65 sash framing

hermetically sealed double glazing factory glazed to pivot sash

65 x 65 sash framing

weather seal to all rebates

Bay Windows ~ these can be defined as any window with side lights which projects in front of the external wall and is supported by a sill height wall. Bay windows not supported by a sill height wall are called oriel windows. They can be of any window type, constructed from any of the usual window materials and are available in three plan formats namely square, splay and circular or segmental. Timber corner posts can be boxed, solid or jointed the latter being the common method.

Typical Examples ~

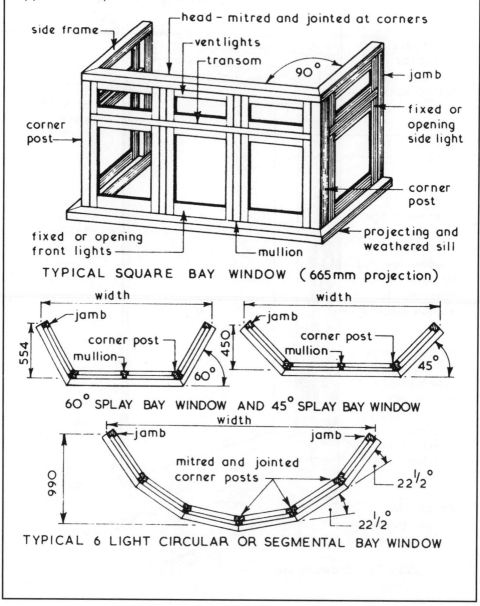

TYPICAL SQUARE BAY WINDOW (665mm projection)

60° SPLAY BAY WINDOW AND 45° SPLAY BAY WINDOW

TYPICAL 6 LIGHT CIRCULAR OR SEGMENTAL BAY WINDOW

Schedules ~ the main function of a schedule is to collect together all the necessary information for a particular group of components such as windows, doors and drainage inspection chambers. There is no standard format for schedules but they should be easy to read, accurate and contain all the necessary information for their purpose. Schedules are usually presented in a tabulated format which can be related to and read in conjunction with the working drawings.

Typical Example ~

WINDOW SCHEDULE – Sheet 1 of 1	Drawn By: RC	Date: 14/4/01	Rev.

Contract Title & Number: Lane End Form – H 341/80	Drg. Nos. C(31) 450–7

Location or Number	Type	Material	Overall Size	Glass	Ironmongery	Sill	
						External	Internal
1 & 2	9 FCV4 – Subframe –	steel softwood	910×1214 970×1275	146×1140 632×553 670×594 3mm float	supplied with casements	2 cos. plain tiles	150×150×15 quarry tiles
3, 4, 5 & 6	240V –	softwood	1206×1206	480×280 580×700 480×1030 3mm float	ditto	ditto	25mm thick softwood
7	Purpose made – Drg. No. C (31)-457	softwood	1770×1600	460×200 1080×300 460×1040 1080×1140 3mm float	1–200mm 1–300mm al stays 1-al alloy fastener	sill of frame	ditto

Glass ~ this material is produced by fusing together soda, lime and silica with other minor ingredients such as magnesia and alumina. A number of glass types are available for domestic work and these include :-

Clear Float ~ used where clear undistorted vision is required. Available thicknesses range from 3 mm to 25 mm.

Clear Sheet ~ suitable for all clear glass areas but because the two faces of the glass are never perfectly flat or parallel some distortion of vision usually occurs. This type of glass is gradually being superseded by the clear float glass. Available thicknesses range from 3 mm to 6 mm.

Translucent Glass ~ these are patterned glasses most having one patterned surface and one relatively flat surface. The amount of obscurity and diffusion obtained depend on the type and nature of pattern. Available thicknesses range from 4 mm to 6 mm for patterned glasses and from 5 mm to 10 mm for rough cast glasses.

Wired Glass ~ obtainable as a clear polished wired glass or as a rough cast wired glass with a nominal thickness of 7 mm. Generally used where a degree of fire resistance is required. Georgian wired glass has a 12 mm square mesh whereas the hexagonally wired glass has a 20 mm mesh.

Choice of Glass ~ the main factors to be considered are :-

1. Resistance to wind loadings. 2. Clear vision required.
3. Privacy. 4. Security. 5. Fire resistance. 6. Aesthetics.

Glazing Terminology ~

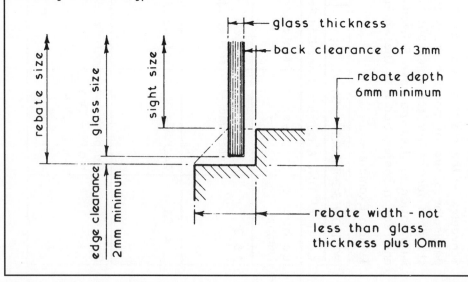

Glazing ~ the act of fixing glass into a frame or surround in domestic work this is usually achieved by locating the glass in a rebate and securing it with putty or beading and should be carried out in accordance with the recommendations contained in BS 6262: Glazing for buildings.

Timber Surrounds ~ linseed oil putty to BS 544 — rebate to be clean, dry and primed before glazing is carried out. Putty should be protected with paint within two weeks of application.

Metal Surrounds ~ metal casement putty if metal surround is to be painted — if surround is not to be painted a non-setting compound should be used.

Typical Glazing Details ~

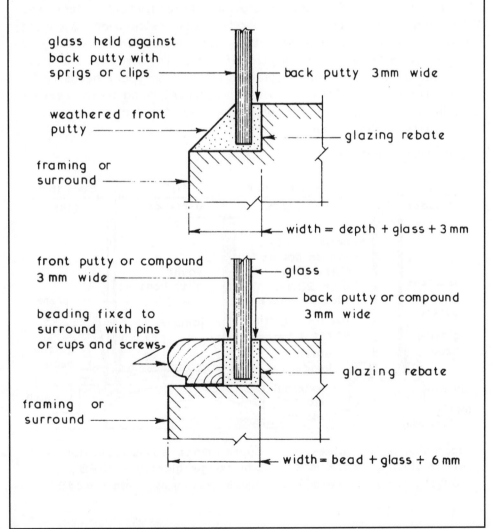

glass held against back putty with sprigs or clips

back putty 3mm wide

weathered front putty

glazing rebate

framing or surround

width = depth + glass + 3 mm

front putty or compound 3 mm wide

glass

back putty or compound 3 mm wide

beading fixed to surround with pins or cups and screws

glazing rebate

framing or surround

width = bead + glass + 6 mm

Double Glazing ~ as its name implies this is where two layers of glass are used instead of the traditional single layer. Double glazing can be used to reduce the rate of heat loss through windows and glazed doors or it can be employed to reduce the sound transmission through windows. In the context of thermal insulation this is achieved by having a small air space within the range of 6 to 20 mm between the two layers of glass. The hermetically sealed double glazing unit with an air space of 20 mm containing dehydrated air to prevent internal misting gives good results. If metal frames are used these should have a thermal break incorporated in their design. All opening sashes in a double glazing system should be fitted with adequate weather seals to reduce the rate of heat loss through the opening clearance gap.

In the context of sound insulation three factors affect the performance of double glazing. Firstly good installation to ensure airtightness, secondly the weight of glass used and thirdly the size of air space between the layers of glass. The heavier the glass used the better the sound insulation and the air space needs to be within the range of 50 to 300 mm. Absorbent lining to the reveals within the air space will also improve the sound insulation properties of the system.

Typical Examples ~

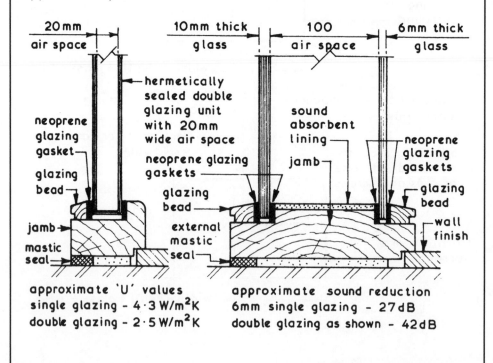

| 20 mm air space | 10 mm thick glass | 100 air space | 6 mm thick glass |

hermetically sealed double glazing unit with 20 mm wide air space

neoprene glazing gasket

glazing bead

jamb

mastic seal

neoprene glazing gaskets

glazing bead

external mastic seal

sound absorbent lining

jamb

neoprene glazing gaskets

glazing bead

wall finish

approximate 'U' values
single glazing - 4·3 W/m²K
double glazing - 2·5 W/m²K

approximate sound reduction
6 mm single glazing - 27 dB
double glazing as shown - 42 dB

Secondary glazing of existing windows is an acceptable method for reducing heat energy losses at wall openings. Providing the existing windows are in a good state of repair, this is a cost effective, simple method for upgrading windows to current energy efficiency standards. In addition to avoiding the disruption of removing existing windows, further advantages of secondary glazing include, retention of the original window features, reduction in sound transmission and elimination of draughts. Applications are manufactured for all types of window, with sliding or hinged variations. The following details are typical of horizontal sliding sashes -

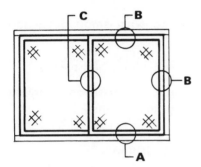

Elevation of frame

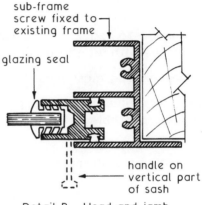

sub-frame screw fixed to existing frame

glazing seal

handle on vertical part of sash

Detail B - Head and jamb

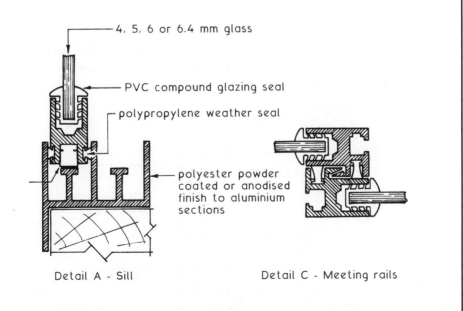

4, 5, 6 or 6.4 mm glass

PVC compound glazing seal

polypropylene weather seal

polyester powder coated or anodised finish to aluminium sections

Detail A - Sill

Detail C - Meeting rails

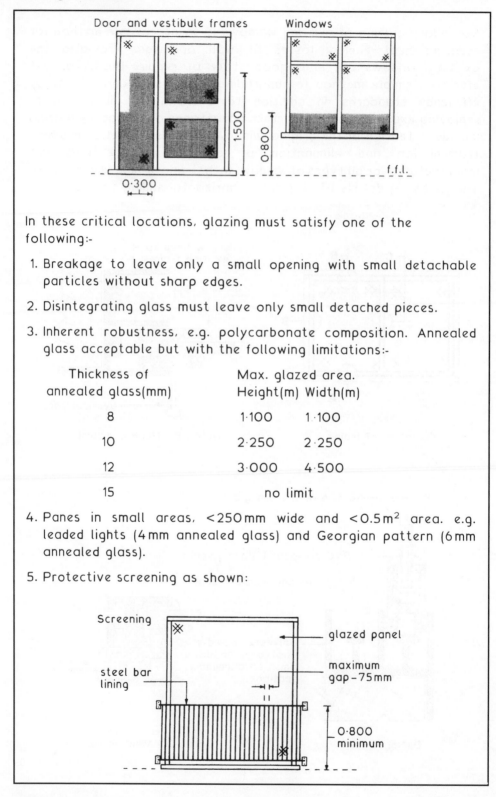

In these critical locations, glazing must satisfy one of the following:-

1. Breakage to leave only a small opening with small detachable particles without sharp edges.

2. Disintegrating glass must leave only small detached pieces.

3. Inherent robustness, e.g. polycarbonate composition. Annealed glass acceptable but with the following limitations:-

Thickness of annealed glass(mm)	Max. glazed area.	
	Height(m)	Width(m)
8	1·100	1·100
10	2·250	2·250
12	3·000	4·500
15	no limit	

4. Panes in small areas, <250mm wide and <0.5m² area. e.g. leaded lights (4mm annealed glass) and Georgian pattern (6mm annealed glass).

5. Protective screening as shown:

Manifestation or Marking of Glass ~ another aspect of the critical location concept which frequently occurs with contemporary glazed features in a building. Commercial premises such as open plan offices, shops and showrooms often incorporate large walled areas of uninterrupted glass to promote visual depth, whilst dividing space or forming part of the exterior envelope. To prevent collision, glazed doors and walls must have prominent framing or intermediate transoms and mullions. An alternative is to position obvious markings 1500 mm above floor level. Glass doors could have large pull/push handles and/or IN and OUT signs in bold lettering. Other areas may be adorned with company logos, stripes, geometric shape, etc.

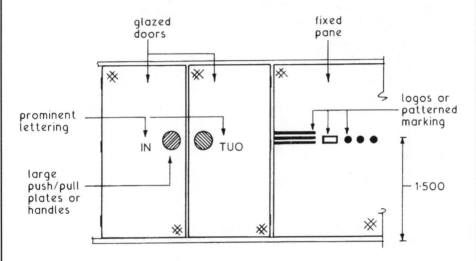

Critical Locations ~ The Building Regulations, Approved Document — N, determines positions where potential personal impact and injury with glazed doors and windows are most critical. In these situations the glazing specification must incorporate a degree of safety such that any breakage would be relatively harmless. Additional measures in British Standard 6206 complement the Building Regulations and provide test requirements and specifications for impact performance for different classes of glazing material. See also BS6262.

Refs. Building Regulations, A.D. N1: Protection against impact.
                        A.D. N2: Manifestation of glazing.
BS 6206: Specification for impact performance requirements for flat safety glass and safety plastics for use in buildings.
BS 6262: Glazing for buildings.

Glass blocks have been used for some time as internal feature partitioning. They now include a variety of applications in external walls, where they combine the benefits of a walling unit with a natural source of light. They have also been used in paving to allow natural light penetration into basements.

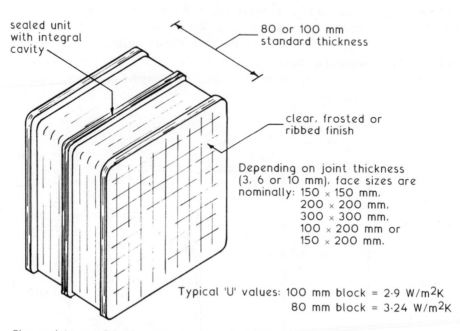

sealed unit with integral cavity

80 or 100 mm standard thickness

clear, frosted or ribbed finish

Depending on joint thickness (3, 6 or 10 mm), face sizes are nominally: 150 × 150 mm,
200 × 200 mm,
300 × 300 mm,
100 × 200 mm or
150 × 200 mm.

Typical 'U' values: 100 mm block = 2·9 W/m^{2}K
80 mm block = 3·24 W/m^{2}K

Fire resistance, BS 476–22 - 1 hour integrity.
Maximum panel size is 9m^2. Maximum panel dimension is 3 m

Laying — glass blocks can be bonded like conventional brickwork, but for aesthetic reasons are usually laid with continuous vertical and horizontal joints.

Jointing — blocks are bedded in mortar with reinforcement from two, 9 gauge galvanised steel wires in horizontal joints. Every 3rd. course for 150 mm units, every 2nd. course for 200 mm units and every course for 300 mm units. First and last course to be reinforced.

Mortar — dryer than for bricklaying as the blocks are non-absorbent. The general specification will include: White Portland Cement (BS 12), High Calcium Lime (BS 890) and Sand (BS 1200 - Table 1). The sand should be white quartzite or silica type. Fine silver sand is acceptable. An integral waterproofing agent should also be provided. Recommended mix ratios — 1 part cement: 0.5 part lime: 4 parts sand.

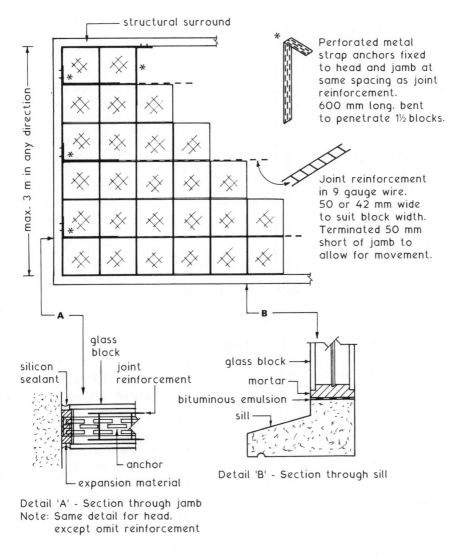

structural surround

max. 3 m in any direction

* Perforated metal strap anchors fixed to head and jamb at same spacing as joint reinforcement. 600 mm long, bent to penetrate 1½ blocks.

Joint reinforcement in 9 gauge wire. 50 or 42 mm wide to suit block width. Terminated 50 mm short of jamb to allow for movement.

A

glass block

silicon sealant

joint reinforcement

anchor

expansion material

Detail 'A' - Section through jamb
Note: Same detail for head, except omit reinforcement

B

glass block

mortar

bituminous emulsion

sill

Detail 'B' - Section through sill

Ref. BS EN 12725: Glass in building — glass block walls — design, dimensions and performance.

Doors—Performance Requirements

Doors ~ can be classed as external or internal. External doors are usually thicker and more robust in design than internal doors since they have more functions to fulfil.

Typical Functions ~

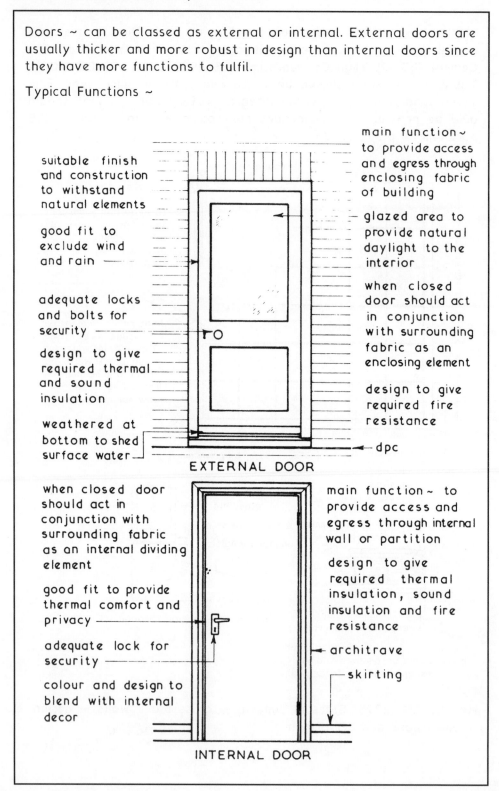

suitable finish and construction to withstand natural elements

good fit to exclude wind and rain

adequate locks and bolts for security

design to give required thermal and sound insulation

weathered at bottom to shed surface water

main function ~ to provide access and egress through enclosing fabric of building

glazed area to provide natural daylight to the interior

when closed door should act in conjunction with surrounding fabric as an enclosing element

design to give required fire resistance

dpc

EXTERNAL DOOR

when closed door should act in conjunction with surrounding fabric as an internal dividing element

good fit to provide thermal comfort and privacy

adequate lock for security

colour and design to blend with internal decor

main function ~ to provide access and egress through internal wall or partition

design to give required thermal insulation, sound insulation and fire resistance

architrave

skirting

INTERNAL DOOR

External Doors ~ these are available in a wide variety of types and styles in timber, aluminium alloy or steel. The majority of external doors are however made from timber, the metal doors being mainly confined to fully glazed doors such as 'patio doors'.

Typical Examples of External Doors ~

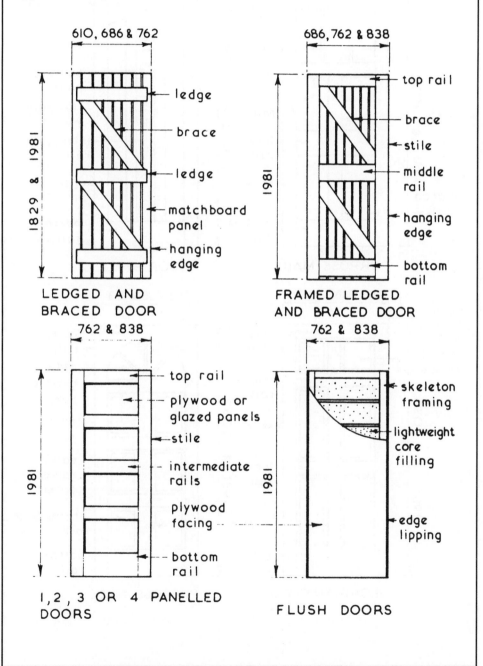

610, 686 & 762

ledge

brace

ledge

matchboard panel

hanging edge

1829 & 1981

LEDGED AND BRACED DOOR

686, 762 & 838

top rail

brace

stile

middle rail

hanging edge

bottom rail

1981

FRAMED LEDGED AND BRACED DOOR

762 & 838

top rail

plywood or glazed panels

stile

intermediate rails

plywood facing

bottom rail

1981

1, 2, 3 OR 4 PANELLED DOORS

762 & 838

skeleton framing

lightweight core filling

edge lipping

1981

FLUSH DOORS

Door Types

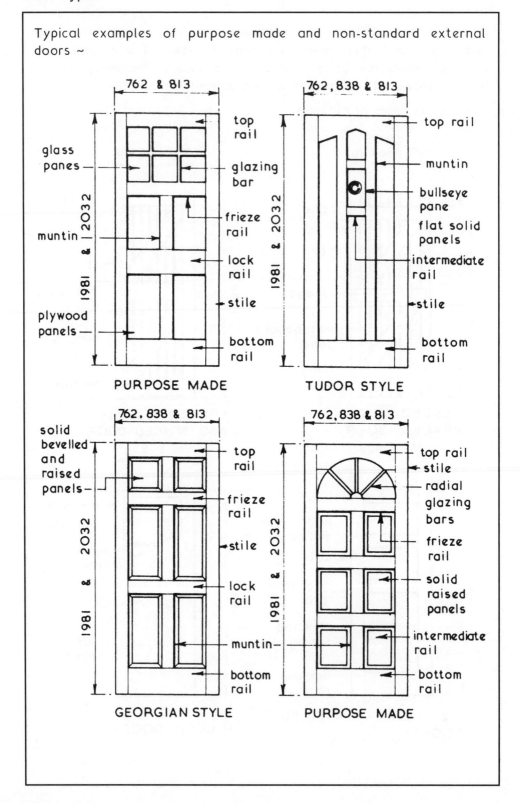

Typical examples of purpose made and non-standard external doors ~

PURPOSE MADE

762 & 813

glass panes

muntin

plywood panels

top rail

glazing bar

frieze rail

lock rail

stile

bottom rail

2032 & 1981

TUDOR STYLE

762, 838 & 813

top rail

muntin

bullseye pane

flat solid panels

intermediate rail

stile

bottom rail

1981 & 2032

GEORGIAN STYLE

762, 838 & 813

solid bevelled and raised panels

top rail

frieze rail

stile

lock rail

muntin

bottom rail

2032 & 1981

PURPOSE MADE

762, 838 & 813

top rail

stile

radial glazing bars

frieze rail

solid raised panels

intermediate rail

bottom rail

1981 & 2032

Door Frames ~ these are available for all standard external doors and can be obtained with a fixed solid or glazed panel above a door height transom. Door frames are available for doors opening inwards or outwards. Most door frames are made to the recommendations set out in BS 4787: Internal and external wood doorsets, door leaves and frames.

Typical Example ~

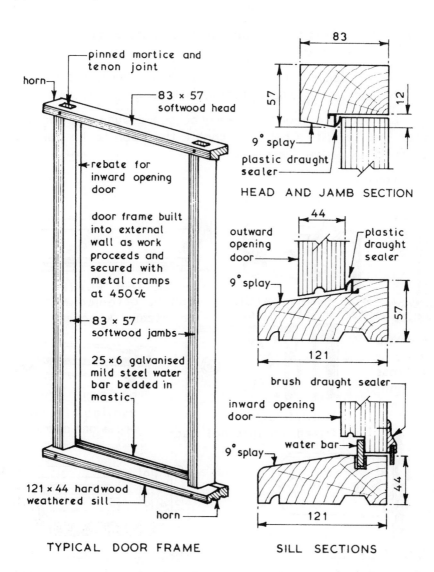

pinned mortice and tenon joint

horn

83 × 57 softwood head

rebate for inward opening door

door frame built into external wall as work proceeds and secured with metal cramps at 450 c/c

83 × 57 softwood jambs

25 × 6 galvanised mild steel water bar bedded in mastic

121 × 44 hardwood weathered sill

horn

TYPICAL DOOR FRAME

83

57

12

9° splay

plastic draught sealer

HEAD AND JAMB SECTION

44

outward opening door

plastic draught sealer

9° splay

57

121

brush draught sealer

inward opening door

water bar

9° splay

44

121

SILL SECTIONS

Door Ironmongery ~ available in a wide variety of materials, styles and finisher's but will consist of essentially the same components:-
Hinges or Butts — these are used to fix the door to its frame or lining and to enable it to pivot about its hanging edge.

Locks, Latches and Bolts ~ the means of keeping the door in its closed position and providing the required degree of security. The handles and cover plates used in conjunction with locks and latches are collectively called door furniture.

Letter Plates — fitted in external doors to enable letters etc., to be deposited through the door.

Other items include Finger and Kicking Plates which are used to protect the door fabric where there is high usage.

Draught Excluders to seal the clearance gap around the edges of the door and Security Chains to enable the door to be partially opened and thus retain some security.

Typical Examples ~

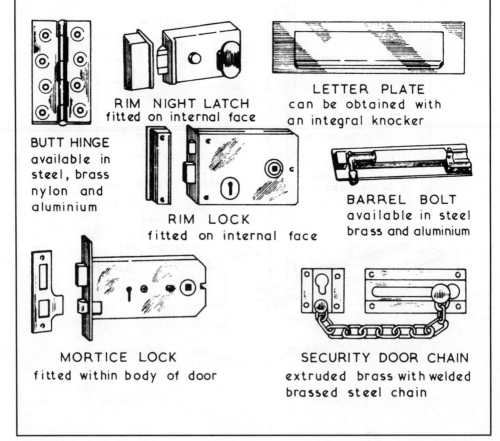

BUTT HINGE
available in
steel, brass
nylon and
aluminium

RIM NIGHT LATCH
fitted on internal face

LETTER PLATE
can be obtained with
an integral knocker

RIM LOCK
fitted on internal face

BARREL BOLT
available in steel
brass and aluminium

MORTICE LOCK
fitted within body of door

SECURITY DOOR CHAIN
extruded brass with welded
brassed steel chain

Industrial Doors ~ these doors are usually classified by their method of operation and construction. There is a very wide range of doors available and the choice should be based on the following considerations:-

1. Movement - vertical or horizontal.

2. Size of opening.

3. Position and purpose of door (s).

4. Frequency of opening and closing door (s).

5. Manual or mechanical operation.

6. Thermal and/or sound insulation requirements.

7. Fire resistance requirements.

Typical Industrial Door Types ~

1. **Straight Sliding -**

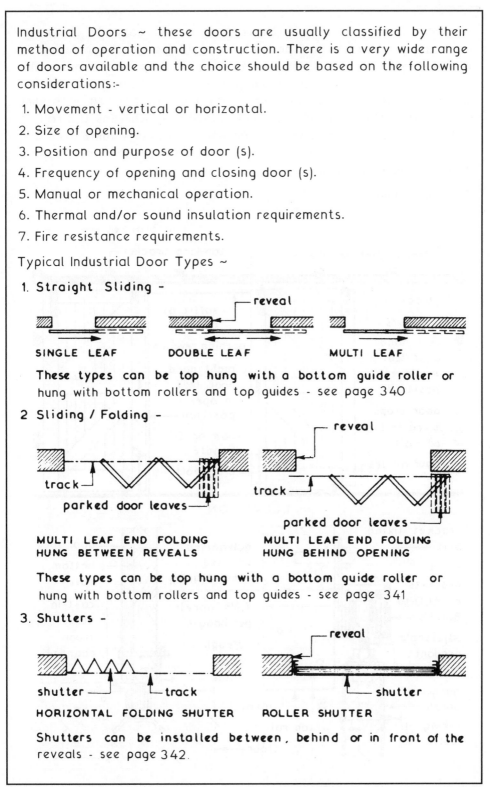

SINGLE LEAF   **DOUBLE LEAF**   **MULTI LEAF**

These types can be top hung with a bottom guide roller or hung with bottom rollers and top guides - see page 340

2 **Sliding / Folding -**

MULTI LEAF END FOLDING HUNG BETWEEN REVEALS   **MULTI LEAF END FOLDING HUNG BEHIND OPENING**

These types can be top hung with a bottom guide roller or hung with bottom rollers and top guides - see page 341

3. **Shutters -**

HORIZONTAL FOLDING SHUTTER   **ROLLER SHUTTER**

Shutters can be installed between, behind or in front of the reveals - see page 342.

Straight Sliding Doors ~ these doors are easy to operate, economic to maintain and present no problems for the inclusion of a wicket gate. They do however take up wall space to enable the leaves to be parked in the open position. The floor guide channel associated with top hung doors can become blocked with dirt causing a malfunction of the sliding movement whereas the rollers in bottom track doors can seize up unless regularly lubricated and kept clean. Straight sliding doors are available with either manual or mechanical operation.

Typical Example ~

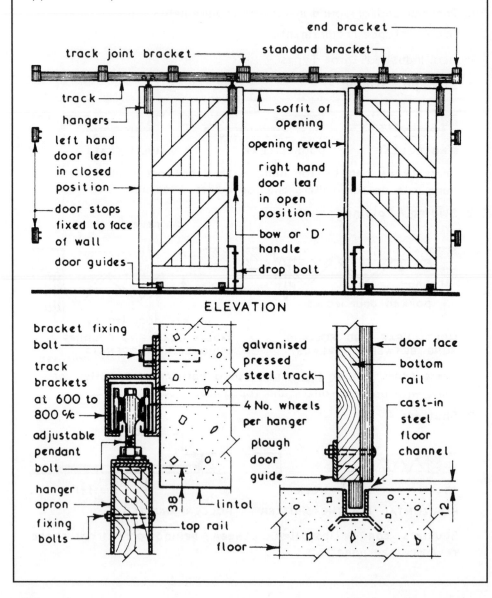

ELEVATION

Sliding/Folding Doors ~ these doors are an alternative format to the straight sliding door types and have the same advantages and disadvantages except that the parking space required for the opened door is less than that for straight sliding doors. Sliding/ folding are usually manually operated and can be arranged in groups of 2 to 8 leaves.

Typical Example ~

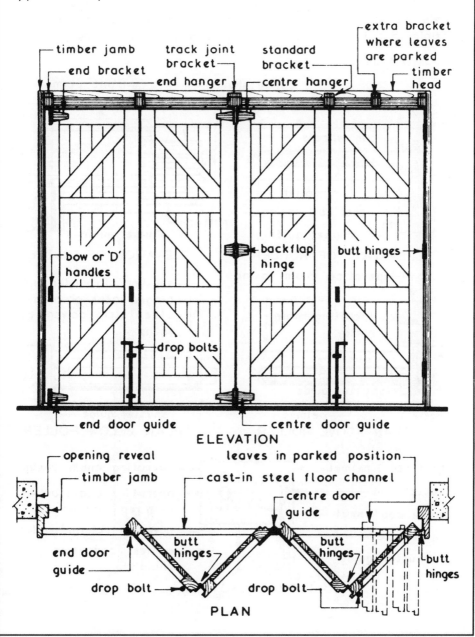

ELEVATION

PLAN

Shutters ~ horizontal folding shutters are similar in operation to sliding/folding doors but are composed of smaller leaves and present the same problems. Roller shutters however do not occupy any wall space but usually have to be fully opened for access. They can be manually operated by means of a pole when the shutters are self coiling, operated by means of an endless chain winding gear or mechanically raised and lowered by an electric motor but in all cases they are slow to open and close. Vision panels cannot be incorporated in the roller shutter but it is possible to include a small wicket gate or door in the design.

Typical Details ~

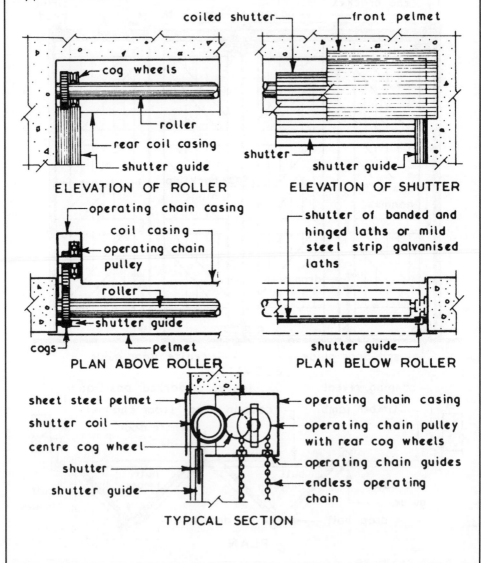

coiled shutter — front pelmet

— cog wheels

— roller

— rear coil casing

— shutter guide

ELEVATION OF ROLLER

shutter — shutter guide

ELEVATION OF SHUTTER

— operating chain casing

coil casing —

— operating chain pulley

— roller

— shutter guide

cogs — — pelmet

PLAN ABOVE ROLLER

— shutter of banded and hinged laths or mild steel strip galvanised laths

shutter guide —

PLAN BELOW ROLLER

sheet steel pelmet — — operating chain casing

shutter coil — — operating chain pulley with rear cog wheels

centre cog wheel — — operating chain guides

shutter — — endless operating chain

shutter guide —

TYPICAL SECTION

Crosswall Construction ~ this is a form of construction where load bearing walls are placed at right angles to the lateral axis of the building, the front and rear walls being essentially non-load bearing cladding. Crosswall construction is suitable for buildings up to 5 storeys high where the floors are similar and where internal separating or party walls are required such as in blocks of flats or maisonettes. The intermediate floors span longitudinally between the crosswalls providing the necessary lateral restraint and if both walls and floors are of cast insitu reinforced concrete the series of 'boxes' so formed is sometimes called box frame construction. Great care must be taken in both design and construction to ensure that the junctions between the non-load bearing claddings and the crosswalls are weathertight. If a pitched roof is to be employed with the ridge parallel to the lateral axis an edge beam will be required to provide a seating for the trussed or common rafters and to transmit the roof loads to the crosswalls.

Typical Crosswall Arrangement Details ~

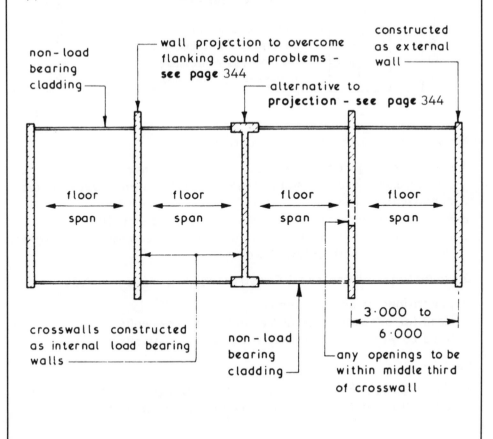

Crosswall Construction

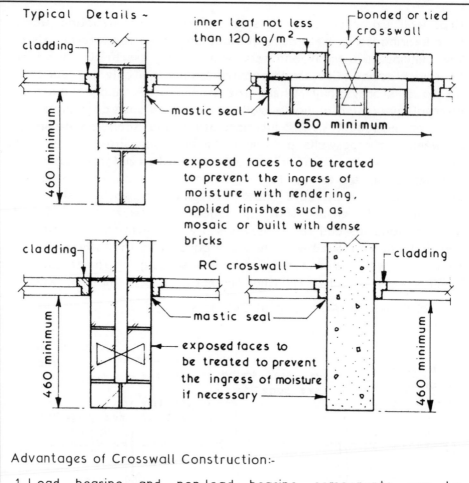

Typical Details ~

cladding

inner leaf not less than 120 kg/m^2

bonded or tied crosswall

460 minimum

mastic seal

650 minimum

exposed faces to be treated to prevent the ingress of moisture with rendering, applied finishes such as mosaic or built with dense bricks

cladding

460 minimum

RC crosswall

cladding

mastic seal

exposed faces to be treated to prevent the ingress of moisture if necessary

460 minimum

Advantages of Crosswall Construction:-

1. Load bearing and non-load bearing components can be standardised and in same cases prefabricated giving faster construction times.
2. Fenestration between crosswalls unrestricted structurally.
3. Crosswalls although load bearing need not be weather resistant as is the case with external walls.

Disadvantages of Crosswall Construction:-

1. Limitations of possible plans.
2. Need for adequate lateral ties between crosswalls.
3. Need to weather adequately projecting crosswalls.

Floors:-

An insitu solid reinforced concrete floor will provide the greatest rigidity, all other form must be adequately tied to walls.

Framing ~ an industry based pre-fabricated house manufacturing process permitting rapid site construction, with considerably fewer site operatives than traditional construction. This technique has a long history of conventional practice in Scandinavia and North America, but has only gained credibility in the UK since the 1960s. Factory-made panels are based on a stud framework of timber, normally ex. 100 × 50 mm, an outer sheathing of plywood, particle-board or similar sheet material, insulation between the framing members and an internal lining of plasterboard. An outer cladding of brickwork weatherproofs the building and provides a traditional appearance.

Assembly techniques are derived from two systems:-

1. Balloon frame

2. Platform frame

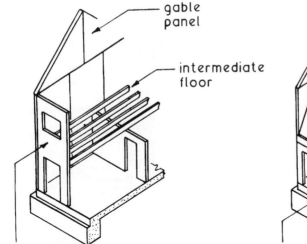

gable panel

intermediate floor

2-storey height panels

BALLOON FRAME

single storey height panels

PLATFORM FRAME

A balloon frame consists of two-storey height panels with an intermediate floor suspended from the framework. In the UK, the platform frame is preferred with intermediate floor support directly on the lower panel. It is also easier to transport, easier to handle on site and has fewer shrinkage and movement problems.

Timber Frame Construction

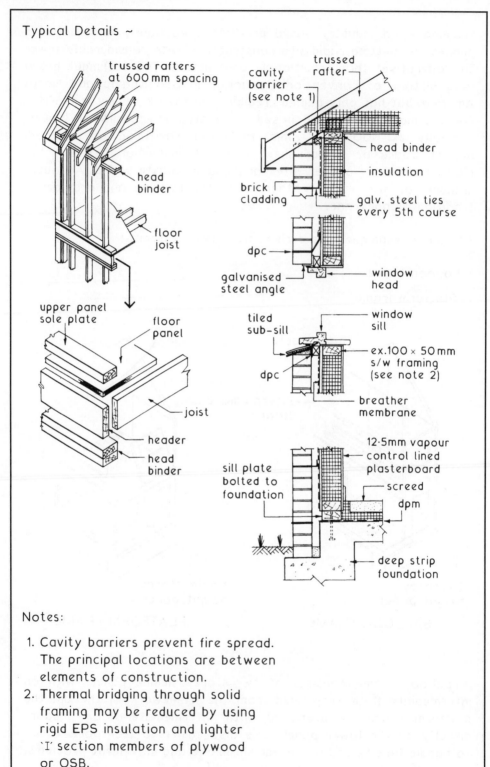

Typical Details ~

trussed rafters at 600 mm spacing

head binder

floor joist

upper panel sole plate

floor panel

joist

header

head binder

trussed rafter

cavity barrier (see note 1)

head binder

insulation

brick cladding

galv. steel ties every 5th course

dpc

galvanised steel angle

window head

tiled sub-sill

window sill

ex.100 × 50 mm s/w framing (see note 2)

dpc

breather membrane

12·5mm vapour control lined plasterboard

sill plate bolted to foundation

screed

dpm

deep strip foundation

Notes:

1. Cavity barriers prevent fire spread. The principal locations are between elements of construction.
2. Thermal bridging through solid framing may be reduced by using rigid EPS insulation and lighter 'I' section members of plywood or OSB.

Simply Supported Slabs ~ these are slabs which rest on a bearing and for design purposes are not considered to be fixed to the support and are therefore, in theory, free to lift. In practice however they are restrained from unacceptable lifting by their own self weight plus any loadings.

Concrete Slabs ~ concrete is a material which is strong in compression and weak in tension and if the member is overloaded its tensile resistance may be exceeded leading to structural failure.

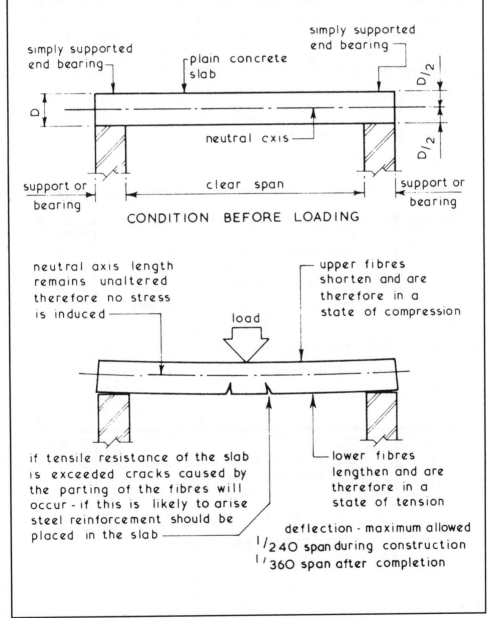

simply supported end bearing

simply supported end bearing

plain concrete slab

D

D/2

D/2

neutral axis

support or bearing

clear span

support or bearing

CONDITION BEFORE LOADING

neutral axis length remains unaltered therefore no stress is induced

upper fibres shorten and are therefore in a state of compression

load

if tensile resistance of the slab is exceeded cracks caused by the parting of the fibres will occur - if this is likely to arise steel reinforcement should be placed in the slab

lower fibres lengthen and are therefore in a state of tension

deflection - maximum allowed
$^{1}/240$ span during construction
$^{1}/360$ span after completion

Simply Supported RC Slabs

Reinforcement ~ generally in the form of steel bars which are used to provide the tensile strength which plain concrete lacks. The number, diameter, spacing, shape and type of bars to be used have to be designed; a basic guide is shown on pages 352 and 353. Reinforcement is placed as near to the outside fibres as practicable, a cover of concrete over the reinforcement is required to protect the steel bars from corrosion and to provide a degree of fire resistance. Slabs which are square in plan are considered to be spanning in two directions and therefore main reinforcing bars are used both ways whereas slabs which are rectangular in plan are considered to span across the shortest distance and main bars are used in this direction only with smaller diameter distribution bars placed at right angles forming a mat or grid.

Typical Details ~

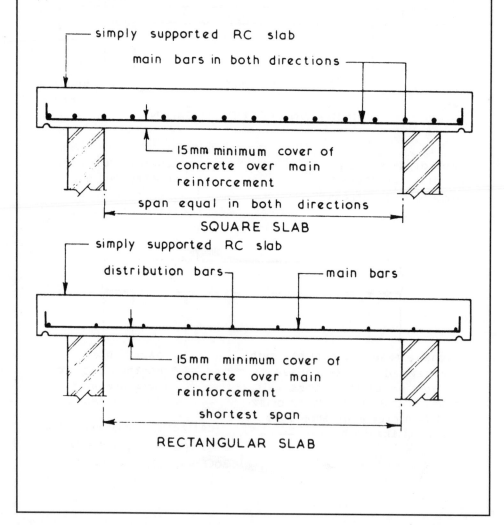

simply supported RC slab

main bars in both directions

15mm minimum cover of concrete over main reinforcement

span equal in both directions

SQUARE SLAB

simply supported RC slab

distribution bars

main bars

15mm minimum cover of concrete over main reinforcement

shortest span

RECTANGULAR SLAB

Construction ~ whatever method of construction is used the construction sequence will follow the same pattern-

1. Assemble and erect formwork.

2. Prepare and place reinforcement.

3. Pour and compact or vibrate concrete.

4. Strike and remove formwork in stages as curing proceeds.

Typical Example ~

edge formwork

concrete poured and compacted or vibrated around reinforcement

main reinforcement - cover maintained by plastic or similar spacers - see Detail 'A'

decking of suitable material such as plywood with all joints sealed or taped to prevent grout loss

distribution bars - position maintained by wire binding or clips - see Detail 'A'

surface finish as specified

adjustable steel or timber props at centres to suit spanning ability of joists

joists supporting decking spaced at centres to suit spanning ability of decking

tying wire or clip

telescopic steel floor centres with sheet steel decking giving clear spans between support walls

plastic spacer

distribution bar

main bars

DETAIL 'A'

ALTERNATIVE DECKING SUPPORT

Profiled galvanised steel decking is a permanent formwork system for construction of composite floor slabs. The steel sheet has surface indentations and deformities to effect a bond with the concrete topping. The concrete will still require reinforcing with steel rods or mesh, even though the metal section will contribute considerably to the tensile strength of the finished slab.

Typical detail -

concrete 30 N/mm^2

mesh or steel rod reinforcement

slab depth 120 to 250 mm*

60 or 80 mm in widths up to 1 m

galv. steel deck and permanent formwork

300-325 mm

* For slab depth and span potential, see BS 5950-4: Code of practice for design of composite slabs with profiled steel sheeting.

Where structural support framing is located at the ends of a section and at intermediate points, studs are through-deck welded to provide resistance to shear -

profiled galv. steel decking

anti-shear studs through-deck welded in pairs to structural support

95 or 120 mm

19 mm dia.

studs 20 mm min. from beam edge

UB support

There are considerable savings in concrete volume compared with standard in-situ reinforced concrete floor slabs. This reduction in concrete also reduces structural load on foundations.

Beams ~ these are horizontal load bearing members which are classified as either main beams which transmit floor and secondary beam loads to the columns or secondary beams which transmit floor loads to the main beams.

Concrete being a material which has little tensile strength needs to be reinforced to resist the induced tensile stresses which can be in the form of ordinary tension or diagonal tension (shear). The calculation of the area, diameter, type, position and number of reinforcing bars required is one of the functions of a structural engineer.

Typical RC Beam Details ~

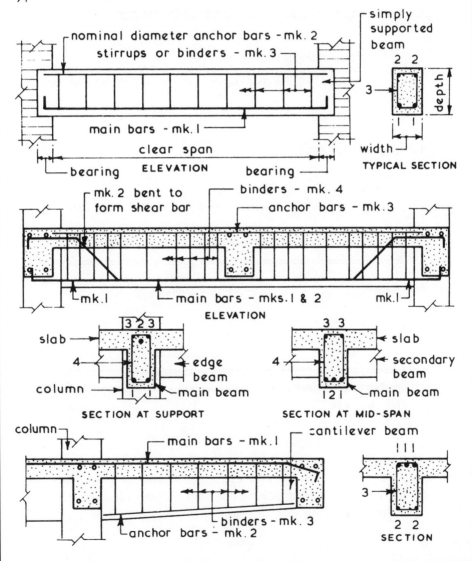

351

Simple Reinforced Concrete Beam and Slab Design (1)

Mild Steel Reinforcement — located in areas where tension occurs in a beam or slab. Concrete specification is normally 25 or 30 N/mm² in this situation.

Simple beam or slab

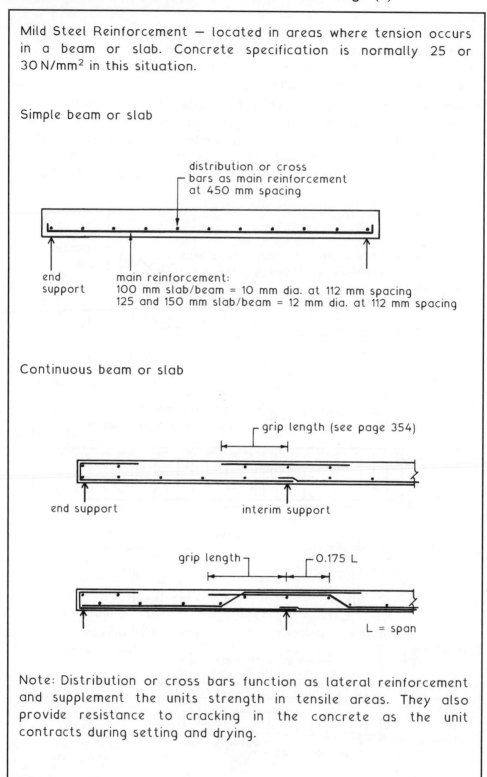

distribution or cross
bars as main reinforcement
at 450 mm spacing

end
support

main reinforcement:
100 mm slab/beam = 10 mm dia. at 112 mm spacing
125 and 150 mm slab/beam = 12 mm dia. at 112 mm spacing

Continuous beam or slab

grip length (see page 354)

end support   interim support

grip length   0.175 L

L = span

Note: Distribution or cross bars function as lateral reinforcement and supplement the units strength in tensile areas. They also provide resistance to cracking in the concrete as the unit contracts during setting and drying.

Guidance — simply supported slabs are capable of the following loading relative to their thickness:

Thickness (mm)	Self weight (kg/m²)	Imposed load* (kg/m²)	Total load (kg/m²)	Total load (kN/m²)	Span (m)
100	240	500	740	7·26	2·4
125	300	500	800	7·85	3·0
150	360	500	860	8·44	3·6

Note: As a *rule of thumb*, it is easy to remember that for general use (as above), thickness of slab equates to 1/24 span.

* Imposed loading varies with application from 1.5 kN/m² (153 kg/m²) for domestic buildings, to over 10 kN/m² (1020 kg/m²) for heavy industrial storage areas. 500 kg/m² is typical for office filing and storage space. See BS 6399—1: Code of practice for dead and imposed loads.

For larger spans — thickness can be increased proportionally to the span, eg. 6 m span will require a 250 mm thickness.

For greater loading — slab thickness is increased proportionally to the square root of the load, eg. for a total load of 1500 kg/m² over a 3 m span:

$$\sqrt{\frac{1500}{800}} \times 125 = 171\cdot2 \text{ i.e. 175 mm}$$

Continuous beams and slabs have several supports, therefore they are stronger than simple beams and slabs. The spans given in the above table may be increased by 20% for interior spans and 10% for end spans.

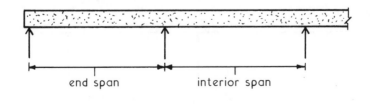

end span   interior span

Grip Length of Reinforcement

Bond Between Concrete and Steel — permissible stress for the bond between concrete and steel can be taken as one tenth of the compressive concrete stress, plus 0.175 N/mm² *. Given the stresses in concrete and steel, it is possible to calculate sufficient grip length.

e.g. concrete working stress of 5 N/mm²
      steel working stress of 125 N/mm²
      sectional area of reinf. bar = $3.142\ r^2$ or $0.7854\ d^2$
      tensile strength of bar = $125 \times 0.7854\ d^2$
      circumference of bar = $3.142\ d$
      area of bar in contact = $3.142 \times d \times L$

Key: r = radius of steel bar
     d = diameter of steel bar
     L = Length of bar in contact

* Conc. bond stress = $(0.10 \times 5\ \text{N/mm}^2) + 0.175 = 0.675\ \text{N/mm}^2$

Total bond stress = $3.142\ d \times L \times 0.675\ \text{N/mm}^2$

Thus, developing the tensile strength of the bar:

$$125 \times 0.7854\ d^2 = 3.142\ d \times L \times 0.675$$

$$98.175\ d = 2.120\ L$$

$$L = 46\ d$$

As a guide to good practice, a margin of 14 d should be added to L. Therefore the bar bond or grip length in this example is equivalent to 60 times the bar diameter.

Columns ~ these are the vertical load bearing members of the structural frame which transmits the beam loads down to the foundations. They are usually constructed in storey heights and therefore the reinforcement must be lapped to provide structural continuity.

Typical RC Column Details ~

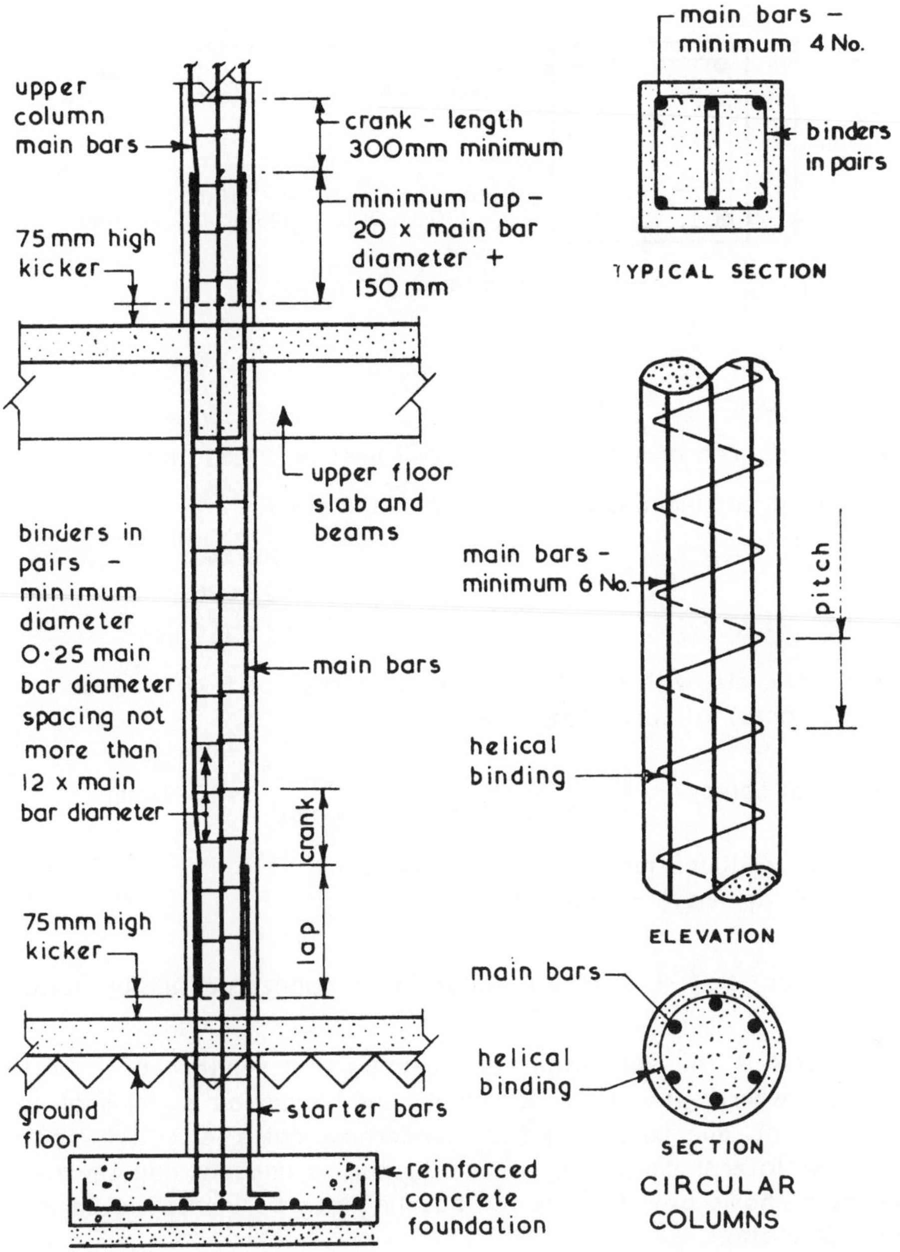

main bars –
minimum 4 No.

binders
in pairs

upper
column
main bars

crank - length
300mm minimum

minimum lap –
20 x main bar
diameter +
150 mm

TYPICAL SECTION

75 mm high
kicker

upper floor
slab and
beams

binders in
pairs –
minimum
diameter
0·25 main
bar diameter
spacing not
more than
12 x main
bar diameter

main bars –
minimum 6 No.

main bars

pitch

crank

helical
binding

lap

75 mm high
kicker

ELEVATION

main bars

helical
binding

ground
floor

starter bars

SECTION

reinforced
concrete
foundation

**CIRCULAR
COLUMNS**

Steel Reinforced Concrete — a modular ratio represents the amount of load that a square unit of steel can safely transmit relative to that of concrete. A figure of 18 is normal, with some variation depending on materials specification and quality.

e.g.

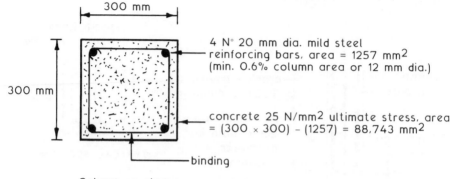

300 mm

300 mm

4 N° 20 mm dia. mild steel reinforcing bars, area = 1257 mm^2 (min. 0.6% column area or 12 mm dia.)

concrete 25 N/mm^2 ultimate stress, area = (300 × 300) – (1257) = 88,743 mm^2

binding

Column on plan

Area of concrete = 88,743 mm^2

Equivalent area of steel = 18 × 1,257 mm^2 = 22,626 mm^2

Equivalent combined area of concrete and steel:

$$\begin{array}{r} 88,743 \\ +22,626 \\ \hline 111,369 \text{ mm}^2 \end{array}$$

Using concrete with a safe or working stress of 5 N/mm^2, derived from a factor of safety of 5, i.e.

Factor of safety $= \dfrac{\text{Ultimate stress}}{\text{Working stress}} = \dfrac{25\,\text{N/mm}^2}{5\,\text{N/mm}^2} = 5\,\text{N/mm}^2$

5 N/mm^2 × 111,369 mm^2 = 556,845 Newtons
kg × 9·81 (gravity) = Newtons

Therefore: $\dfrac{556,845}{9\cdot81}$ = 56,763 kg or 56·76 tonnes permissible load

Note: This is the safe load calculation for a reinforced concrete column where the load is axial and bending is minimal or nonexistant, due to a very low slenderness ratio (effective length to least lateral dimension). In reality this is unusual and the next example shows how factors for buckling can be incorporated into the calculation.

Buckling or Bending Effect — the previous example assumed total rigidity and made no allowance for column length and attachments such as floor beams.

The working stress unit for concrete may be taken as 0.8 times the maximum working stress of concrete where the effective length of column (see page 377) is less than 15 times its least lateral dimension. Where this exceeds 15, a further factor for buckling can be obtained from the following:

Effective length ÷ Least lateral dimension	Buckling factor
15	1·0
18	0·9
21	0·8
24	0·7
27	0·6
30	0·5
33	0·4
36	0·3
39	0·2
42	0·1
45	0

Using the example from the previous page, with a column effective length of 9 metres and a modular ratio of 18:

Effective length÷Least lateral dimension = 9000÷300 = 30

From above table the buckling factor = 0·5

Concrete working stress = 5N/mm^2

Equivalent combined area of concrete and steel = 111,369 mm^2

Therefore: 5 × 0·8 × 0·5 × 111,369 = 222,738 Newtons

$\dfrac{222,738}{9·81}$ = 22,705 kg or 22·7 tonnes permissible load

Bar Coding ~ a convenient method for specifying and coordinating the prefabrication of steel reinforcement in the assembly area. It is also useful on site, for checking deliveries and locating materials relative to project requirements. BS 4466 provides guidance for a simplified coding system, such that bars can be manufactured and labelled without ambiguity for easy recognition and application on site.

A typical example is the beam shown on page 351, where the lower longitudinal reinforcement (mk.1) could be coded:~

2T20-1-200B

    2 = number of bars
    T = deformed high yield steel (460 N/mm², 8—40 mm dia.)
   20 = diameter of bar (mm)
    1 = bar mark or ref. no.
  200 = spacing (mm)
    B = located in bottom of member

Other common notation:-

    R = plain round mild steel (250 N/mm², 8—16 mm dia.)
    S = stainless steel
    W = wire reinforcement (4—12 mm dia.)
    T (at the end) = located in top of member
  abr = alternate bars reversed (useful for offsets)

Thus, bar mk.2 = 2R10-2-200T
     and mk.3 = 10R8-3-270

All but the most obscure reinforcement shapes are illustrated in the British Standard. For the beam referred to on page 351, the standard listing is :-

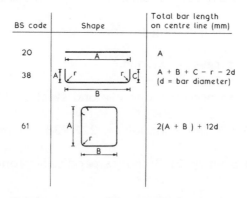

BS code	Shape	Total bar length on centre line (mm)
20	A	A
38	A, B, C (shape)	A + B + C − r − 2d (d = bar diameter)
61	A, B (shape)	2(A + B) + 12d

Ref. BS 4466: Specification for scheduling, dimensioning, bending and cutting of steel reinforcement for concrete.

Bar Schedule ~ this can be derived from the coding explained on the previous page. Assuming 10 No. beams are required:-

Site ref ....   Schedule ref ....

Prepared by ....   Date ....

Member	Bar mark	Type and size	No. of members	No. of bars in each	Total No.	Bar length (mm)	Shape code	A	B	C	D	E/r
Beam	1	T20	10	2	20	3080	38	200	2700	200		
	2	R10	10	2	20	2700	20	2700				
	3	R8	10	10	100	1336	61	400	220			

Note: r = 2 × d for mild steel
　　　　3 × d for high yield steel

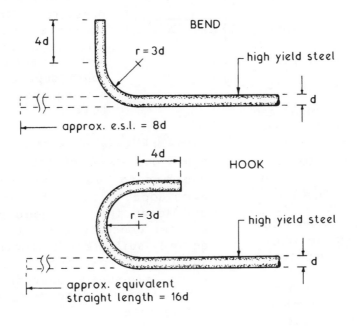

Basic Formwork ~ concrete when first mixed is a fluid and therefore to form any concrete member the wet concrete must be placed in a suitable mould to retain its shape, size and position as it sets. It is possible with some forms of concrete foundations to use the sides of the excavation as the mould but in most cases when casting concrete members a mould will have to be constructed on site. These moulds are usually called formwork. It is important to appreciate that the actual formwork is the reverse shape of the concrete member which is to be cast.

Basic Principles ~

formwork · sides can be designed to offer all the necessary resistance to the imposed pressures as a single member or alternatively they can be designed to use a thinner material which is adequately strutted — for economic reasons the latter method is usually employed

grout tight joints ———

formwork soffits can be designed to offer all the necessary resistance to the imposed loads as a single member or alternatively they can be designed to a thinner material which is adequately propped — for economic reasons the latter method is usually employed

wet concrete - density is greater than that of the resultant set and dry concrete

formwork sides — limits width and shape of wet concrete and has to resist the hydrostatic pressure of the wet concrete which will diminish to zero within a matter of hours depending on · setting and curing rate

formwork base or soffit – limits depth and shape of wet concrete and has to resist the initial dead load of the wet concrete and later the dead load of the dry set concrete until it has gained sufficient strength to support its own dead weight which is usually several days after casting depending on curing rate.

Typical Simple Beam Formwork Details ~

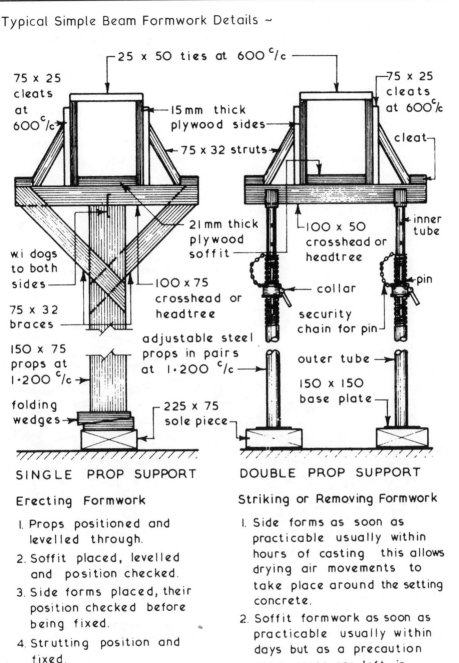

75 x 25 cleats at 600 ^c/c

25 x 50 ties at 600 ^c/c

15 mm thick plywood sides

75 x 32 struts

75 x 25 cleats at 600 ^c/c

cleat

21 mm thick plywood soffit

w.i dogs to both sides

75 x 32 braces

100 x 75 crosshead or headtree

100 x 50 crosshead or headtree

inner tube

150 x 75 props at 1·200 ^c/c

adjustable steel props in pairs at 1·200 ^c/c

collar

security chain for pin

pin

folding wedges

225 x 75 sole piece

outer tube

150 x 150 base plate

SINGLE PROP SUPPORT

DOUBLE PROP SUPPORT

Erecting Formwork

1. Props positioned and levelled through.
2. Soffit placed, levelled and position checked.
3. Side forms placed, their position checked before being fixed.
4. Strutting position and fixed.
5. Final check before casting.

Suitable Formwork Materials~ timber, steel and special plastics.

Striking or Removing Formwork

1. Side forms as soon as practicable usually within hours of casting this allows drying air movements to take place around the setting concrete.
2. Soffit formwork as soon as practicable usually within days but as a precaution some props are left in position until concrete member is self supporting.

Beam Formwork ~ this is basically a three sided box supported and propped in the correct position and to the desired level. The beam formwork sides have to retain the wet concrete in the required shape and be able to withstand the initial hydrostatic pressure of the wet concrete whereas the formwork soffit apart from retaining the concrete has to support the initial load of the wet concrete and finally the set concrete until it has gained sufficient strength to be self supporting. It is essential that all joints in the formwork are constructed to prevent the escape of grout which could result in honeycombing and/or feather edging in the cast beam. The removal time for the formwork will vary with air temperature, humidity and consequent curing rate.

Typical Details ~

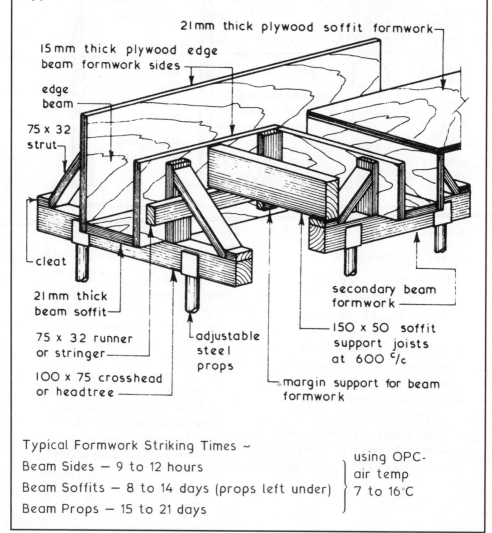

21mm thick plywood soffit formwork

15mm thick plywood edge beam formwork sides

edge beam

75 x 32 strut

cleat

21mm thick beam soffit

75 x 32 runner or stringer

100 x 75 crosshead or headtree

adjustable steel props

secondary beam formwork

150 x 50 soffit support joists at 600 c/c

margin support for beam formwork

Typical Formwork Striking Times ~

Beam Sides — 9 to 12 hours

Beam Soffits — 8 to 14 days (props left under)

Beam Props — 15 to 21 days

using OPC-
air temp
7 to 16°C

Column Formwork ~ this consists of a vertical mould of the desired shape and size which has to retain the wet concrete and resist the initial hydrostatic pressure caused by the wet concrete. To keep the thickness of the formwork material to a minimum horizontal clamps or yokes are used at equal centres for batch filling and at varying centres for complete filling in one pour. The head of the column formwork can be used to support the incoming beam formwork which gives good top lateral restraint but results in complex formwork. Alternatively the column can be cast to the underside of the beams and at a later stage a collar of formwork can be clamped around the cast column to complete casting and support the incoming beam formwork. Column forms are located at the bottom around a 75 to 100 mm high concrete plinth or kicker which has the dual function of location and preventing grout loss from the bottom of the column formwork.

Typical Details ~

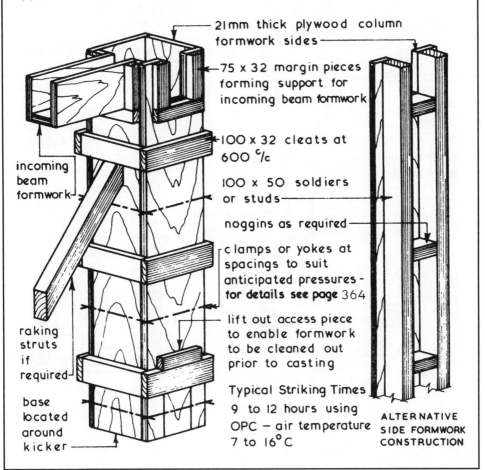

21mm thick plywood column formwork sides

75 x 32 margin pieces forming support for incoming beam formwork

100 x 32 cleats at 600 $^c/c$

100 x 50 soldiers or studs

noggins as required

clamps or yokes at spacings to suit anticipated pressures - for details see page 364

lift out access piece to enable formwork to be cleaned out prior to casting

incoming beam formwork

raking struts if required

base located around kicker

Typical Striking Times 9 to 12 hours using OPC – air temperature 7 to 16°C

ALTERNATIVE SIDE FORMWORK CONSTRUCTION

Column Yokes ~ these are obtainable as a metal yoke or clamp or they can be purpose made from timber.

Typical Examples ~

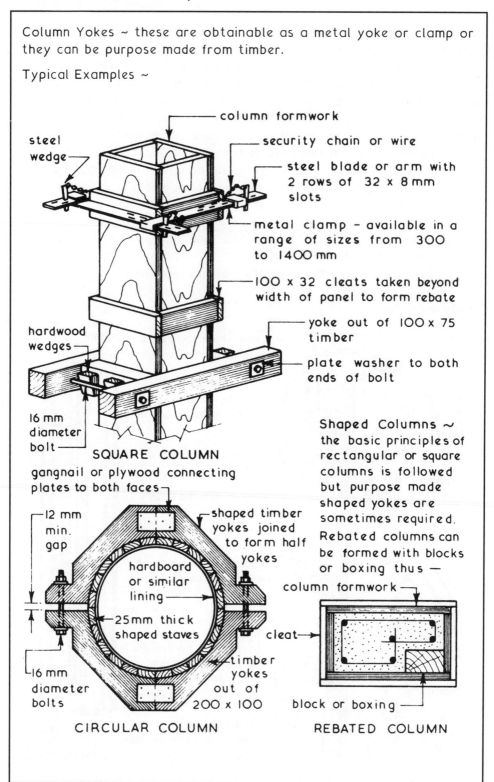

column formwork

security chain or wire

steel blade or arm with 2 rows of 32 x 8mm slots

metal clamp - available in a range of sizes from 300 to 1400 mm

100 x 32 cleats taken beyond width of panel to form rebate

yoke out of 100 x 75 timber

plate washer to both ends of bolt

steel wedge

hardwood wedges

16 mm diameter bolt

SQUARE COLUMN

gangnail or plywood connecting plates to both faces

12 mm min. gap

shaped timber yokes joined to form half yokes

hardboard or similar lining

25mm thick shaped staves

16 mm diameter bolts

timber yokes out of 200 x 100

CIRCULAR COLUMN

Shaped Columns ~ the basic principles of rectangular or square columns is followed but purpose made shaped yokes are sometimes required. Rebated columns can be formed with blocks or boxing thus —

column formwork

cleat

block or boxing

REBATED COLUMN

Precast Concrete Frames ~ these frames are suitable for single storey and low rise applications, the former usually in the form of portal frames which are normally studied separately. Precast concrete frames provide the skeleton for the building and can be clad externally and finished internally by all the traditional methods. The frames are usually produced as part of a manufacturer's standard range of designs and are therefore seldom purpose made due mainly to the high cost of the moulds.

Advantages :-

1. Frames are produced under factory controlled conditions resulting in a uniform product of both quality and accuracy.

2. Repetitive casting lowers the cost of individual members.

3. Off site production releases site space for other activities.

4. Frames can be assembled in cold weather and generally by semi-skilled labour.

Disadvantages :-

1. Although a wide choice of frames is available from various manufacturer's these systems lack the design flexibility of cast insitu purpose made frames.

2. Site planning can be limited by manufacturer's delivery and unloading programmes and requirements.

3. Lifting plant of a type and size not normally required by traditional construction methods may be needed.

Typical Site Activities ~

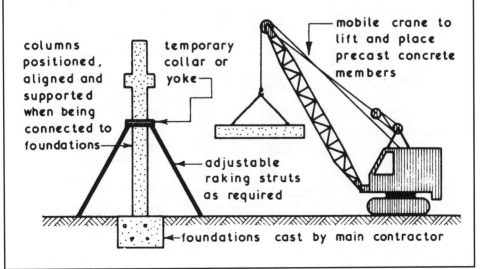

columns positioned, aligned and supported when being connected to foundations

temporary collar or yoke

mobile crane to lift and place precast concrete members

adjustable raking struts as required

foundations cast by main contractor

Precast Concrete Frames

Foundation Connections ~ the preferred method of connection is to set the column into a pocket cast into a reinforced concrete pad foundation and is suitable for light to medium loadings. Where heavy column loadings are encountered it may be necessary to use a steel base plate secured to the reinforced concrete pad foundation with holding down bolts.

Typical Details ~

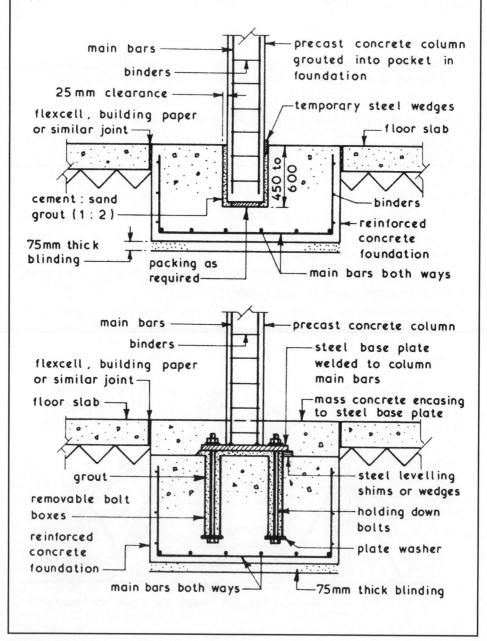

Column to Column Connection ~ precast columns are usually cast in one length and can be up to four storeys in height. They are either reinforced with bar reinforcement or they are prestressed according to the loading conditions. If column to column are required they are usually made at floor levels above the beam to column connections and can range from a simple dowel connection to a complex connection involving insitu concrete.

Typical Details ~

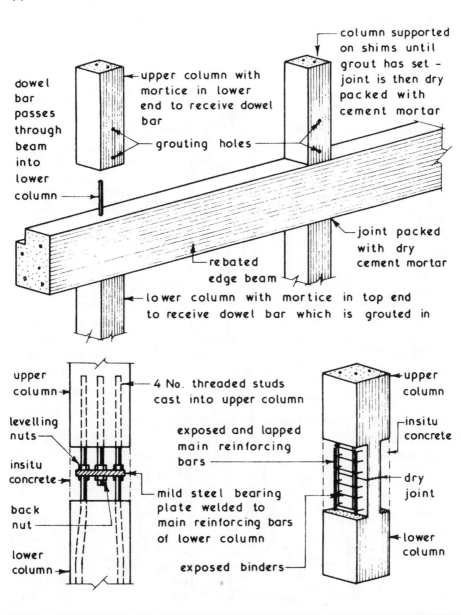

dowel bar passes through beam into lower column

upper column with mortice in lower end to receive dowel bar

grouting holes

column supported on shims until grout has set - joint is then dry packed with cement mortar

joint packed with dry cement mortar

rebated edge beam

lower column with mortice in top end to receive dowel bar which is grouted in

upper column

levelling nuts

insitu concrete

back nut

lower column

4 No. threaded studs cast into upper column

exposed and lapped main reinforcing bars

mild steel bearing plate welded to main reinforcing bars of lower column

exposed binders

upper column

insitu concrete

dry joint

lower column

Beam to Column Connections ~ as with the column to column connections (see page 367) the main objective is to provide structural continuity at the junction. This is usually achieved by one of two basic methods:-

1. Projecting bearing haunches cast onto the columns with a projecting dowel or stud bolt to provide both location and fixing.
2. Steel to steel fixings which are usually in the form of a corbel or bracket projecting from the column providing a bolted connection to a steel plate cast into the end of the beam.

Typical Details ~

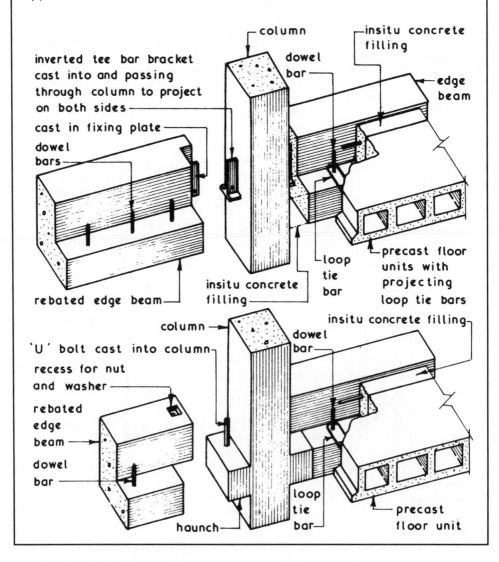

Principles ~ the well known properties of concrete are that it has high compressive strength and low tensile strength. The basic concept of reinforced concrete is to include a designed amount of steel bars in a predetermined pattern to give the concrete a reasonable amount of tensile strength. In prestressed concrete a precompression is induced into the member to make full use of its own inherent compressive strength when loaded. The design aim is to achieve a balance of tensile and compressive forces so that the end result is a concrete member which is resisting only stresses which are compressive. In practice a small amount of tension may be present but providing this does not exceed the tensile strength of the concrete being used tensile failure will not occur.

Comparison of Reinforced and Prestressed Concrete ~

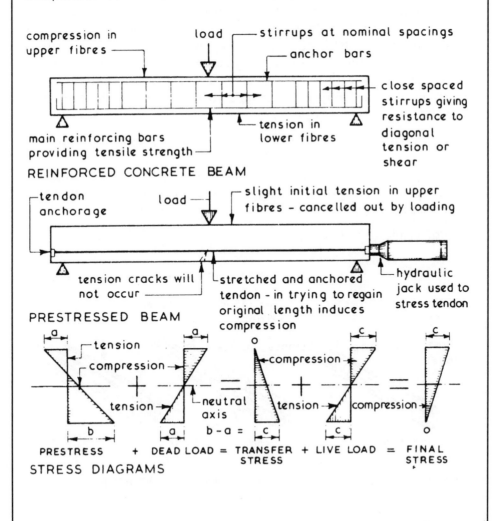

compression in upper fibres

load

stirrups at nominal spacings

anchor bars

close spaced stirrups giving resistance to diagonal tension or shear

main reinforcing bars providing tensile strength

tension in lower fibres

REINFORCED CONCRETE BEAM

tendon anchorage

load

slight initial tension in upper fibres - cancelled out by loading

tension cracks will not occur

stretched and anchored tendon - in trying to regain original length induces compression

hydraulic jack used to stress tendon

PRESTRESSED BEAM

tension

compression

tension

neutral axis

compression

tension

compression

PRESTRESS + DEAD LOAD = TRANSFER STRESS + LIVE LOAD = FINAL STRESS

STRESS DIAGRAMS

Materials ~ concrete will shrink whilst curing and it can also suffer sectional losses due to creep when subjected to pressure. The amount of shrinkage and creep likely to occur can be controlled by designing the strength and workability of the concrete, high strength and low workability giving the greatest reduction in both shrinkage and creep. Mild steel will suffer from relaxation losses which is where the stresses in steel under load decrease to a minimum value after a period of time and this can be overcome by increasing the initial stress in the steel. If mild steel is used for prestressing the summation of shrinkage, creep and relaxation losses will cancel out any induced compression, therefore special alloy steels must be used to form tendons for prestressed work.

Tendons — these can be of small diameter wires (2 to 7 mm) in a plain round, crimped or indented format, these wires may be individual or grouped to form cables. Another form of tendon is strand which consists of a straight core wire around which is helically wound further wires to give formats such as 7 wire (6 over 1) and 19 wire (9 over 9 over 1) and like wire tendons strand can be used individually or in groups to form cables. The two main advantages of strand are:-

1. A large prestressing force can be provided over a restricted area.
2. Strand can be supplied in long flexible lengths capable of being stored on drums thus saving site storage and site fabrication space.

Typical Tendon Formats ~

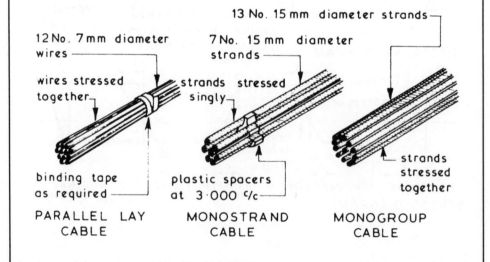

12 No. 7 mm diameter wires

wires stressed together

binding tape as required

PARALLEL LAY CABLE

7 No. 15 mm diameter strands

strands stressed singly

plastic spacers at 3·000 c/c

MONOSTRAND CABLE

13 No. 15 mm diameter strands

strands stressed together

MONOGROUP CABLE

Pre-tensioning ~ this method is used mainly in the factory production of precast concrete components such as lintels, floor units and small beams. Many of these units are formed by the long line method where precision steel moulds up to 120·000 long are used with spacer or dividing plates to form the various lengths required. In pre-tensioning the wires are stressed within the mould before the concrete is placed around them. Steam curing is often used to accelerate this process to achieve a 24 hour characteristic strength of 28 N/mm^2 with a typical 28 day cube strength of 40N/mm^2. Stressing of the wires is carried out by using hydraulic jacks operating from one or both ends of the mould to achieve an initial 10% overstress to counteract expected looses. After curing the wires are released or cut and the bond between the stressed wires and the concrete prevents the tendons from regaining their original length thus maintaining the precompression or prestress.

At the extreme ends of the members the bond between the stressed wires and concrete is not fully developed due to low frictional resistance. This results in a small contraction and swelling at the ends of the wire forming in effect a cone shape anchorage. The distance over which this contraction occurs is called the transfer length and is equal to 80 to 120 times the wire diameter. To achieve a greater total surface contact area it is common practice to use a larger number of small diameter wires rather than a smaller number of large diameter wires giving the same total cross sectional area.

Typical Pre-tensioning Arrangement ~

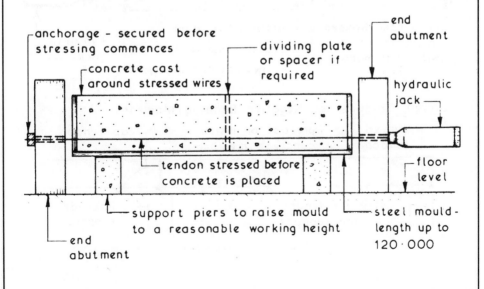

371

Prestressed Concrete

Post-tensioning ~ this method is usually employed where stressing is to be carried out on site after casting an insitu component or where a series of precast concrete units are to be joined together to form the required member. It can also be used where curved tendons are to be used to overcome negative bending moments. In post-tensioning the concrete is cast around ducts or sheathing in which the tendons are to be housed. Stressing is carried out after the concrete has cured by means of hydraulic jacks operating from one or both ends of the member. The anchorages (see page 373) which form part of the complete component prevent the stressed tendon from regaining its original length thus maintaining the precompression or prestress. After stressing the annular space in the tendon ducts should be filled with grout to prevent corrosion of the tendons due to any entrapped moisture and to assist in stress distribution. Due to the high local stresses at the anchorage positions it is usual for a reinforcing spiral to be included in the design.

Typical Post-tensioning Arrangement ~

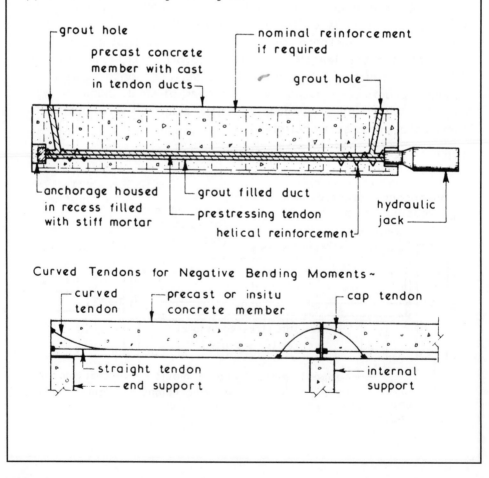

Curved Tendons for Negative Bending Moments ~

Anchorages ~ the formats for anchorages used in conjunction with post-tensioned prestressed concrete works depends mainly on whether the tendons are to be stressed individually or as a group, but most systems use a form of split cone wedges or jaws acting against a form of bearing or pressure plate.

Typical Anchorage Details ~

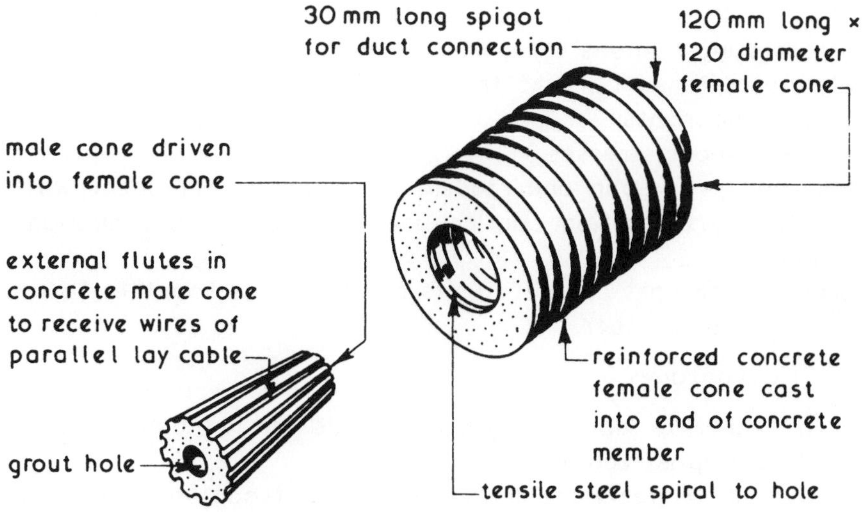

30 mm long spigot for duct connection

120 mm long × 120 diameter female cone

male cone driven into female cone

external flutes in concrete male cone to receive wires of parallel lay cable

grout hole

reinforced concrete female cone cast into end of concrete member

tensile steel spiral to hole

FREYSSINET ANCHORAGE

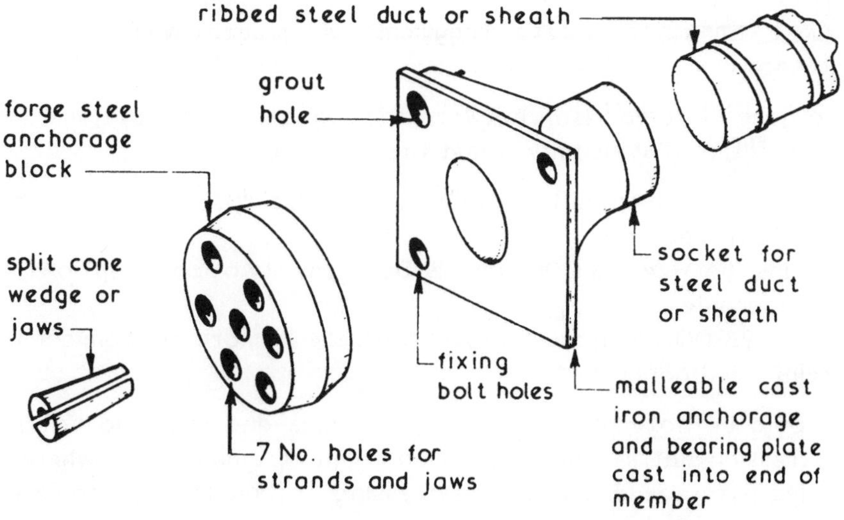

ribbed steel duct or sheath

grout hole

forge steel anchorage block

split cone wedge or jaws

socket for steel duct or sheath

fixing bolt holes

7 No. holes for strands and jaws

malleable cast iron anchorage and bearing plate cast into end of member

TYPICAL MONOSTRAND ANCHORAGE

Comparison with Reinforced Concrete ~ when comparing prestressed concrete with conventional reinforced concrete the main advantages and disadvantages can be enumerated but in the final analysis each structure and/or component must be decided on its own merit.

Main advantages :-

1. Makes full use of the inherent compressive strength of concrete.
2. Makes full use of the special alloy steels used to form the prestressing tendons.
3. Eliminates tension cracks thus reducing the risk of corrosion of steel components.
4. Reduces shear stresses.
5. For any given span and loading condition a component with a smaller cross section can be used thus giving a reduction in weight.
6. Individual precast concrete units can be joined together to form a composite member.

Main Disadvantages :-

1. High degree of control over materials, design and quality of workmanship is required.
2. Special alloy steels are dearer than most traditional steels used in reinforced concrete.
3. Extra cost of special equipment required to carry out the prestressing activities.
4. Cost of extra safety requirements needed whilst stressing tendons.

As a general comparison between the two structural options under consideration it is usually found that :-

1. Up to 6·000 span traditional reinforced concrete is the most economic method.
2. Spans between 6·000 and 9·000 the two cost options are comparable.
3. Over 9·000 span prestressed concrete is more economical than reinforced concrete.

It should be noted that generally columns and walls do not need prestressing but in tall columns and high retaining walls where the bending stresses are high, prestressing techniques can sometimes be economically applied.

Ground Anchors ~ these are a particular application of post-tensioning prestressing techniques and can be used to form ground tie backs to cofferdams, retaining walls and basement walls. They can also be used as vertical tie downs to basement and similar slabs to prevent flotation during and after construction. Ground anchors can be of a solid bar format (rock anchors) or of a wire or cable format for granular and cohesive soils. A lined or unlined bore hole must be drilled into the soil to the design depth and at the required angle to house the ground anchor. In clay soils the bore hole needs to be underreamed over the anchorage length to provide adequate bond. The tail end of the anchor is pressure grouted to form a bond with the surrounding soil, the remaining length being unbonded so that it can be stressed and anchored at head thus inducing the prestress. The void around the unbonded or elastic length is gravity grouted after completion of the stressing operation.

Typical Ground Anchor Details ~

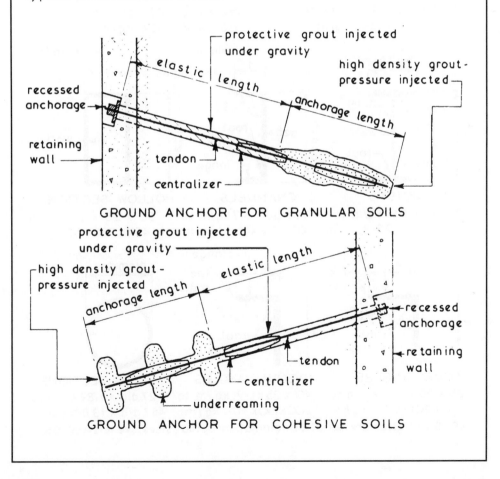

GROUND ANCHOR FOR GRANULAR SOILS

GROUND ANCHOR FOR COHESIVE SOILS

Structural Steelwork ~ the standard sections available for use in structural steelwork are given in BS 4 and BS EN's 10056 and 10210. These standards give a wide range of sizes and weights to enable the designer to formulate an economic design.

Typical Standard Steelwork Sections ~

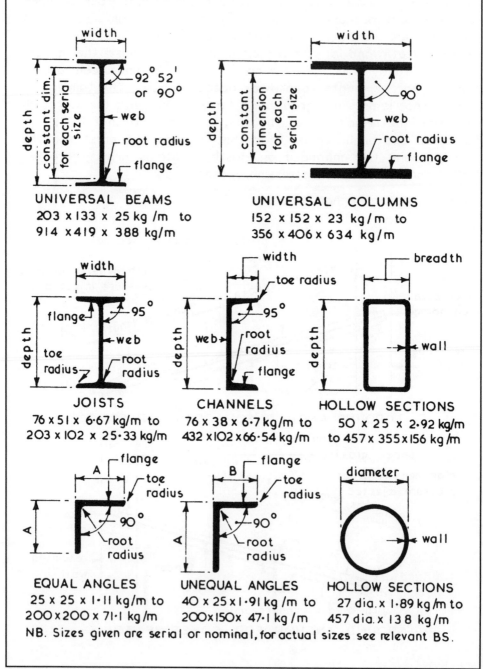

UNIVERSAL BEAMS
203 x 133 x 25 kg /m to
914 x 419 x 388 kg/m

UNIVERSAL COLUMNS
152 x 152 x 23 kg/m to
356 x 406 x 634 kg/m

JOISTS
76 x 51 x 6·67 kg/m to
203 x 102 x 25·33 kg/m

CHANNELS
76 x 38 x 6·7 kg/m to
432 x 102 x 66·54 kg/m

HOLLOW SECTIONS
50 x 25 x 2·92 kg/m
to 457 x 355 x 156 kg/m

EQUAL ANGLES
25 x 25 x 1·11 kg/m to
200 x 200 x 71·1 kg/m

UNEQUAL ANGLES
40 x 25 x 1·91 kg/m to
200 x 150 x 47·1 kg /m

HOLLOW SECTIONS
27 dia. x 1·89 kg/m to
457 dia. x 13 8 kg/m

NB. Sizes given are serial or nominal, for actual sizes see relevant BS.

Compound Sections — these are produced by welding together standard sections. Various profiles are possible, which can be designed specifically for extreme situations such as very high loads and long spans, where standard sections alone would be insufficient. Some popular combinations of standard sections include:

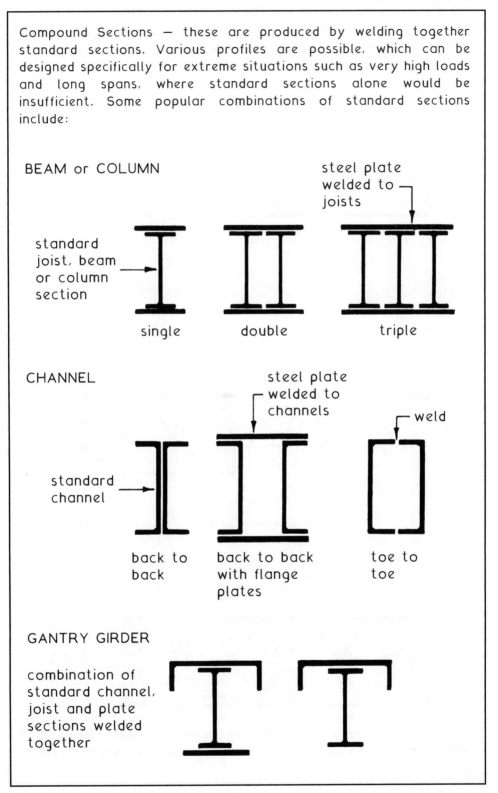

BEAM or COLUMN

standard joist, beam or column section

steel plate welded to joists

single   double   triple

CHANNEL

standard channel

steel plate welded to channels

weld

back to back

back to back with flange plates

toe to toe

GANTRY GIRDER

combination of standard channel, joist and plate sections welded together

Open Web Beams — these are particularly suited to long spans with light to moderate loading. The relative increase in depth will help resist deflection and voids in the web will reduce structural dead load.

Perforated Beam — a standard beam section with circular voids cut about the neutral axis.

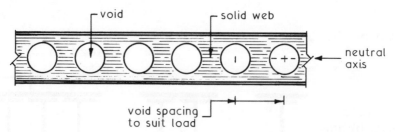

Castellated Beam — a standard beam section web is profile cut into two by oxy-acetylene torch. The projections on each section are welded together to create a new beam 50% deeper than the original.

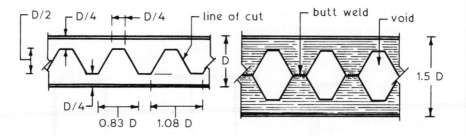

Litzka Beam — a standard beam cut as the castellated beam, but with overall depth increased further by using spacer plates welded to the projections. Minimal increase in weight.

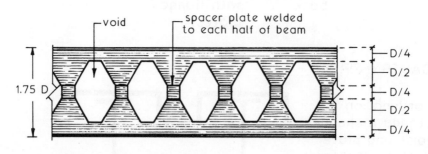

Note: Voids at the end of open web beams should be filled with a welded steel plate, as this is the area of maximum shear stress in a beam.

Lattices — these are an alternative type of open web beam, using standard steel sections to fabricate high depth to weight ratio units capable of spans up to about 15 m. The range of possible components is extensive and some examples are shown below:

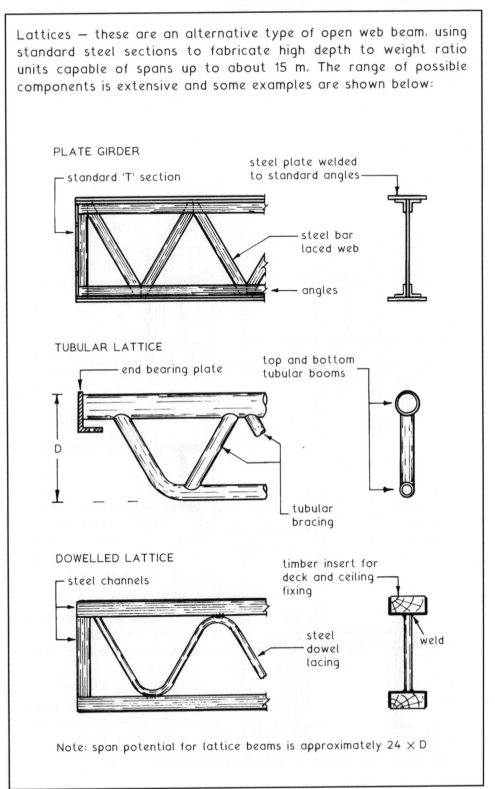

PLATE GIRDER

standard 'T' section

steel plate welded to standard angles

steel bar laced web

angles

TUBULAR LATTICE

end bearing plate

top and bottom tubular booms

D

tubular bracing

DOWELLED LATTICE

steel channels

timber insert for deck and ceiling fixing

steel dowel lacing

weld

Note: span potential for lattice beams is approximately 24 × D

Structural Steelwork Connections ~ these are either shop or site connections according to where the fabrication takes place. Most site connections are bolted whereas shop connections are very often carried out by welding. The design of structural steelwork members and their connections is the province of the structural engineer who selects the type and number of bolts or the size and length of weld to be used according to the connection strength to be achieved.

Typical Connection Examples ~

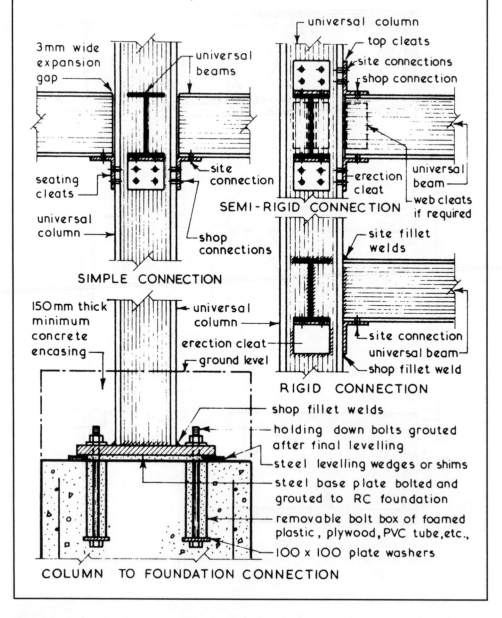

SIMPLE CONNECTION

SEMI-RIGID CONNECTION

RIGID CONNECTION

COLUMN TO FOUNDATION CONNECTION

Typical Connection Examples ~

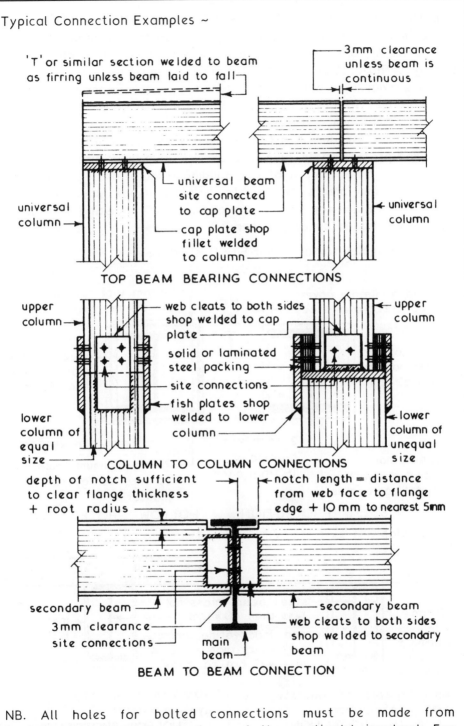

'T' or similar section welded to beam as firring unless beam laid to fall

3mm clearance unless beam is continuous

universal column

universal beam site connected to cap plate

cap plate shop fillet welded to column

universal column

TOP BEAM BEARING CONNECTIONS

upper column

web cleats to both sides shop welded to cap plate

solid or laminated steel packing

site connections

fish plates shop welded to lower column

upper column

lower column of equal size

lower column of unequal size

COLUMN TO COLUMN CONNECTIONS

depth of notch sufficient to clear flange thickness + root radius

notch length = distance from web face to flange edge + 10 mm to nearest 5mm

secondary beam

3mm clearance

site connections

main beam

secondary beam

web cleats to both sides shop welded to secondary beam

BEAM TO BEAM CONNECTION

NB. All holes for bolted connections must be made from backmarking the outer surface of the section(s) involved. For actual positions see structural steelwork manuals.

Fire Resistance of Structural Steelwork ~ although steel is a non-combustible material with negligible surface spread of flame properties it does not behave very well under fire conditions. During the initial stages of a fire the steel will actually gain in strength but this reduces to normal at a steel temperature range of 250 to 400°C and continues to decrease until the steel temperature reaches 550°C when it has lost most of its strength. Since the temperature rise during a fire is rapid, most structural steelwork will need protection to give it a specific degree of fire resistance in terms of time. Part B of the Building Regulations sets out the minimum requirements related to building usage and size, BRE report 'Guidelines for the construction of fire resisting structural elements' gives acceptable methods.

Typical Examples for a 2 Hour Fire Resistance ~

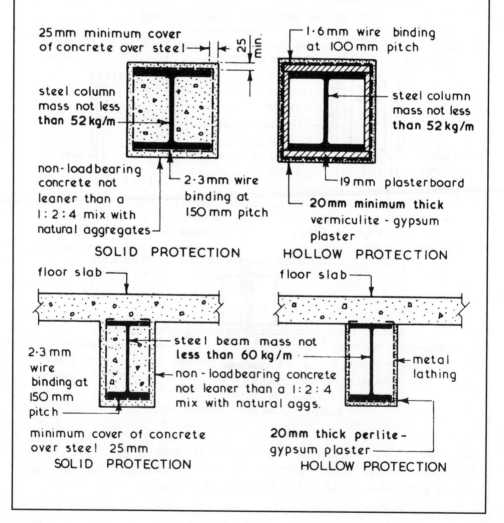

25 mm minimum cover of concrete over steel

25 min.

steel column mass not less than 52 kg/m

non-loadbearing concrete not leaner than a 1:2:4 mix with natural aggregates

2·3 mm wire binding at 150 mm pitch

SOLID PROTECTION

1·6 mm wire binding at 100 mm pitch

steel column mass not less than 52 kg/m

19 mm plasterboard

20 mm minimum thick vermiculite - gypsum plaster

HOLLOW PROTECTION

floor slab

2·3 mm wire binding at 150 mm pitch

steel beam mass not less than 60 kg/m

non-loadbearing concrete not leaner than a 1:2:4 mix with natural aggs.

minimum cover of concrete over steel 25 mm

SOLID PROTECTION

floor slab

metal lathing

20 mm thick perlite-gypsum plaster

HOLLOW PROTECTION

Section Factors — these are criteria found in tabulated fire protection data such as the Loss Prevention Council's 'Design Guide for the Fire Protection of Buildings'. These factors can be used to establish the minimum thickness or cover of protective material for structural sections. This interpretation is usually preferred by buildings insurance companies, as it often provides a standard in excess of the building regulations. Section factors are categorised: < 90, $90 - 140$ and > 140. They can be calculated by the following formula:

Section Factor $= Hp/A$ (m^{-1})
Hp = Perimeter of section exposed to fire (m)
A = Cross sectional area of steel (m^2) [see BS4 or Structural Steel Tables]
Examples:

UB serial size, $305 \times 127 \times 42$ kg/m

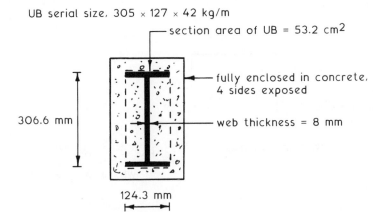

section area of UB = 53.2 cm^2

fully enclosed in concrete, 4 sides exposed

306.6 mm

web thickness = 8 mm

124.3 mm

$Hp = (2 \times 124.3) + (2 \times 306.6) + 2(124.3 - 8) = 1.0944$ m
$A = 53.2$ cm^2 or 0.00532 m^2
Section Factor, $Hp/A = 1.0944/0.00532 = 205$

As beam above, but 3 sides only exposed

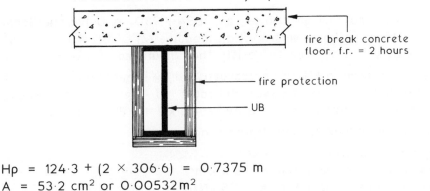

fire break concrete floor, f.r. = 2 hours

fire protection

UB

$Hp = 124.3 + (2 \times 306.6) = 0.7375$ m
$A = 53.2$ cm^2 or 0.00532 m^2
Section Factor, $Hp/A = 0.7375/0.00532 = 138$

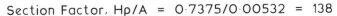

References: BS4: Structural steel sections

BS449: Specification for the use of structural steel in building

BS5950: Structural use of steelwork in building

Simple beam design (Bending)

Formula:

$$Z = \frac{M}{f}$$

where: Z = section or elastic modulus (BS4)

M = moment of resistance > or = max. bending moment

f = fibre stress of the material, (normally 165 N/mm² for rolled steel sections)

In simple situations the bending moment can be calculated:-

(a) Point loads

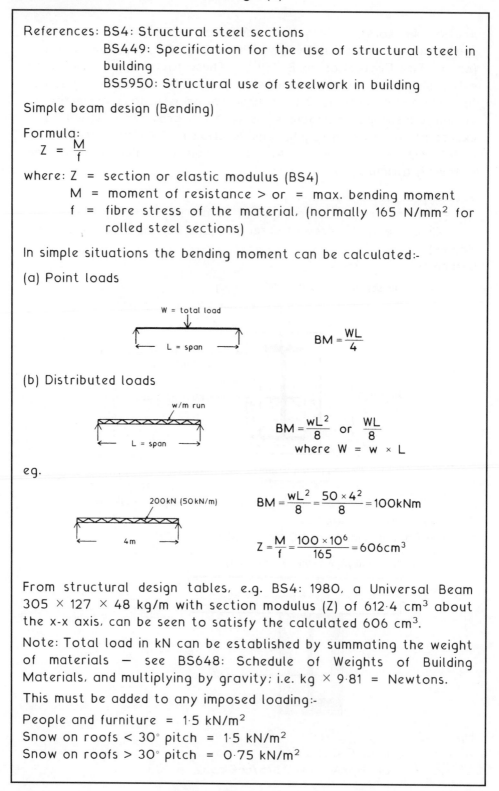

W = total load

L = span

$$BM = \frac{WL}{4}$$

(b) Distributed loads

w/m run

L = span

$$BM = \frac{wL^2}{8} \text{ or } \frac{WL}{8}$$

where W = w × L

eg.

200kN (50kN/m)

4m

$$BM = \frac{wL^2}{8} = \frac{50 \times 4^2}{8} = 100\,kNm$$

$$Z = \frac{M}{f} = \frac{100 \times 10^6}{165} = 606\,cm^3$$

From structural design tables, e.g. BS4: 1980, a Universal Beam 305 × 127 × 48 kg/m with section modulus (Z) of 612·4 cm³ about the x-x axis, can be seen to satisfy the calculated 606 cm³.

Note: Total load in kN can be established by summating the weight of materials — see BS648: Schedule of Weights of Building Materials, and multiplying by gravity; i.e. kg × 9·81 = Newtons.

This must be added to any imposed loading:-

People and furniture = 1·5 kN/m²

Snow on roofs < 30° pitch = 1·5 kN/m²

Snow on roofs > 30° pitch = 0·75 kN/m²

Simple beam design (Shear)
From the previous example, the section profile is:-

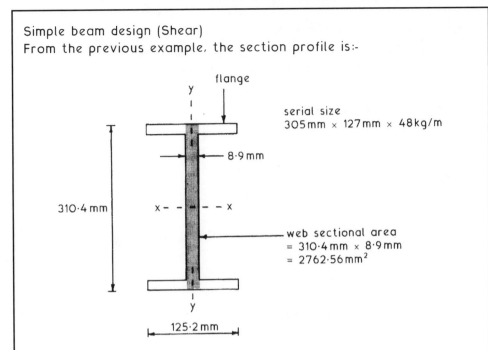

flange

serial size
305mm × 127mm × 48kg/m

8·9mm

310·4 mm

x — — — — — x

web sectional area
= 310·4mm × 8·9mm
= 2762·56mm^2

125·2 mm

Maximum shear force normally occurs at the support points, i.e. near the end of the beam. Calculation is made of the average stress value on the web sectional area.

Using the example of 200 kN load distributed over the beam, the maximum shear force at each end support will be 100 kN.

Therefore, the average shear stress $= \dfrac{\text{shear force}}{\text{web sectional area}}$

$$= \frac{100 \times 10^3}{2762 \cdot 56}$$

$$= 36 \cdot 20 \,\text{N/mm}^2$$

Reference to BS449 indicates that if using Grade 43 steel[*], i.e. 430 N/ mm^2 tensile strength, this has an allowable shear stress in the web of 110 N/ mm^2. Therefore the example section of serial size: 305 mm × 127 mm × 48 kg/m with only 36·20 N/mm^2 calculated average shear stress is more than capable of resisting the applied forces.

* Note: Details of grading steel for structural applications is found in BS 7668, BS EN's 10029 and 10113. Grades of 40, 43, 50 and 55 correspond to minimum tensile strength, e.g. 43 = 43 × 10^7 N/m^2, but the preferred specification is now 430 N/mm^2.

Simple beam design (Deflection)

The deflection due to loading, other than the weight of the structure, should not exceed 1/360 of the span.

The formula to determine the extent of deflection varies, depending on:-

(a) Point loading

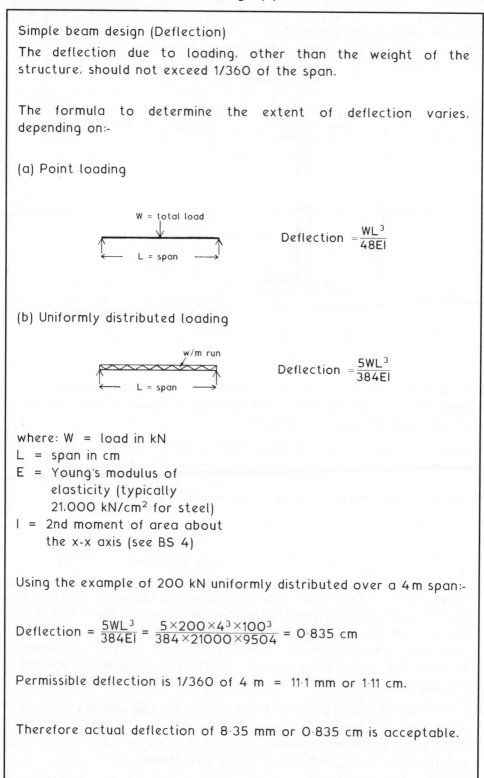

$$\text{Deflection } = \frac{WL^3}{48EI}$$

(b) Uniformly distributed loading

$$\text{Deflection } = \frac{5WL^3}{384EI}$$

where: W = load in kN
L = span in cm
E = Young's modulus of
    elasticity (typically
    21,000 kN/cm^2 for steel)
I = 2nd moment of area about
    the x-x axis (see BS 4)

Using the example of 200 kN uniformly distributed over a 4 m span:-

$$\text{Deflection} = \frac{5WL^3}{384EI} = \frac{5 \times 200 \times 4^3 \times 100^3}{384 \times 21000 \times 9504} = 0.835 \text{ cm}$$

Permissible deflection is 1/360 of 4 m = 11.1 mm or 1.11 cm.

Therefore actual deflection of 8.35 mm or 0.835 cm is acceptable.

Simple column design

Steel columns or stanchions have a tendency to buckle or bend under extreme loading. This can be attributed to:

(a) length,

(b) cross sectional area,

(c) method of end fixing, and

(d) the shape of section.

(b) and (d) are incorporated into a geometric property of section, known as the radius of gyration (r). It can be calculated:-

$$r = \sqrt{\frac{I}{A}}$$

where: I = 2nd moment of area

A = cross sectional area

Note: r,I and A are all listed in steel design tables, eg. BS4:1980.

The radius of gyration about the y-y axis is used for calculation, as this is normally the weaker axis.

Effective length of columns

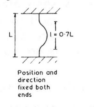
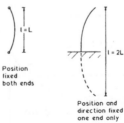

| Position and direction fixed both ends | Position fixed both ends. direction fixed at one end | Position fixed both ends | Position and direction fixed one end only |

Position and direction fixed is location at specific points by beams or other means of retention. Position fixed only means hinged or pinned. eg. A Universal Column 203 mm × 203 mm × 46 kg/m, 10 m long, position and direction fixed both ends. Determine the maximum axial loading.

Effective length (l) = 0.7 × 10 m = 7 m

(r) from BS4 = 51.1 mm

Slenderness ratio = $\frac{l}{r}$ = $\frac{7 \times 10^3}{51.1}$ = 137

Maximum allowable stress for grade 43 steel = 49 N/mm^2 (BS449)

Cross sectional area of stanchion (UC) = 5880 mm^2 (BS4)

The total axial load = $\frac{49 \times 5880}{10^3}$ = 288kN (approx. 29 tonnes)

Portal Frames ~ these can be defined as two dimensional rigid frames which have the basic characteristic of a rigid joint between the column and the beam. The main objective of this form of design is to reduce the bending moment in the beam thus allowing the frame to act as one structural unit. The transfer of stresses from the beam to the column can result in a rotational movement at the foundation which can be overcome by the introduction of a pin or hinge joint. The pin or hinge will allow free rotation to take place at the point of fixity whilst transmitting both load and shear from one member to another. In practice a true 'pivot' is not always required but there must be enough movement to ensure that the rigidity at the point of connection is low enough to overcome the tendency of rotational movement.

Typical Single Storey Portal Frame Formats ~

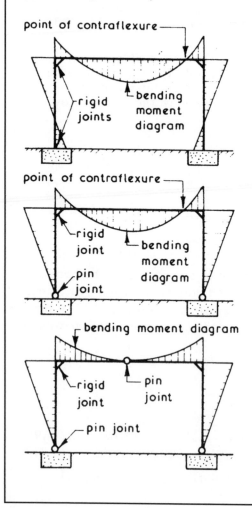

point of contraflexure

rigid joints

bending moment diagram

point of contraflexure

rigid joint

bending moment diagram

pin joint

bending moment diagram

rigid joint

pin joint

pin joint

FIXED or RIGID PORTAL FRAME :-

all joints or connections are rigid giving lower bending moments than other formats. Used for small to medium span frames where moments at foundations are not excessive.

TWO PIN PORTAL FRAME:-

pin joints or hinges used at foundation connections to eliminate tendency of base to rotate. Used where high base moments and weak ground are encountered.

THREE PIN PORTAL FRAME:-

pin joints or hinges used at foundation connections and at centre of beam which reduces bending moment in beam but increases deflection. Used as an alternative to a 2 pin frame.

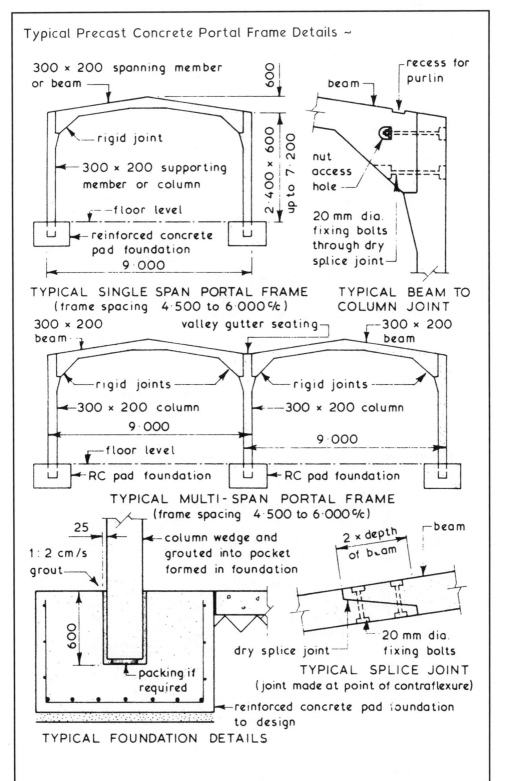

Typical Precast Concrete Portal Frame Details ~

300 × 200 spanning member or beam

600

rigid joint

300 × 200 supporting member or column

floor level

reinforced concrete pad foundation

9·000

2·400 × 600   up to 7·200

TYPICAL SINGLE SPAN PORTAL FRAME
(frame spacing 4·500 to 6·000%)

recess for purlin

beam

nut access hole

20 mm dia. fixing bolts through dry splice joint

TYPICAL BEAM TO COLUMN JOINT

300 × 200 beam

valley gutter seating

300 × 200 beam

rigid joints

rigid joints

300 × 200 column

9·000

300 × 200 column

9·000

floor level

RC pad foundation

RC pad foundation

TYPICAL MULTI-SPAN PORTAL FRAME
(frame spacing 4·500 to 6·000%)

25

1:2 cm/s grout

column wedge and grouted into pocket formed in foundation

600

packing if required

reinforced concrete pad foundation to design

TYPICAL FOUNDATION DETAILS

2 × depth of beam

beam

dry splice joint

20 mm dia. fixing bolts

TYPICAL SPLICE JOINT
(joint made at point of contraflexure)

Typical Precast Concrete Portal Frame Hinge Details ~

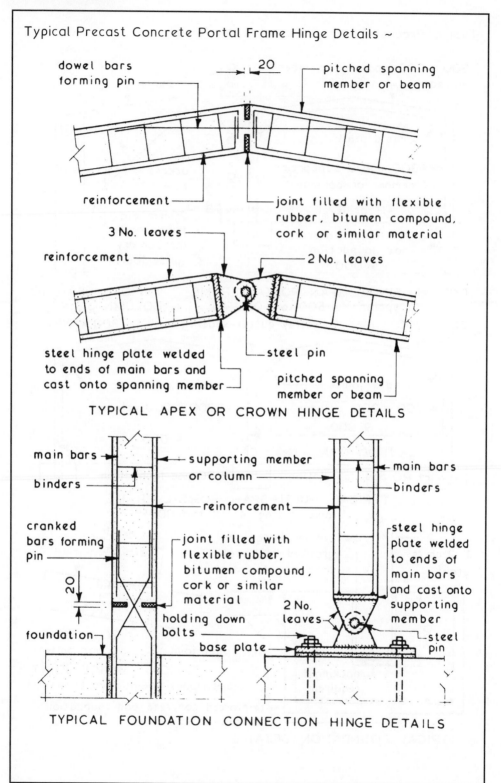

TYPICAL APEX OR CROWN HINGE DETAILS

TYPICAL FOUNDATION CONNECTION HINGE DETAILS

Typical Steel Portal Frame Details ~

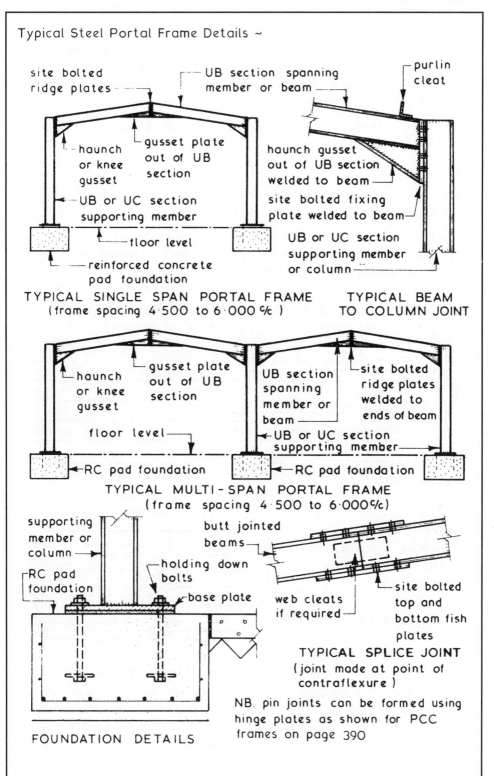

site bolted
ridge plates

UB section spanning
member or beam

purlin
cleat

haunch
or knee
gusset

gusset plate
out of UB
section

haunch gusset
out of UB section
welded to beam

site bolted fixing
plate welded to beam

UB or UC section
supporting member

UB or UC section
supporting member
or column

floor level

reinforced concrete
pad foundation

TYPICAL SINGLE SPAN PORTAL FRAME
(frame spacing 4·500 to 6·000 ℅)

**TYPICAL BEAM
TO COLUMN JOINT**

haunch
or knee
gusset

gusset plate
out of UB
section

UB section
spanning
member or
beam

site bolted
ridge plates
welded to
ends of beam

floor level

UB or UC section
supporting member

RC pad foundation

RC pad foundation

RC pad foundation

TYPICAL MULTI-SPAN PORTAL FRAME
(frame spacing 4·500 to 6·000℅)

supporting
member or
column

RC pad
foundation

holding down
bolts

base plate

butt jointed
beams

web cleats
if required

site bolted
top and
bottom fish
plates

TYPICAL SPLICE JOINT
(joint made at point of
contraflexure)

NB. pin joints can be formed using
hinge plates as shown for PCC
frames on page 390

FOUNDATION DETAILS

Laminated Timber ~ sometimes called `Gluelam` and is the process of building up beams, ribs, arches, portal frames and other structural units by gluing together layers of timber boards so that the direction of the grain of each board runs parallel with the longitudinal axis of the member being fabricated.

Laminates ~ these are the layers of board and may be jointed in width and length.

Joints ~

Width — joints in consecutive layers should lap twice the board thickness or one quarter of its width whichever is the greater.

Length — scarf and finger joints can be used. Scarf joints should have a minimum slope of 1 in 12 but this can be steeper (say 1 in 6) in the compression edge of a beam :-

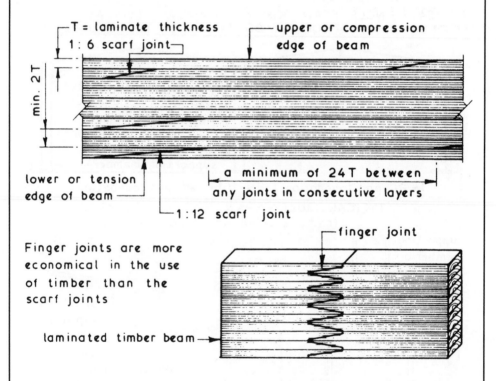

Moisture Content ~ timber should have a moisture content equal to that which the member will reach in service and this is known as its equilibrium moisture content; for most buildings this will be between 11 and 15%. Generally at the time of gluing timber should not exceed 15 ± 3% in moisture content.

Vertical Laminations ~ not often used for structural laminated timber members and is unsatisfactory for curved members.

vertical laminates ———→

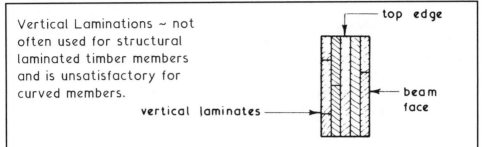

Horizontal Laminations ~ most popular method for all types of laminated timber members. The stress diagrams below show that laminates near the upper edge are subject to a compressive stress whilst those near the lower edge to a tensile stress and those near the neutral axis are subject to shear stress.

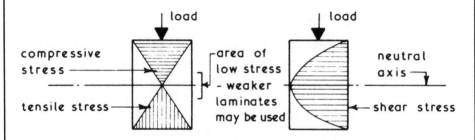

compressive stress ———→

area of low stress - weaker laminates may be used

tensile stress ———→

neutral axis ⌐

shear stress ←———

Flat sawn timber shrinks twice as much as quarter sawn timber therefore flat and quarter sawn timbers should not be mixed in the same member since the different shrinkage rates will cause unacceptable stresses to occur on the glue lines.

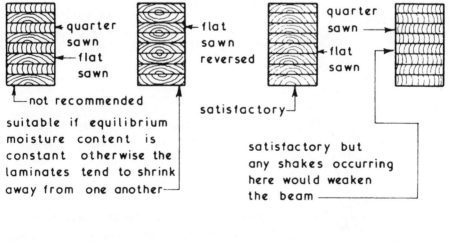

←quarter sawn
←flat sawn

—not recommended

←flat sawn reversed

quarter sawn ———→
←flat sawn

suitable if equilibrium moisture content is constant otherwise the laminates tend to shrink away from one another—

satisfactory—

satisfactory but any shakes occurring here would weaken the beam ———

Planing ~ before gluing, laminates should be planed so that the depth of the planer cutter marks are not greater than 0·025 mm.

Gluing ~ this should be carried out within 48 hours of the planing operation to reduce the risk of the planed surfaces becoming contaminated or case hardened (for suitable adhesives see page 395). Just before gluing up the laminates they should be checked for 'cupping.' The amount of cupping allowed depends upon the thickness and width of the laminates and has a range of 0·75 mm to 1·5 mm.

Laminate Thickness ~ no laminate should be more than 50 mm thick since seasoning up to this thickness can be carried out economically and there is less chance of any individual laminate having excessive cross grain strength.

Straight Members — laminate thickness is determined by the depth of the member, there must be enough layers to allow the end joints (i.e. scarf or finger joints — see page 392) to be properly staggered.

Curved Members — laminate thickness is determined by the radius to which the laminate is to be bent and the species together with the quality of the timber being used. Generally the maximum laminate thickness should be 1/150 of the sharpest curve radius although with some softwoods 1/100 may be used.

Typical Laminated Timber Curved Member ~

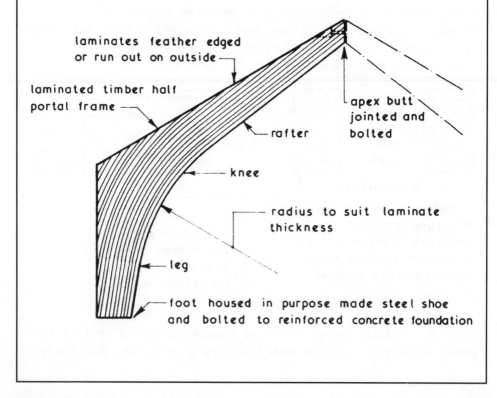

laminates feather edged or run out on outside

laminated timber half portal frame

apex butt jointed and bolted

rafter

knee

radius to suit laminate thickness

leg

foot housed in purpose made steel shoe and bolted to reinforced concrete foundation

Adhesives ~ although timber laminates are carefully machined, the minimum of cupping permitted and efficient cramping methods employed it is not always possible to obtain really tight joints between the laminates. One of the important properties of the adhesive is therefore that it should be gap filling. The maximum permissible gap being 1.25 mm.

There are four adhesives suitable for laminated timber work which have the necessary gap filling property and they are namely :-

1. Casein — this is made from sour milk to the requirements of BS 5442. It is a cold setting adhesive in the form of a powder which is mixed with water, it has a tendency to stain timber and is only suitable for members used in dry conditions of service.

2. Urea Formaldehyde — this is a cold setting resin glue formulated to BS 1204 type MR/GF (moisture resistant/gap filling). Although moisture resistant it is not suitable for prolonged exposure in wet conditions and there is a tendency for the glue to lose its strength in temperatures above 40°C such as when exposed to direct sunlight. The use of this adhesive is usually confined to members used in dry, unexposed conditions of service. This adhesive will set under temperatures down to 10°C.

3. Resorcinol Formaldehyde — this is a cold setting glue formulated to BS 1204 type WBP/GF (weather and boilproof/gap filling). It is suitable for members used in external situations but is relatively expensive. This adhesive will set under temperatures down to 15°C and does not lose its strength at high temperatures.

4. Phenol Formaldehyde — this is a similar glue to resorcinol formaldehyde but is a warm setting adhesive requiring a temperature of above 86°C in order to set. A mixture called phenol/resorcinol formaldehyde is available and is sometimes used having similar properties to but less expensive than resorcinol formaldehyde but needs a setting temperature of at least 23°C.

Preservative Treatment — this can be employed if required, provided that the pressure impregnated preservative used is selected with regard to the adhesive being employed. See also page 435.

Composite Timber Beams

Composite Beams ~ stock sizes of structural softwood have sectional limitations of about 225mm and corresponding span potential in the region of 6m. At this distance, even modest loadings could interpose with the maximum recommended deflection of 0·003 × span.

Fabricated softwood box, lattice and plywood beams are an economic consideration for medium spans. They are produced with adequate depth to resist deflection and with sufficient strength for spans into double figures. The high strength to weight ratio and simple construction provides advantages in many situations otherwise associated with steel or reinforced concrete, e.g. frames, trusses, beams and purlins in gymnasia, workshops, garages, churches, shops, etc. They are also appropriate as purlins in loft conversion.

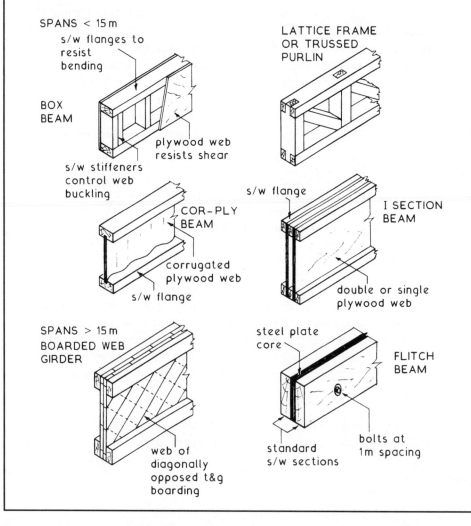

SPANS < 15m

s/w flanges to resist bending

LATTICE FRAME OR TRUSSED PURLIN

BOX BEAM

plywood web resists shear

s/w stiffeners control web buckling

COR-PLY BEAM

s/w flange

I SECTION BEAM

corrugated plywood web

s/w flange

double or single plywood web

SPANS > 15m
BOARDED WEB GIRDER

steel plate core

FLITCH BEAM

web of diagonally opposed t&g boarding

standard s/w sections

bolts at 1m spacing

Multi-storey Structures ~ these buildings are usually designed for office, hotel or residential use and contain the means of vertical circulation in the form of stairs and lifts occupying up to 20% of the floor area. These means of circulation can be housed within a core inside the structure and this can be used to provide a degree of restraint to sway due to lateral wind pressures (see page 398).

Typical Basic Multi-storey Structure Types ~

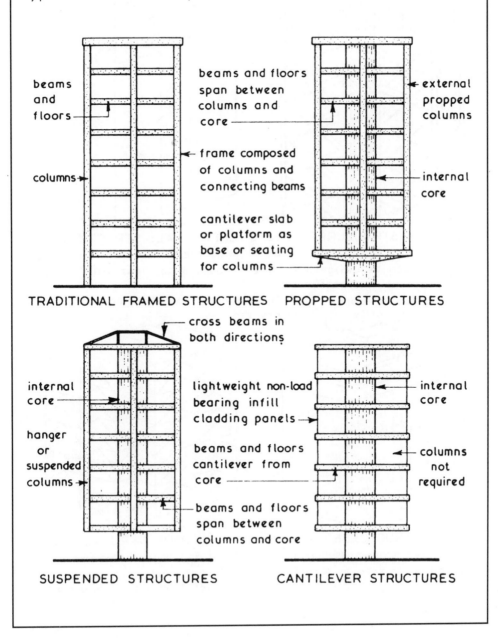

beams and floors

beams and floors span between columns and core

external propped columns

columns

frame composed of columns and connecting beams

cantilever slab or platform as base or seating for columns

internal core

TRADITIONAL FRAMED STRUCTURES   PROPPED STRUCTURES

cross beams in both directions

internal core

lightweight non-load bearing infill cladding panels

internal core

hanger or suspended columns

beams and floors cantilever from core

columns not required

beams and floors span between columns and core

SUSPENDED STRUCTURES   CANTILEVER STRUCTURES

Typical Multi-storey Structures ~ the formats shown below are designed to provide lateral restraint against wind pressures.

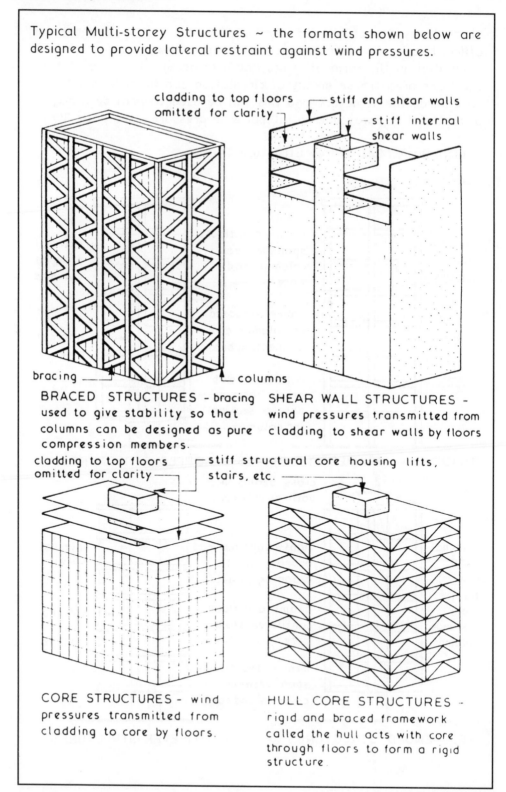

cladding to top floors omitted for clarity

stiff end shear walls

stiff internal shear walls

bracing ___

columns

BRACED STRUCTURES - bracing used to give stability so that columns can be designed as pure compression members.

SHEAR WALL STRUCTURES - wind pressures transmitted from cladding to shear walls by floors

cladding to top floors omitted for clarity

stiff structural core housing lifts, stairs, etc.

CORE STRUCTURES - wind pressures transmitted from cladding to core by floors.

HULL CORE STRUCTURES - rigid and braced framework called the hull acts with core through floors to form a rigid structure.

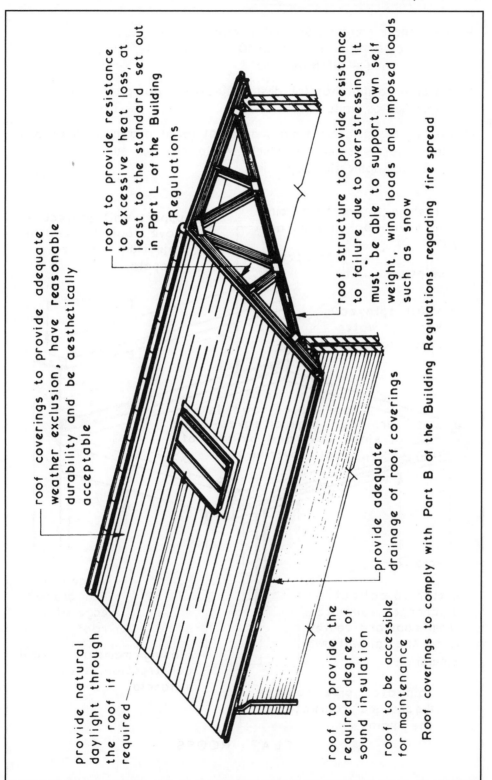

roof to provide resistance to excessive heat loss, at least to the standard set out in Part L of the Building Regulations

roof structure to provide resistance to failure due to overstressing. It must be able to support own self weight, wind loads and imposed loads such as snow

roof coverings to provide adequate weather exclusion, have reasonable durability and be aesthetically acceptable

provide adequate drainage of roof coverings

provide natural daylight through the roof if required

provide the required degree of sound insulation

roof to be accessible for maintenance

Roof coverings to comply with Part B of the Building Regulations regarding fire spread

Roofs ~ these can be classified as either:-
    Flat — pitch from 0° to 10°
    Pitched — pitch over 10°

It is worth noting that for design purposes roof pitches over 70° are classified as walls.

Roofs can be designed in many different forms and in combinations of these forms some of which would not be suitable and/or economic for domestic properties.

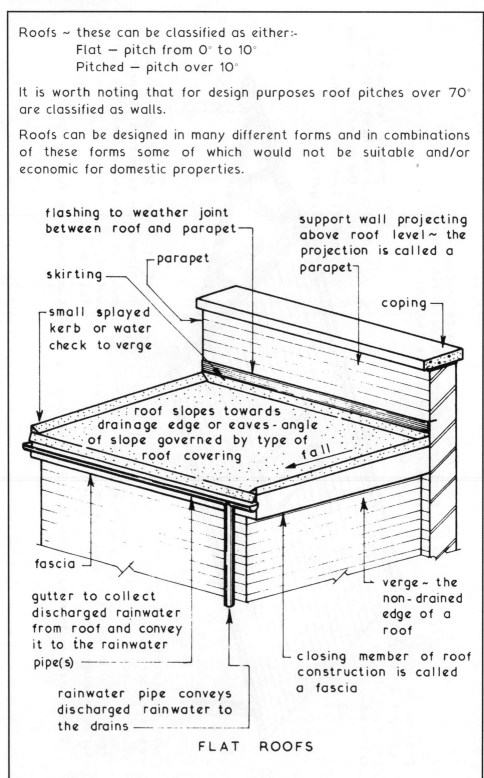

flashing to weather joint between roof and parapet

support wall projecting above roof level~ the projection is called a parapet

parapet

skirting

coping

small splayed kerb or water check to verge

roof slopes towards drainage edge or eaves - angle of slope governed by type of roof covering

fall

fascia

gutter to collect discharged rainwater from roof and convey it to the rainwater pipe(s)

verge ~ the non-drained edge of a roof

closing member of roof construction is called a fascia

rainwater pipe conveys discharged rainwater to the drains

FLAT ROOFS

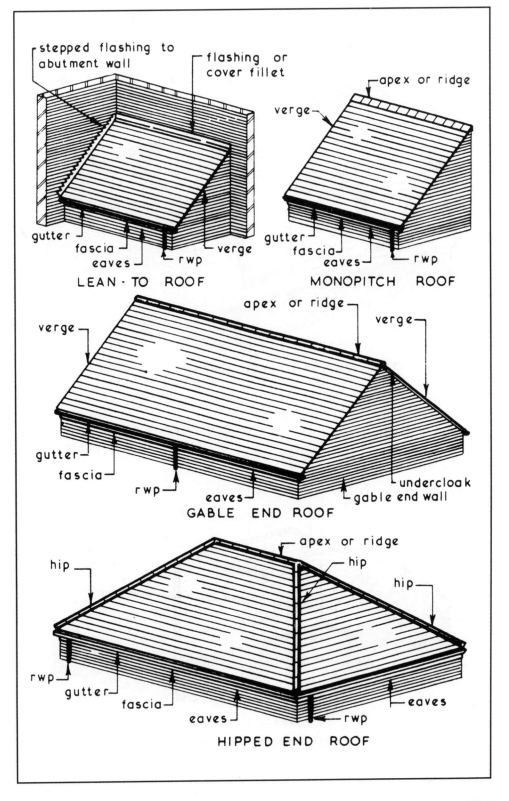

stepped flashing to
abutment wall

flashing or
cover fillet

apex or ridge

verge

gutter
fascia
eaves
rwp
verge

LEAN · TO ROOF

gutter
fascia
eaves
rwp

MONOPITCH ROOF

verge

apex or ridge

verge

gutter
fascia
rwp
eaves

undercloak
gable end wall

GABLE END ROOF

apex or ridge

hip

hip

hip

hip

rwp
gutter
fascia
eaves
eaves
rwp

HIPPED END ROOF

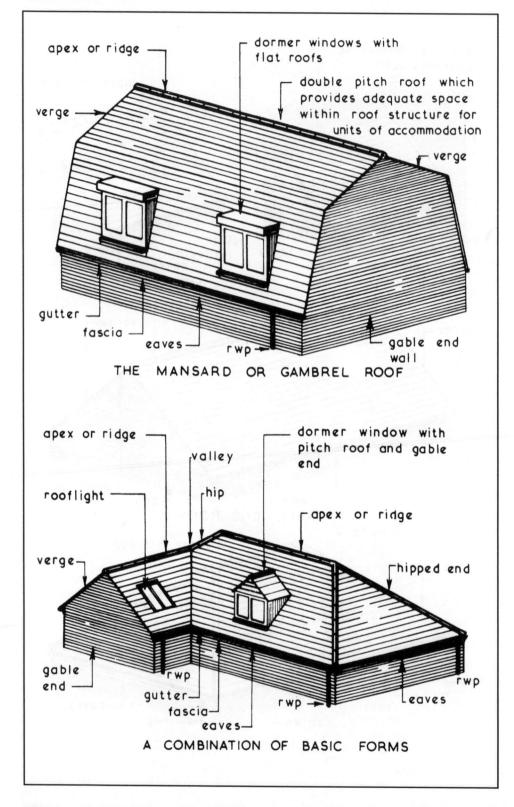

apex or ridge

dormer windows with flat roofs

double pitch roof which provides adequate space within roof structure for units of accommodation

verge

verge

gutter

fascia

eaves

rwp

gable end wall

THE MANSARD OR GAMBREL ROOF

apex or ridge

dormer window with pitch roof and gable end

valley

rooflight

hip

apex or ridge

verge

hipped end

gable end

rwp

gutter

fascia

eaves

rwp

rwp

eaves

A COMBINATION OF BASIC FORMS

Pitched Roofs ~ the primary functions of any domestic roof are to:-

1. Provide an adequate barrier to the penetration of the elements.
2. Maintain the internal environment by providing an adequate resistance to heat loss.

A roof is in a very exposed situation and must therefore be designed and constructed in such a manner as to:-

1. Safely resist all imposed loadings such as snow and wind.
2. Be capable of accommodating thermal and moisture movements.
3. Be durable so as to give a satisfactory performance and reduce maintenance to a minimum.

Component Parts of a Pitched Roof ~

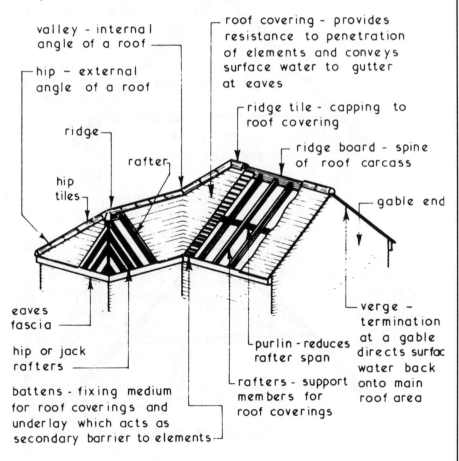

valley - internal angle of a roof

hip - external angle of a roof

roof covering - provides resistance to penetration of elements and conveys surface water to gutter at eaves

ridge tile - capping to roof covering

ridge

rafter

ridge board - spine of roof carcass

hip tiles

gable end

eaves fascia

hip or jack rafters

battens - fixing medium for roof coverings and underlay which acts as secondary barrier to elements

purlin - reduces rafter span

rafters - support members for roof coverings

verge - termination at a gable directs surface water back onto main roof area

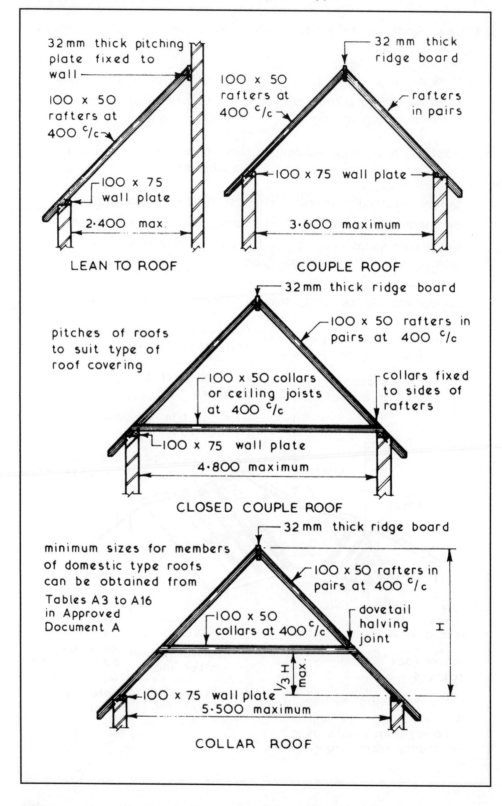

32 mm thick pitching plate fixed to wall

100 x 50 rafters at 400 c/c

100 x 75 wall plate

2·400 max.

LEAN TO ROOF

32 mm thick ridge board

100 x 50 rafters at 400 c/c

rafters in pairs

100 x 75 wall plate

3·600 maximum

COUPLE ROOF

32 mm thick ridge board

pitches of roofs to suit type of roof covering

100 x 50 rafters in pairs at 400 c/c

100 x 50 collars or ceiling joists at 400 c/c

collars fixed to sides of rafters

100 x 75 wall plate

4·800 maximum

CLOSED COUPLE ROOF

32 mm thick ridge board

minimum sizes for members of domestic type roofs can be obtained from Tables A3 to A16 in Approved Document A

100 x 50 rafters in pairs at 400 c/c

100 x 50 collars at 400 c/c

dovetail halving joint

H

100 x 75 wall plate

$1/3$ H max.

5·500 maximum

COLLAR ROOF

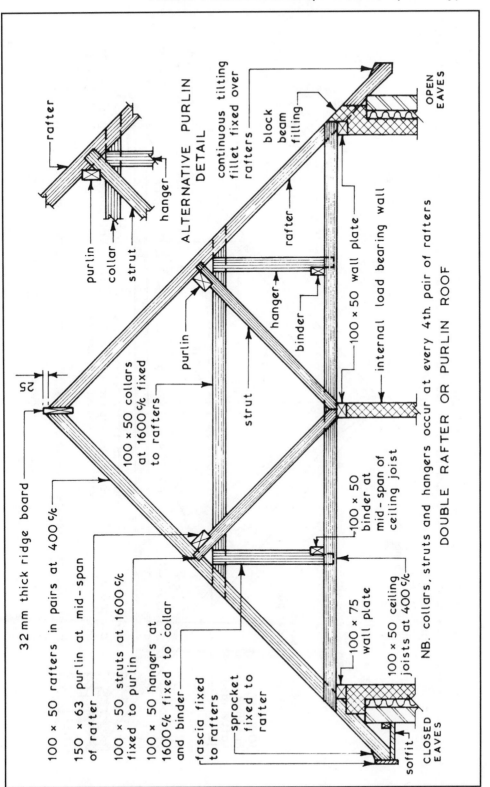

rafter

ALTERNATIVE PURLIN DETAIL

purlin

collar

strut

hanger

continuous tilting fillet fixed over rafters

block

beam filling

OPEN EAVES

rafter

hanger

binder

100 × 50 wall plate

internal load bearing wall

NB. collars, struts and hangers occur at every 4th. pair of rafters

DOUBLE RAFTER OR PURLIN ROOF

25

32 mm thick ridge board

100 × 50 rafters in pairs at 400 ℅

150 × 63 purlin at mid-span of rafter

100 × 50 struts at 1600 ℅ fixed to purlin

100 × 50 hangers at 1600 ℅ fixed to collar and binder

fascia fixed to rafters

sprocket fixed to rafter

purlin

strut

100 × 50 collars at 1600 ℅ fixed to rafters

100 × 50 binder at mid-span of ceiling joist

100 × 75 wall plate

100 × 50 ceiling joists at 400 ℅

CLOSED EAVES

soffit

Roof Trusses ~ these are triangulated plane roof frames designed to give clear spans between the external supporting walls. They are usually prefabricated or partially prefabricated off site and are fixed at 1·800 centres to support purlins which accept loads from the infill rafters.

24 No. triangular teeth bent alternately at 90° to face of steel plate

bolt hole

BS 1579 TOOTHED PLATE TIMBER CONNECTOR

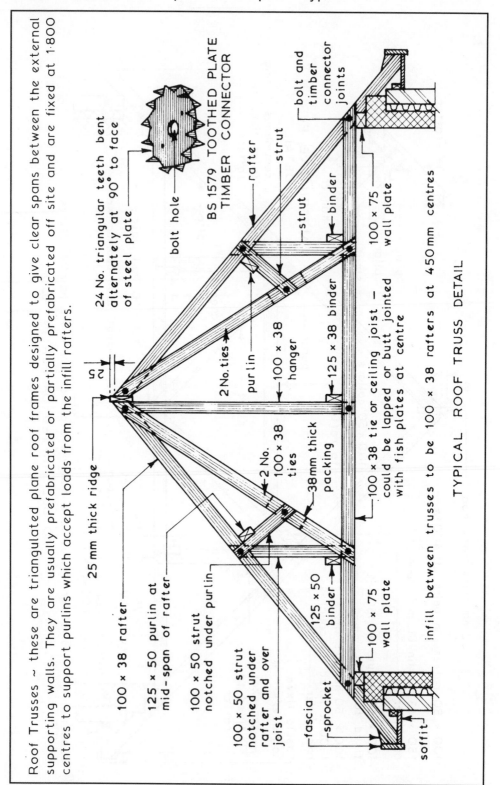

25 mm thick ridge

100 × 38 rafter

125 × 50 purlin at mid-span of rafter

100 × 50 strut notched under purlin

100 × 50 strut notched under rafter and over joist

fascia

sprocket

soffit

100 × 75 wall plate

125 × 50 binder

2 No. 100 × 38 ties

38 mm thick packing

2 No. ties

purlin

100 × 38 hanger

125 × 38 binder

rafter

strut

strut

binder

bolt and timber connector joints

100 × 75 wall plate

100 × 38 tie or ceiling joist — could be lapped or butt jointed with fish plates at centre

infill between trusses to be 100 × 38 rafters at 450 mm centres

TYPICAL ROOF TRUSS DETAIL

Trussed Rafters ~ these are triangulated plane roof frames designed to give clear spans between the external supporting walls. They are delivered to site as a prefabricated component where they are fixed to the wall plates at 600mm centres. Trussed rafters do not require any ridge board or purlins since they receive their lateral stability by using larger tiling battens than those used on traditional roofs.

7mm thick external quality plywood gussets glued to both faces of trussed rafter at all butt joints

ALTERNATIVE JOINT DETAIL

galvanised steel truss plate connectors formed by cutting, punching and bending a series of spikes at right angles to the face. Plates are inserted under heavy pressure to both faces of trussed rafter at all butt joints

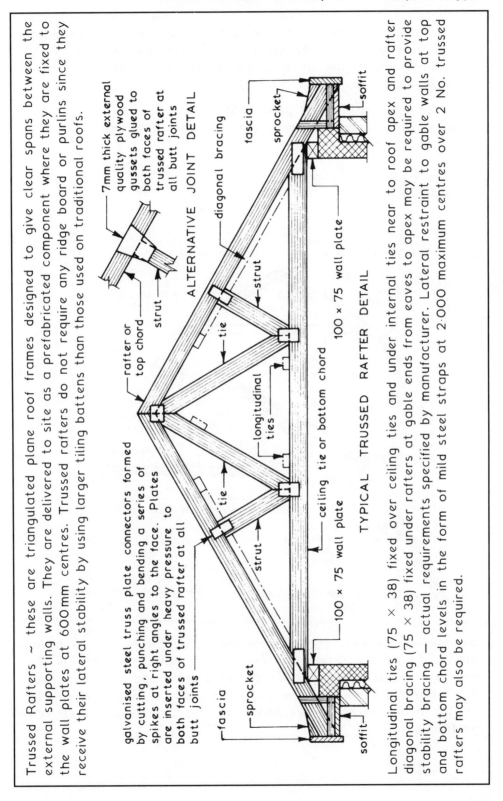

rafter or top chord

strut

diagonal bracing

fascia

sprocket

soffit

tie

strut

100 × 75 wall plate

longitudinal ties

ceiling tie or bottom chord

tie

100 × 75 wall plate

TYPICAL TRUSSED RAFTER DETAIL

strut

fascia

sprocket

soffit

Longitudinal ties (75 × 38) fixed over ceiling ties and under internal ties near to roof apex and rafter diagonal bracing (75 × 38) fixed under rafters at gable ends from eaves to apex may be required to provide stability bracing — actual requirements specified by manufacturer. Lateral restraint to gable walls at top and bottom chord levels in the form of mild steel straps at 2·000 maximum centres over 2 No. trussed rafters may also be required.

Gambrel roofs are double pitched with a break in the roof slope. The pitch angle above the break is less than 45° relative to the horizontal, whilst the pitch angle below the break is greater. Generally, these angles are 30° and 60°.

Gambrels are useful in providing more attic headroom and frequently incorporate dormers and rooflights. They have a variety of constructional forms.

Typically —

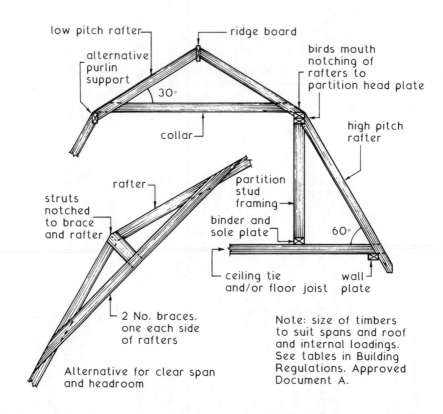

low pitch rafter

ridge board

alternative purlin support

birds mouth notching of rafters to partition head plate

30°

high pitch rafter

collar

rafter

partition stud framing

struts notched to brace and rafter

binder and sole plate

60°

2 No. braces, one each side of rafters

ceiling tie and/or floor joist

wall plate

Alternative for clear span and headroom

Note: size of timbers to suit spans and roof and internal loadings. See tables in Building Regulations, Approved Document A.

Intermediate support can be provided in various ways as shown above. To create headroom for accommodation in what would otherwise be attic space, a double head plate and partition studing is usual. The collar beam and rafters can conveniently locate on the head plates or prefabricated trusses can span between partitions.

Roof Underlays ~ sometimes called sarking or roofing felt provides the barrier to the entry of snow, wind and rain blown between the tiles or states, it also prevents the entry of water from capillary action.

Suitable Materials ~

Bitumen fibre based felts ⎫ supplied in rolls 1m wide × 10 or
Bitumen glass fibre based felts ⎭ 20m long to BS 747

Sheathing and Hair felts — supplied in rolls 810mm wide × 25m long to the recommendations of BS 747: Specification for roofing felts.

Plastic Sheeting underlays — these are lighter, require less storage space, have greater flexibility at low temperatures and high resistance to tearing but have a greater risk to the formation of condensation than the BS 747 felts and should not be used on roof pitches below 20°. Materials permeable to water vapour are preferred as these do not need to be perforated to ventilate the roof space. See BS 4016: Specification for flexible building membranes (breather type).

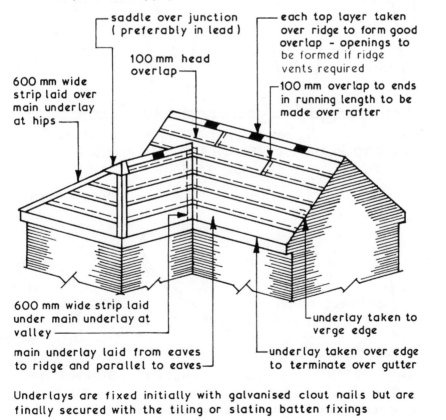

saddle over junction (preferably in lead)

100 mm head overlap

each top layer taken over ridge to form good overlap - openings to be formed if ridge vents required

600 mm wide strip laid over main underlay at hips

100 mm overlap to ends in running length to be made over rafter

600 mm wide strip laid under main underlay at valley

underlay taken to verge edge

main underlay laid from eaves to ridge and parallel to eaves

underlay taken over edge to terminate over gutter

Underlays are fixed initially with galvanised clout nails but are finally secured with the tiling or slating batten fixings

Double Lap Tiles ~ these are the traditional tile covering for pitched roofs and are available made from clay and concrete and are usually called plain tiles. Plain tiles have a slight camber in their length to ensure that the tail of the tile will bed and not ride on the tile below. There is always at least two layers of tiles covering any part of the roof. Each tile has at least two nibs on the underside of its head so that it can be hung on support battens nailed over the rafters. Two nail holes provide the means of fixing the tile to the batten, in practice only every 4th course of tiles is nailed unless the roof exposure is high. Double lap tiles are laid to a bond so that the edge joints between the tiles are in the centre of the tiles immediately below and above the course under consideration.

Typical Plain Tile Details ~

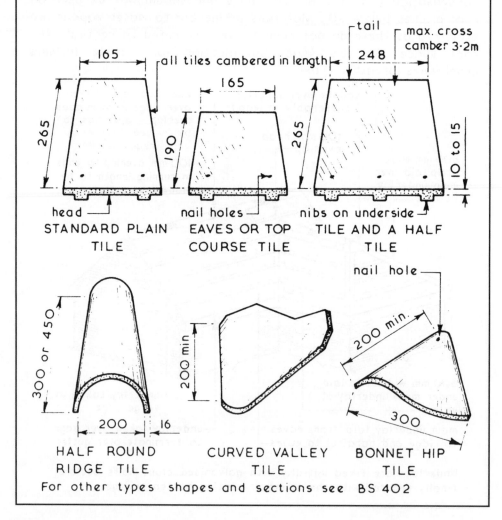

STANDARD PLAIN TILE

EAVES OR TOP COURSE TILE

TILE AND A HALF TILE

HALF ROUND RIDGE TILE

CURVED VALLEY TILE

BONNET HIP TILE

For other types shapes and sections see BS 402

Typical Details ~

ridge vents spaced to provide equivalent of 5 mm continuous gap alternatively, purpose made tile vents may be positioned at high level

half round ridge tiles bedded in cm. mt. (1:3) butt jointed in length, with end of ridge tile filled with cm. mt. and tile slip inserts

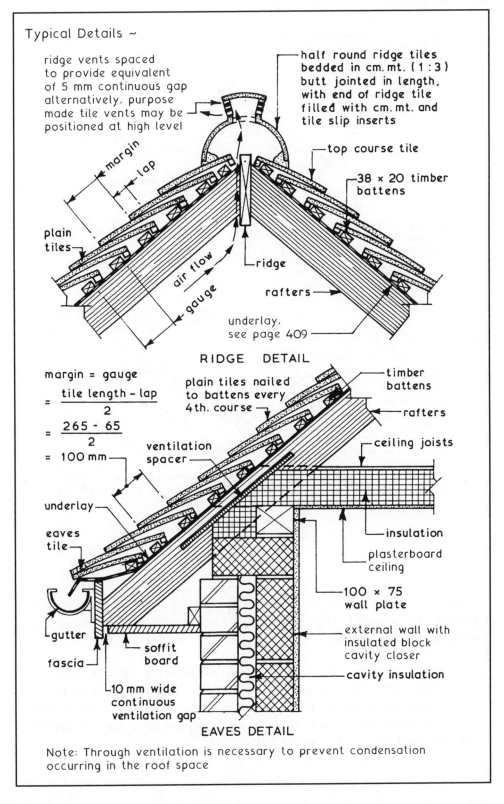

margin

lap

plain tiles

air flow

gauge

ridge

top course tile

38 × 20 timber battens

rafters

underlay, see page 409

RIDGE DETAIL

margin = gauge

$$= \frac{\text{tile length} - \text{lap}}{2}$$

$$= \frac{265 - 65}{2}$$

$$= 100 \text{ mm}$$

plain tiles nailed to battens every 4th. course

ventilation spacer

underlay

eaves tile

gutter

fascia

soffit board

10 mm wide continuous ventilation gap

timber battens

rafters

ceiling joists

insulation

plasterboard ceiling

100 × 75 wall plate

external wall with insulated block cavity closer

cavity insulation

EAVES DETAIL

Note: Through ventilation is necessary to prevent condensation occurring in the roof space

Eaves and Ridge—Alternative Treatment

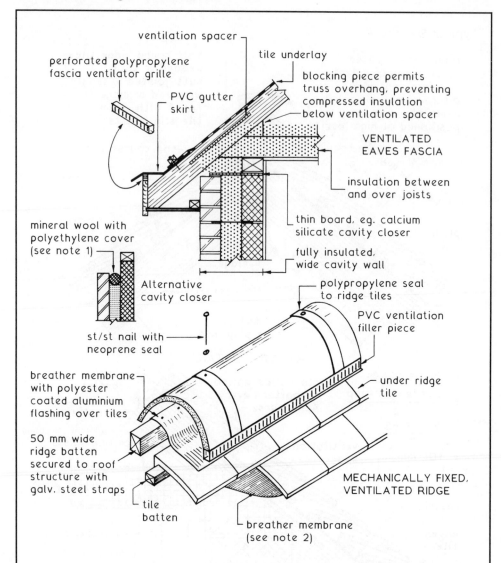

Note 1. If a cavity closer is also required to function as a cavity barrier to prevent fire spread, it should provide at least 30 minutes fire resistance.

Note 2. A breather membrane is an alternative to conventional bituminous felt as an under-tiling layer. It has the benefit of restricting liquid water penetration whilst allowing water vapour transfer from within the roof space. This permits air circulation without perforating the under-tiling layer.

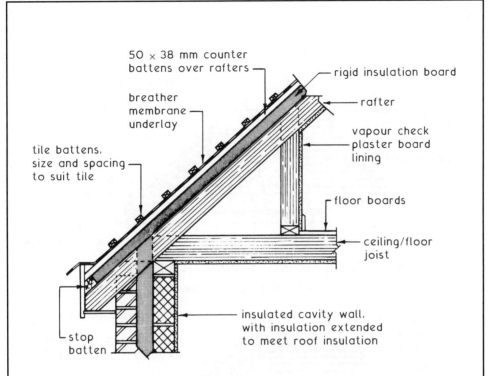

50 × 38 mm counter battens over rafters

rigid insulation board

breather membrane underlay

rafter

tile battens, size and spacing to suit tile

vapour check plaster board lining

floor boards

ceiling/floor joist

stop batten

insulated cavity wall, with insulation extended to meet roof insulation

Where a roof space is used for habitable space, insulation must be provided within the roof slope. Insulation above the rafters (as shown) creates a 'warm roof', eliminating the need for continuous ventilation. Insulation placed between the rafters creates a 'cold roof', requiring a continuous 50 mm ventilation void above the insulation to prevent the possible occurrence of interstitial condensation.

Suitable rigid insulants include; low density polyisocyanurate (PIR) foam, reinforced with long strand glass fibres, both faces bonded to aluminium foil with joints aluminium foil taped on the upper surface; high density mineral wool slabs over rafters with less dense mineral wool between rafters.

An alternative location for the breather membrane is under the counter battens. This is often preferred as the insulation board will provide uniform support for the underlay. Otherwise, extra insulation could be provided between the counter battens.

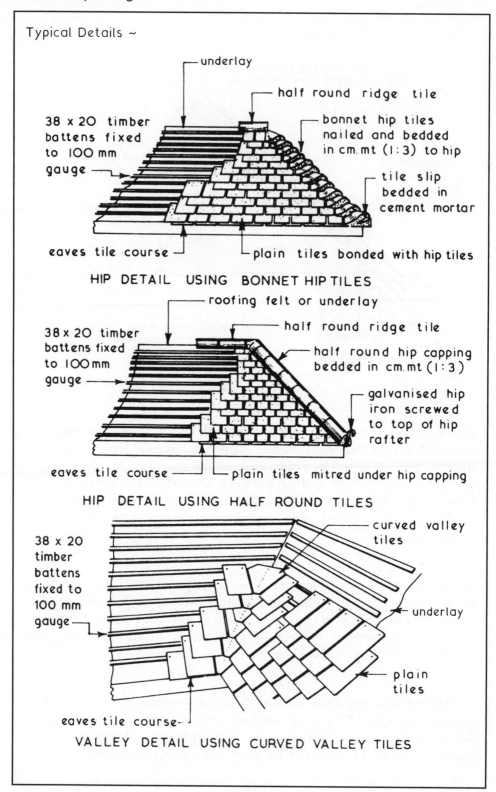

Typical Details ~

HIP DETAIL USING BONNET HIP TILES

- underlay
- half round ridge tile
- 38 x 20 timber battens fixed to 100 mm gauge
- bonnet hip tiles nailed and bedded in cm.mt (1:3) to hip
- tile slip bedded in cement mortar
- eaves tile course
- plain tiles bonded with hip tiles

HIP DETAIL USING HALF ROUND TILES

- roofing felt or underlay
- half round ridge tile
- 38 x 20 timber battens fixed to 100 mm gauge
- half round hip capping bedded in cm.mt (1:3)
- galvanised hip iron screwed to top of hip rafter
- eaves tile course
- plain tiles mitred under hip capping

VALLEY DETAIL USING CURVED VALLEY TILES

- curved valley tiles
- 38 x 20 timber battens fixed to 100 mm gauge
- underlay
- plain tiles
- eaves tile course

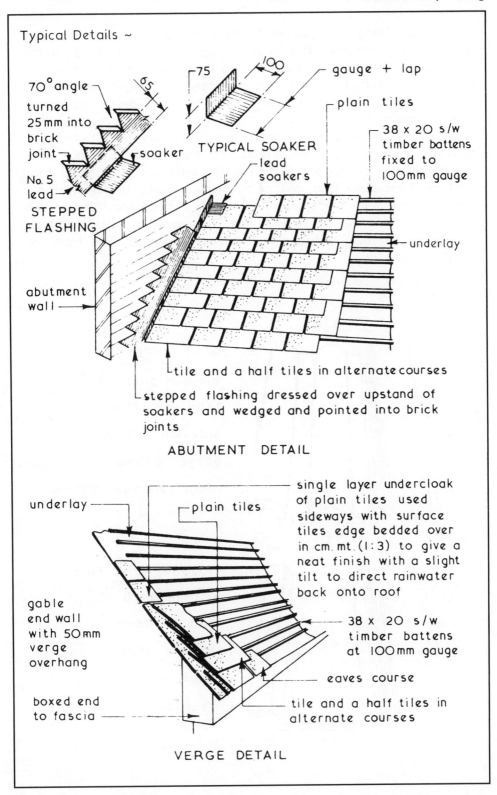

Typical Details ~

70° angle turned 25 mm into brick joint

65

soaker

No. 5 lead

STEPPED FLASHING

75

100

TYPICAL SOAKER

gauge + lap

plain tiles

38 x 20 s/w timber battens fixed to 100 mm gauge

lead soakers

abutment wall

underlay

tile and a half tiles in alternate courses

stepped flashing dressed over upstand of soakers and wedged and pointed into brick joints

ABUTMENT DETAIL

underlay

plain tiles

single layer undercloak of plain tiles used sideways with surface tiles edge bedded over in cm. mt. (1:3) to give a neat finish with a slight tilt to direct rainwater back onto roof

gable end wall with 50 mm verge overhang

boxed end to fascia

38 x 20 s/w timber battens at 100 mm gauge

eaves course

tile and a half tiles in alternate courses

VERGE DETAIL

Single Lap Tiling ~ so called because the single lap of one tile over another provides the weather tightness as opposed to the two layers of tiles used in double lap tiling. Most of the single lap tiles produced in clay and concrete have a tongue and groove joint along their side edges and in some patterns on all four edges which forms a series of interlocking joints and therefore these tiles are called single lap interlocking tiles. Generally there will be an overall reduction in the weight of the roof covering when compared with double lap tiling but the batten size is larger than that used for plain tiles and as a minimum every tile in alternate courses should be twice nailed although a good specification will require every tile to be twice nailed. The gauge or batten spacing for single lap tiling is found by subtracting the end lap from the length of the tile.

Typical Single Lap Tiles ~

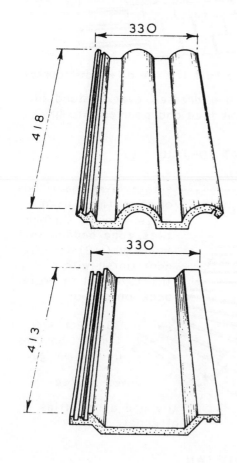

ROLL TYPE TILE

minimum pitch 30°

head lap 75mm

side lap 30 mm

gauge 343 mm

linear coverage 300 mm

TROUGH TYPE TILE

minimum pitch 15°

head lap 75 mm

side lap 38 mm

gauge 338 mm

linear coverage 292 mm

Typical Details ~

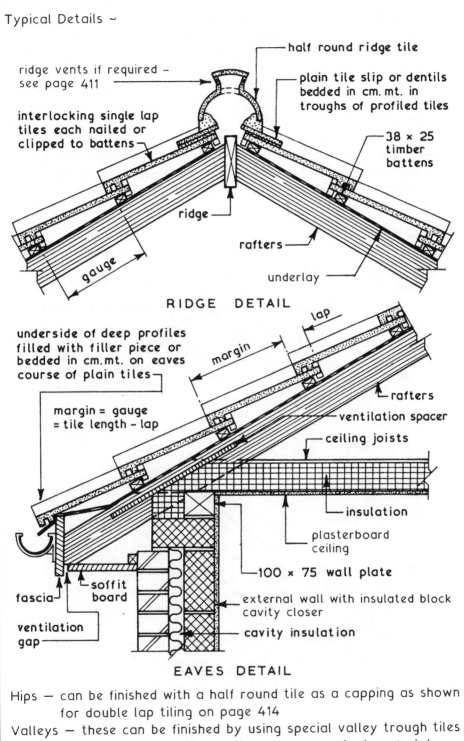

RIDGE DETAIL

ridge vents if required – see page 411

interlocking single lap tiles each nailed or clipped to battens

half round ridge tile

plain tile slip or dentils bedded in cm. mt. in troughs of profiled tiles

38 × 25 timber battens

ridge

rafters

gauge

underlay

EAVES DETAIL

underside of deep profiles filled with filler piece or bedded in cm.mt. on eaves course of plain tiles

margin = gauge = tile length – lap

lap

margin

rafters

ventilation spacer

ceiling joists

insulation

plasterboard ceiling

100 × 75 wall plate

external wall with insulated block cavity closer

cavity insulation

fascia

soffit board

ventilation gap

Hips – can be finished with a half round tile as a capping as shown for double lap tiling on page 414

Valleys – these can be finished by using special valley trough tiles or with a lead lined gutter – see manufacturer's data.

Roof Slating

Slates ~ slate is a natural dense material which can be split into thin sheets and cut to form a small unit covering suitable for pitched roofs in excess of 25° pitch. Slates are graded according to thickness and texture, the thinnest being known as 'Bests'. These are of 4mm nominal thickness. Slates are laid to the same double lap principles as plain tiles. Ridges and hips are normally covered with half round or angular tiles whereas valley junctions are usually of mitred slates over soakers. Unlike plain tiles every course is fixed to the battens by head or centre nailing, the latter being used on long slates and on pitches below 35° to overcome the problem of vibration caused by the wind which can break head nailed long slates.

Typical Details ~

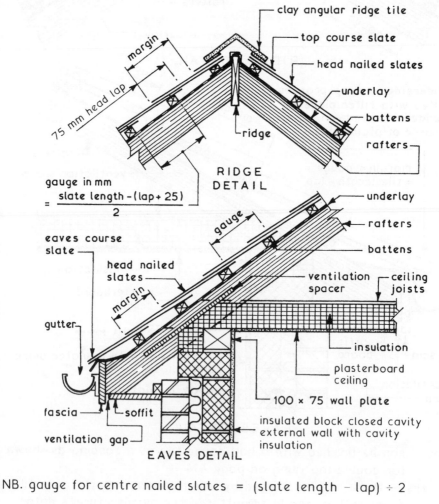

$$\text{gauge in mm} = \frac{\text{slate length} - (\text{lap} + 25)}{2}$$

NB. gauge for centre nailed slates = (slate length – lap) ÷ 2

The UK has been supplied with its own slate resources from quarries in Wales, Cornwall and Westermorland. Imported slate is also available from Spain, Argentina and parts of the Far East.

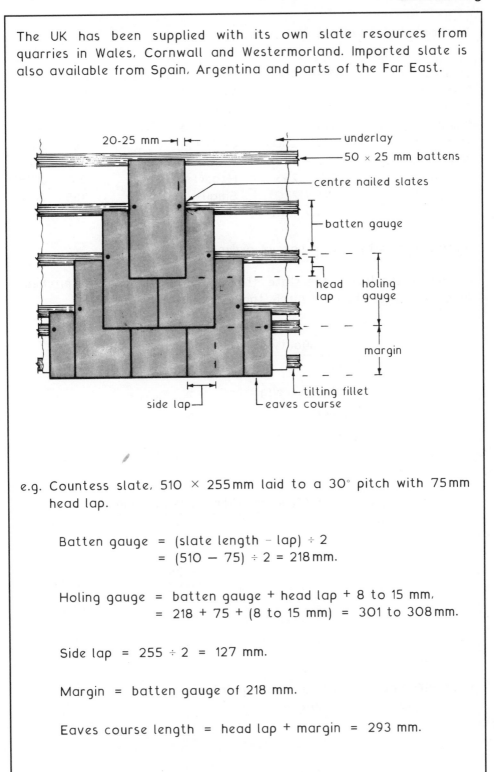

e.g. Countess slate, 510 × 255mm laid to a 30° pitch with 75mm head lap.

Batten gauge = (slate length − lap) ÷ 2
= (510 − 75) ÷ 2 = 218 mm.

Holing gauge = batten gauge + head lap + 8 to 15 mm,
= 218 + 75 + (8 to 15 mm) = 301 to 308 mm.

Side lap = 255 ÷ 2 = 127 mm.

Margin = batten gauge of 218 mm.

Eaves course length = head lap + margin = 293 mm.

Roof Slating

Traditional slate names and sizes (mm) —

Empress	650 × 400	Wide Viscountess	460 × 255
Princess	610 × 355	Viscountess	460 × 230
Duchess	610 × 305	Wide Ladies	405 × 255
Small Duchess	560 × 305	Broad ladies	405 × 230
Marchioness	560 × 280	Ladies	405 × 205
Wide Countess	510 × 305	Wide Headers	355 × 305
Countess	510 × 255	Headers	355 × 255
.. ..	510 × 230	Small Ladies	355 × 203
.. ..	460 × 305	Narrow Ladies	355 × 180

Sizes can also be cut to special order.

Generally, the larger the slate, the lower the roof may be pitched. Also, the lower the roof pitch, the greater the head lap.

Slate quality	Thickness (mm)
Best	4
Medium strong	5
Heavy	6
Extra heavy	9

Roof pitch (degrees)	Min. head lap (mm)
20	115
25	85
35	75
45	65

See also:

1. BS 680: Specification for roofing slates.
2. Slate producers catalogues.

Roof hip examples —

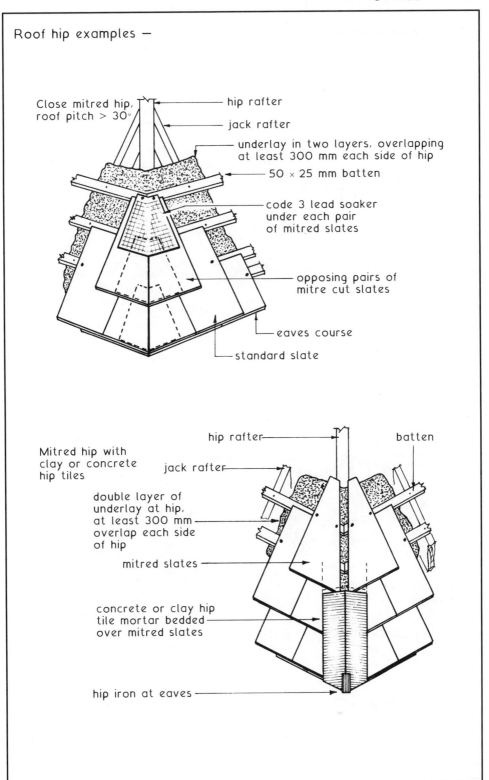

Close mitred hip, roof pitch > 30°

hip rafter

jack rafter

underlay in two layers, overlapping at least 300 mm each side of hip

50 × 25 mm batten

code 3 lead soaker under each pair of mitred slates

opposing pairs of mitre cut slates

eaves course

standard slate

Mitred hip with clay or concrete hip tiles

hip rafter

batten

jack rafter

double layer of underlay at hip, at least 300 mm overlap each side of hip

mitred slates

concrete or clay hip tile mortar bedded over mitred slates

hip iron at eaves

Roof valley examples —

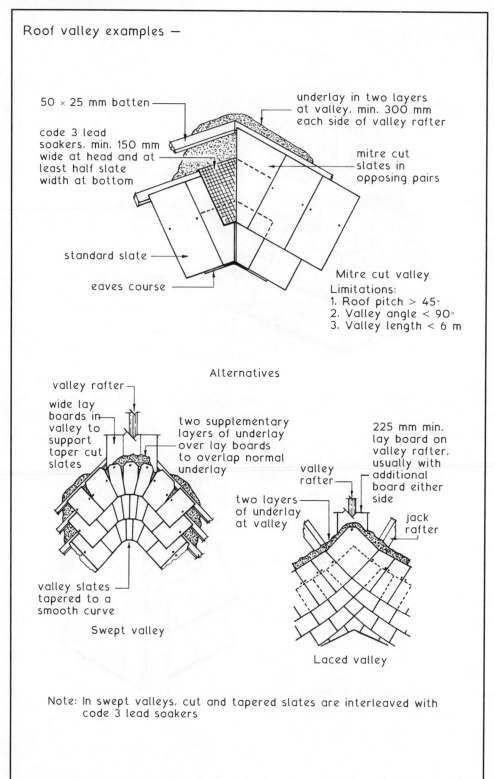

50 × 25 mm batten

code 3 lead soakers, min. 150 mm wide at head and at least half slate width at bottom

underlay in two layers at valley, min. 300 mm each side of valley rafter

mitre cut slates in opposing pairs

standard slate

eaves course

Mitre cut valley
Limitations:
1. Roof pitch > 45°
2. Valley angle < 90°
3. Valley length < 6 m

Alternatives

valley rafter

wide lay boards in valley to support taper cut slates

two supplementary layers of underlay over lay boards to overlap normal underlay

225 mm min. lay board on valley rafter, usually with additional board either side

valley rafter

two layers of underlay at valley

jack rafter

valley slates tapered to a smooth curve

Swept valley

Laced valley

Materials — water reed (Norfolk reed), wheat straw (Spring or Winter), Winter being the most suitable. Wheat for thatch is often known as wheat reed, long straw or Devon reed. Other thatches include rye and oat straws, and sedge. Sedge is harvested every fourth year to provide long growth, making it most suitable as a ridging material.

There are various patterns and styles of thatching, relating to the skill of the thatcher and local traditions.

Typical detail —

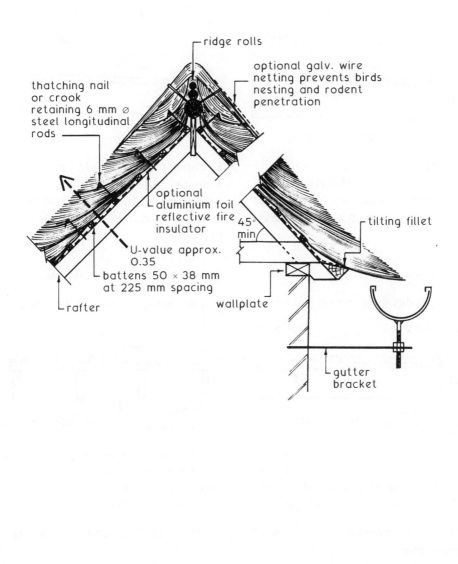

ridge rolls

optional galv. wire netting prevents birds nesting and rodent penetration

thatching nail or crook retaining 6 mm ⌀ steel longitudinal rods

optional aluminium foil reflective fire insulator

45° min

tilting fillet

U-value approx. 0.35

battens 50 × 38 mm at 225 mm spacing

wallplate

rafter

gutter bracket

Flat Roofs ~ these roofs are very seldom flat with a pitch of 0° but are considered to be flat if the pitch does not exceed 10°. The actual pitch chosen can be governed by the roof covering selected and/or by the required rate of rainwater discharge off the roof. As a general rule the minimum pitch for smooth surfaces such as asphalt should be 1:80 or 0°—43' and for sheet coverings with laps 1:60 or 0°—57'.

Methods of Obtaining Falls ~

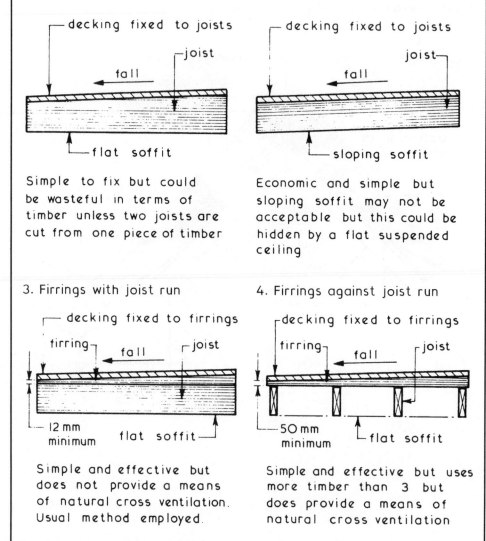

1. Joists cut to falls

decking fixed to joists
—joist
fall
flat soffit

Simple to fix but could be wasteful in terms of timber unless two joists are cut from one piece of timber

2. Joists laid to falls

decking fixed to joists
joist
fall
sloping soffit

Economic and simple but sloping soffit may not be acceptable but this could be hidden by a flat suspended ceiling

3. Firrings with joist run

decking fixed to firrings
firring
fall
joist
12 mm minimum
flat soffit

Simple and effective but does not provide a means of natural cross ventilation. Usual method employed.

4. Firrings against joist run

decking fixed to firrings
firring
fall
joist
50 mm minimum
flat soffit

Simple and effective but uses more timber than 3 but does provide a means of natural cross ventilation

Wherever possible joists should span the shortest distance of the roof plan

Timber Roof Joists ~ the spacing and sizes of joists is related to the loadings and span, actual dimensions for domestic loadings can be taken direct from Tables A17 to A22 of Approved Document A or they can be calculated from first principles in the same manner as used for timber upper floors. Strutting between joists should be used if the span exceeds 2·400 to restrict joist movements and twisting.

Typical Eaves Details ~

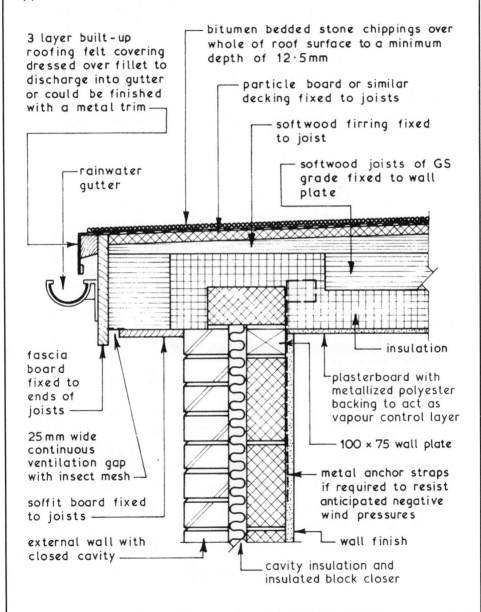

3 layer built-up roofing felt covering dressed over fillet to discharge into gutter or could be finished with a metal trim

rainwater gutter

bitumen bedded stone chippings over whole of roof surface to a minimum depth of 12·5mm

particle board or similar decking fixed to joists

softwood firring fixed to joist

softwood joists of GS grade fixed to wall plate

fascia board fixed to ends of joists

25 mm wide continuous ventilation gap with insect mesh

soffit board fixed to joists

external wall with closed cavity

insulation

plasterboard with metallized polyester backing to act as vapour control layer

100 × 75 wall plate

metal anchor straps if required to resist anticipated negative wind pressures

wall finish

cavity insulation and insulated block closer

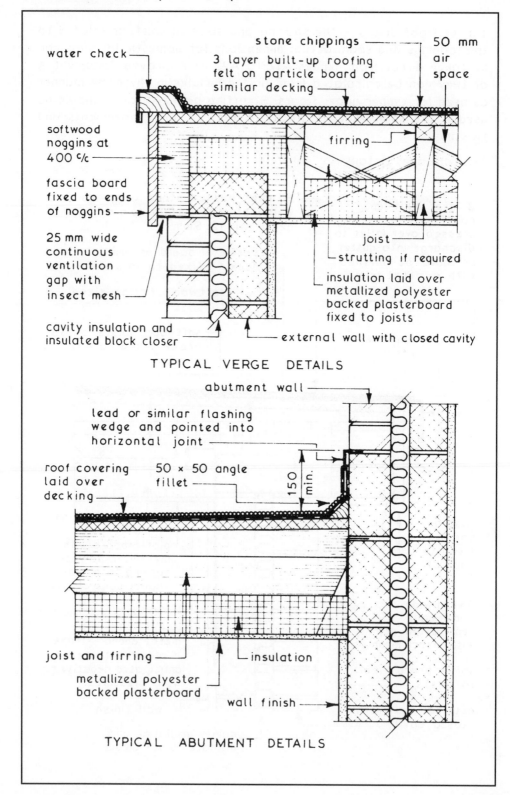

water check

stone chippings

3 layer built-up roofing
felt on particle board or
similar decking

50 mm
air
space

softwood
noggins at
400 c/c

firring

fascia board
fixed to ends
of noggins

25 mm wide
continuous
ventilation
gap with
insect mesh

joist

strutting if required

insulation laid over
metallized polyester
backed plasterboard
fixed to joists

cavity insulation and
insulated block closer

external wall with closed cavity

TYPICAL VERGE DETAILS

abutment wall

lead or similar flashing
wedge and pointed into
horizontal joint

roof covering
laid over
decking

50 × 50 angle
fillet

150 min.

joist and firring

insulation

metallized polyester
backed plasterboard

wall finish

TYPICAL ABUTMENT DETAILS

A dormer is the framework for a vertical window constructed from the roof slope. It may be used as a feature, but is more likely as an economical and practical means for accessing light and ventilation to an attic room. Dormers are normally external with the option of a flat or pitched roof. Frame construction is typical of the following illustrations, with connections made by traditional housed and tenoned joints or simpler galvanized steel brackets and hangers.

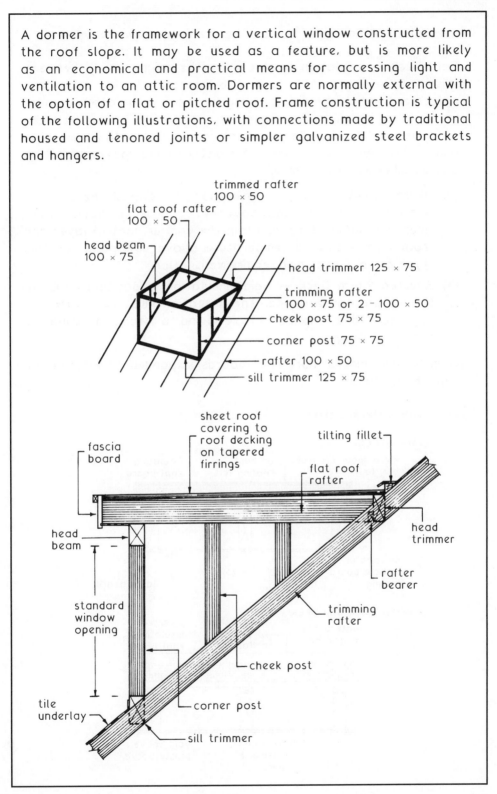

Conservation of Energy ~ this can be achieved in two ways:

1. Cold Deck — insulation is placed on the ceiling lining, between joists. See pages 425 and 426 for details. A metallized polyester lined plasterboard ceiling functions as a vapour control layer, with a minimum 50 mm air circulation space between insulation and decking. The air space corresponds with eaves vents and both provisions will prevent moisture build-up, condensation and possible decay of timber.

2. (a) Warm Deck — rigid* insulation is placed below the waterproof covering and above the roof decking. The insulation must be sufficient to maintain the vapour control layer and roof members at a temperature above dew point, as this type of roof does not require ventilation.

   (b) Inverted Warm Deck — rigid* insulation is positioned above the waterproof covering. The insulation must be unaffected by water and capable of receiving a stone dressing or ceramic pavings.

* Resin bonded mineral fibre roof boards, expanded polystyrene or polyurethane slabs.

Typical Warm Deck Details ~

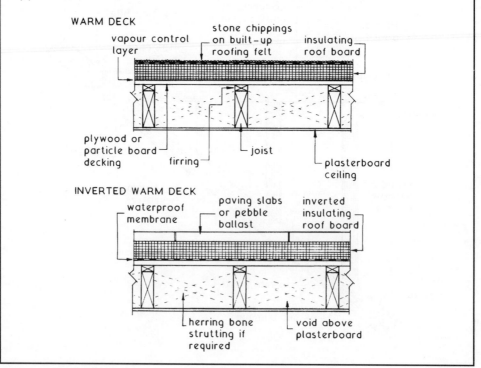

WARM DECK

vapour control layer — stone chippings on built-up roofing felt — insulating roof board

plywood or particle board decking — firring — joist — plasterboard ceiling

INVERTED WARM DECK

waterproof membrane — paving slabs or pebble ballast — inverted insulating roof board

herring bone strutting if required — void above plasterboard

Built-up Roofing Felt ~ this consists of three layers of bitumen roofing felt to BS 747 and should be laid to the recommendations of BS 8217. The layers of felt are bonded together with hot bitumen and should have staggered laps of 50 mm minimum for side laps and 75 mm minimum for end laps — for typical details see pages 425 & 426

Other felt materials which could be used are the two layer polyester based roofing felts which use a non-woven polyester base instead of the woven base used in the BS 747 felts.

Mastic Asphalt ~ this consists of two layers of mastic asphalt laid breaking joints and built up to a minimum thickness of 20 mm and should be laid to the recommendations of BS 8218. The mastic asphalt is laid over an isolating membrane of black sheathing felt complying with BS 747A (i) which should be laid loose with 50 mm minimum overlaps.

Typical Details ~

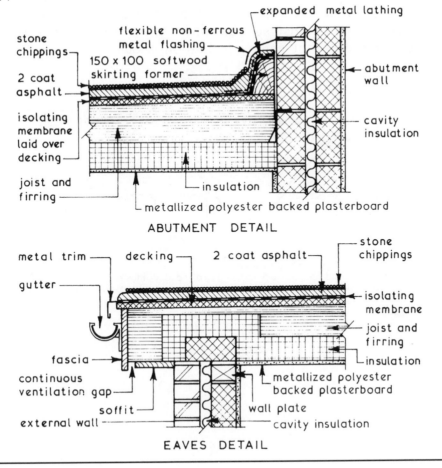

stone chippings
2 coat asphalt
isolating membrane laid over decking
joist and firring

flexible non-ferrous metal flashing
150 x 100 softwood skirting former
expanded metal lathing
abutment wall
cavity insulation
insulation
metallized polyester backed plasterboard

ABUTMENT DETAIL

metal trim
gutter
fascia
continuous ventilation gap
soffit
external wall

decking
2 coat asphalt
stone chippings
isolating membrane
joist and firring
insulation
metallized polyester backed plasterboard
wall plate
cavity insulation

EAVES DETAIL

429

Air carries water vapour, the amount increasing proportionally with the air temperature. As the water vapour increases so does the pressure and this causes the vapour to migrate from warmer to cooler parts of a building. As the air temperature reduces, so does its ability to hold water and this manifests as condensation on cold surfaces. Insulation between living areas and roof spaces increases the temperature differential and potential for condensation in the roof void.

Condensation can be prevented by either of the following:

* Providing a vapour control layer on the warm side of any insulation.
* Removing the damp air by ventilating the colder area.

The most convenient form of vapour layer is vapour check plasterboard which has a moisture resistant lining bonded to the back of the board. A typical patented product is a foil or metal-lized polyester backed plasterboard in 9·5 and 12·5 mm standard thicknesses. This is most suitable where there are rooms in roofs and for cold deck flat roofs. Ventilation is appropriate to larger roof spaces.

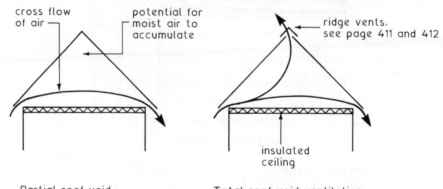

cross flow of air

potential for moist air to accumulate

ridge vents, see page 411 and 412

insulated ceiling

Partial roof void ventilation through the eaves

Total roof void ventilation through eaves and high level vents

Roof ventilation — provision of eaves ventilation alone should allow adequate air circulation in most situations. However, in some climatic conditions and where the air movement is not directly at right angles to the building, moist air can be trapped in the roof apex. Therefore, supplementary ridge ventilation is recommended.

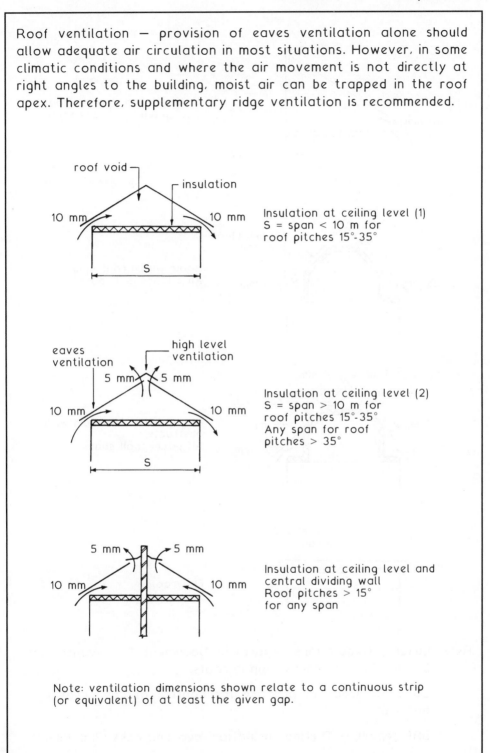

Insulation at ceiling level (1)
S = span < 10 m for
roof pitches 15°-35°

Insulation at ceiling level (2)
S = span > 10 m for
roof pitches 15°-35°
Any span for roof
pitches > 35°

Insulation at ceiling level and
central dividing wall
Roof pitches > 15°
for any span

Note: ventilation dimensions shown relate to a continuous strip (or equivalent) of at least the given gap.

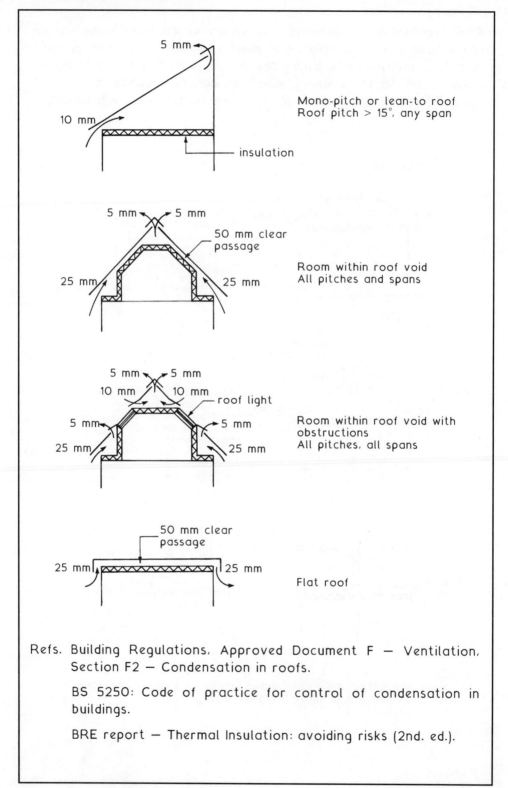

5 mm

10 mm

insulation

Mono-pitch or lean-to roof
Roof pitch > 15°, any span

5 mm   5 mm

50 mm clear passage

25 mm   25 mm

Room within roof void
All pitches and spans

5 mm   5 mm

10 mm   10 mm

roof light

5 mm   5 mm

25 mm   25 mm

Room within roof void with obstructions
All pitches, all spans

50 mm clear passage

25 mm   25 mm

Flat roof

Refs. Building Regulations, Approved Document F — Ventilation, Section F2 — Condensation in roofs.

BS 5250: Code of practice for control of condensation in buildings.

BRE report — Thermal Insulation: avoiding risks (2nd. ed.).

Lateral Restraint — stability of gable walls and construction at the eaves, plus integrity of the roof structure during excessive wind forces, requires complementary restraint and continuity through 30 × 5mm cross sectional area galvanised steel straps.

Exceptions may occur if the roof:-

1. exceeds 15° pitch, and
2. is tiled or slated, and
3. has the type of construction known locally to resist gusts, and
4. has ceiling joists and rafters bearing onto support walls at not more than 1·2 m centres.

Applications ~

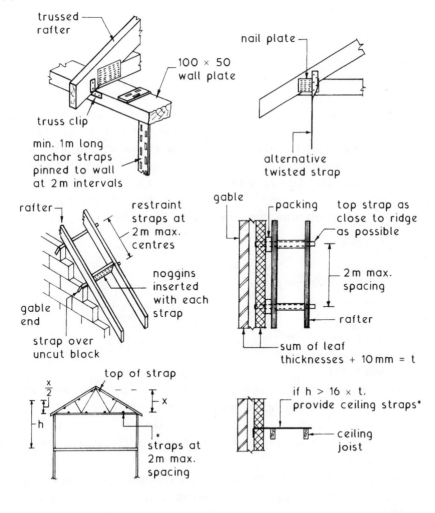

trussed rafter

100 × 50 wall plate

truss clip

min. 1m long anchor straps pinned to wall at 2m intervals

nail plate

alternative twisted strap

rafter

restraint straps at 2m max. centres

noggins inserted with each strap

gable end

strap over uncut block

gable

packing

top strap as close to ridge as possible

2m max. spacing

rafter

sum of leaf thicknesses + 10mm = t

top of strap

$\frac{x}{2}$

x

h

straps at 2m max. spacing

if h > 16 × t, provide ceiling straps*

ceiling joist

Preservation ~ ref. Building Regulations: Materials and Workmanship. Approved Document to support Regulation 7.

Woodworm infestation of untreated structural timbers is common. However, the smaller woodborers such as the abundant Furniture beetle are controllable. It is the threat of considerable damage potential from the House Longhorn beetle that has forced many local authorities in Surrey and the fringe areas of adjacent counties to seek timber preservation listing in the Building Regulations (see Table 1 in the above reference). Prior to the introduction of pretreated timber (c. 1960s), the House Longhorn beetle was once prolific in housing in the south of England, establishing a reputation for destroying structural roof timbers, particularly in the Camberley area.

House Longhorn beetle data:-

Latin name — Hylotrupes bajulus
Life cycle — Mature beetle lays up to 200 eggs on rough surface of untreated timber.
        After 2-3 weeks, larvae emerge and bore into wood, preferring sapwood to denser growth areas. Up to 10 years in the damaging larval stage. In 3 weeks, larvae change to chrysalis to emerge as mature beetles in summer to reproduce.
Timber appearance — powdery deposits (frass) on the surface and the obvious mature beetle flight holes.

Beetle appearance —

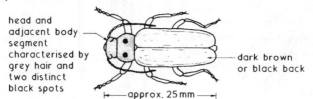

head and adjacent body segment characterised by grey hair and two distinct black spots — approx. 25 mm — dark brown or black back

Other woodborers:-

Furniture beetle — dark brown, 6—8 mm long, lays 20—50 eggs on soft or hardwoods. Bore holes only 1—2 mm diameter.

Lyctus powder post beetle — reddish brown, 10—15 mm long, lays 70—200 eggs on sapwood of new hardwood. Bore holes only 1—2 mm in diameter.

Death Watch beetle — dark brown, sometimes speckled in lighter shades. Lays 40—80 eggs on hardwood. Known for preferring the oak timbers used in old churches and similar buildings.

Bore holes about 3 mm diameter.

Preservation ~ treatment of timber to prevent damage from House Longhorn beetle.

In the areas specified (see previous page), all softwood used in roof structures including ceiling joists and any other softwood fixings should be treated with insecticide prior to installation. Specific chemicals and processes have not been listed in the Building Regulations since the 1976 issue, although the processes detailed then should suffice:-

1. Treatment to BS 4072.
2. Diffusion with sodium borate (boron salts).
3. Steeping for at least 10 mins in an organic solvent wood preservative.

NB. Steeping or soaking in creosote will be effective, but problems of local staining are likely.

BS 4072 provides guidance on an acceptable blend of copper, chromium and arsenic known commercially as Tanalizing. Application is at specialist timber yards by vacuum/pressure impregnation in large cylindrical containers.

Insect treatment adds about 10% to the cost of timber and also enhances its resistance to moisture. Other parts of the structure, e.g. floors and partitions are less exposed to woodworm damage as they are enclosed. Also, there is a suggestion that if these areas received treated timber, the toxic fumes could be harmful to the health of building occupants. Current requirements for through ventilation in roofs has the added benefit of discouraging wood boring insects, as they prefer draught-free damp areas.

Refs. BS 4072: Copper/chromium/arsenic preparations for wood preservation.
BS 4261: Wood preservation. Vocabulary.
BS 5589: Code of practice for preservation of timber
DD 239: Recommendations for the preservation of timber.

Steel Roof Trusses ~ these are triangulated plane frames which carry purlins to which the roof coverings can be fixed. Steel is stronger than timber and will not spread fire over its surface and for these reasons it is often preferred to timber for medium and long span roofs. The rafters are restrained from spreading by being connected securely at their feet by a tie member. Struts and ties are provided within the basic triangle to give adequate bracing. Angle sections are usually employed for steel truss members since they are economic and accept both tensile and compressive stresses. The members of a steel roof truss are connected together with bolts or by welding to shaped plates called gussets. Steel trusses are usually placed at 3·000 to 4·500 centres which gives an economic purlin size.

Typical Steel Roof Truss Formats ~

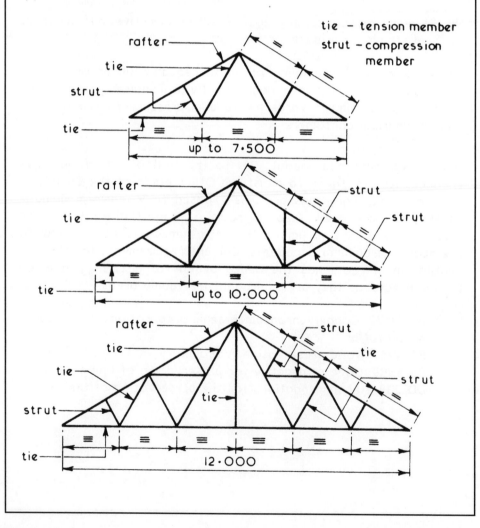

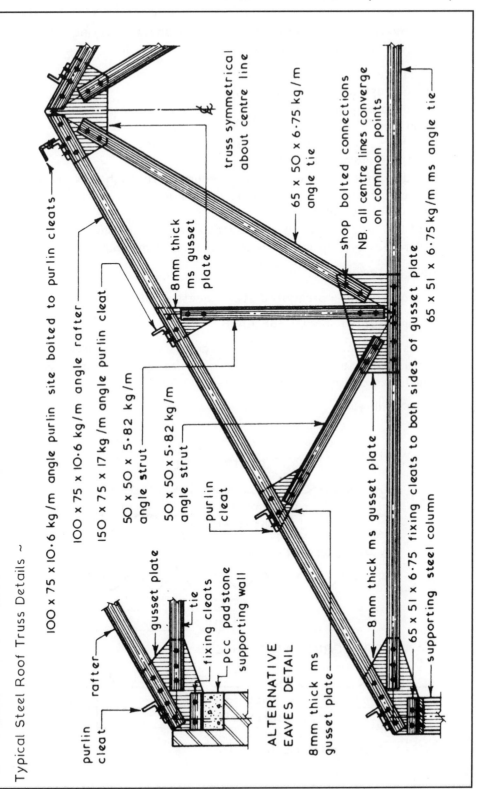

Typical Steel Roof Truss Details ~

100 x 75 x 10·6 kg/m angle purlin site bolted to purlin cleats

100 x 75 x 10·6 kg/m angle rafter

150 x 75 x 17kg/m angle purlin cleat

50 x 50 x 5·82 kg/m angle strut

50 x 50 x 5·82 kg/m angle strut

purlin cleat

truss symmetrical about centre line

8mm thick ms gusset plate

65 x 50 x 6·75 kg/m angle tie

shop bolted connections
NB. all centre lines converge on common points

65 x 51 x 6·75 kg/m ms angle tie

8mm thick ms gusset plate

65 x 51 x 6·75 fixing cleats to both sides of gusset plate

supporting steel column

purlin cleat

rafter

gusset plate

tie

fixing cleats

pcc padstone supporting wall

ALTERNATIVE EAVES DETAIL

8mm thick ms gusset plate

Sheet Coverings ~ the basic functions of sheet coverings used in conjunction with steel roof trusses are to :-

1. Provide resistance to penetration by the elements.
2. Provide restraint to wind and snow loads.
3. Provide a degree of thermal insulation of not less than that set out in Part L of the Building Regulations.
4. Provide resistance to surface spread of flame as set out in Part B of the Building Regulations.
5. Provide any natural daylight required through the roof in accordance with the maximum permitted areas set out in Part L of the Building Regulations.
6. Be of low self weight to give overall design economy.
7. Be durable to keep maintenance needs to a minimum.

Suitable Materials ~

Hot-dip galvanised corrugated steel sheets — BS 3083

Aluminium profiled sheets —BS 4868.

Asbestos free profiled sheets — various manufacturers whose products are usually based on a mixture of Portland cement, mineral fibres and density modifiers — ISO 9383.

Typical Profiles ~

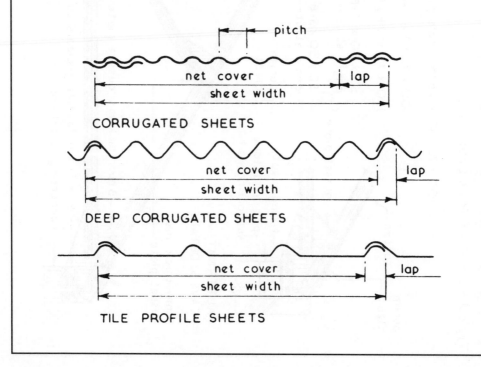

CORRUGATED SHEETS

DEEP CORRUGATED SHEETS

TILE PROFILE SHEETS

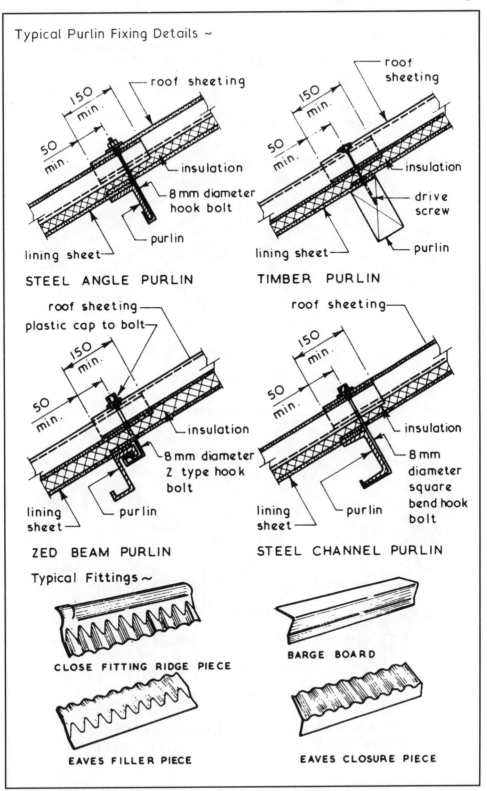

Typical Purlin Fixing Details ~

STEEL ANGLE PURLIN
- roof sheeting
- 150 min.
- 50 min.
- insulation
- 8 mm diameter hook bolt
- purlin
- lining sheet

TIMBER PURLIN
- roof sheeting
- 150 min.
- 50 min.
- insulation
- drive screw
- purlin
- lining sheet

ZED BEAM PURLIN
- roof sheeting
- plastic cap to bolt
- 150 min.
- 50 min.
- insulation
- 8 mm diameter Z type hook bolt
- purlin
- lining sheet

STEEL CHANNEL PURLIN
- roof sheeting
- 150 min.
- 50 min.
- insulation
- 8 mm diameter square bend hook bolt
- purlin
- lining sheet

Typical Fittings ~

CLOSE FITTING RIDGE PIECE

BARGE BOARD

EAVES FILLER PIECE

EAVES CLOSURE PIECE

Typical Details ~

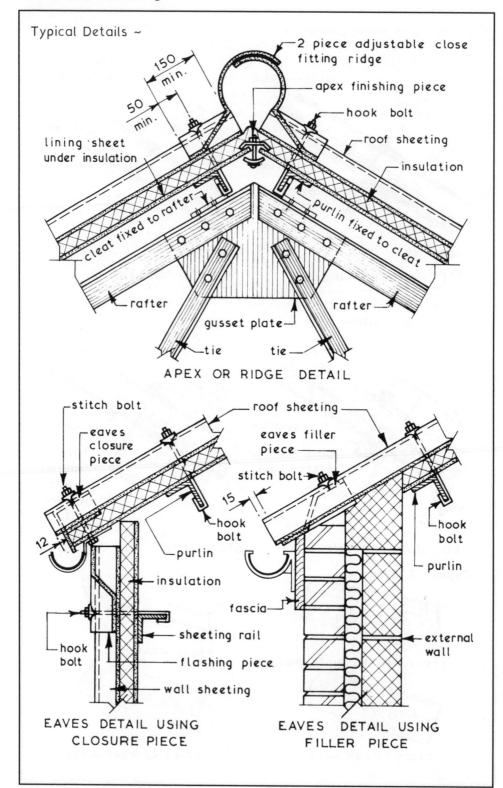

2 piece adjustable close fitting ridge

apex finishing piece

hook bolt

roof sheeting

insulation

150 min.

50 min.

lining sheet under insulation

cleat fixed to rafter

purlin fixed to cleat

rafter

rafter

gusset plate

tie

tie

APEX OR RIDGE DETAIL

stitch bolt

roof sheeting

eaves closure piece

eaves filler piece

stitch bolt

15

hook bolt

12

purlin

hook bolt

purlin

insulation

fascia

external wall

hook bolt

sheeting rail

flashing piece

wall sheeting

EAVES DETAIL USING CLOSURE PIECE

EAVES DETAIL USING FILLER PIECE

Double Skin, Energy Roof systems ~ apply to industrial and commercial use buildings. In addition to new projects constructed to current thermal insulation standards, these systems can be specified to upgrade existing sheet profiled roofs with superimposed supplementary insulation and protective decking. Thermal performance with resin bonded mineral wool fibre of up to 250 mm overall depth may provide 'U' values as low as $0.13\,W/m^2K$.

Typical Details ~

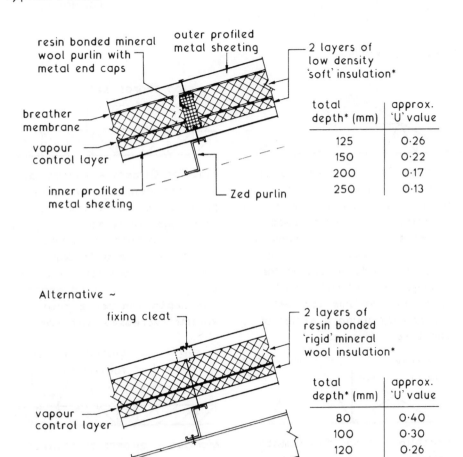

resin bonded mineral wool purlin with metal end caps

outer profiled metal sheeting

2 layers of low density 'soft' insulation*

breather membrane

vapour control layer

inner profiled metal sheeting

Zed purlin

total depth* (mm)	approx. 'U' value
125	0.26
150	0.22
200	0.17
250	0.13

Alternative ~

fixing cleat

2 layers of resin bonded 'rigid' mineral wool insulation*

vapour control layer

portal frame

total depth* (mm)	approx. 'U' value
80	0.40
100	0.30
120	0.26
140	0.22

441

Long Span Roofs ~ these can be defined as those exceeding 12·000 in span. They can be fabricated in steel, aluminium alloy, timber, reinforced concrete and prestressed concrete. Long span roofs can be used for buildings such as factories. Large public halls and gymnasiums which require a large floor area free of roof support columns. The primary roof functions of providing weather protection, thermal insulation, sound insulation and restricting spread of fire over the roof surface are common to all roof types but these roofs may also have to provide strength sufficient to carry services lifting equipment and provide for natural daylight to the interior by means of rooflights.

Basic Roof Forms ~

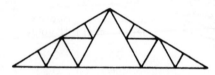

Pitched Trusses - spaced at suitable centres to carry purlins to which the roof coverings are fixed. Good rainwater run off - reasonable daylight spread from rooflights - high roof volume due to the triangulated format - on long spans roof volume can be reduced by using a series of short span trusses.

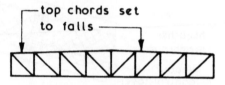

Flat Top Girders - spaced at suitable centres to carry purlins to which the roof coverings are fixed. Low pitch to give acceptable rainwater run off - reasonable daylight spread from rooflights - can be designed for very long spans but depth and hence roof volume increases with span.

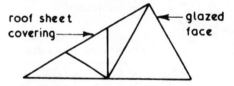

Northlight - spaced at suitable centres to carry purlins to which roof sheeting is fixed. Good rainwater run off - if correctly orientated solar glare is eliminated - long spans can be covered by a series of short span frames

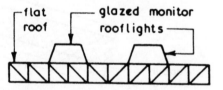

Monitor - girders or cranked beams at centres to suit low pitch decking used. Good even daylight spread from monitor lights which is not affected by orientation of building.

Pitched Trusses ~ these can be constructed with a symmetrical outline (as shown on pages 436 to 440) or with an asymmetrical outline (Northlight — see detail below). They are usually made from standard steel sections with shop welded or bolted connections, alternatively they can be fabricated using timber members joined together with bolts and timber connectors or formed as a precast concrete portal frame.

Typical Multi-span Northlight Roof Details ~

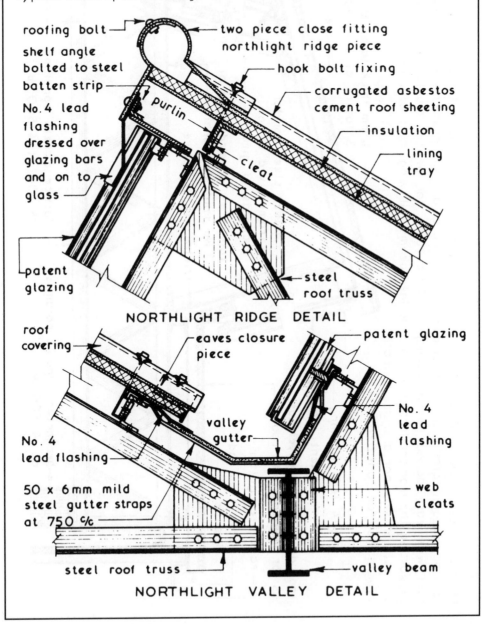

NORTHLIGHT RIDGE DETAIL

NORTHLIGHT VALLEY DETAIL

Monitor Roofs ~ these are basically a flat roof with raised glazed portions called monitors which forms a roof having a uniform distribution of daylight with no solar glare problems irrespective of orientation and a roof with easy access for maintenance. These roofs can be constructed with light long span girders supporting the monitor frames, cranked welded beams following the profile of the roof or they can be of a precast concrete portal frame format.
Typical Monitor Roof Details ~

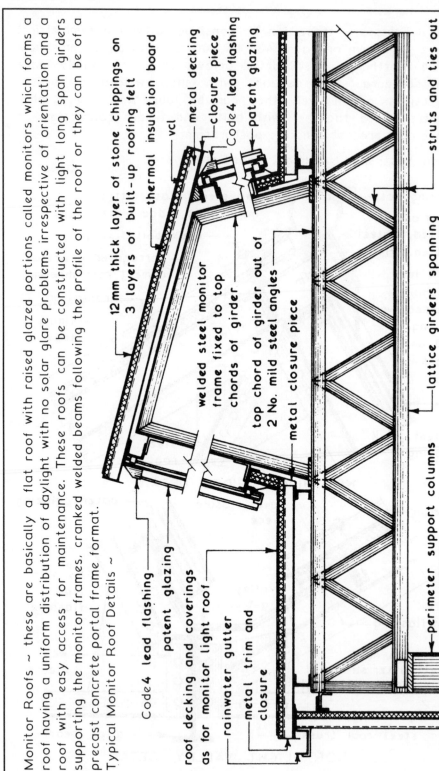

12 mm thick layer of stone chippings on 3 layers of built-up roofing felt

thermal insulation board

vcl

metal decking

closure piece

Code 4 lead flashing

patent glazing

welded steel monitor frame fixed to top chords of girder

top chord of girder out of 2 No. mild steel angles

metal closure piece

struts and ties out of tee section bar

lattice girders spanning 15·000 at 4·500 %

perimeter support columns at 4·500 %

Code 4 lead flashing

patent glazing

roof decking and coverings as for monitor light roof

rainwater gutter

metal trim and closure

Flat Top Girders ~ these are suitable for roof spans ranging from 15·000 to 45·000 and are basically low pitched lattice beams used to carry purlins which support the roof coverings. One of the main advantages of this form of roof is the reduction in roof volume. The usual materials employed in the fabrication of flat top girders are timber and steel.

Typical Flat Top Girder Details ~

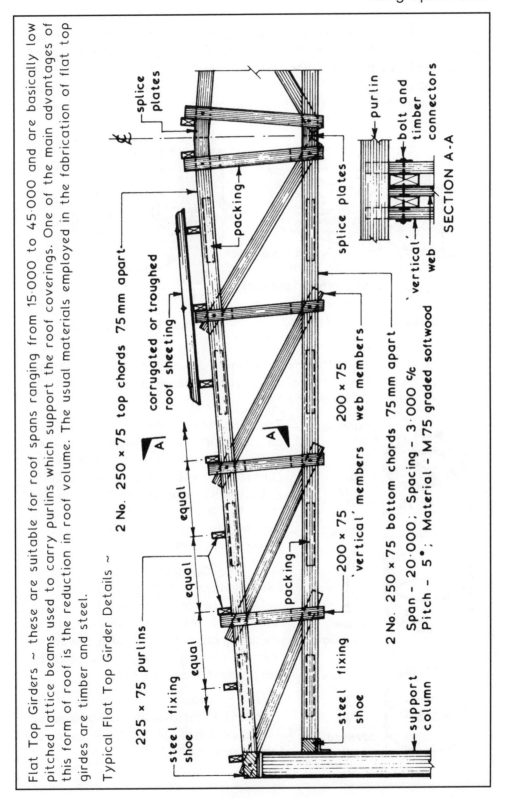

splice plates

𝄴

packing

corrugated or troughed roof sheeting

A

225 × 75 purlins

steel fixing shoe

equal equal equal equal

A

2 No. 250 × 75 top chords 75 mm apart

packing

200 × 75 'vertical' members

splice plates

200 × 75 web members

steel fixing shoe

support column

2 No. 250 × 75 bottom chords 75 mm apart

Span – 20·000; Spacing – 3·000 c/c
Pitch – 5°; Material – M 75 graded softwood

purlin

bolt and timber connectors

'vertical' web

SECTION A·A

445

Connections ~ nails, screws and bolts have their limitations when used to join structural timber members. The low efficiency of joints made with a rigid bar such as a bolt is caused by the usual low shear strength of timber parallel to the grain and the non-uniform distribution of bearing stress along the shank of the bolt —

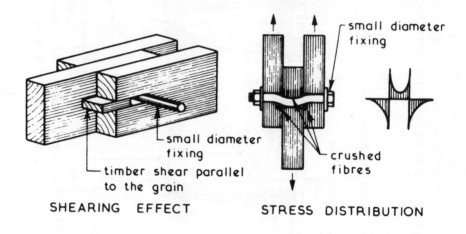

SHEARING EFFECT

STRESS DISTRIBUTION

Timber Connectors ~ these are designed to ovecome the problems of structural timber connections outlined above by increasing the effective bearing area of the bolts.

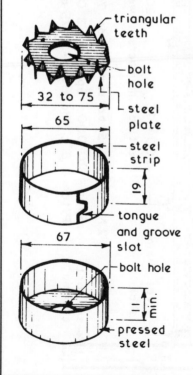

Toothed Plate Connector — provides an efficient joint without special tools or equipment — suitable for all connections especially small sections — bolt holes are drilled 2 mm larger than the bolt diameter, the timbers forming the joint being held together whilst being drilled.

Split Ring Connector — very efficient and develops a high joint strength — suitable for all connections — split ring connectors are inserted into a precut groove formed with a special tool making the connector independent from the bolt.

Shear Plate Connector — counterpart of a split ring connector — housed flush into timber — used for temporary joints.

Space Deck ~ this is a structural roofing system based on a simple repetitive pyramidal unit to give large clear spans of up to 22·000 for single spanning designs and up to 33·000 for two way spanning designs. The steel units are easily transported to site before assembly into beams and the complete space deck at ground level before being hoisted into position on top of the perimeter supports. A roof covering of wood wool slabs with built-up roofing felt could be used, although any suitable structural lightweight decking is appropriate. Rooflights can be mounted directly onto the square top space deck units

Typical Details ~

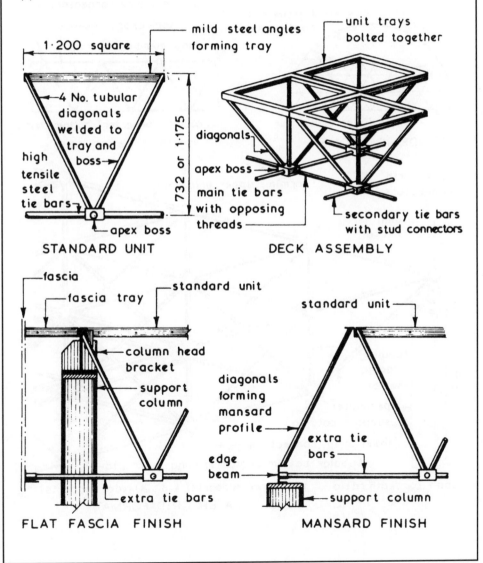

mild steel angles forming tray

unit trays bolted together

1·200 square

4 No. tubular diagonals welded to tray and boss

high tensile steel tie bars

732 or 1·175

diagonals

apex boss

main tie bars with opposing threads

apex boss

STANDARD UNIT

secondary tie bars with stud connectors

DECK ASSEMBLY

fascia

fascia tray

standard unit

column head bracket

support column

extra tie bars

FLAT FASCIA FINISH

standard unit

diagonals forming mansard profile

extra tie bars

edge beam

support column

MANSARD FINISH

Space Frames ~ these are roofing systems which consist of a series of connectors which joins together the chords and bracing members of the system. Single or double layer grids are possible, the former usually employed in connection with small domes or curved roofs. Space frames are similar in concept to space decks but they have greater flexibility in design and layout possibilities. Most space frames are fabricated from structural steel tubes or tubes of aluminium alloy although any suitable structural material could be used.

Typical Examples ~

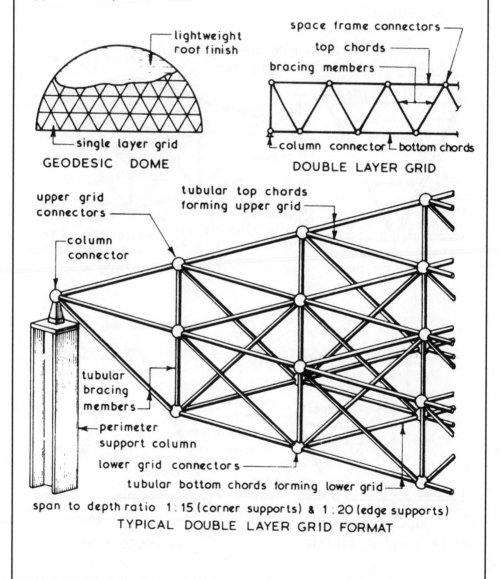

GEODESIC DOME

DOUBLE LAYER GRID

span to depth ratio 1:15 (corner supports) & 1:20 (edge supports)
TYPICAL DOUBLE LAYER GRID FORMAT

Shell Roofs ~ these can be defined as a structural curved skin covering a given plan shape and area where the forces in the shell or membrane are compressive and in the restraining edge beams are tensile. The usual materials employed in shell roof construction are insitu reinforced concrete and timber. Concrete shell roofs are constructed over formwork which in itself is very often a shell roof making this format expensive since the principle of use and reuse of formwork can not normally be applied. The main factors of shell roofs are:-

1. The entire roof is primarily a structural element.
2. Basic strength of any particular shell is inherent in its geometrical shape and form.
3. Comparatively less material is required for shell roofs than other forms of roof construction.

Domes ~ these are double curvature shells which can be rotationally formed by any curved geometrical plane figure rotating about a central vertical axis. Translation domes are formed by a curved line moving over another curved line whereas pendentive domes are formed by inscribing within the base circle a regular polygon and vertical planes through the true hemispherical dome.

Typical Examples ~

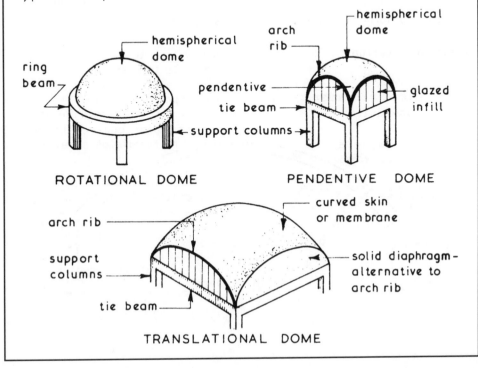

ROTATIONAL DOME   PENDENTIVE DOME

TRANSLATIONAL DOME

Shell Roof Construction

Barrel Vaults ~ these are single curvature shells which are essentially a cut cylinder which must be restrained at both ends to overcome the tendency to flatten. A barrel vault acts as a beam whose span is equal to the length of the roof. Long span barrel vaults are those whose span is longer than its width or chord length and conversely short barrel vaults are those whose span is shorter than its width or chord length. In every long span barrel vaults thermal expansion joints will be required at 30·000 centres which will create a series of abutting barrel vault roofs weather sealed together (see page 451).

Typical Single Barrel Vault Principles ~

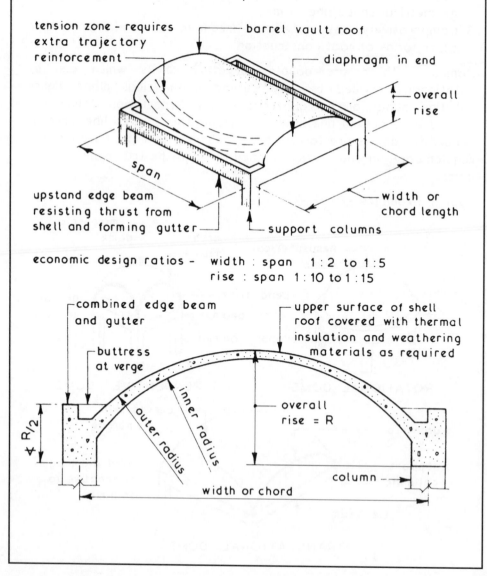

economic design ratios – width : span   1:2 to 1:5
rise : span 1:10 to 1:15

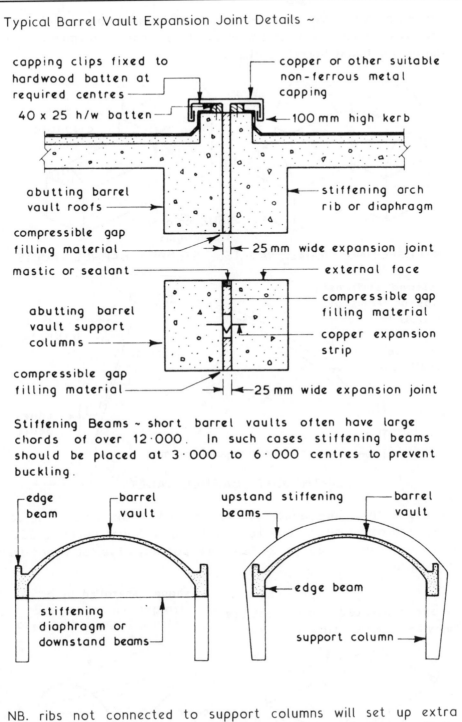

Typical Barrel Vault Expansion Joint Details ~

capping clips fixed to hardwood batten at required centres

copper or other suitable non-ferrous metal capping

40 x 25 h/w batten

100 mm high kerb

abutting barrel vault roofs

stiffening arch rib or diaphragm

compressible gap filling material

25 mm wide expansion joint

mastic or sealant

external face

abutting barrel vault support columns

compressible gap filling material

copper expansion strip

compressible gap filling material

25 mm wide expansion joint

Stiffening Beams ~ short barrel vaults often have large chords of over 12·000. In such cases stiffening beams should be placed at 3·000 to 6·000 centres to prevent buckling.

edge beam

barrel vault

upstand stiffening beams

barrel vault

edge beam

stiffening diaphragm or downstand beams

support column

NB. ribs not connected to support columns will set up extra stresses within the shell roof therefore extra reinforcement will be required at the stiffening rib or beam positions.

451

Other Forms of Barrel Vault ~ by cutting intersecting and placing at different levels the basic barrel vault roof can be formed into a groin or northlight barrel vault roof :-

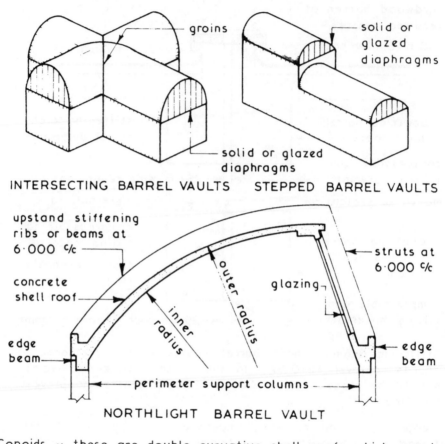

INTERSECTING BARREL VAULTS   STEPPED BARREL VAULTS

NORTHLIGHT BARREL VAULT

Conoids ~ these are double curvative shell roofs which can be considered as an alternative to barrel vaults. Spans up to 12·000 with chord lengths up to 24·000 are possible. Typical chord to span ratio 2:1.

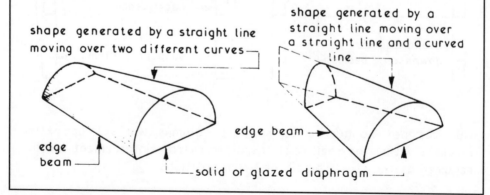

Hyperbolic Paraboloids ~ the true hyperbolic paraboloid shell roof shape is generated by moving a vertical parabola (the generator) over another vertical parabola (the directrix) set at right angles to the moving parabola. This forms a saddle shape where horizontal sections taken through the roof are hyperbolic in format and vertical sections are parabolic. The resultant shape is not very suitable for roofing purposes therefore only part of the saddle shape is used and this is formed by joining the centre points thus :-

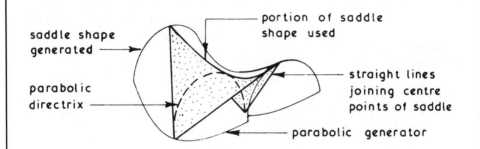

To obtain a more practical shape than the true saddle a straight line limited hyperbolic paraboloid is used. This is formed by raising or lowering one or more corners of a square forming a warped parallelogram thus :-

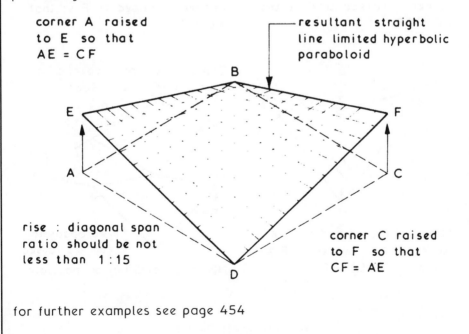

for further examples see page 454

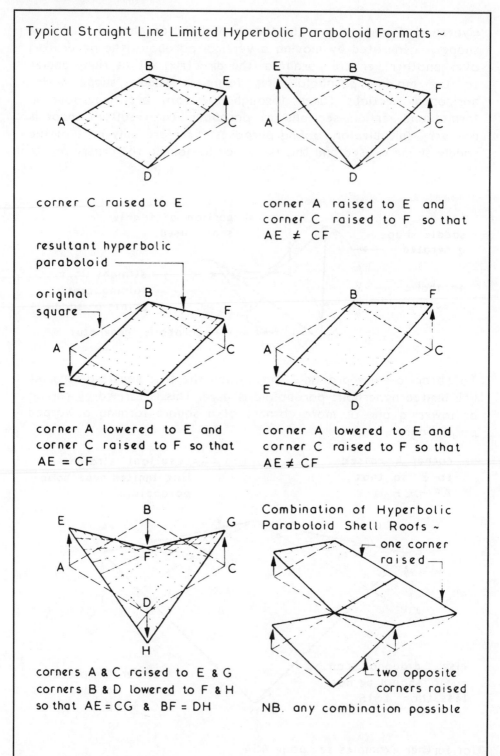

Typical Straight Line Limited Hyperbolic Paraboloid Formats ~

corner C raised to E

corner A raised to E and
corner C raised to F so that
AE ≠ CF

resultant hyperbolic
paraboloid

original
square

corner A lowered to E and
corner C raised to F so that
AE = CF

corner A lowered to E and
corner C raised to F so that
AE ≠ CF

corners A & C raised to E & G
corners B & D lowered to F & H
so that AE = CG & BF = DH

Combination of Hyperbolic
Paraboloid Shell Roofs ~

one corner
raised

two opposite
corners raised

NB. any combination possible

Concrete Hyperbolic Paraboloid Shell Roofs ~ these can be constructed in reinforced concrete (characteristic strength 25 or 30N/mm²) with a minimum shell thickness of 50mm with diagonal spans up to 35·000. These shells are cast over a timber form in the shape of the required hyperbolic paraboloid format. In practice therefore two roofs are constructed and it is one of the reasons for the popularity of timber versions of this form of shell roof.

Timber Hyperbolic Paraboloid Shell Roofs ~ these are usually constructed using laminated edge beams and layers of t & g boarding to form the shell membrane. For roofs with a plan size of up to 6·000 × 6·000 only 2 layers of boards are required and these are laid parallel to the diagonals with both layers running in opposite directions. Roofs with a plan size of over 6·000 × 6·000 require 3 layers of board as shown below. The weather protective cover can be of any suitable flexible material such as built-up roofing felt, copper and lead. During construction the relatively lightweight roof is tied down to a framework of scaffolding until the anchorages and wall infilling have been completed. This is to overcome any negative and positive wind pressures due to the open sides.

Typical Details ~

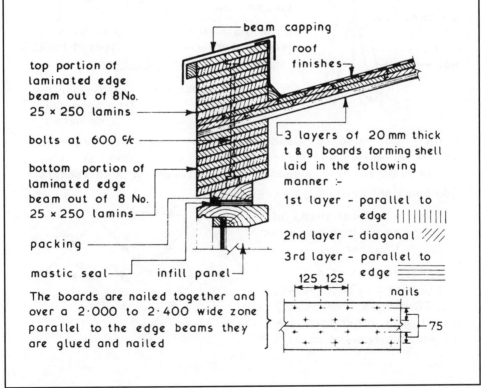

beam capping

roof finishes

top portion of laminated edge beam out of 8 No. 25 × 250 lamins

bolts at 600 ℅

bottom portion of laminated edge beam out of 8 No. 25 × 250 lamins

packing

mastic seal

infill panel

3 layers of 20 mm thick t & g boards forming shell laid in the following manner :-

1st layer – parallel to edge ||||||||||

2nd layer – diagonal ////

3rd layer – parallel to edge ═══

125 125

nails

75

The boards are nailed together and over a 2·000 to 2·400 wide zone parallel to the edge beams they are glued and nailed

Support Considerations ~ in timber hyperbolic paraboloid shell roofs only two supports are required :-

Edge beams are in compression forces P are transmitted to B and D resulting in a vertical force V and a horizontal force H at both positions therefore support columns are required at B and D.

Vertical force V is transmitted directly down the columns to a suitable foundation. The outward or horizontal force H can be accommodated in one of two ways :-

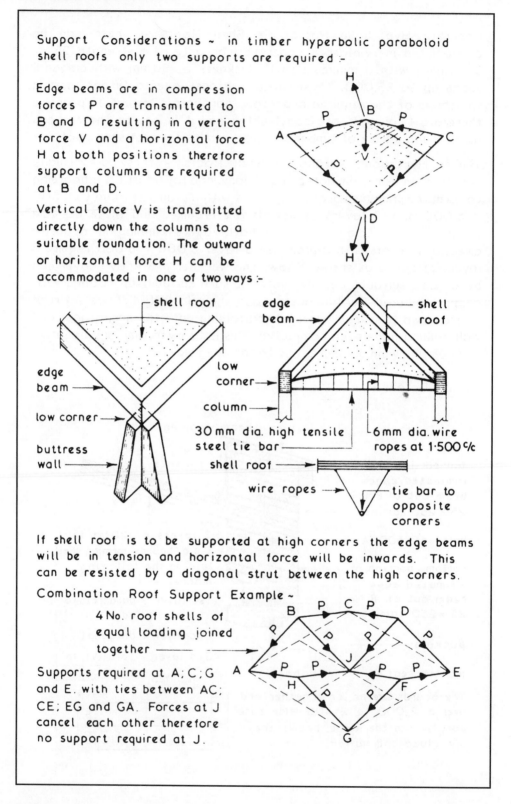

If shell roof is to be supported at high corners the edge beams will be in tension and horizontal force will be inwards. This can be resisted by a diagonal strut between the high corners.

Combination Roof Support Example ~

4 No. roof shells of equal loading joined together ———— ———→

Supports required at A; C; G and E. with ties between AC; CE; EG and GA. Forces at J cancel each other therefore no support required at J.

Rooflights ~ the useful penetration of daylight through the windows in external walls of buildings is from 6·000 to 9·000 depending on the height and size of the window. In buildings with spans over 18·000 side wall daylighting needs to be supplemented by artificial lighting or in the case of top floors or single storey buildings by rooflights. The total maximum area of wall window openings and rooflights for the various purpose groups is set out in the Building Regulations with allowances for increased areas if double or triple glazing is used. In pitched roofs such as northlight and monitor roofs the rooflights are usually in the form of patent glazing (see Long Span Roofs on pages 443 and 444). In flat roof construction natural daylighting can be provided by one or more of the following methods :-

1. Lantern lights — see page 459

2. Lens lights — see page 459

3. Dome, pyramid and similar rooflights — see page 460

Patent Glazing ~ these are systems of steel or aluminium alloy glazing bars which span the distance to be glazed whilst giving continuous edge support to the glass. They can be used in the roof forms noted above as well as in pitched roofs with profiled coverings where the patent glazing bars are fixed above and below the profiled sheets — see page 458

Typical Patent Glazing Bar Sections ~

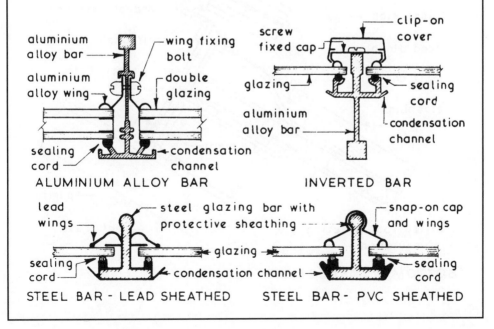

ALUMINIUM ALLOY BAR   INVERTED BAR

STEEL BAR - LEAD SHEATHED   STEEL BAR - PVC SHEATHED

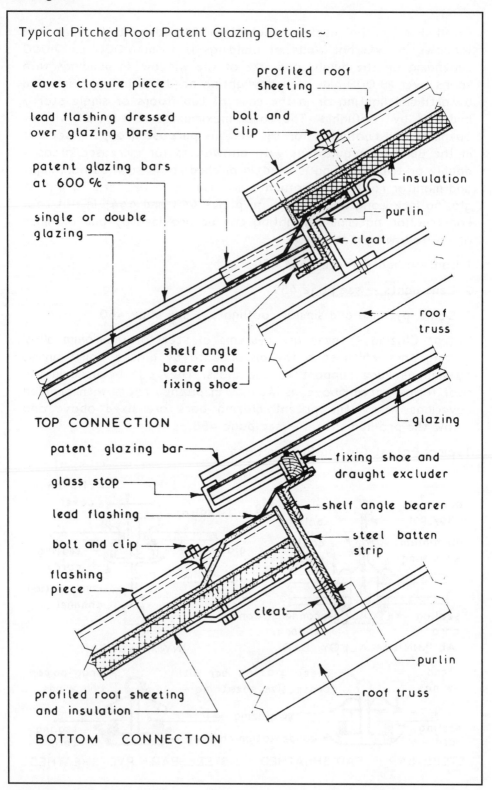

Typical Pitched Roof Patent Glazing Details ~

eaves closure piece

lead flashing dressed over glazing bars

patent glazing bars at 600 ℀

single or double glazing

profiled roof sheeting

bolt and clip

insulation

purlin

cleat

roof truss

shelf angle bearer and fixing shoe

TOP CONNECTION

glazing

patent glazing bar

glass stop

lead flashing

bolt and clip

flashing piece

fixing shoe and draught excluder

shelf angle bearer

steel batten strip

cleat

purlin

roof truss

profiled roof sheeting and insulation

BOTTOM CONNECTION

Lantern Lights ~ these are a form of rooflight used in conjuction with flat roofs. They consist of glazed vertical sides and fully glazed pitched roof which is usually hipped at both ends. Part of the glazed upstand sides is usually formed as an opening light or alternatively glazed with louvres to provide a degree of controllable ventilation. They can be constructed of timber, metal or a combination of these two materials. Lantern lights in the context of new buildings have been generally superseded by the various forms of dome light (see page 460)

Typical Lantern Light Details ~

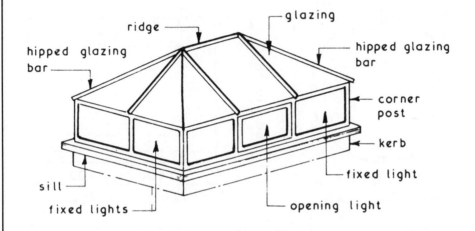

Lens Lights ~ these are small square or round blocks of translucent toughened glass especially designed for casting into concrete and are suitable for use in flat roofs and curved roofs such as barrel vaults. They can also be incorporated in precast concrete frames for inclusion into a cast insitu roof.

Typical Detail ~

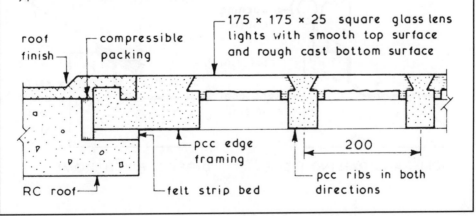

Dome, Pyramid and Similar Rooflights ~ these are used in conjuction with flat roofs and may be framed or unframed. The glazing can be of glass or plastics such as polycarbonate, acrylic, PVC and glass fibre reinforced polyester resin (grp). The whole component is fixed to a kerb and may have a raising piece containing hit and miss ventilators, louvres or flaps for controllable ventilation purposes.

Typical Details ~

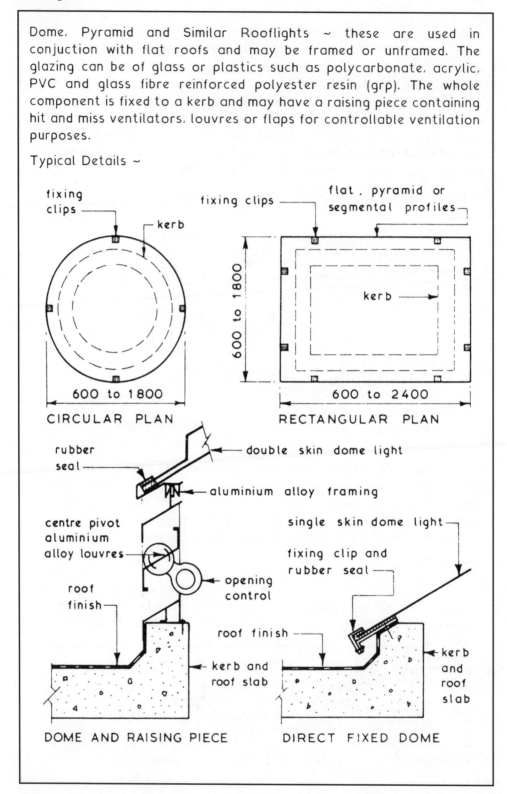

CIRCULAR PLAN

RECTANGULAR PLAN

DOME AND RAISING PIECE

DIRECT FIXED DOME

Claddings to External Walls ~ external walls of block or timber frame construction can be clad with tiles, timber boards or plastic board sections. The tiles used are plain roofing tiles with either a straight or patterned bottom edge. They are applied to the vertical surface in the same manner as tiles laid on a sloping surface (see pages 410 to 411) except that the gauge can be wider and each tile is twice nailed. External and internal angles can be formed using special tiles or they can be mitred. Timber boards such as matchboarding and shiplap can be fixed vertically to horizontal battens or horizontally to vertical battens. Plastic moulded board claddings can be applied in a similar manner. The battens to which the claddings are fixed should be treated with a preservative against fungi and beetle attack and should be fixed with corrosion resistant nails.

Typical Details ~

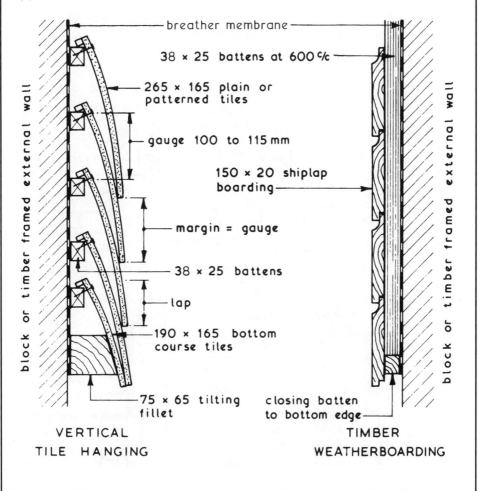

breather membrane

38 × 25 battens at 600 c/c

265 × 165 plain or patterned tiles

gauge 100 to 115 mm

150 × 20 shiplap boarding

margin = gauge

38 × 25 battens

lap

190 × 165 bottom course tiles

75 × 65 tilting fillet

closing batten to bottom edge

block or timber framed external wall

block or timber framed external wall

VERTICAL TILE HANGING

TIMBER WEATHERBOARDING

Non-load Bearing Brick Panel Walls ~ these are used in conjunction with framed structures as an infill between the beams and columns. They are constructed in the same manner as ordinary brick walls with the openings being formed by traditional methods.

Basic Requirements ~

1. To be adequately supported by and tied to the structural frame.
2. Have sufficient strength to support own self weight plus any attached finishes and imposed loads such as wind pressures.
3. Provide the necessary resistance to penetration by the natural elements.
4. Provide the required degree of thermal insulation, sound insulation and fire resistance.
5. Have sufficient durability to reduce maintenance costs to a minimum.
6. Provide for movements due to moisture and thermal expansion of the panel and for contraction of the frame.

Typical Details ~

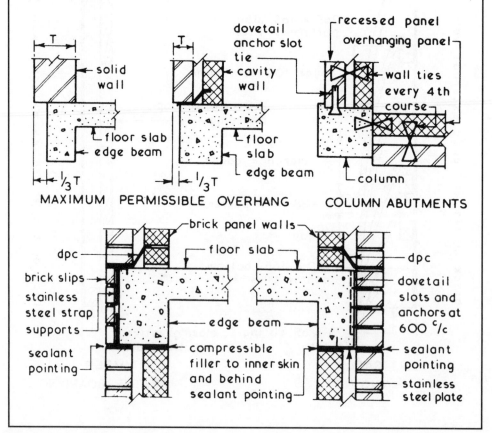

MAXIMUM PERMISSIBLE OVERHANG   COLUMN ABUTMENTS

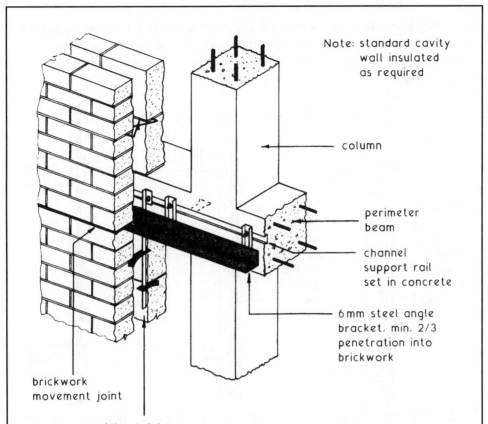

Note: standard cavity wall insulated as required

column

perimeter beam

channel support rail set in concrete

6mm steel angle bracket, min. 2/3 penetration into brickwork

brickwork movement joint

sliding brick anchor~(ties fit loosely over guide)

Application — multi-storey buildings, where a traditional brick facade is required.

Brickwork movement — to allow for climatic changes and differential movement between the cladding and main structure, a 'soft' joint (cellular polyethylene, cellular polyurethane, expanded rubber or sponge rubber with polysulphide or silicon pointing) should be located below the support angle. Vertical movement joints may also be required at a maximum of 12m spacing.

Lateral restraint — provided by normal wall ties between inner and outer leaf of masonry, plus sliding brick anchors below the support angle.

Infill Panel Walls ~ these can be used between the framing members of a building to provide the cladding and division between the internal and external environments and are distinct from claddings and facing :-

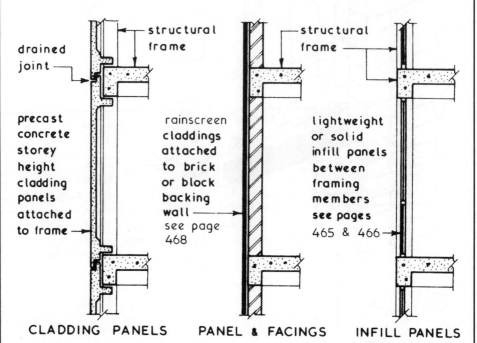

drained joint —

structural frame

structural frame

precast concrete storey height cladding panels attached to frame →

rainscreen claddings attached to brick or block backing wall → see page 468

lightweight or solid infill panels between framing members see pages 465 & 466 →

CLADDING PANELS   PANEL & FACINGS   INFILL PANELS

Functional Requirements ~ all forms of infill panel should be designed and constructed to fulfil the following functional requirements :-

1. Self supporting between structural framing members.
2. Provide resistance to the penetration of the elements.
3. Provide resistance to positive and negative wind pressures.
4. Give the required degree of thermal insulation.
5. Give the required degree of sound insulation.
6. Give the required degree of fire resistance.
7. Have sufficient openings to provide the required amount of natural ventilation.
8. Have sufficient glazed area to fulfil the natural daylight and vision out requirements.
9. Be economic in the context of construction and maintenance.
10. Provide for any differential movements between panel and structural frame.

Brick infill Panels ~ these can be constructed in a solid or cavity format the latter usually having an inner skin of blockwork to increase the thermal insulation properties of the panel. All the fundamental construction processes and detail of solid and cavity walls (bonding, lintels over openings, wall ties, damp-proof courses etc.,) apply equally to infill panel walls. The infill panel walls can be tied to the columns by means of wall ties cast into the columns at 300mm centres or located in cast-in dovetail anchor slots. The head of every infill panel should have a compressible joint to allow for any differential movements between the frame and panel.

Typical Details

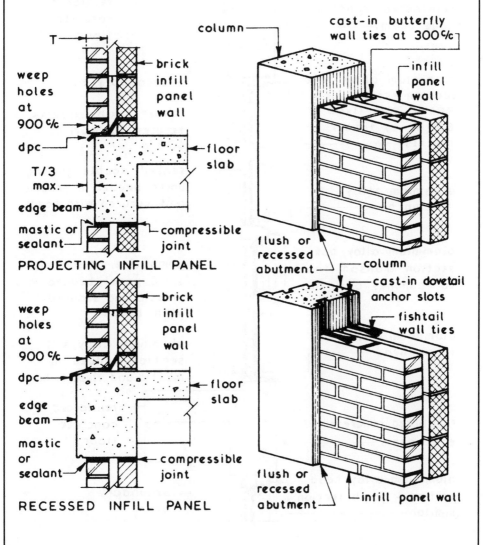

Infill Panel Walls

Lightweight Infill Panels ~ these can be constructed from a wide variety or combination of materials such as timber, metals and plastics into which single or double glazing can be fitted. If solid panels are to be used below a transom they are usually of a composite or sandwich construction to provide the required sound insulation, thermal insulation and fire resistance properties.

Typical Example ~

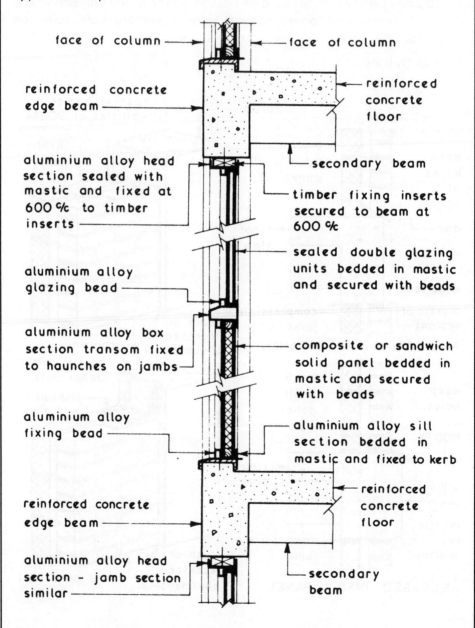

face of column

face of column

reinforced concrete edge beam

reinforced concrete floor

aluminium alloy head section sealed with mastic and fixed at 600 % to timber inserts

secondary beam

timber fixing inserts secured to beam at 600 %

sealed double glazing units bedded in mastic and secured with beads

aluminium alloy glazing bead

aluminium alloy box section transom fixed to haunches on jambs

composite or sandwich solid panel bedded in mastic and secured with beads

aluminium alloy fixing bead

aluminium alloy sill section bedded in mastic and fixed to kerb

reinforced concrete edge beam

reinforced concrete floor

aluminium alloy head section – jamb section similar

secondary beam

Lightweight Infill Panels ~ these can be fixed between the structural horizontal and vertical members of the frame or fixed to the face of either the columns or beams to give a grid, horizontal or vertical emphasis to the façade thus —

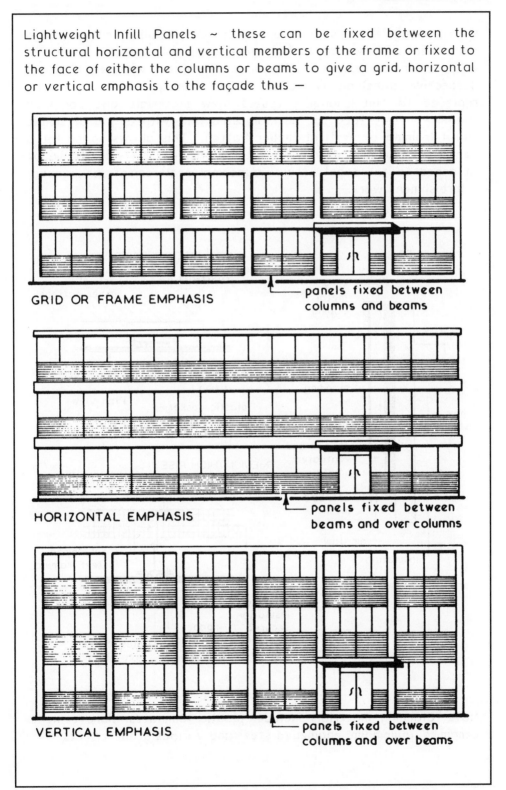

GRID OR FRAME EMPHASIS

panels fixed between columns and beams

HORIZONTAL EMPHASIS

panels fixed between beams and over columns

VERTICAL EMPHASIS

panels fixed between columns and over beams

Overcladding ~ a superficial treatment, applied either as a component of new construction work, or as a façade and insulation enhancement to existing structures. The outer weather resistant decorative panelling is 'loose fit' in concept, which is easily replaced to suit changing tastes, new materials and company image. Panels attach to the main structure with a grid of simple metal framing or vertical timber battens. This allows space for a ventilated and drained cavity, with provision for insulation to be attached to the substructure; a normal requirement in upgrade/refurbishment work.

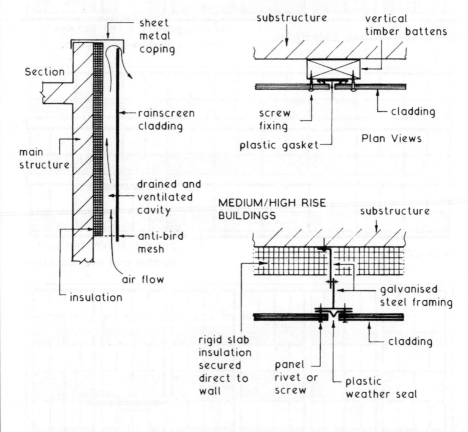

Note: Cladding materials include, plastic laminates, fibre cement, ceramics, aluminium, enamelled steel and various stone effects.

Glazed facades have been associated with hi-tech architecture since the 1970s. The increasing use of this type of cladding is largely due to developments in toughened glass and improved qualities of elastomeric silicone sealants. The properties of the latter must incorporate a resilience to varying atmospheric conditions as well as the facility to absorb structural movement without loss of adhesion.

Systems — two edge and four edge.

The two edge system relies on conventional glazing beads/fixings to the head and sill parts of a frame, with sides silicone bonded to mullions and styles.

The four edge system relies entirely on structural adhesion, using silicone bonding between glazing and support frame — see details.

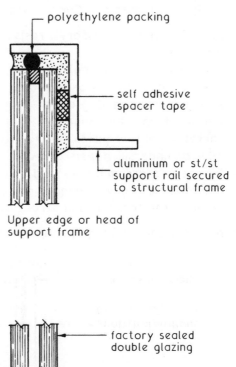

polyethylene packing

self adhesive
spacer tape

aluminium or st/st
support rail secured
to structural frame

Upper edge or head of
support frame

factory sealed
double glazing

structural
sealant

precured silicon
spacer block

silicon seal
and pointing

Lower edge of support frame to sill
Note: Sides of frame as head.

Curtain Walling ~ this is a form of lightweight non-load bearing external cladding which forms a complete envelope or sheath around the structural frame. In low rise structures the curtain wall framing could be of timber or patent glazing but in the usual high rise context, box or solid members of steel or aluminium alloy are normally employed.

Basic Requirements for Curtain Walls ~

1. Provide the necessary resistance to penetration by the elements.
2. Have sufficient strength to carry own self weight and provide resistance to both positive and negative wind pressures.
3. Provide required degree of fire resistance — glazed areas are classified in the Building Regulations as unprotected areas therefore any required fire resistance must be obtained from the infill or undersill panels and any backing wall or beam.
4. Be easy to assemble, fix and maintain.
5. Provide the required degree of sound and thermal insulation.
6. Provide for thermal and structural movements.

Typical Curtain Walling Arrangement ~

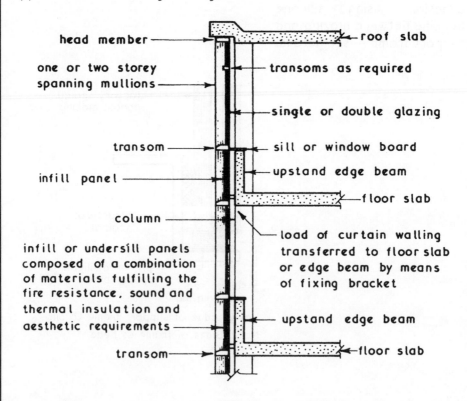

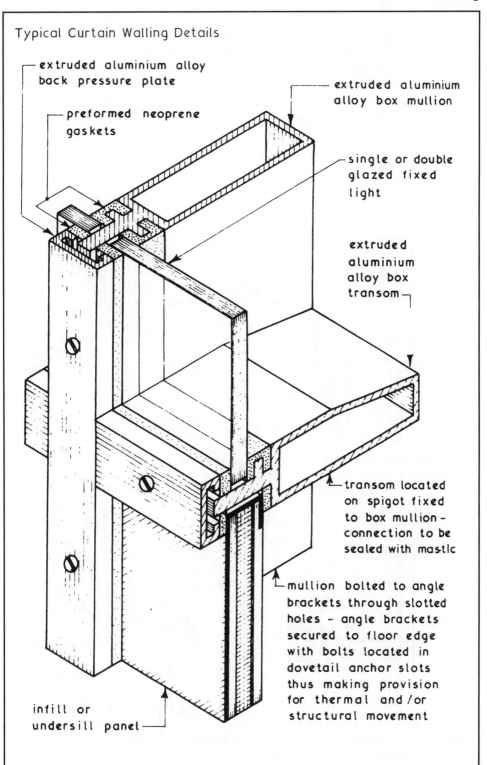

Typical Curtain Walling Details

extruded aluminium alloy back pressure plate

preformed neoprene gaskets

extruded aluminium alloy box mullion

single or double glazed fixed light

extruded aluminium alloy box transom ⌐

transom located on spigot fixed to box mullion - connection to be sealed with mastic

mullion bolted to angle brackets through slotted holes - angle brackets secured to floor edge with bolts located in dovetail anchor slots thus making provision for thermal and / or structural movement

infill or undersill panel

Curtain Walling

Fixing Curtain Walling to the Structure ~ in curtain walling systems it is the main vertical component or mullion which carries the loads and transfers them to the structural frame at every or alternate floor levels depending on the spanning ability of the mullion. At each fixing point the load must be transferred and an allowance made for thermal expansion and differential movement between the structural frame and curtain walling. The usual method employed is slotted bolt fixings.

Typical Examples ~

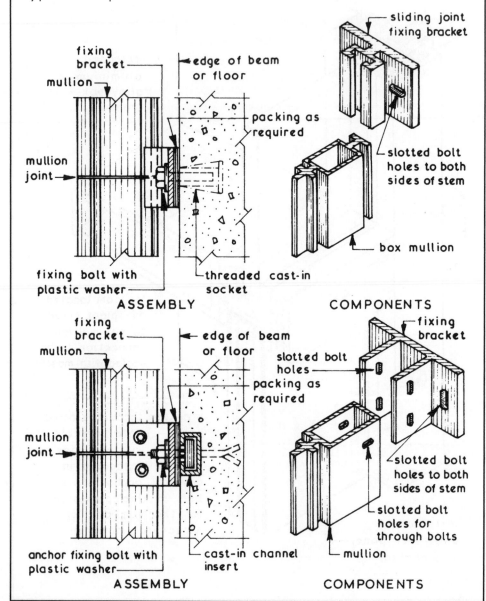

Loadbearing Concrete Panels ~ this form of construction uses storey height loadbearing precast reinforced concrete perimeter panels. The width and depth of the panels is governed by the load(s) to be carried, the height and exposure of the building. Panels can be plain or fenestrated providing the latter leaves sufficient concrete to transmit the load(s) around the opening. The cladding panels, being structural, eliminate the need for perimeter columns and beams and provide an internal surface ready to receive insulation, attached services and decorations. In the context of design these structures must be formed in such a manner that should a single member be removed by an internal explosion, wind pressure or similar force progressive or structural collapse will not occur, the minimum requirements being set out in Part A of the Building Regulations. Loadbearing concrete panel construction can be a cost effective method of building.

Typical Details ~

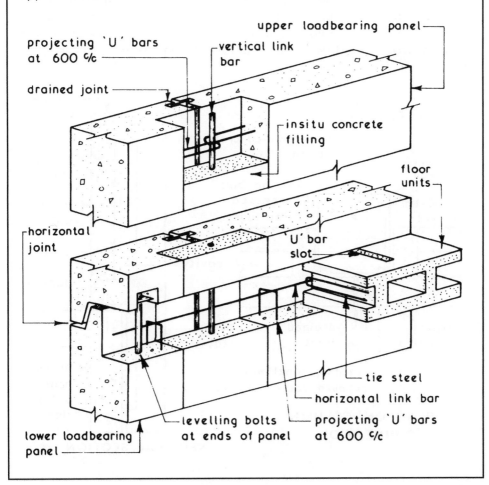

Concrete Cladding Panels ~ these are usually of reinforced precast concrete to an undersill or storey height format, the former being sometimes called apron panels. All precast concrete cladding panels should be designed and installed to fulfil the following functions:-

1. Self supporting between framing members.
2. Provide resistance to penetration by the natural elements.
3. Resist both positive and negative wind pressures.
4. Provide required degree of fire resistance.
5. Provide required degree of thermal insulation by having the insulating material incorporated within the body of the cladding or alternatively allow the cladding to act as the outer leaf of cavity wall panel.
6. Provide required degree of sound insulation.

Undersill or Apron Cladding Panels ~ these are designed to span from column to column and provide a seating for the windows located above. Levelling is usually carried out by wedging and packing from the lower edge before being fixed with grouted dowels.

Typical Details ~

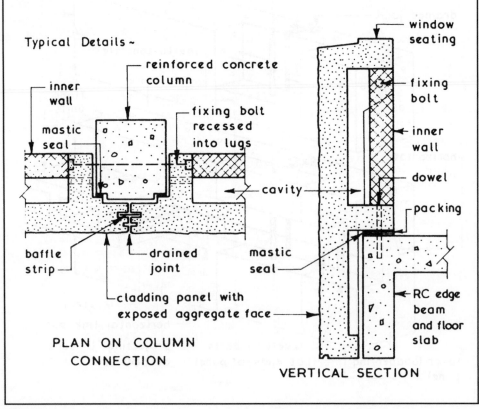

PLAN ON COLUMN
CONNECTION

VERTICAL SECTION

Storey Height Cladding Panels~ these are designed to span vertically from beam to beam and can be fenestrated if required. Levelling is usually carried out by wedging and packing from floor level before being fixed by bolts or grouted dowels.

Typical Details ~

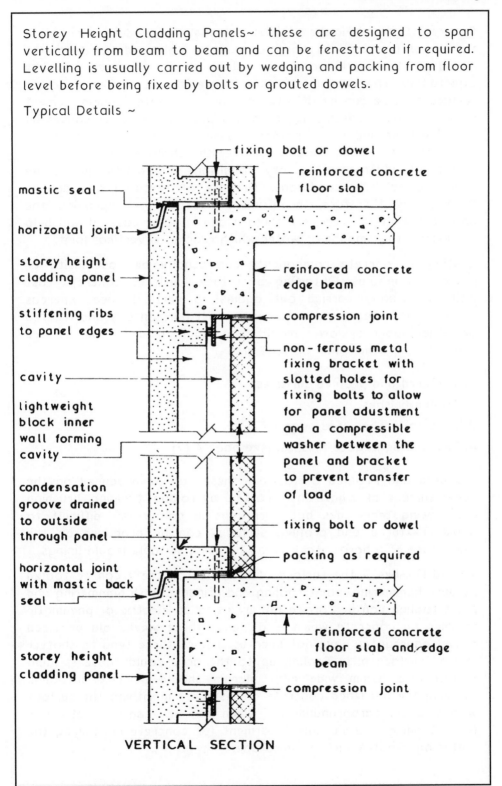

mastic seal

horizontal joint

storey height cladding panel

stiffening ribs to panel edges

cavity

lightweight block inner wall forming cavity

condensation groove drained to outside through panel

horizontal joint with mastic back seal

storey height cladding panel

fixing bolt or dowel

reinforced concrete floor slab

reinforced concrete edge beam

compression joint

non-ferrous metal fixing bracket with slotted holes for fixing bolts to allow for panel adustment and a compressible washer between the panel and bracket to prevent transfer of load

fixing bolt or dowel

packing as required

reinforced concrete floor slab and edge beam

compression joint

VERTICAL SECTION

Concrete Surface Finishes ~ it is not easy to produce a concrete surface with a smooth finish of uniform colour direct from the mould or formwork since the colour of the concrete can be affected by the cement and fine aggregate used. The concrete surface texture can be affected by the aggregate grading, cement content, water content, degree of compaction, pin holes caused by entrapped air and rough patches caused by adhesion to parts of the formworks. Complete control over the above mentioned causes is difficult under ideal factory conditions and almost impossible under normal site conditions. The use of textured and applied finishes has therefore the primary function of improving the appearance of the concrete surface and in some cases it will help to restrict the amount of water which reaches a vertical joint.

Casting ~ concrete components can usually be cast insitu or precast in moulds. Obtaining a surface finish to concrete cast insitu is usually carried out against a vertical face, whereas precast concrete components can be cast horizontally and treated on either upper or lower mould face. Apart from a plain surface concrete the other main options are :-

1. Textured and profiled surfaces.

2. Tooled finishes.

3. Cast-on finishes. (see page 477)

4. Exposed aggregate finishes. (see page 477)

Textured and Profiled Surfaces ~ these can be produced on the upper surface of a horizontal casting by rolling, tamping, brushing and sawing techniques but variations in colour are difficult to avoid. Textured and profiled surfaces can be produced on the lower face of a horizontal casting by using suitable mould linings.

Tooled Finishes ~ the surface of hardened concrete can be tooled by bush hammering, point tooling and grinding. Bush hammering and point tooling can be carried out by using an electric or pneumatic hammer on concrete which is at least three weeks old provided gravel aggregates have not been used since these tend to shatter leaving surface pits. Tooling up to the arris could cause spalling therefore a 10 mm wide edge margin should be left untooled. Grinding the hardened concrete consists of smoothing the surface with a rotary carborundum disc which may have an integral water feed. Grinding is a suitable treatment for concrete containing the softer aggregates such as limestone.

Cast-on Finishes ~ these finishes include split blocks, bricks, stone, tiles and mosaic. Cast-on finishes to the upper surface of a horizontal casting are not recommended although such finishes could be bedded onto the fresh concrete. Lower face treatment is by laying the materials with sealed or grouted joints onto the base of mould or alternatively the materials to be cast-on may be located in a sand bed spread over the base of the mould.

Exposed Aggregate Finishes ~ attractive effects can be obtained by removing the skin of hardened cement paste or surface matrix, which forms on the surface of concrete, to expose the aggregate. The methods which can be employed differ with the casting position.

Horizontal Casting — treatment to the upper face can consist of spraying with water and brushing some two hours after casting, trowelling aggregate into the fresh concrete surface or by using the felt-float method. This method consists of trowelling 10 mm of dry mix fine concrete onto the fresh concrete surface and using the felt pad to pick up the cement and fine particles from the surface leaving a clean exposed aggregate finish.

Treatment to the lower face can consist of applying a retarder to the base of the mould so that the partially set surface matrix can be removed by water and/or brushing as soon as the castings are removed from the moulds. When special face aggregates are used the sand bed method could be employed.

Vertical Casting — exposed aggregate finishes to the vertical faces can be obtained by tooling the hardened concrete or they can be cast-on by the aggregate transfer process. This consists of sticking the selected aggregate onto the rough side of pegboard sheets with a mixture of water soluble cellulose compounds and sand fillers. The cream like mixture is spread evenly over the surface of the pegboard to a depth of one third the aggregate size and the aggregate sprinkled or placed evenly over the surface before being lightly tamped into the adhesive. The prepared board is then set aside for 36 hours to set before being used as a liner to the formwork or mould.

The liner is used in conjunction with a loose plywood or hardboard baffle placed against the face of the aggregate. The baffle board is removed as the concrete is being placed.

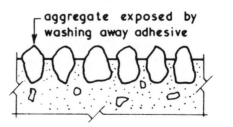

aggregate exposed by washing away adhesive

Thermal insulation of external elements of construction is measured in terms of thermal transmittance rate, otherwise known as the U-value. It is the amount of heat energy in watts transmitted through one square metre of construction for every one degree Kelvin between external and internal air temperature, i.e. W/m^2K.

U-values are unlikely to be entirely accurate, due to:

* the varying effects of solar radiation, atmospheric dampness and prevailing winds.

* inconsistencies in construction, even with the best of supervision.

* 'bridging' where different structural components meet, e.g. dense mortar in lightweight blockwork.

Nevertheless, calculation of the U-value for a particular element of construction will provide guidance as to whether the structure is thermally acceptable. The Building Regulations, Approved Document L, Conservation of fuel and power, determines acceptable energy efficiency standards for modern buildings, with the objective of limiting the emission of carbon dioxide and other burnt gases into the atmosphere.

The U-value is calculated by taking the reciprocal of the summed thermal resistances (R) of the component parts of an element of construction:

$$U = \frac{1}{\Sigma R} = W/m^2K$$

R is expressed in m^2K/W. The higher the value, the better a component's insulation. Conversely, the lower the value of U, the better the insulative properties of the structure.

Thermal resistances (R) are a combination of the different structural, surface and air space components which make up an element of construction. Typically:

$$U = \frac{1}{R_{so} + R_1 + R_2 + R_a + R_3 + R_4 \text{ etc} \cdots + R_{si}(m^2 K/W)}$$

Where: R_{so} = Outside or external surface resistance.
R_1, R_2, etc. = Thermal resistance of structural components.
R_a = Air space resistance, eg. wall cavity.
R_{si} = Internal surface resistance.

The thermal resistance of a structural component (R_1, R_2, etc.) is calculated by dividing its thickness (L) by its thermal conductivity (λ), i.e.

$$R(m^2 K/W) = \frac{L(m)}{\lambda(W/mK)}$$

eg. 1. A 102mm brick with a conductivity of 0·84 W/mK has a thermal resistance (R) of: 0·102 ÷ 0·84 = 0·121 m²K/W.

eg. 2.

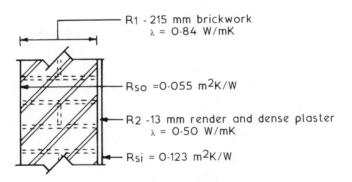

R1 - 215 mm brickwork
λ = 0·84 W/mK

Rso =0·055 m²K/W

R2 -13 mm render and dense plaster
λ = 0·50 W/mK

Rsi = 0·123 m²K/W

Note: the effect of mortar joints in the brickwork can be ignored, as both components have similar density and insulative properties.

$$U = \frac{1}{R_{so} + R_1 + R_2 + R_{si}}$$

$R_1 = 0·215 ÷ 0·84 = 0·256$
$R_2 = 0·013 ÷ 0·50 = 0·026$

$$U = \frac{1}{0·055 + 0·256 + 0·026 + 0·123} = 2·17 W/m^2 K$$

Thermal Insulation, Surface and Air Spaces Resistances

Typical values in: m²K/W

Internal surface resistances (R$_{si}$):

  Walls — 0·123
  Floors or ceilings for upward heat flow — 0·104
  Floors or ceilings for upward heat flow — 0·148
  Roofs (flat or pitched) — 0·104

External surface resistances (R$_{so}$):

Surface	Exposure		
	Sheltered	Normal	Severe
Wall — high emissivity	0·080	0·055	0·030
Wall — low emissivity	0·110	0·070	0·030
Roof — high emissivity	0·070	0·045	0·020
Roof — low emissivity	0·090	0·050	0·020
Floor — high emissivity	0·070	0·040	0·020

  Sheltered — town buildings to 3 storeys.
  Normal — town buildings 4 to 8 storeys and most suburban
         premises.
  Severe — > 9 storeys in towns.
         > 5 storeys elsewhere and any buildings on exposed
         coasts and hills.

Air space resistances (R$_a$):

  Pitched or flat roof space — 0·180
  Behind vertical tile hanging — 0·120
  Cavity wall void — 0·180
  Between high and low emissivity surfaces — 0·300
  Unventilated/sealed — 0·180

Emissivity relates to the heat transfer across and from surfaces by radiant heat emission and absorption effects. The amount will depend on the surface texture, the quantity and temperature of air movement across it, the surface position or orientation and the temperature of adjacent bodies or materials. High surface emissivity is appropriate for most building materials. An example of low emissivity would be bright aluminium foil on one or both sides of an air space.

Typical values —

Material	Density (kg/m³)	Conductivity (λ) (W/mK)
WALLS:		
Boarding (hardwood)	700	0·18
.. .. .. (softwood)	500	0·13
Brick outer leaf	1700	0·84
.. .. inner leaf	1700	0·62
Calcium silicate board	875	0·17
Ceramic tiles	2300	1·30
Concrete	2400	1·93
.. ..	2200	1·59
.. ..	2000	1·33
.. ..	1800	1·13
.. .. (lightweight)	1200	0·38
.. .. (reinforced)	2400	2·50
Concrete block (lightweight)	600	0·18
.. .. (mediumweight)	1400	0·53
Cement mortar (protected)	1750	0·88
.. .. (exposed)	1750	0·94
Fibreboard	350	0·08
Gypsum plaster (dense)	1300	0·57
Gypsum plaster (lightweight)	600	0·16
Plasterboard	950	0·16
Tile hanging	1900	0·84
Rendering	1300	0·57
Sandstone	2600	2·30
Wall ties (st/st)	7900	17·00
ROOFS:		
Aerated concrete slab	500	0·16
Asphalt	1900	0·60
Bituminous felt in 3 layers	1700	0·50
Sarking felt	1700	0·50
Stone chippings	1800	0·96
Tiles (clay)	2000	1·00
.. .. (concrete)	2100	1·50
Wood wool slab	500	0·10

Typical values —

Material	Density (kg/m³)	Conductivity (λ) (W/mK)
FLOORS:		
Cast concrete	2000	1·33
Hardwood block/strip	700	0·18
Plywood/particle board	650	0·14
Screed	1200	0·41
Softwood board	500	0·13
Steel tray	7800	50·00
INSULATION:		
Expanded polystyrene board	20	0·035
Mineral wool batt/slab	25	0·038
Mineral wool quilt	12	0·042
Phenolic foam board	30	0·025
Polyurethane board	30	0·025
Urea formaldehyde foam	10	0·040

Notes:

1. For purposes of calculating U-values, the effect of mortar in external brickwork is usually ignored as the density and thermal properties of bricks and mortar are similar.

2. Where butterfly wall ties are used at normal spacing no adjustment is required to calculations. If vertical twist ties are used in very wide cavities, 0·020 W/m²K should be added to the U-value.

* Tables and charts — Approved Document L to the Building Regs.
* Calculation using the Proportional Area Method.
* Calculation using the Combined Method — BS EN ISO 6946.

Tables and charts — these apply where specific U-values are required and standard forms of construction are adopted. The appendices to Approved Document L contain an extensive range of tables for application to floors, walls and roofs, with indicative values for windows, doors and rooflights. The values contain appropriate allowances for variable heat transfer due to different components in the construction, ie. thermal bridging. The example below shows the principle applied to a solid ground floor with embedded insulation of $\lambda = 0.03$ W/mK.

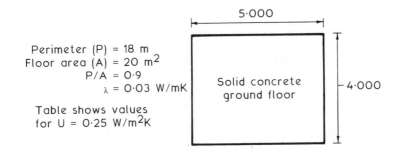

Perimeter (P) = 18 m
Floor area (A) = 20 m^2
P/A = 0.9
λ = 0.03 W/mK

Table shows values
for U = 0.25 W/m^{2}K

5.000

Solid concrete
ground floor

4.000

Typical table for floor insulation:

P/A	0.020	0.025	0.030*	0.035	0.040	0.045	W/mK	
1.0	61	76	91	107	122	137	mm	ins.
0.9*	60	75	90	105	120	135	..	..
0.8	58	73	88	102	117	132	..	..
0.7	57	71	85	99	113	128	..	..
0.6	54	68	82	95	109	122	..	..
0.5	51	64	77	90	103	115	..	..

90 mm of insulation required.

Proportional Area Method (Wall)

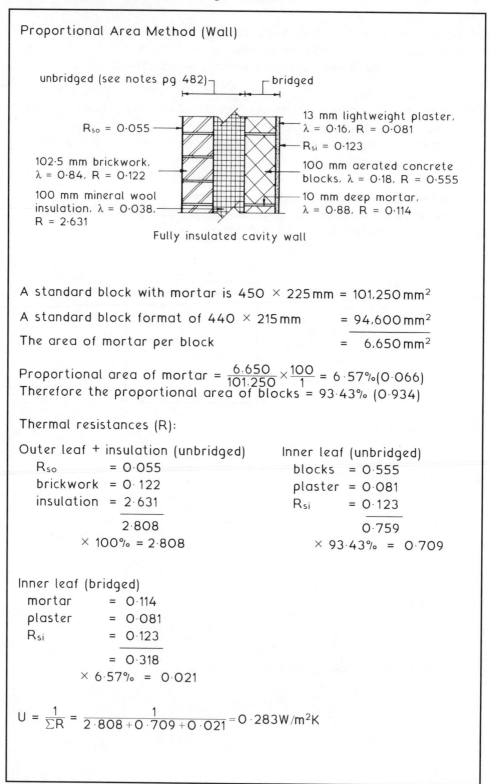

unbridged (see notes pg 482)   bridged

$R_{so} = 0.055$

13 mm lightweight plaster, $\lambda = 0.16$, $R = 0.081$

$R_{si} = 0.123$

102.5 mm brickwork, $\lambda = 0.84$, $R = 0.122$

100 mm aerated concrete blocks, $\lambda = 0.18$, $R = 0.555$

100 mm mineral wool insulation, $\lambda = 0.038$, $R = 2.631$

10 mm deep mortar, $\lambda = 0.88$, $R = 0.114$

Fully insulated cavity wall

A standard block with mortar is $450 \times 225\,mm = 101{,}250\,mm^2$

A standard block format of $440 \times 215\,mm = 94{,}600\,mm^2$

The area of mortar per block $= 6{,}650\,mm^2$

Proportional area of mortar $= \dfrac{6{,}650}{101{,}250} \times \dfrac{100}{1} = 6.57\%(0.066)$

Therefore the proportional area of blocks $= 93.43\%\ (0.934)$

Thermal resistances (R):

Outer leaf + insulation (unbridged)

R_{so}	$= 0.055$
brickwork	$= 0.122$
insulation	$= 2.631$
	$\underline{2.808}$
$\times 100\% = 2.808$	

Inner leaf (unbridged)

blocks	$= 0.555$
plaster	$= 0.081$
R_{si}	$= 0.123$
	$\underline{0.759}$
$\times 93.43\% = 0.709$	

Inner leaf (bridged)

mortar	$= 0.114$
plaster	$= 0.081$
R_{si}	$= 0.123$
	$\underline{0.318}$
$\times 6.57\% = 0.021$	

$$U = \frac{1}{\Sigma R} = \frac{1}{2.808 + 0.709 + 0.021} = 0.283\,W/m^2K$$

Combined Method (BS EN ISO 6946)

This method considers the upper and lower thermal resistance (R) limits of an element of structure. The average of these is reciprocated to provide the U-value.

Formula for upper and lower resistances = $\dfrac{1}{\Sigma(F_x \div R_x)}$

Where: F_x = Fractional area of a section
$\quad\quad$ R_x = Total thermal resistance of a section

Using the wall example from the previous page:

Upper limit of resistance (R) through section containing blocks —
(R_{so}, 0·055) + (brkwk, 0·122) + (ins, 2·631) + (blocks, 0·555) + (plstr, 0·081) + (R_{si}, 0·123) = 3·567 m²K/W

Fractional area of section (F) = 93·43% or 0·934

Upper limit of resistance (R) through section containing mortar —
(R_{so} 0·055) + (brkwk, 0·122) + (ins, 2·631) + (mortar, 0·114) + (plstr, 0·081) + (R_{si}, 0·123) = 3·126 m²K/W

Fractional area of section (F) = 6·57% or 0·066

The upper limit of resistance =

$$\frac{1}{\Sigma(0\cdot943 \div 3\cdot567) + (0\cdot066 \div 3\cdot126)} = 3\cdot533\, m^2K/W$$

Lower limit of resistance (R) is obtained by summating the resistance of all the layers —
(R_{so}, 0·055) + (brkwk, 0·122) + (ins, 2·631) + (bridged layer, 1÷ [0·934 ÷ 0·555] + [0·066 ÷ 0·114] = 0·442) + (plstr, 0·081) + (R_{si}, 0·123) = 3·454 m²K/W

Total resistance (R) of wall is the average of upper and lower limits = (3·533 + 3·454) ÷ 2 = 3·493 m²K/W

U-value = $\dfrac{1}{R}$ = $\dfrac{1}{3\cdot493}$ = 0·286 W/m²K

Note: Both proportional area and combined method calculations require an addition of 0·020 W/m²K to the calculated U value. This is for vertical twist type wall ties in the wide cavity. See page 291 and note 2 on page 482.

Proportional Area Method (Roof)

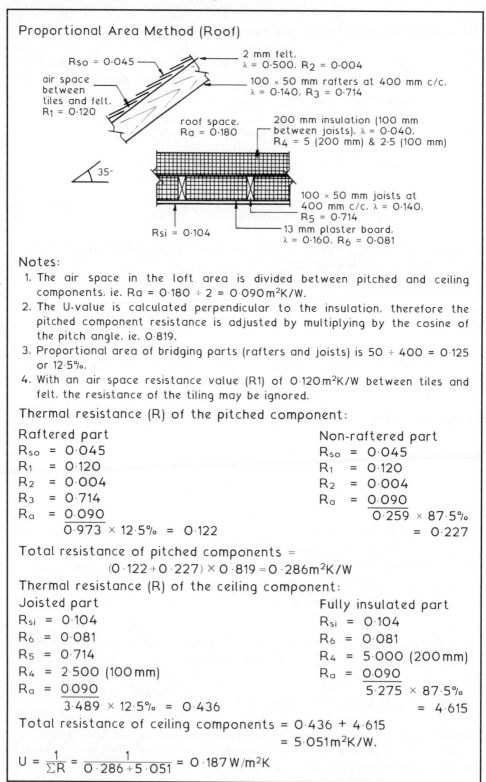

Rso = 0·045

2 mm felt, λ = 0·500, R2 = 0·004

air space between tiles and felt, R1 = 0·120

100 × 50 mm rafters at 400 mm c/c, λ = 0·140, R3 = 0·714

roof space, Ra = 0·180

200 mm insulation (100 mm between joists), λ = 0·040, R4 = 5 (200 mm) & 2·5 (100 mm)

35°

100 × 50 mm joists at 400 mm c/c, λ = 0·140, R5 = 0·714

Rsi = 0·104

13 mm plaster board, λ = 0·160, R6 = 0·081

Notes:

1. The air space in the loft area is divided between pitched and ceiling components, ie. $Ra = 0·180 \div 2 = 0·090\,m^2K/W$.
2. The U-value is calculated perpendicular to the insulation, therefore the pitched component resistance is adjusted by multiplying by the cosine of the pitch angle, ie. 0·819.
3. Proportional area of bridging parts (rafters and joists) is $50 \div 400 = 0·125$ or 12·5%.
4. With an air space resistance value (R1) of $0·120\,m^2K/W$ between tiles and felt, the resistance of the tiling may be ignored.

Thermal resistance (R) of the pitched component:

Raftered part

Rso	= 0·045
R1	= 0·120
R2	= 0·004
R3	= 0·714
Ra	= 0·090

$\dfrac{}{0·973} \times 12·5\% = 0·122$

Non-raftered part

Rso	= 0·045
R1	= 0·120
R2	= 0·004
Ra	= 0·090

$\dfrac{}{0·259} \times 87·5\% = 0·227$

Total resistance of pitched components =
$$(0·122 + 0·227) \times 0·819 = 0·286\,m^2K/W$$

Thermal resistance (R) of the ceiling component:

Joisted part

Rsi	= 0·104
R6	= 0·081
R5	= 0·714
R4	= 2·500 (100 mm)
Ra	= 0·090

$\dfrac{}{3·489} \times 12·5\% = 0·436$

Fully insulated part

Rsi	= 0·104
R6	= 0·081
R4	= 5·000 (200 mm)
Ra	= 0·090

$\dfrac{}{5·275} \times 87·5\% = 4·615$

Total resistance of ceiling components = 0·436 + 4·615
$$= 5·051\,m^2K/W.$$

$$U = \frac{1}{\Sigma R} = \frac{1}{0·286 + 5·051} = 0·187\,W/m^2K$$

Standard Assessment Procedure — the Approved Document to Part L of the Building Regulations emphasises the importance of quantifying the energy costs of running new homes. For this purpose it uses the Government's Standard Assessment Procedure (SAP). SAP has a numerical scale of 1 to 120 and takes into account the fabric losses, ventilation, boiler efficiency and incidentals such as solar gains.

Builders must submit energy rating (SAP) calculations to the local building control authority. Whilst there is no obligation to achieve a particular SAP value, the following may be used for guidance:

Dwelling floor area (m²)	SAP energy rating
80 or less	80
81–90	81
91–100	82
101–110	83
111–120	84
Over 120	85

SAP ratings are required to provide prospective home purchasers or tenants with an indication of the expected fuel costs for hot water and heating. This must be documented and included with the property conveyance.

The calculation involves tables and work sheets found in Approved Document L.

Additionally, new dwellings must comply with any one of the following assessments for limiting heat losses through the structure:

* Elemental method
* Target U-value method
* Carbon index method

Thermal Insulation, Elemental Method

Domestic buildings —

Element of Construction	Maximum U-value (W/m²K)
Pitched roof (insulation between rafters)	0·20
Pitched roof (insulation between joists)	0·16
Flat roof	0·25
Wall	0·35
Floor	0·25
Windows, doors and rooflights (average)	2·20 (metal frames) 2.00 (wood or uPVC frames)

Note: Maximum area of windows, doors and rooflights not greater than 25% of the total floor area.

Energy source — gas or oil fired central heating boiler with a minimum SEDBUK as follows:

Mains natural gas — 78%; LPG — 80%; Oil — 85%.

SEDBUK = Seasonal Efficiency of a Domestic Boiler in the United Kingdom. SEDBUK values are defined in the Government's Standard Assessment Procedure for the Energy Rating of Dwellings, 1998 edition. There is also a SEDBUK website, www.sedbuk.com.

Extensions and alterations to existing dwellings can be treated in the same manner as a new dwelling, by applying the standard area provision for windows, doors and rooflights occupying not more than 25% of the total floor area of the extension.

The average U-value for windows, doors and rooflights is area weighted. This will depend on individual U-values of glazed and frame components and the area they occupy.

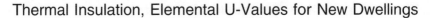

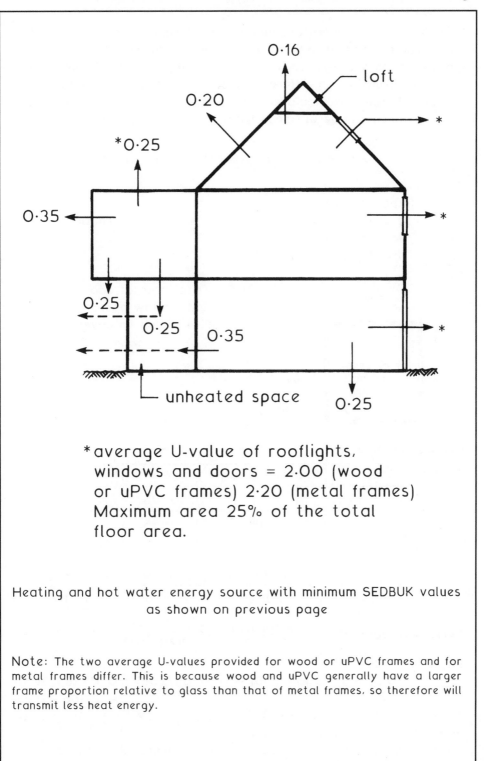

0·16

loft

0·20

*0·25

0·35

0·25

0·25   0·35

unheated space   0·25

*average U-value of rooflights,
windows and doors = 2·00 (wood
or uPVC frames) 2·20 (metal frames)
Maximum area 25% of the total
floor area.

Heating and hot water energy source with minimum SEDBUK values
as shown on previous page

Note: The two average U-values provided for wood or uPVC frames and for
metal frames differ. This is because wood and uPVC generally have a larger
frame proportion relative to glass than that of metal frames, so therefore will
transmit less heat energy.

New dwellings — this procedure is less design restrictive than the elemental method. It allows for a number of variables, including glazing/opening areas, solar gains, heating system efficiency and levels of insulation in different elements of the construction. The average U-value is calculated and it must be less than the target U-value, which can be obtained from the following formula:

Target U-value =
$$0{\cdot}35 - [0{\cdot}19\,(A_R/A_T) - 0{\cdot}10\,(A_{GF}/A_T) + 0{\cdot}413\,(A_F/A_T)]$$

where:

A_R = Exposed roof area
A_T = Total area of exposed elements of the construction
A_{GF} = Ground floor area
A_F = Total floor area

eg.

single storey building

8 m

8 m

floor area = 64 m²

Note: Total area of door and window openings = 12 m²
Floor level to eaves = 2·5 m

Average U-value:

Element	Exposed area (m²)	U (W/m²K)	Heat loss (W/K)
Floor	64	0·25	16
Wall	68	0·35	23·8
Door/windows (ave·)	12	2·20	26·4
Roof	64	0·20	12·8
total	208 m²	total	79 W/K

$$\text{Average U-value} = \frac{\text{total heat loss}}{\text{total exposed surface area}} = \frac{79}{208} = 0{\cdot}379\,\text{W/m}^2\text{k}$$

Target U-value:
$$0{\cdot}35 - [0{\cdot}19\,(64/208) - 0{\cdot}10\,(64/208) + 0{\cdot}413\,(64/208)]$$
$$= 0{\cdot}388\,\text{W/m}^2\text{K}$$

As the average U-value of 0·379 is less than the target U-value of 0·388, the proposal is satisfactory.

Note: This target U-value calculation assumes that factors for boiler rating and solar gains are unnecessary — see next page.

* Boiler rating — no adjustment is necessary when the SEDBUK is the same as the quoted percentage figures for the elemental method. If the proposed boiler is better or worse, the target U-value should be multiplied by the value calculated:

$$\frac{\text{Proposed boiler SEDBUK}(\%)}{\text{Reference boiler SEDBUK}(\%) see\ page\ 488}$$

eg. Proposed mains gas boiler SEDBUK = 80%
Reference boiler SEDBUK = 78%
80% ÷78% = 1·0256

* If electricity or solid fuel is used, the basic target U-value should be divided by 1·15. This is to improve the insulation properties of construction, as a counter measure against the potential for greater carbon dioxide emissions.

* Solar gains — it is assumed that glazing areas to both north and south elevations are similar. Where the glazing is greater to the south, solar benefits can be calculated and added to the basic target U-value after any adjustments for boiler rating:

$$U = 0.04[(A_S - A_N) \div A_{TG}]$$

Where: A_S = Glazed area (inc. frame) facing south (+ or −30°)
A_N = Glazed area (inc. frame) facing north (+ or −30°)
A_{TG} = Total area of glazed elements of the dwelling

eg. A_S = 4·5m², A_N = 2·5m² and A_{TG} = 12m²
$U = 0.04\ [(4·5 - 2·5) \div 12]$
$U = 0.0067$

Note: when incorporating high energy efficiency systems/features in a dwelling, it may be possible to use less demanding U-values in construction. However, consideration must be given to the possibility of condensation occurring in these areas and Approved Document L to the Building Regulations provides guidance on the poorest values — see note on next page.

The Government's Standard Assessment Procedure (SAP) for energy rating dwellings includes an optional facility to calculate carbon dioxide (CO_2) emissions in kilograms or tonnes per year. This is adjusted for dwelling floor area to obtain a carbon factor (CF):

$$CF = CO_2 \div (\text{total floor area} + 45)$$

The carbon index (CI) $= 17 \cdot 7 - (9 \log. CF)$

Note: log. = logarithm to the base 10.

eg. A dwelling of total floor area $100 \, m^2$, with CO_2 emissions of 2900 kg/yr.

$$CF = 2900 \div (100 + 45) = 20$$

$$CI = 17 \cdot 7 - (9 \log. 20) = 6$$

The carbon index (CI) is expressed on a scale of 0 to 10. The higher the number the better. Every new dwelling should have a CI value of a least 8, therefore the example above is unacceptable and will require some modification. Some examples of dwelling construction with a CI of at least 8 are shown in Appendix G to Approved Document L of the Building Regulations.

Note: When using the carbon index or the target U-value methods of assessment, consideration should also be given to avoiding U-values poorer than the following:

    pitched and flat roofs 0·35,
    exposed walls and floors 0·70
    windows, doors and rooflights, 3·30 (average)

Approved Document L to the Building Regulations has guidance on this and may require further reductions depending on the heating system efficiency.

Elemental method — establishes a standard of insulation for each component of construction:

Element	Maximum U-value
Pitched Roof with horizontal insulation between or over joists	0·16
Pitched Roof with integral insulation	0·20
Wall	0·35
Exposed floor and ground floor	0·25
Windows, doors & rooflights (ave.)	2·20 (Metal frames)
Windows, doors & rooflights (ave.)	2.00 (Wood or uPVC frames)
Vehicle access and other large doors	0·70
Flat roof	0·25

Windows, doors and rooflights — compliance will be satisfied by

the following:

Building type	Max. % of window and door to exposed wall area	Max. % of rooflight to roof area
Residential	30	20
Assembly places, offices and shops	40	20
Industrial and storage	15	20

Note: vehicle access doors as required

To provide a degree of design flexibility the given U-values and glazed areas can be varied or traded off. This is provided the rate of heat loss does not exceed that of an equivalent building complying with the criteria and the U-values for parts of specific elements do not exceed: roof 0·35, wall and floor 0·70.

Alternative methods of energy assessment —

* Offices may be assessed on the basis of the Whole Building Carbon Index Method. To comply, the service systems comprising heating, ventilation, air conditioning and lighting must operate within the carbon emissions per square metre, per annum benchmark, based on ECON 19 data. See the Energy Consumption Guide No. 19, (DETR 1998).

* Any non-domestic building can be assessed by the Carbon Emissions Calculation Method. To comply, it must be shown that the annual carbon emissions will not exceed that of an equivalent notional building which satisfies the criteria defined in the elemental method. Calculations must be in accordance with the benchmark tests indicated by the Chartered Institution of Building Services Engineers in their publication, Building Energy and Environmental Modelling, ref. AM11 1998.

Note 1: For all methods of energy efficiency assessment a maximum building air leakage standard Of $10^3/h/m^2$ of external surface at an applied pressure differential of 50 Pascals (N/m^2) can be used for guidance.

Note 2: When using either of the alternative methods, consideration should be given to the guidance on the poorest acceptable U-values for the following elements:

Roofs with loft space and insulation between or over joist   0·25

Roofs with integral insulation (residential buildings)   0·35

Roofs with integral insulation (non-residential buildings)   0·45

Walls   0·45

Exposed floors and ground floor   0·45

Windows, doors and rooflights (ave.)   3·30

It is also possible to trade off between construction elements and heating system efficiency by formula adjustment as shown in Approved Document L to the Building Regulations

Thermal Insulation ~ this is required in most roofs to reduce the heat loss from the interior of the building which will create a better internal environment reducing the risk of condensation and give a saving on heating costs.

Part L of the Building Regulations when dealing with dwellings gives the need to make reasonable provision for the conservation of fuel and power in buildings. To satisfy this requirement Approved Document L gives a maximum allowable thermal transmittance coefficient or U value of 0·16 W/m²K for roofs and 0·20 W/m²K where they form a sloping wall in a loft room. This is usually achieved by placing thermal insulating material(s) at ceiling level creating a cold roof void. Alternatively the insulation can be placed above rafter level thus creating a warm roof void — see page 413.

Typical Details ~

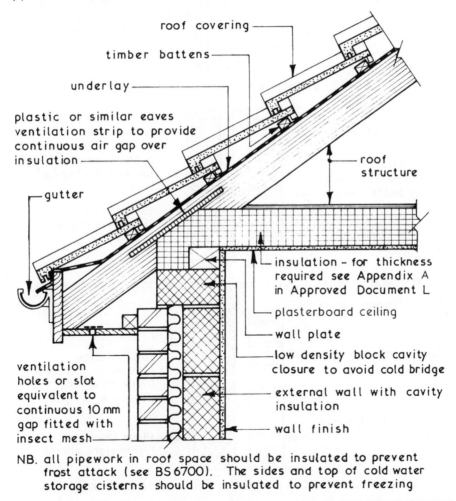

roof covering

timber battens

underlay

plastic or similar eaves ventilation strip to provide continuous air gap over insulation

gutter

roof structure

ventilation holes or slot equivalent to continuous 10 mm gap fitted with insect mesh

insulation – for thickness required see Appendix A in Approved Document L

plasterboard ceiling

wall plate

low density block cavity closure to avoid cold bridge

external wall with cavity insulation

wall finish

NB. all pipework in roof space should be insulated to prevent frost attack (see BS 6700). The sides and top of cold water storage cisterns should be insulated to prevent freezing

Thermal insulation to Walls ~ the minimum performance standards for exposed walls set out in Approved Document L to meet the requirements of Part L of the Building Regulations can be achieved in several ways (see pages 484 and 485). The usual methods require careful specification, detail and construction of the wall fabric, insulating material(s) and/or applied finishes.

Typical Examples of existing construction that would require upgrading to satisfy contemporary UK standards

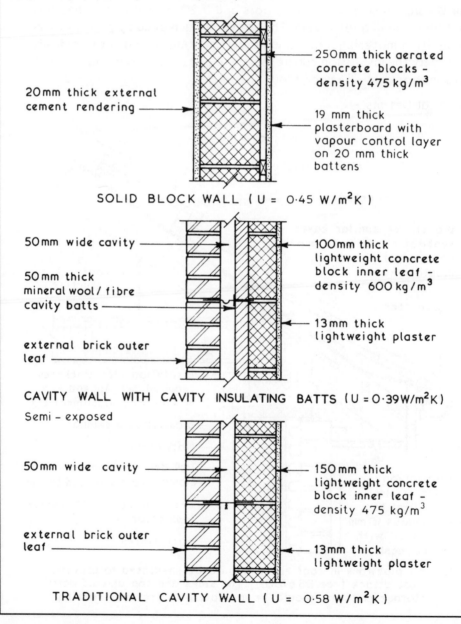

20mm thick external cement rendering

250mm thick aerated concrete blocks - density 475 kg/m³

19 mm thick plasterboard with vapour control layer on 20 mm thick battens

SOLID BLOCK WALL (U = 0·45 W/m²K)

50mm wide cavity

50mm thick mineral wool/fibre cavity batts

external brick outer leaf

100mm thick lightweight concrete block inner leaf - density 600 kg/m³

13mm thick lightweight plaster

CAVITY WALL WITH CAVITY INSULATING BATTS (U = 0·39W/m²K)

Semi - exposed

50mm wide cavity

external brick outer leaf

150mm thick lightweight concrete block inner leaf - density 475 kg/m³

13mm thick lightweight plaster

TRADITIONAL CAVITY WALL (U = 0·58 W/m²K)

Thermal or Cold Bridging ~ this is heat loss and possible condensation, occurring mainly around window and door openings and at the junction between ground floor and wall. Other opportunities for thermal bridging occur where uniform construction is interrupted by unspecified components, e.g. occasional use of bricks and/or tile slips to make good gaps in thermal block inner leaf construction.

NB. This practice was quite common, but no longer acceptable by current legislative standards in the UK.

Prime areas for concern —

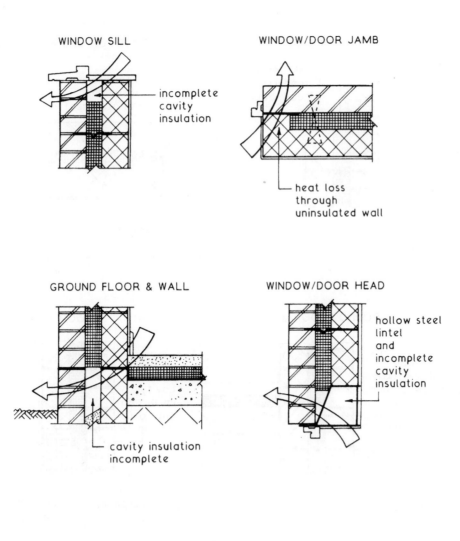

WINDOW SILL

incomplete cavity insulation

WINDOW/DOOR JAMB

heat loss through uninsulated wall

GROUND FLOOR & WALL

cavity insulation incomplete

WINDOW/DOOR HEAD

hollow steel lintel and incomplete cavity insulation

Thermal Bridging — for dwellings the significance can be calculated as:

$$0.3 \times \frac{\text{total length of opening surrounds}}{\text{total exposed surface areas}} + \text{`average U value'}$$

If the figure is below the 'target U value', thermal bridging is insignificant.

Nevertheless, it is better if all construction conforms without applying exempting calculations and the following details should be observed:

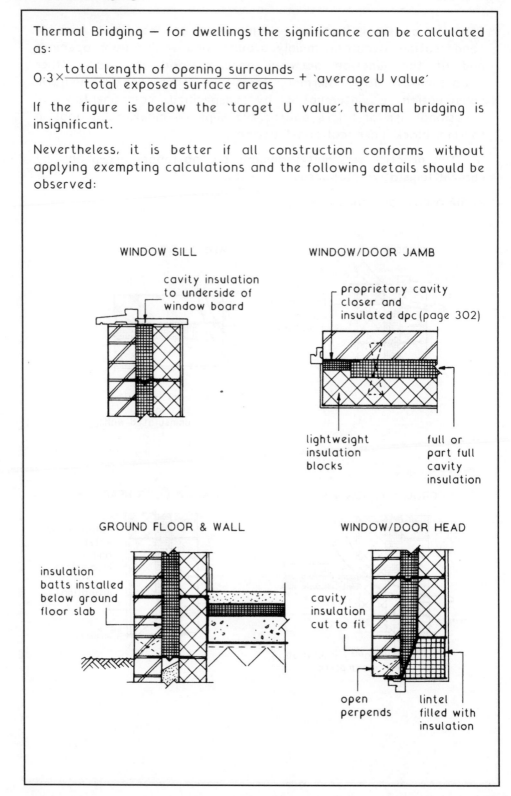

WINDOW SILL

cavity insulation to underside of window board

WINDOW/DOOR JAMB

proprietory cavity closer and insulated dpc (page 302)

lightweight insulation blocks

full or part full cavity insulation

GROUND FLOOR & WALL

insulation batts installed below ground floor slab

WINDOW/DOOR HEAD

cavity insulation cut to fit

open perpends

lintel filled with insulation

Air Infiltration ~ heating costs will increase if cold air is allowed to penetrate peripheral gaps and breaks in the continuity of construction. Furthermore, heat energy will escape through structural breaks and the following are prime situations for treatment :-

1. Loft hatch
2. Services penetrating the structure
3. Opening components in windows, doors and rooflights
4. Gaps between dry lining and masonry walls

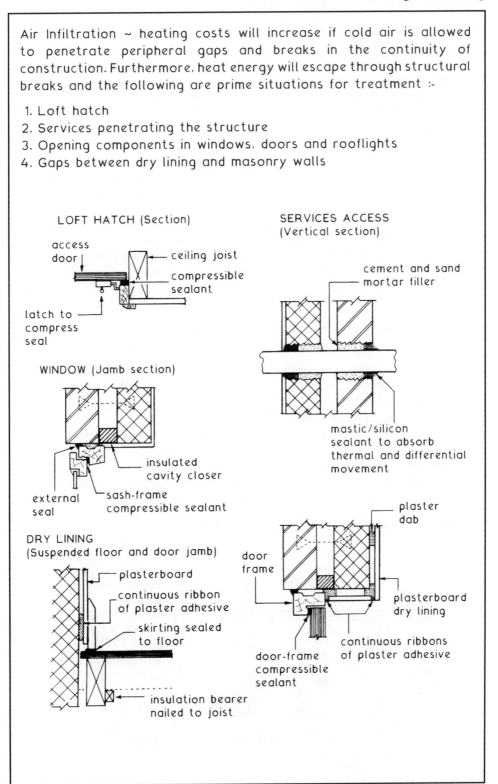

LOFT HATCH (Section)

access door
ceiling joist
compressible sealant
latch to compress seal

SERVICES ACCESS (Vertical section)

cement and sand mortar filler

mastic/silicon sealant to absorb thermal and differential movement

WINDOW (Jamb section)

insulated cavity closer
external seal
sash-frame compressible sealant

DRY LINING (Suspended floor and door jamb)

plasterboard
continuous ribbon of plaster adhesive
skirting sealed to floor
insulation bearer nailed to joist

door frame
plaster dab
plasterboard dry lining
door-frame compressible sealant
continuous ribbons of plaster adhesive

Sound Insulation

Sound Insulation ~ sound can be defined as vibrations of air which are registered by the human ear. All sounds are produced by a vibrating object which causes tiny particles of air around it to move in unison. These displaced air particles collide with adjacent air particles setting them in motion and in unison with the vibrating object. This continuous chain reaction creates a sound wave which travels through the air until at some distance the air particle movement is so small that it is inaudible to the human ear. Sounds are defined as either impact or airborne sound, the definition being determined by the source producing the sound. Impact sounds are created when the fabric of structure is vibrated by direct contact whereas airborne sound only sets the structural fabric vibrating in unison when the emitted sound wave reaches the enclosing structural fabric. The vibrations set up by the structural fabric can therefore transmit the sound to adjacent rooms which can cause annoyance, disturbance of sleep and of the ability to hold a normal conservation. The objective of sound insulation is to reduce transmitted sound to an acceptable level, the intensity of which is measured in units of decibels (dB).

The Building Regulations, Approved Document E: Resistance to the passage of sound, establishes sound insulation standards as follows:

E1: Between dwellings and between dwellings and other buildings.
E2: Within a dwelling, ie. between rooms, particularly WC and habitable rooms, and bedrooms and other rooms.
E3: From external noise and dwellings.
E4: Control of reverberation noise in common parts (stairwells and corridors) of buildings containing dwellings, ie. flats.
E5: Specific applications to acoustic conditions in schools.
Note: E1 includes, hotels, hostels, student accommodation, nurses' homes and homes for the elderly, but not hospitals and prisons.

Typical Sources and Transmission of Sound ~

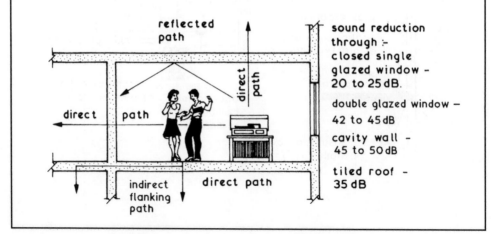

reflected path

direct path

sound reduction through :-
closed single glazed window –
20 to 25 dB.

double glazed window –
42 to 45 dB

cavity wall –
45 to 50 dB

tiled roof –
35 dB

direct   path

indirect flanking path

direct path

Separating Walls ~ types :-

    1. Solid masonry
    2. Cavity masonry
    3. Masonry between isolating panels
    4. Timber frame

Type 1 — relies on mass

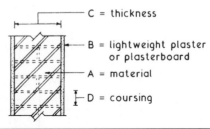

C = thickness

B = lightweight plaster or plasterboard

A = material

D = coursing

Material A	Density of A [Kg/m³]	Finish B	Combined mass A + B (Kg/m²)	Thickness C [mm]	Coursing D [mm]
brickwork	1610	13 mm lwt. pl.	375	215	75
.. ..	.. ..	12·5 mm pl. brd.	.. ..	.. ..	.. ..
Concrete block	1840	13 mm lwt. pl	415	.. ..	110
.. ..	1840	12·5 mm pl. brd	.. ..	.. ..	150
Insitu concrete	2200	Optional	415	190	n/a

Type 2 — relies on mass and isolation

C = leaf thickness

A = material

B = lightweight plaster or plasterboard

butterfly type ties only

D = coursing

E = cavity width

Material A	Density of A [Kg/m³]	Finish B	Mass A + B (Kg/m²)	Thickness C [mm]	Coursing D [mm]	Cavity E [mm]
bkwk.	1970	13 mm lwt. pl.	415	102	75	50
concrete block	1990	..	..	100	225	..
lwt. conc. block	1375	.. or 12.5 mm pl. brd.	300	100	225	75

Type 3 ~ relies on: (a) core material type and mass,
(b) isolation, and
(c) mass of isolated panels.

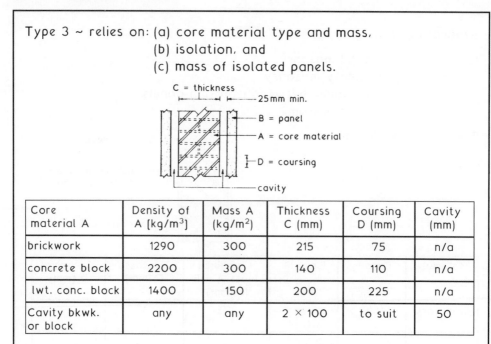

Core material A	Density of A [kg/m³]	Mass A (kg/m²)	Thickness C (mm)	Coursing D (mm)	Cavity (mm)
brickwork	1290	300	215	75	n/a
concrete block	2200	300	140	110	n/a
lwt. conc. block	1400	150	200	225	n/a
Cavity bkwk. or block	any	any	2 × 100	to suit	50

Panel materials — B

(i) Plasterboard with cellular core plus plaster finish, mass 18 kg/m². All joints taped. Fixed floor and ceiling only.

(ii) 2 No. plasterboard sheets, 12.5 mm each, with joints staggered. Frame support or 30 mm overall thickness.

Type 4 — relies on mass, frame separation and absorption of sound.

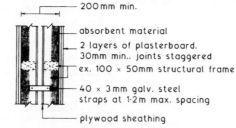

Absorbent material — quilting of unfaced mineral fibre batts with a minimum density of 10 kg/m³, located in the cavity or frames.

Thickness (mm)	Location
25	Suspended in cavity
50	Fixed within one frame
2 × 25	Each quilt fixed within each frame

Separating Floors ~ types:-

      1. Concrete with soft covering
      2. Concrete with floating layer
      3. Timber with floating layer

Type 1. Airborne resistance depends on mass of concrete and ceiling.
      Impact resistance depends on softness of covering.

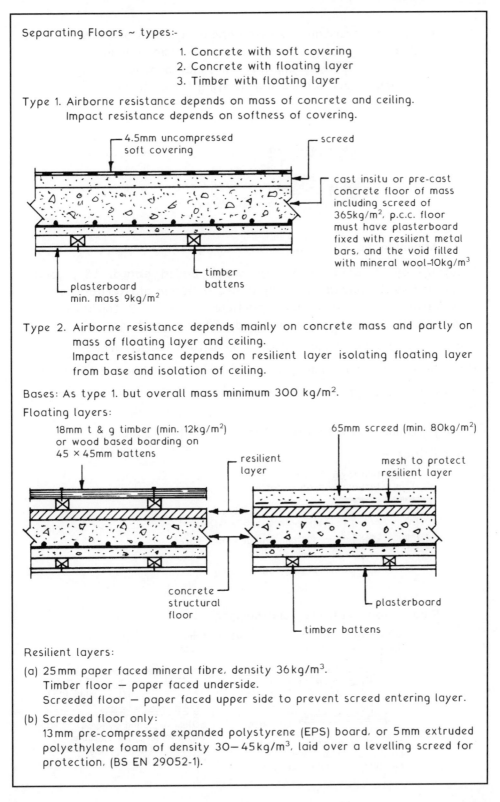

Type 2. Airborne resistance depends mainly on concrete mass and partly on mass of floating layer and ceiling.
      Impact resistance depends on resilient layer isolating floating layer from base and isolation of ceiling.

Bases: As type 1. but overall mass minimum 300 kg/m^2.

Floating layers:

18mm t & g timber (min. 12kg/m^2) or wood based boarding on 45 × 45mm battens

65mm screed (min. 80kg/m^2)

Resilient layers:

(a) 25mm paper faced mineral fibre, density 36kg/m^3.
    Timber floor — paper faced underside.
    Screeded floor — paper faced upper side to prevent screed entering layer.

(b) Screeded floor only:
    13mm pre-compressed expanded polystyrene (EPS) board, or 5mm extruded polyethylene foam of density 30—45kg/m^3, laid over a levelling screed for protection, (BS EN 29052-1).

Type 3. Airborne resistance varies depending on floor construction, absorbency of materials, extent of pugging and partly on the floating layer. Impact resistance depends mainly on the resilient layer separating floating from structure.

Platform floor ~

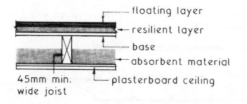

floating layer
resilient layer
base
absorbent material
45mm min. wide joist
plasterboard ceiling

Note: Minimum mass per unit area = 25 kg/m²

Floating layer: 18 mm timber or wood based board, t&g joints glued and spot bonded to a sub-strate of 19 mm plasterboard. Alternatively, cement bonded particle board in 2 thicknesses — 24 mm total, joints staggered, glued and screwed together.

Resilient layer: 25 mm mineral fibre, density 60–100 kg/m³.

Base: 12 mm timber boarding or wood based board nailed to joists.

Absorbent material: 100 mm unfaced rock fibre, minimum density 10 kg/m³.

Ceiling: 30 mm plasterboard in 2 layers, joints staggered.

Ribbed floor ~

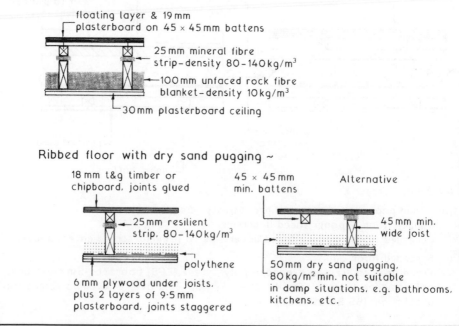

floating layer & 19 mm plasterboard on 45 × 45 mm battens
25 mm mineral fibre strip–density 80–140 kg/m³
100 mm unfaced rock fibre blanket–density 10 kg/m³
30 mm plasterboard ceiling

Ribbed floor with dry sand pugging ~

18 mm t&g timber or chipboard, joints glued
45 × 45 mm min. battens
Alternative
25 mm resilient strip, 80–140 kg/m³
45 mm min. wide joist
polythene
6 mm plywood under joists, plus 2 layers of 9·5 mm plasterboard, joints staggered
50 mm dry sand pugging, 80 kg/m² min. not suitable in damp situations, e.g. bathrooms, kitchens, etc.

Main features —

* Site entrance or car parking space to building entrance to be firm and level, with a 900 mm min. width. A gentle slope is acceptable with a gradient up to 1 in 20 and up to 1 in 40 in cross falls. A slightly steeper ramped access or easy steps should satisfy A.D. M2: Sections 6·14 & 6·15, and 6·16 & 6·17 respectively.
* An accessible threshold for wheelchairs is required at the principal entrance — see illustration.
* Entrance door — minimum clear opening width of 775 mm.
* Corridors, passageways and internal doors of adequate width for wheelchair circulation. Minimum 750 mm — see also table 1 in A.D. M2: Section 7.
* Stair minimum clear width of 900 mm, with provision of handrails both sides. Other requirements as A.D. K for private stairs.
* Accessible light switches, power, telephone and aerial sockets between 450 and 1200 mm above floor level.
* WC provision in the entrance storey or first habitable storey. Door to open outwards. Clear wheelchair space of at least 750 mm in front of WC and a preferred dimension of 500 mm either side of the WC as measured from its centre.
* Special provisions are required for passenger lifts and stairs in blocks of flats, to enable disabled people to access other storeys. See A.D. M2: Section 9 for details.

Note: A.D. refers to the Building Regulations, Approved Document.

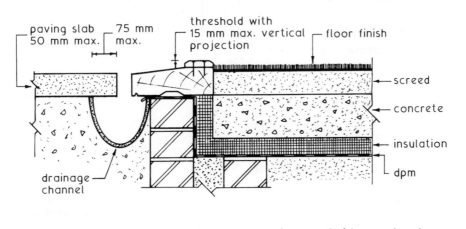

Refs. Accessible thresholds in new housing — Guidance for house builders and designers. The Stationery Office.
BS 5619: Code of practice for design of housing for the convenience of disabled people.

Main features —

* Site entrance, or car parking space to building entrance to be firm and level, ie. maximum gradient 1 in 20 and minimum width 1200 mm. Ramped and easy stepped approaches are also acceptable, see A.D. M2, Sections 1·12 to 1·24.

* Access to include tactile warnings, ie. profiled (blistered or ribbed) pavings for the benefit of people with impaired vision. Dropped kerbs are required to ease wheelchair use.

* Special provision for handrails is necessary for those who may have difficulty in negotiating changes in level. See A.D. M2, Sections 1·25 and 1·26

* Guarding and warning to be provided where projections or obstructions occur, eg. tactile paving could be used around window opening areas.

* Sufficient space for wheelchair manoeuvrability in entrances.

Minimum entrance width of 800 mm. Unobstructed space of at least 300 mm to the leading (opening) edge of door. Glazed panel in the door to provide visibility from 900 to 1500 mm above floor level. Entrance lobby space as A.D. M2, Sections 1·40 and 1·41.

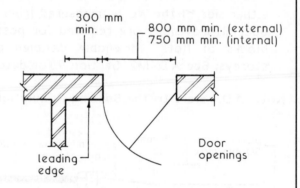

* Internal door openings, minimum width 750 mm. Unobstructed space of at least 300 mm to the leading edge. Visibility panel as above.

continued······

* Corridors and passageways, minimum unobstructed width 1200 mm. Internal lobbies as A.D. M2, Sections 2·7 and 2·8.

* Lift dimensions and capacities to suit the building size. See A.D M2, Sections 2·12 to 2·14. Alternative vertical access may be by wheelchair stairlift — BS 5776: *Specification for powered stairlifts,* or a platform lift — BS 6440: *Powered lifting platforms for use by disabled people.*

* Stair minimum width 1000 mm, with step nosings brightly distinguished. Rise maximum 1800 mm between landings. Landings to have 1200 mm of clear space from any door swings. Step rise, maximum 170 mm and uniform throughout. Step going, minimum 250 mm and uniform throughout. Tapered treads acceptable if 250 mm going is measured 270 mm from the inside of the stair. No open risers. Handrail to each side of the stair.

* Number and location of WC's to reflect ease of access for wheelchair users. In no case should a wheelchair user have to travel more than one storey. Provision may be 'unisex' which is generally more suitable, or 'integral' with specific sex conveniences. Particular provision is outlined in A.D. M3 (3), Section 4.

* A.D. M3 (3), Section 3 should be consulted for special provisions for restaurants, bars and hotel bedrooms. Section 5 has special provisions for spectator seating in theatres and stadia.

Refs. Building Regulations, Approved Document M: Access and facilities for disabled people.
Disability Discrimination Act.
BS 5810: Code of practice for access for the disabled to buildings.
BS 5588-8: Code of practice for means of escape for disabled people.
BS 8300-1: Code of practice for the design of buildings for the convenience and use of people with disabilities. (Currently in draft form for public comment).
PD 6523: Information on access to and movement within and around buildings and on certain facilities for disabled people.

6 INTERNAL CONSTRUCTION AND FINISHES

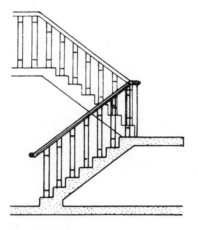

INTERNAL ELEMENTS
INTERNAL WALLS
CONSTRUCTION JOINTS
PARTITIONS
PLASTERS AND PLASTERING
DRY LINING TECHNIQUES
WALL TILING
DOMESTIC FLOORS AND FINISHES
LARGE CAST INSITU GROUND FLOORS
CONCRETE FLOOR SCREEDS
TIMBER SUSPENDED FLOORS
TIMBER BEAM DESIGN
REINFORCED CONCRETE SUSPENDED FLOORS
PRECAST CONCRETE FLOORS
RAISED ACCESS FLOORS
TIMBER, CONCRETE AND METAL STAIRS
INTERNAL DOORS
FIRE RESISTING DOORS
PLASTERBOARD CEILINGS
SUSPENDED CEILINGS
PAINTS AND PAINTING
JOINERY PRODUCTION
COMPOSITE BOARDING
PLASTICS IN BUILDING

NB. roof coverings, roof insulation and guttering not shown

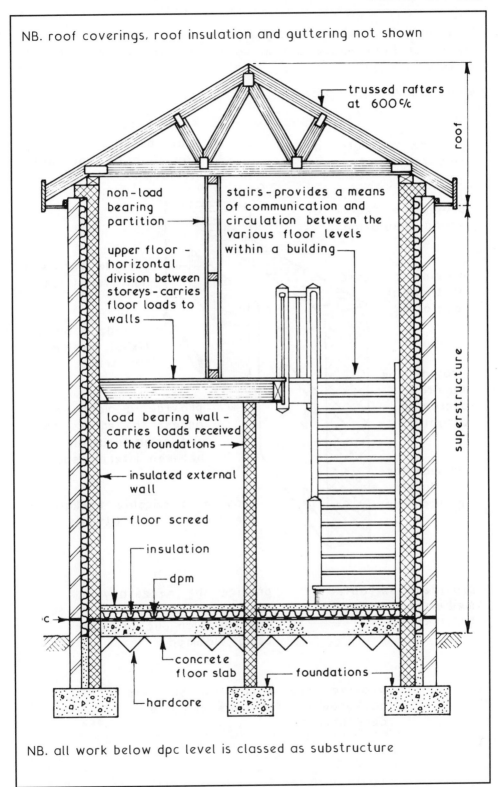

trussed rafters at 600 c/c

roof

non-load bearing partition

stairs - provides a means of communication and circulation between the various floor levels within a building

upper floor - horizontal division between storeys - carries floor loads to walls

load bearing wall - carries loads received to the foundations

insulated external wall

floor screed

insulation

dpm

superstructure

concrete floor slab

foundations

hardcore

NB. all work below dpc level is classed as substructure

511

Internal Walls ~ their primary function is to act as a vertical divider of floor space and in so doing form a storey height enclosing element.

Other Possible Functions:-

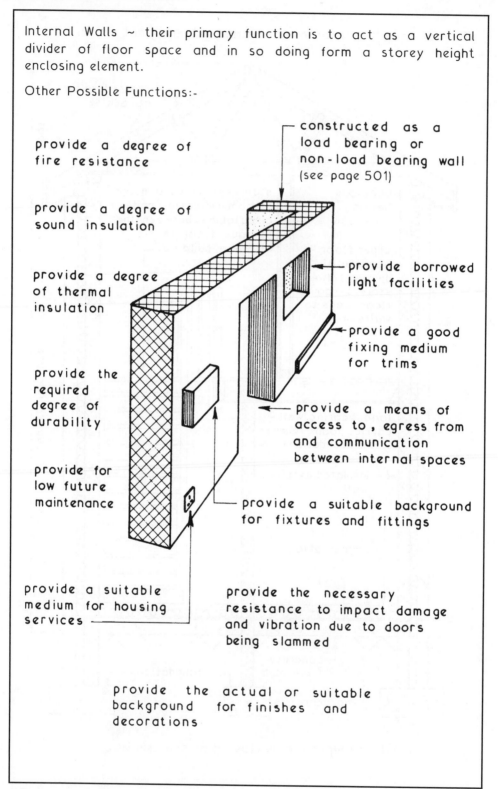

provide a degree of fire resistance

provide a degree of sound insulation

provide a degree of thermal insulation

provide the required degree of durability

provide for low future maintenance

provide a suitable medium for housing services

constructed as a load bearing or non-load bearing wall (see page 501)

provide borrowed light facilities

provide a good fixing medium for trims

provide a means of access to, egress from and communication between internal spaces

provide a suitable background for fixtures and fittings

provide the necessary resistance to impact damage and vibration due to doors being slammed

provide the actual or suitable background for finishes and decorations

Internal Walls ~ there are two basic design concepts for internal walls those which accept and transmit structural loads to the foundations are called Load Bearing Walls and those which support only their own self-weight and do not accept any structural loads are called Non-load Bearing Walls or Partitions.

Typical Examples ~

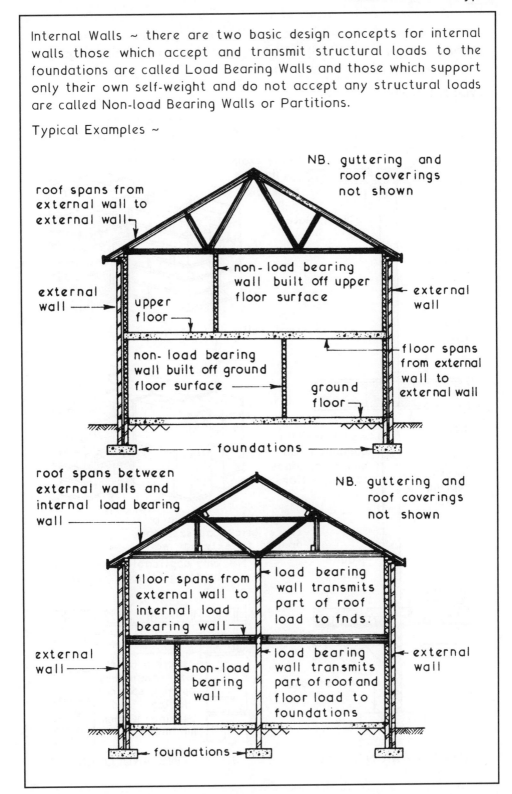

NB. guttering and roof coverings not shown

roof spans from external wall to external wall

external wall

upper floor

non-load bearing wall built off upper floor surface

external wall

non-load bearing wall built off ground floor surface

ground floor

floor spans from external wall to external wall

foundations

roof spans between external walls and internal load bearing wall

NB. guttering and roof coverings not shown

floor spans from external wall to internal load bearing wall

load bearing wall transmits part of roof load to fnds.

external wall

non-load bearing wall

load bearing wall transmits part of roof and floor load to foundations

external wall

foundations

513

Internal Brick Walls ~ these can be load bearing or non-load bearing (see previous 501) and for most two storey buildings are built in half brick thickness in stretcher bond.

Typical Details ~

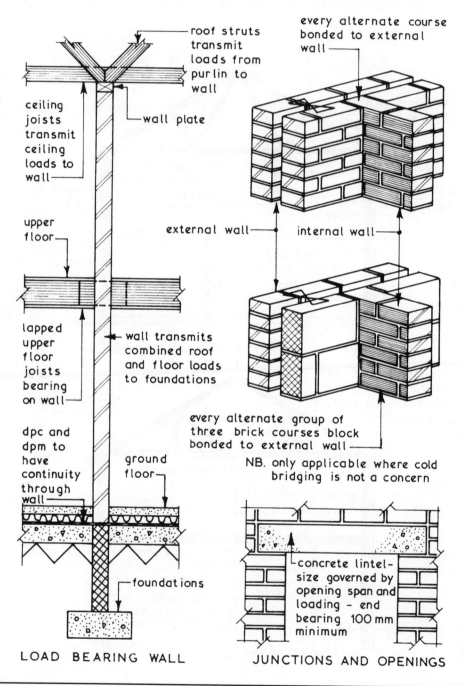

roof struts transmit loads from purlin to wall

wall plate

ceiling joists transmit ceiling loads to wall

upper floor

lapped upper floor joists bearing on wall

wall transmits combined roof and floor loads to foundations

dpc and dpm to have continuity through wall

ground floor

foundations

LOAD BEARING WALL

every alternate course bonded to external wall

external wall

internal wall

every alternate group of three brick courses block bonded to external wall

NB. only applicable where cold bridging is not a concern

concrete lintel size governed by opening span and loading - end bearing 100 mm minimum

JUNCTIONS AND OPENINGS

Internal Block Walls ~ these can be load bearing or non-load bearing (see page 501) the thickness and type of block to be used will depend upon the loadings it has to carry.

Typical Details ~

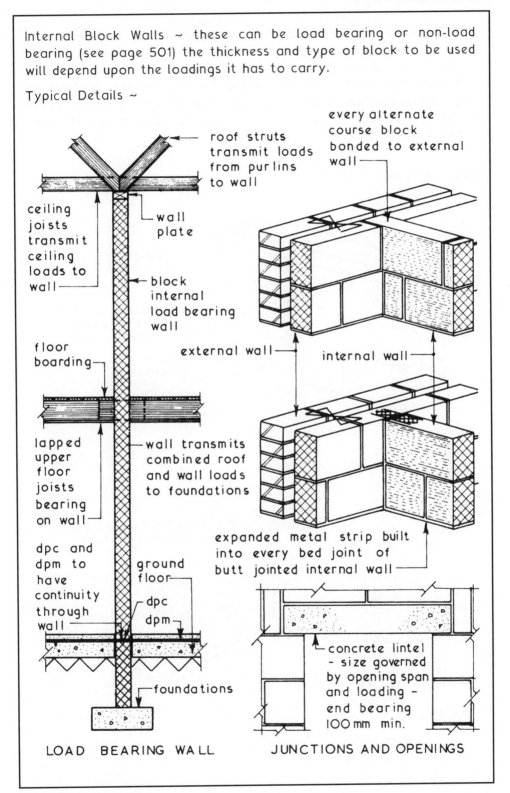

roof struts transmit loads from purlins to wall

ceiling joists transmit ceiling loads to wall

wall plate

block internal load bearing wall

floor boarding

lapped upper floor joists bearing on wall

dpc and dpm to have continuity through wall

ground floor

dpc

dpm

foundations

wall transmits combined roof and wall loads to foundations

LOAD BEARING WALL

every alternate course block bonded to external wall

external wall

internal wall

expanded metal strip built into every bed joint of butt jointed internal wall

concrete lintel – size governed by opening span and loading – end bearing 100 mm min.

JUNCTIONS AND OPENINGS

515

Internal Walls ~ an alternative to brick and block bonding shown on the preceding two pages is application of wall profiles. These are quick and simple to install, provide adequate lateral stability, sufficient movement flexibility and will overcome the problem of thermal bridging where a brick partition would otherwise bond into a block inner leaf. They are also useful for attaching extension walls at right angles to existing masonry.

Application ~

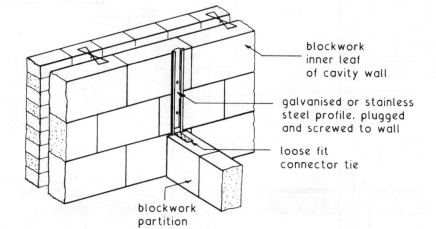

blockwork inner leaf of cavity wall

galvanised or stainless steel profile, plugged and screwed to wall

loose fit connector tie

blockwork partition

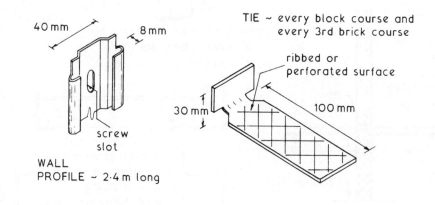

40 mm

8 mm

screw slot

WALL PROFILE ~ 2·4 m long

TIE ~ every block course and every 3rd brick course

ribbed or perforated surface

30 mm

100 mm

Movement or Construction Joints ~ provide an alternative to ties or mesh reinforcement in masonry bed joints. Even with reinforcement, lightweight concrete block walls are renowned for producing unsightly and possibly unstable shrinkage cracks. Galvanised or stainless steel formers and ties are built in at approximately 6m horizontal spacing to accommodate initial drying, shrinkage movement and structural settlement. One side of the former is fitted with profiled or perforated ties to bond into bed joints and the other has plastic sleeved ties. The sleeved tie maintains continuity, but restricts bonding to allow for controlled movement.

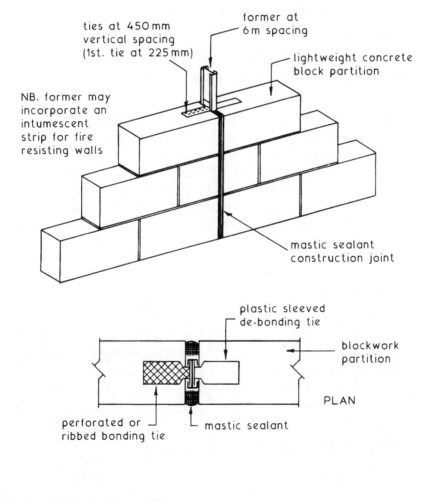

ties at 450mm vertical spacing (1st. tie at 225mm)

former at 6m spacing

lightweight concrete block partition

NB. former may incorporate an intumescent strip for fire resisting walls

mastic sealant construction joint

plastic sleeved de-bonding tie

blockwork partition

PLAN

perforated or ribbed bonding tie

mastic sealant

Ref. BS 5628—3: Use of Masonry

517

Internal Partitions ~ these are vertical dividers which are used to separate the internal space of a building into rooms and circulation areas such as corridors. Partitions which give support to a floor or roof are classified as load bearing whereas those which give no such support are called non-load bearing.

Load Bearing Partitions ~ these walls can be constructed of bricks, blocks or insitu concrete by traditional methods and have the design advantages of being capable of having good fire resistance and/or high sound insulation. Their main disadvantage is permanence giving rise to an inflexible internal layout.

Non-load Bearing Partitions ~ the wide variety of methods available makes it difficult to classify the form of partition but most can be placed into one of three groups:-

1. Masonry partitions.

2. Stud partitions — see pages 519 & 520.

3. Demountable partitions — see pages 521 & 522.

Masonry Partitions ~ these are usually built with blocks of clay or lightweight concrete which are readily available and easy to construct thus making them popular. These masonry partitions should be adequately tied to the structure or load bearing walls to provide continuity as a sound barrier, provide edge restraint and to reduce the shrinkage cracking which inevitably occurs at abutments. Wherever possible openings for doors should be in the form of storey height frames to provide extra stiffness at these positions.

Typical Details ~

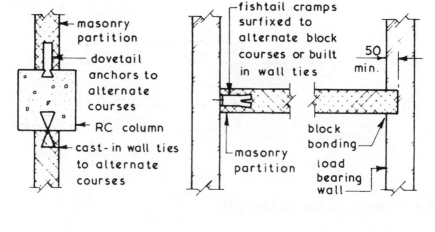

masonry partition

dovetail anchors to alternate courses

RC column

cast-in wall ties to alternate courses

fishtail cramps surfixed to alternate block courses or built in wall ties

masonry partition

50 min.

block bonding

load bearing wall

Timber Stud Partitions ~ these are non-load bearing internal dividing walls which are easy to construct, lightweight, adaptable and can be clad and infilled with various materials to give different finishes and properties. The timber studs should be of prepared or planed material to ensure that the wall is of constant thickness with parallel faces. Stud spacings will be governed by the size and spanning ability of the facing or cladding material.

Typical Details ~

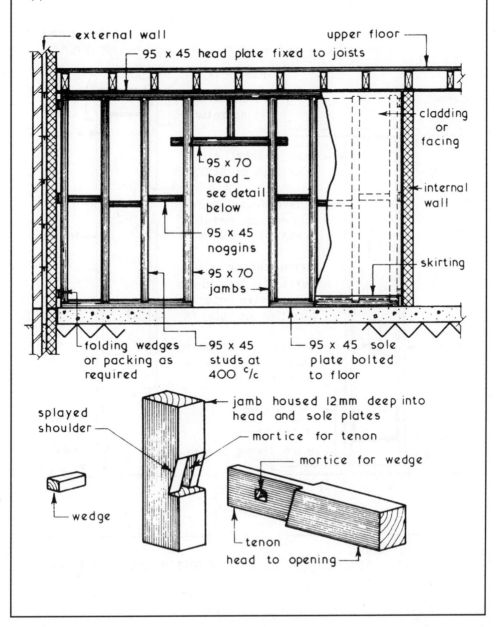

Metal Stud Partitions

Stud Partitions ~ these non-load bearing partitions consist of a framework of vertical studs to which the facing material can be attached. The void between the studs created by the two faces can be infilled to meet specific design needs. The traditional material for stud partitions is timber (see Timber Stud Partitions on page 519) but a similar arrangement can be constructed using metal studs faced on both sides with plasterboard.

Typical Metal Stud Partition Details ~

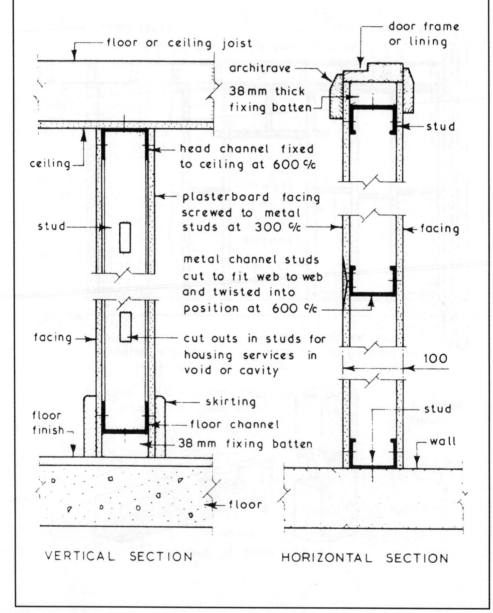

VERTICAL SECTION   HORIZONTAL SECTION

Partitions ~ these can be defined as vertical internal space dividers and are usually non-loadbearing. They can be permanent, constructed of materials such as bricks or blocks or they can be demountable constructed using lightweight materials and capable of being taken down and moved to a new location incurring little or no damage to the structure or finishes. There is a wide range of demountable partitions available constructed from a variety of materials giving a range that will be suitable for most situations. Many of these partitions have a permanent finish which requires no decoration and only periodic cleaning in the context of planned maintenance.

Typical Example ~

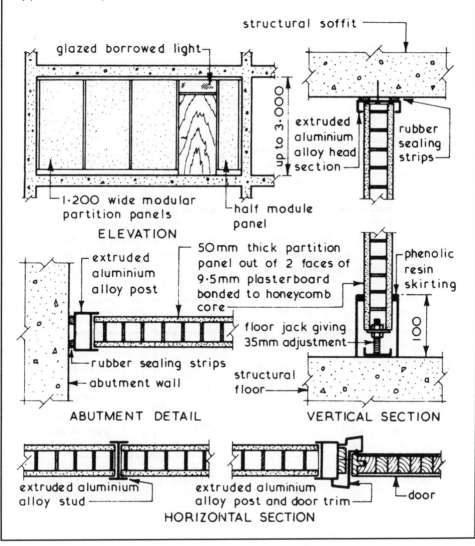

Demountable Partitions ~ it can be argued that all internal non-load bearing partitions are demountable and therefore the major problem is the amount of demountability required in the context of ease of moving and the possible frequency anticipated. The range of partitions available is very wide including stud partitions, framed panel partitions (see Demountable Partitions on page 521) panel to panel partitions and sliding/folding partitions which are similar in concept to industrial doors (see industrial Doors on pages 339 and 341) The latter type is often used where movement of the partition is required frequently. The choice is therefore based on the above stated factors taking into account finish and glazing requirements together with any personal preference for a particular system but in all cases the same basic problems will have to be considered:-

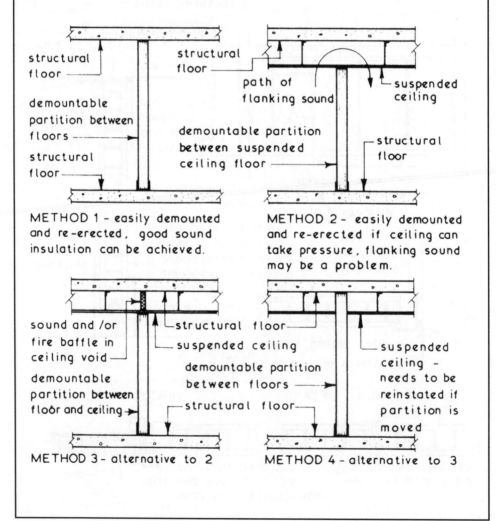

structural floor

demountable partition between floors

structural floor

METHOD 1 - easily demounted and re-erected, good sound insulation can be achieved.

structural floor

path of flanking sound

suspended ceiling

demountable partition between suspended ceiling floor

structural floor

METHOD 2 - easily demounted and re-erected if ceiling can take pressure, flanking sound may be a problem.

sound and /or fire baffle in ceiling void

demountable partition between floor and ceiling

structural floor

suspended ceiling

METHOD 3 - alternative to 2

demountable partition between floors

structural floor

suspended ceiling - needs to be reinstated if partition is moved

METHOD 4 - alternative to 3

Plaster ~ this is a wet mixed material applied to internal walls as a finish to fill in any irregularities in the wall surface and to provide a smooth continuous surface suitable for direct decoration. The plaster finish also needs to have a good resistance to impact damage. The material used to fulfil these requirements is gypsum plaster. Gypsum is a crystalline combination of calcium sulphate and water. The raw material is crushed, screened and heated to dehydrate the gypsum and this process together with various additives defines its type as set out in BS 1191: Specification for gypsum building plasters.

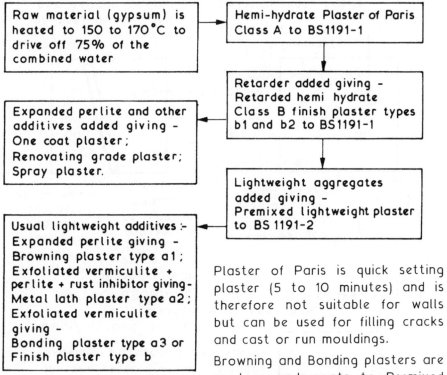

| Raw material (gypsum) is heated to 150 to 170°C to drive off 75% of the combined water | → | Hemi-hydrate Plaster of Paris Class A to BS1191-1 |

| Expanded perlite and other additives added giving - One coat plaster; Renovating grade plaster; Spray plaster. | ← | Retarder added giving - Retarded hemi hydrate Class B finish plaster types b1 and b2 to BS1191-1 |

| Usual lightweight additives :- Expanded perlite giving - Browning plaster type a1; Exfoliated vermiculite + perlite + rust inhibitor giving - Metal lath plaster type a2; Exfoliated vermiculite giving - Bonding plaster type a3 or Finish plaster type b | ← | Lightweight aggregates added giving - Premixed lightweight plaster to BS 1191-2 |

Plaster of Paris is quick setting plaster (5 to 10 minutes) and is therefore not suitable for walls but can be used for filling cracks and cast or run mouldings.

Browning and Bonding plasters are used as undercoats to Premixed lightweight plasters.

All plaster should be stored in dry conditions since any absorption of moisture before mixing may shorten the normal setting time of about one and a half hours which can reduce the strength of the set plaster. Gypsum plasters are not suitable for use in temperatures exceeding 43°C and should not be applied to frozen backgrounds.

A good key to the background and between successive coats is essential for successful plastering. Generally brick and block walls provide the key whereas concrete unless cast against rough formwork will need to be treated to provide the key.

Internal Wall Finishes ~ these can be classified as wet or dry. The traditional wet finish is plaster which is mixed and applied to the wall in layers to achieve a smooth and durable finish suitable for decorative treatments such as paint and wallpaper.

Most plasters are supplied in 25kg paper sacks and require only the addition of clean water or sand and clean water according to the type of plaster being used.

Typical Method of Application ~

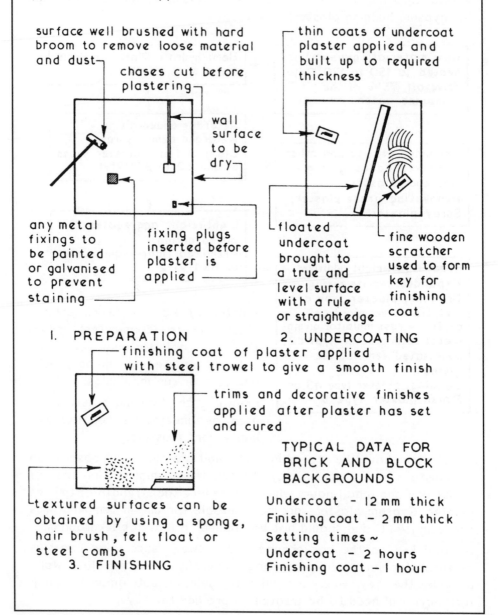

surface well brushed with hard broom to remove loose material and dust

chases cut before plastering

wall surface to be dry

thin coats of undercoat plaster applied and built up to required thickness

any metal fixings to be painted or galvanised to prevent staining

fixing plugs inserted before plaster is applied

floated undercoat brought to a true and level surface with a rule or straightedge

fine wooden scratcher used to form key for finishing coat

1. PREPARATION

2. UNDERCOATING

finishing coat of plaster applied with steel trowel to give a smooth finish

trims and decorative finishes applied after plaster has set and cured

TYPICAL DATA FOR BRICK AND BLOCK BACKGROUNDS

textured surfaces can be obtained by using a sponge, hair brush, felt float or steel combs

3. FINISHING

Undercoat - 12 mm thick
Finishing coat - 2 mm thick
Setting times ~
Undercoat - 2 hours
Finishing coat - 1 hour

Plasterboard ~ a board material made of two sheets of thin mill - board with gypsum plaster between — three edge profiles are available:-

Tapered Edge —

joint filling

A flush seamless surface is obtained by filling the joint with a special filling plaster, applying a joint tape over the filling and finishing with a thin layer of joint filling plaster the edge of which is feathered out using a slightly damp jointing sponge.

Square Edge — edges are close butted and finished with a cover fillet or the joint is covered with a jute scrim before being plastered.

Bevelled Edge — edges are close butted forming a vee-joint which becomes a feature of the lining.

Typical Details ~

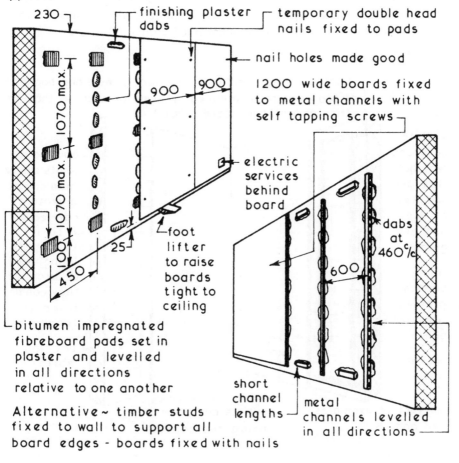

230

finishing plaster dabs

temporary double head nails fixed to pads

nail holes made good

1070 max.

900   900

1200 wide boards fixed to metal channels with self tapping screws

1070 max.

electric services behind board

100   25   450

foot lifter to raise boards tight to ceiling

dabs at 460 °/c

600

bitumen impregnated fibreboard pads set in plaster and levelled in all directions relative to one another

short channel lengths

metal channels levelled in all directions

Alternative~ timber studs fixed to wall to support all board edges - boards fixed with nails

Dry Linings ~ the internal surfaces of walls and partitions are usually covered with a wet finish (plaster or rendering) or with a dry lining such as plasterboard, insulating fibre board, hardboard, timber boards, and plywood, all of which can be supplied with a permanent finish or they can be supplied to accept an applied finish such as paint or wallpaper. The main purpose of any applied covering to an internal wall surface is to provide an acceptable but not necessarily an elegant or expensive wall finish. It is also very difficult and expensive to build a brick or block wall which has a fair face to both sides since this would involve the hand selection of bricks and blocks to ensure a constant thickness together with a high degree of skill to construct a satisfactory wall. The main advantage of dry lining walls is that the drying out period required with wet finishes is eliminated. By careful selection and fixing of some dry lining materials it is possible to improve the thermal insulation properties of a wall. Dry linings can be fixed direct to the backing by means of a recommended adhesive or they can be fixed to a suitable arrangement of wall battens.

Typical Example ~

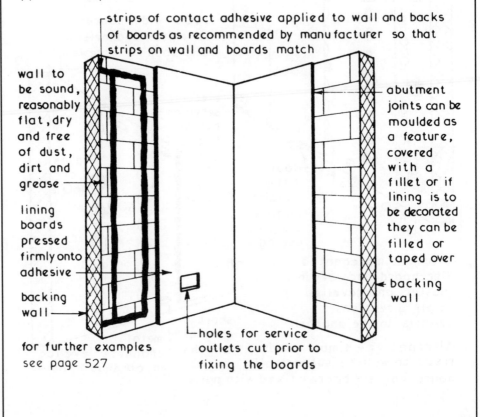

strips of contact adhesive applied to wall and backs of boards as recommended by manufacturer so that strips on wall and boards match

wall to be sound, reasonably flat, dry and free of dust, dirt and grease

lining boards pressed firmly onto adhesive

backing wall

abutment joints can be moulded as a feature, covered with a fillet or if lining is to be decorated they can be filled or taped over

backing wall

holes for service outlets cut prior to fixing the boards

for further examples see page 527

Typical Examples ~

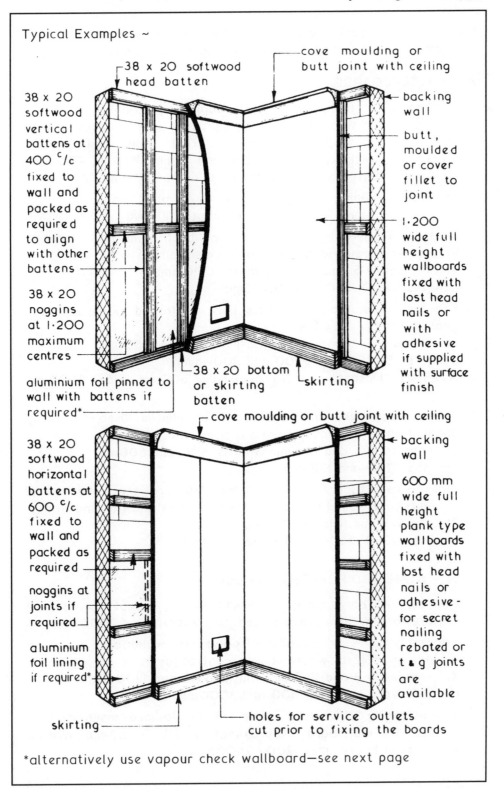

38 x 20 softwood
head batten

cove moulding or
butt joint with ceiling

38 x 20
softwood
vertical
battens at
400 ᶜ/c
fixed to
wall and
packed as
required
to align
with other
battens

38 x 20
noggins
at 1·200
maximum
centres

aluminium foil pinned to
wall with battens if
required*

backing
wall

butt,
moulded
or cover
fillet to
joint

1·200
wide full
height
wallboards
fixed with
lost head
nails or
with
adhesive
if supplied
with surface
finish

38 x 20 bottom
or skirting
batten

skirting

cove moulding or butt joint with ceiling

38 x 20
softwood
horizontal
battens at
600 ᶜ/c
fixed to
wall and
packed as
required

noggins at
joints if
required

aluminium
foil lining
if required*

skirting

backing
wall

600 mm
wide full
height
plank type
wallboards
fixed with
lost head
nails or
adhesive -
for secret
nailing
rebated or
t & g joints
are
available

holes for service outlets
cut prior to fixing the boards

*alternatively use vapour check wallboard—see next page

Plasterboard Types ~ to BS 1230-1: Specification for plasterboard excluding materials submitted to secondary operations.

BS PLASTERBOARDS:~

1. Wallboard — ivory faced for taping, jointing and direct decoration; grey faced for finishing plaster or wall adhesion with dabs. General applications, i.e. internal walls, ceilings and partitions. Thicknesses: 9·5, 12·5 and 15mm. Widths: 900 and 1200mm. Lengths: vary between 1800 and 3000mm.

2. Baseboard — lining ceilings requiring direct plastering.
Thickness: 9·5mm. Width: 900mm. Length: 1219mm.

3. Moisture Resistant — wallboard for bathrooms and kitchens. Pale green colour, ideal base for ceramic tiling.
Thicknesses: 9·5 and 12·5mm. Width: 1200mm.
Lengths: 2400, 2700 and 3000mm.

4. Firecheck — wallboard of glass fibre reinforced vermiculite and gypsum for fire cladding.
Thicknesses: 12·5 and 15mm. Widths: 900 and 1200mm.
Lengths: 1800, 2400, 2700 and 3000mm.
A 25mm thickness is also produced, 600mm wide × 3000mm long.

5. Lath — rounded edge wallboard of limited area for easy application to ceilings requiring a direct plaster finish.
Thicknesses: 9·5 and 12·5mm. Widths: 400 and 600mm.
Lengths: 1200 and 1219mm.

6. Plank — used as fire protection for structural steel and timber, in addition to sound insulation in wall panels and floating floors.
Thickness: 19mm. Width: 600mm.
Lengths: 2350, 2400, 2700 and 3000mm.

NON — STANDARD PLASTERBOARDS:~

1. Contour — only 6mm in thickness to adapt to curved featurework. Width: 1200mm. Lengths: 2400m and 3000mm.

2. Vapourcheck — a metallized polyester wallboard lining to provide an integral water vapour control layer.
Thicknesses: 9·5 and 12·5mm. Widths: 900 and 1200mm.
Lengths: vary between 1800 and 3000mm.

3. Thermalcheck — various expanded or foamed insulants are bonded to wallboard. Approximately 25 — 50mm overall thickness in board sizes 1200 × 2400mm.

Glazed Wall Tiles ~ internal glazed wall tiles are usually made to the recommendations of BS 6431. External glazed wall tiles made from clay or clay/ceramic mixtures are manufactured but there is no British Standard available.

Internal Glazed Wall Tiles ~ the body of the tile can be made from ball-clay, china clay, china stone, flint and limestone. The material is usually mixed with water to the desired consistency, shaped and then fired in a tunnel oven at a high temperature (1150°C) for several days to form the unglazed biscuit tile. The glaze, pattern and colour can now be imparted onto to the biscuit tile before the final firing process at a temperature slightly lower than that of the first firing (1050°C) for about two days.

Typical Internal Glazed Wall Tiles and Fittings ~

Sizes — Modular   100 × 100 × 5mm thick and
                      200 × 100 × 6·5mm thick.

      Non-modular   152 × 152 × 5 to 8mm thick and
                      108 × 108 × 4 and 6·5mm thick.

Fittings — wide range available particularly in the non-modular format.

lugs to maintain joint spacing

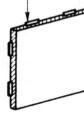

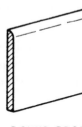

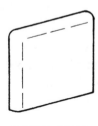

SPACER TILES   ROUND EDGE   ROUND EDGE   ROUND EDGE
                                  EXTERNAL   SQUARE HEAD
                                  CORNER

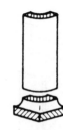

ROUND EDGE AND   ATTACHED ANGLE TILE FITTINGS   BEAD
COVE FOOT   FITTINGS

Bedding of Internal Wall Tiles ~ generally glazed internal wall tiles are considered to be inert in the context of moisture and thermal movement, therefore if movement of the applied wall tile finish is to be avoided attention must be given to the background and the method of fixing the tiles.

Backgrounds ~ these are usually of a cement rendered or plastered surface and should be flat, dry, stable, firmly attached to the substrate and sufficiently old enough for any initial shrinkage to have taken place. The flatness of the background should be not more than 3mm in 2·000 for the thin bedding of tiles and not more than 6mm in 2·000 for thick bedded tiles.

Fixing Wall Tiles ~ two methods are in general use:-

1. Thin Bedding — lightweight internal glazed wall tiles fixed dry using a recommended adhesive which is applied to wall in small areas 1m² at a time with a notched trowel, the tile being pressed or tapped into the adhesive.

2. Thick Bedding — cement mortar within the mix range of 1:3 to 1:4 is used as the adhesive either by buttering the backs of the tiles which are then pressed or tapped into position or by rendering the wall surface to a thickness of approximately 10mm and then applying the lightly buttered tiles (1:2 mix) to the rendered wall surface within two hours. It is usually necessary to soak the wall tiles in water to reduce suction before they are placed in position.

Grouting ~ when the wall tiles have set, the joints can be grouted by rubbing into the joints a grout paste either using a sponge or brush. Most grouting materials are based on cement with inert fillers and are used neat.

Typical Example ~

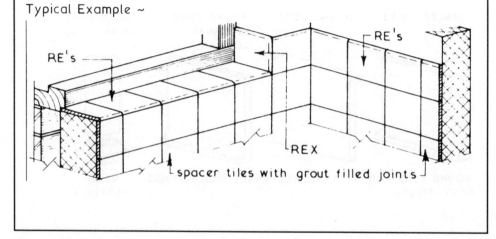

Primary Functions ~

1. Provide a level surface with sufficient strength to support the imposed loads of people and furniture.

2. Exclude the passage of water and water vapour to the interior of the building.

3. Provide resistance to unacceptable heat loss through the floor.

4. Provide the correct type of surface to receive the chosen finish.

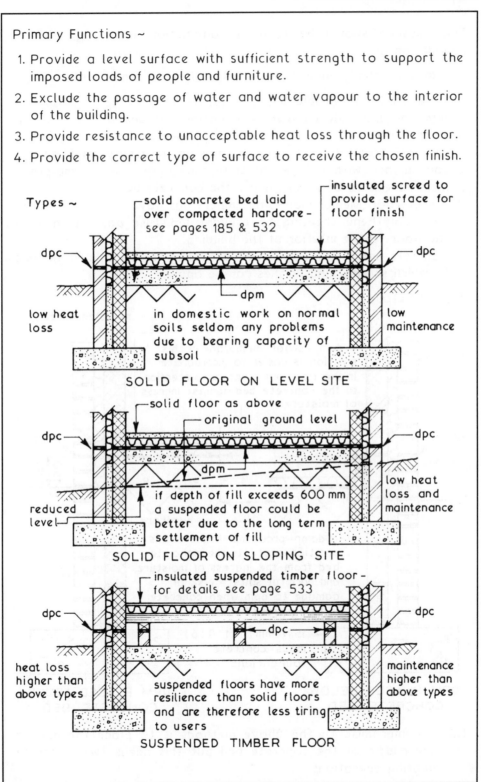

Types ~

SOLID FLOOR ON LEVEL SITE

solid concrete bed laid over compacted hardcore – see pages 185 & 532

insulated screed to provide surface for floor finish

dpc

dpc

dpm

low heat loss

low maintenance

in domestic work on normal soils seldom any problems due to bearing capacity of subsoil

SOLID FLOOR ON SLOPING SITE

solid floor as above

original ground level

dpc

dpc

dpm

reduced level

if depth of fill exceeds 600 mm a suspended floor could be better due to the long term settlement of fill

low heat loss and maintenance

SUSPENDED TIMBER FLOOR

insulated suspended timber floor – for details see page 533

dpc

dpc

dpc

heat loss higher than above types

maintenance higher than above types

suspended floors have more resilience than solid floors and are therefore less tiring to users

Domestic Solid Ground Floors

This drawing should be read in conjunction with page 185 — Foundation Beds.

A domestic solid ground floor consists of three components:-

1. Hardcore — a suitable filling material to make up the top soil removal and reduced level excavations. It should have a top surface which can be rolled out to ensure that cement grout is not lost from the concrete. It may be necessary to blind the top surface with a layer of sand especially if the damp-proof membrane is to be placed under the concrete bed.
2. Damp-proof Membrane — an impervious layer such as heavy duty polythene sheeting to prevent moisture passing through the floor to the interior of the building.
3. Concrete Bed — the component providing the solid level surface to which screeds and finishes can be applied.

Typical Details ~

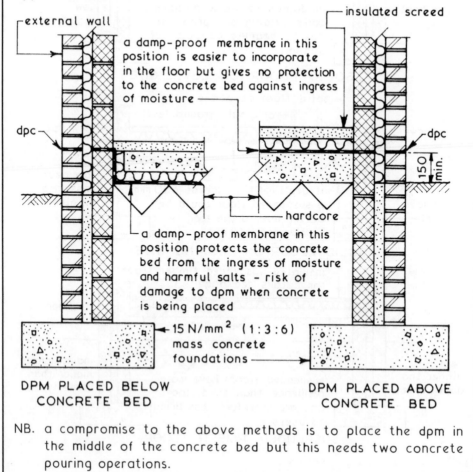

a damp-proof membrane in this position is easier to incorporate in the floor but gives no protection to the concrete bed against ingress of moisture

a damp-proof membrane in this position protects the concrete bed from the ingress of moisture and harmful salts - risk of damage to dpm when concrete is being placed

external wall

insulated screed

dpc

dpc

150 min.

hardcore

15 N/mm^2 (1:3:6) mass concrete foundations

DPM PLACED BELOW CONCRETE BED

DPM PLACED ABOVE CONCRETE BED

NB. a compromise to the above methods is to place the dpm in the middle of the concrete bed but this needs two concrete pouring operations.

Suspended Timber Ground Floors ~ these need to have a well ventilated space beneath the floor construction to prevent the moisture content of the timber rising above an unacceptable level (i.e. not more than 20%) which would create the conditions for possible fungal attack.

Typical Details ~

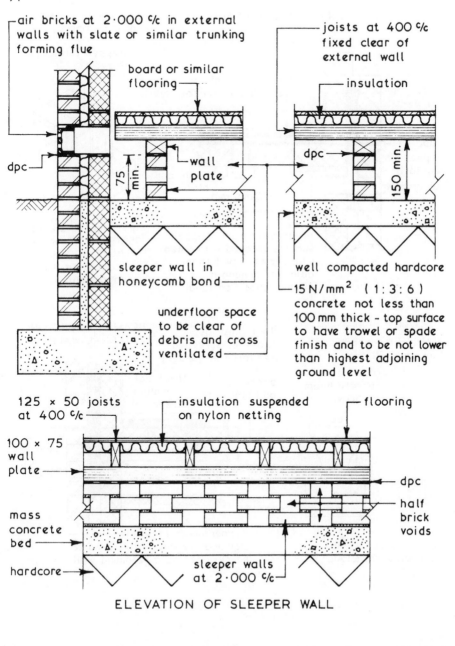

air bricks at 2·000 c/c in external walls with slate or similar trunking forming flue

board or similar flooring

joists at 400 c/c fixed clear of external wall

insulation

dpc

75 min.

wall plate

dpc

150 min.

sleeper wall in honeycomb bond

well compacted hardcore

underfloor space to be clear of debris and cross ventilated

15 N/mm² (1:3:6) concrete not less than 100 mm thick – top surface to have trowel or spade finish and to be not lower than highest adjoining ground level

125 × 50 joists at 400 c/c

insulation suspended on nylon netting

flooring

100 × 75 wall plate

dpc

half brick voids

mass concrete bed

hardcore

sleeper walls at 2·000 c/c

ELEVATION OF SLEEPER WALL

Precast Concrete Floors ~ these have been successfully adapted from commercial building practice (see pages 551 to 553), as an economic alternative construction technique for suspended timber and solid concrete domestic ground (and upper) floors. See also page 304 for special situations.

Typical Details ~

BEAMS PARALLEL WITH EXTERNAL WALL

insulated cavity wall

cavity tray over vent

18 mm t&g chipboard over vapour control layer

insulation

dpc

stepped ventilator

GL

coursing slip

150 mm min.

beam and block

dpc

organic material stripped; surface treated with weed killer; lower level than adjacent ground if free draining (not Scotland)

POLYPROPYLENE VENTILATOR

stepped telescopic sleeve

grill clips to sleeve; 1500 mm^2/m run of wall OR 500 mm^2/m^2 of floor area (take greater value)

lightweight concrete block, min. 7N/mm^2

prestressed concrete beam min. 50 N/mm^2

60

100

60

130

TYPICAL BEAM/RIB AND BLOCK DETAIL

floor finish as above or screeded (reinforce in garage)

beam and block

this block to wall strength

BEAMS BEARING ON EXTERNAL WALL

dpc

vent

GL

dpc

if inner ground level is significantly lower, this area to be designed as a retaining wall

Floor Finishes ~ these are usually applied to a structural base but may form part of the floor structure as in the case of floor boards. Most finishes are chosen to fulfil a particular function such as:-

1. Appearance — chosen mainly for their aesthetic appeal or effect but should however have reasonable wearing properties. Examples are carpets; carpet tiles and wood blocks.

2. High Resistance — chosen mainly for their wearing and impact resistance properties and for high usage areas such as kitchens. Examples are quarry tiles and granolithic pavings.

3. Hygiene — chosen to provide an impervious easy to clean surface with reasonable aesthetic appeal. Examples are quarry tiles and polyvinyl chloride (PVC) sheets and tiles.

Carpets and Carpet Tiles — made from animal hair, mineral fibres and man made fibres such as nylon and acrylic. They are also available in mixtures of the above. A wide range of patterns; sizes and colours are available. Carpets and carpet tiles can be laid loose, stuck with a suitable adhesive or in the case of carpets edge fixed using special grip strips.

PVC Tiles — made from a blended mix of thermoplastic binders; fillers and pigments in a wide variety of colours and patterns to the recommendations of BS 3261. PVC tiles are usually 305 × 305 × 1·6 mm thick and are stuck to a suitable base with special adhesives as recommended by the manufacturer.

Quarry Tiles ~

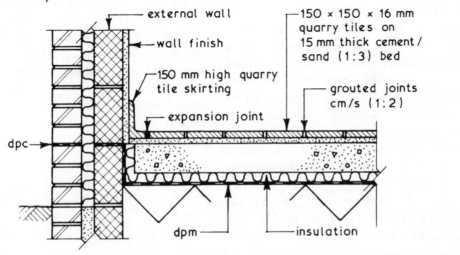

external wall

wall finish

150 mm high quarry tile skirting

expansion joint

dpc

150 × 150 × 16 mm quarry tiles on 15 mm thick cement / sand (1:3) bed

grouted joints cm/s (1:2)

dpm

insulation

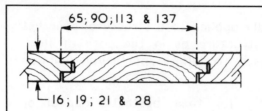

65; 90; 113 & 137

16; 19; 21 & 28

Tongue and Groove Boarding ~ prepared from softwoods to the recommendations of BS 1297. Boards are laid at right angles to the joists and are fixed with 2 No. 65mm long cut floor brads per joists. The ends of board lengths are butt jointed on the centre line of the supporting joist.

Maximum board spans are:-

16 mm thick — 505 mm
19 mm thick — 600 mm
21 mm thick — 635 mm
28 mm thick — 790 mm

75

22

secret
nailing

Timber Strip Flooring ~ strip flooring is usually considered to be boards under 100 mm face width. In good class work hardwoods would be specified the boards being individually laid and secret nailed. Strip flooring can be obtained treated with a spirit-based fungicide. Spacing of supports depends on type of timber used and applied loading. After laying the strip flooring should be finely sanded and treated with a seal or wax. In common with all timber floorings a narrow perimeter gap should be left for moisture movement.

Chipboard ~ sometimes called Particle Board is made from particles of wood bonded with a synthetic resin and/or other organic binders to the recommendations of BS 5669. It can be obtained with a rebated or tongue and groove joint in 600mm wide boards 19mm thick. The former must be supported on all the longitudinal edges whereas the latter should be supported at all cross joints.

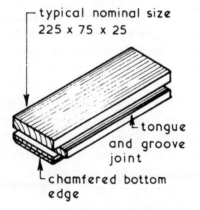

typical nominal size
225 x 75 x 25

tongue
and groove
joint

chamfered bottom
edge

Wood Blocks ~ prepared from hardwoods and softwoods to the recommendations of BS 1187. Wood blocks can be laid to a variety of patterns, also different timbers can be used to create colour and grain effects. Laid blocks should be finely sanded and sealed or polished.

Large Cast-Insitu Ground Floors ~ these are floors designed to carry medium to heavy loadings such as those used in factories, warehouses, shops, garages and similar buildings. Their design and construction is similar to that used for small roads. (see pages 108 to 111). Floors of this type are usually laid in alternate 4·500 wide strips running the length of the building or in line with the anticipated traffic flow where applicable. Transverse joints will be required to control the tensile stresses due to the thermal movement and contraction of the slab. The spacing of these joints will be determined by the design and the amount of reinforcement used. Such joints can either be formed by using a crack inducer or by sawing a 20 to 25mm deep groove into the upper surface of the slab within 20 to 30 hours of casting.

Typical Layout ~

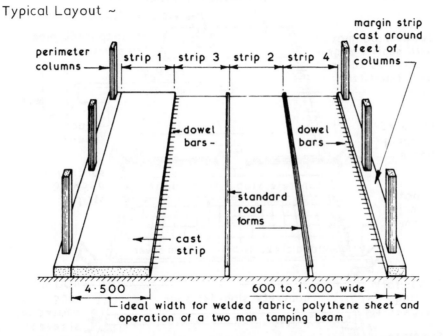

Surface Finishing ~ the surface of the concrete may be finished by power floating or trowelling which is carried out whilst the concrete is still plastic but with sufficient resistance to the weight of machine and operator whose footprint should not leave a depression of more than 3mm. Power grinding of the surface is an alternative method which is carried out within a few days of the concrete hardening. The wet concrete having been surface finished with a skip float after the initial levelling with a tamping bar has been carried out. Power grinding removes 1 to 2mm from the surface and is intended to improve surface texture and not to make good deficiencies in levels.

Vacuum Dewatering ~ if the specification calls for a power float surface finish vacuum dewatering could be used to shorten the time delay between tamping the concrete and power floating the surface. This method is suitable for slabs up to 300mm thick. The vacuum should be applied for approximately 3 minutes for every 25mm depth of concrete which will allow power floating to take place usually within 20 to 30 minutes of the tamping operation. The applied vacuum forces out the surplus water by compressing the slab and this causes a reduction in slab depth of approximately 2% therefore packing strips should be placed on the side forms before tamping to allow for sufficient surcharge of concrete.

Typical Details ~

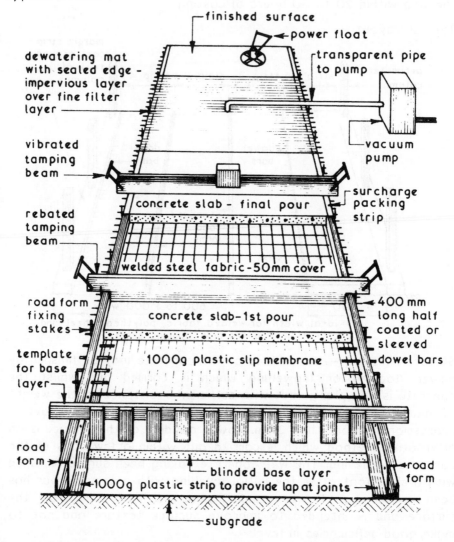

Concrete Floor Screeds ~ these are used to give a concrete floor a finish suitable to receive the floor finish or covering specified. It should be noted that it is not always necessary or desirable to apply a floor screed to receive a floor covering, techniques are available to enable the concrete floor surface to be prepared at the time of casting to receive the coverings at a later stage.

Typical Screed Mixes ~

Screed Thickness	Cement	Dry Fine Aggregate <5mm	Coarse Aggregate >5mm <10mm
up to 40mm	I	3 to 4 1/2	–
40 to 75mm	I	3 to 4 1/2	–
	I	1 1/2	3

Laying Floor Screeds ~ floor screeds should not be laid in bays since this can cause curling at the edges, screeds can however be laid in 3·000 wide strips to receive thin coverings. Levelling of screeds is achieved by working to levelled timber screeding batten or alternatively a 75mm wide band of levelled screed with square edges can be laid to the perimeter of the floor prior to the general screed laying operation.

Screed Types ~

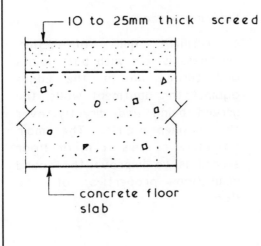

— 10 to 25mm thick screed

— concrete floor slab

Monolithic Screeds —

screed laid directly on concrete floor slab within three hours of placing concrete — before any screed is placed all surface water should be removed — all screeding work should be carried out from scaffold board runways to avoid walking on the 'green' concrete slab.

Concrete Floor Screeds

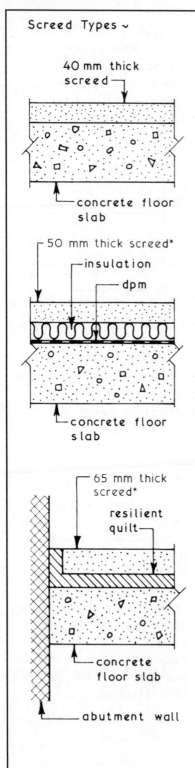

Screed Types ˅

40 mm thick screed

concrete floor slab

50 mm thick screed*

insulation

dpm

concrete floor slab

65 mm thick screed*

resilient quilt

concrete floor slab

abutment wall

Separate Screeds —

screed is laid onto the concrete floor slab after it has cured. The floor surface must be clean and rough enough to ensure an adequate bond unless the floor surface is prepared by applying a suitable bonding agent or by brushing with a cement/water grout of a thick cream like consistency just before laying the screed.

Unbonded Screeds —

screed is laid directly over a damp-proof membrane or over a damp-proof membrane and insulation. A rigid form of floor insulation is required where the concrete floor slab is in contact with the ground. Care must be taken during this operation to ensure that the damp-proof membrane is not damaged.

Floating Screeds —

a resilient quilt of 25mm thickness is laid with butt joints and turned up at the edges against the abutment walls, the screed being laid directly over the resilient quilt. The main objective of this form of floor screed is to improve the sound insulation properties of the floor.

*preferably wire mesh reinforced

Primary Functions ~

1. Provide a level surface with sufficient strength to support the imposed loads of people and furniture plus the dead loads of flooring and ceiling.
2. Reduce heat loss from lower floor as required.
3. Provide required degree of sound insulation.
4. Provide required degree of fire resistance.

Basic Construction — a timber suspended upper floor consists of a series of beams or joists support by load bearing walls sized and spaced to carry all the dead and imposed loads.

Joist Sizing — three methods can be used:-

1. Building Regs. Imposed load to be not more than 1·5 kN/m². Calculate dead load in kg/m² supported by joist excluding the mass of the joist and select suitable size from Tables A1 and A2 in Approved Document A

2. Calculation formula:-

$$BM = \frac{fbd^2}{6}$$

where

BM = bending moment
f = fibre stress
b = breadth
d = depth in mm must be assumed

3. Empirical formula:-

$$D = \frac{\text{span in mm}}{24} + 50$$

where

D = depth of joist in mm

above assumes that joists have a breadth of 50 mm and are at 400 c/c spacing

Support and Restraint ~

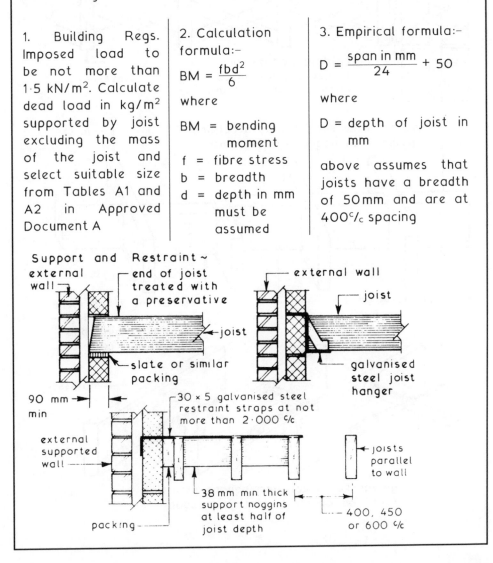

90 mm min

external supported wall

packing

end of joist treated with a preservative

joist

slate or similar packing

30 × 5 galvanised steel restraint straps at not more than 2·000 c/c

38 mm min thick support noggins at least half of joist depth

external wall

joist

galvanised steel joist hanger

joists parallel to wall

400, 450 or 600 c/c

Timber Suspended Upper Floors

Strutting ~ used in timber suspended floors to restrict the movements due to twisting and vibration which could damage ceiling finishes. Strutting should be included if the span of the floor joists exceeds 2·500 and is positioned on the centre line of the span.

Typical Details ~

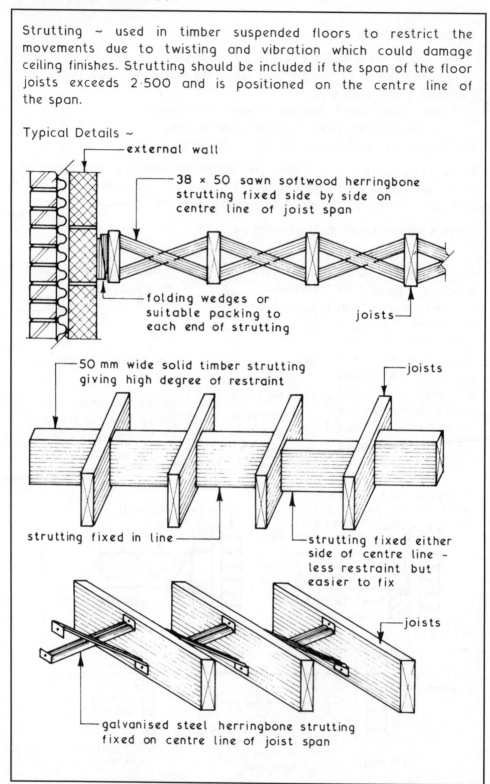

external wall

38 × 50 sawn softwood herringbone strutting fixed side by side on centre line of joist span

folding wedges or suitable packing to each end of strutting

joists

50 mm wide solid timber strutting giving high degree of restraint

joists

strutting fixed in line

strutting fixed either side of centre line - less restraint but easier to fix

joists

galvanised steel herringbone strutting fixed on centre line of joist span

Lateral Restraint ~ external, compartment (fire), separating (party) and internal loadbearing walls must have horizontal support from adjacent floors, to restrict movement. Exceptions occur when the wall is less than 3m long.

Methods:

1. 90mm end bearing of floor joists, spaced not more than 1·2m apart — see page 541

2. Galvanised steel straps spaced at intervals not exceeding 2m and fixed square to joists — see page 541

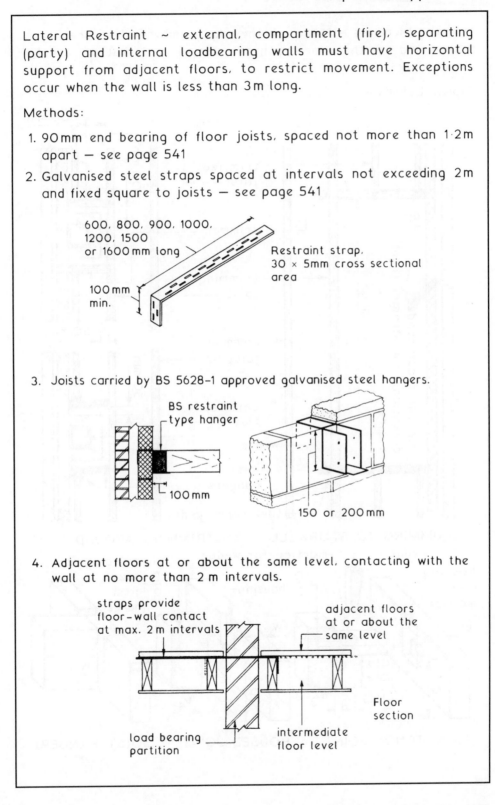

600, 800, 900, 1000, 1200, 1500 or 1600mm long

100mm min.

Restraint strap, 30 × 5mm cross sectional area

3. Joists carried by BS 5628-1 approved galvanised steel hangers.

BS restraint type hanger

100mm

150 or 200mm

4. Adjacent floors at or about the same level, contacting with the wall at no more than 2 m intervals.

straps provide floor-wall contact at max. 2 m intervals

adjacent floors at or about the same level

load bearing partition

intermediate floor level

Floor section

Trimming Members ~ these are the edge members of an opening in a floor and are the same depth as common joists but are usually 25mm wider.

Typical Details ~

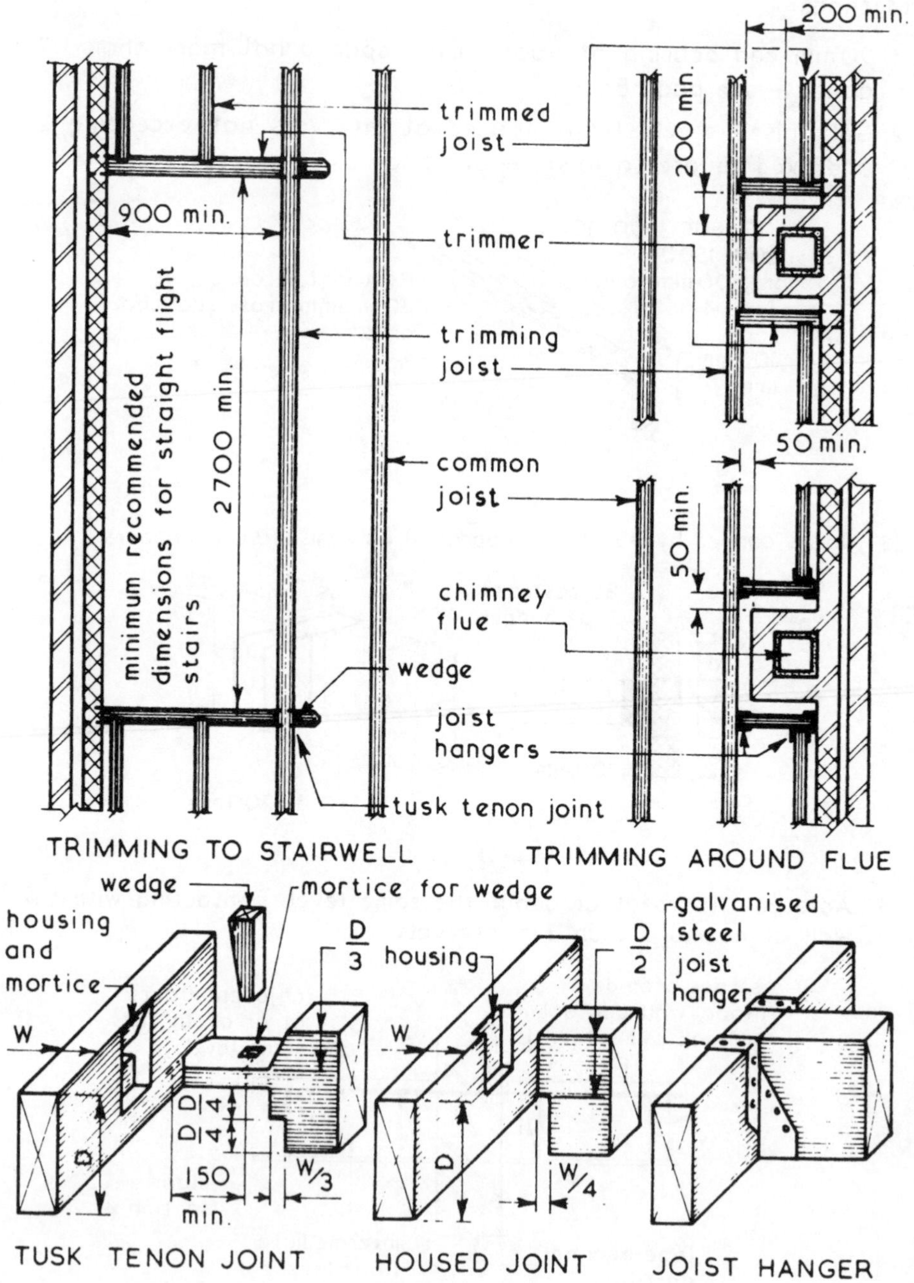

TRIMMING TO STAIRWELL

TRIMMING AROUND FLUE

TUSK TENON JOINT

HOUSED JOINT

JOIST HANGER

Joist and Beam Sizing ~ design tables and formulae have limitations, therefore where loading, span and/or conventional joist spacings are exceeded, calculations are required. BS 5268: Structural Use Of Timber and BS EN 338: Structural Timber — Strength Classes, are both useful resource material for detailed information on a variety of timber species. The following example serves to provide guidance on the design process for determining joist size, measurement of deflection, safe bearing and resistance to shear force:-

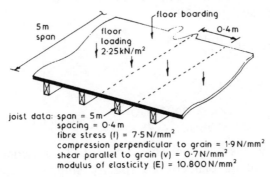

joist data: span = 5 m
spacing = 0·4 m
fibre stress (f) = 7·5 N/mm²
compression perpendicular to grain = 1·9 N/mm²
shear parallel to grain (v) = 0·7 N/mm²
modulus of elasticity (E) = 10.800 N/mm²

Total load (W) per joist = 5 m × 0·4 m × 2·25 kN/m² = 4·5 kN

$$\text{or: } \frac{4.5\,\text{kN}}{5\,\text{m span}} = 0.9\,\text{kN/m}$$

Resistance to bending ~

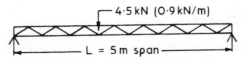

L = 5 m span

Bending moment formulae are shown on page 384

$$BM = \frac{WL}{8} = \frac{fbd^2}{6}$$

Where: W = total load, 4·5 kN (4500 N)
     L = span, 5 m (5000 mm)
     f = fibre stress of timber, 7·5 N/mm²
     b = breadth of joist, try 50 mm
     d = depth of joist, unknown

Transposing:

$$\frac{WL}{8} = \frac{fbd^2}{6}$$

Becomes:

$$d = \sqrt{\frac{6WL}{8fb}} = \sqrt{\frac{6 \times 4500 \times 5000}{8 \times 7.5 \times 50}} = 212\,\text{mm}$$

Nearest commercial size: 50 mm × 225 mm

Timber Beam Design

Joist and Beam Sizing ~ calculating overall dimensions alone is insufficient, checks should also be made to satisfy: resistance to deflection, adequate safe bearing and resistance to shear.

Deflection — should be minimal to prevent damage to plastered ceilings. An allowance of up to $0.003 \times span$ is normally acceptable; for the preceding example this will be:-

$0.003 \times 5000\,mm = 15\,mm$

The formula for calculating deflection due to a uniformly distributed load (see page 386) is: ~

$$\frac{5WL^3}{384EI} \quad where \quad I = \frac{bd^3}{12}$$

$$I = \frac{50 \times (225)^3}{12} = 4.75 \times (10)^7$$

So, deflection $= \dfrac{5 \times 4500 \times (5000)^3}{384 \times 10800 \times 4.75 \times (10)^7} = 14.27\,mm$

NB. This is only just within the calculated allowance of 15 mm, therefore it would be prudent to specify slightly wider or deeper joists to allow for unknown future use.

Safe Bearing ~

$$= \frac{load\ at\ the\ joist\ end.\ W/2}{compression\ perpendicular\ to\ grain \times breadth}$$

$$= \frac{4500/2}{1.9 \times 50} = 24\,mm.$$

therefore full support from masonry (90 mm min.) or joist hangers will be more than adequate.

Shear Strength ~

$$V = \frac{2bdv}{3}$$

where: V = vertical loading at the joist end, W/2

v = shear strength parallel to the grain, 0.7 N/mm²
Transposing:-

$$bd = \frac{3V}{2v} = \frac{3 \times 2250}{2 \times 0.7} = 4821\,mm^2\ minimum$$

Actual bd = 50 mm $\times$ 225 mm = 11,250 mm²

Resistance to shear is satisfied as actual is well above the minimum.

Reinforced Concrete Suspended Floors ~ a simple reinforced concrete flat slab cast to act as a suspended floor is not usually economical for spans over 5·000. To overcome this problem beams can be incorporated into the design to span in one or two directions. Such beams usually span between columns which transfers their loads to the foundations. The disadvantages of introducing beams are the greater overall depth of the floor construction and the increased complexity of the formwork and reinforcement. To reduce the overall depth of the floor construction flat slabs can be used where the beam is incorporated with the depth of the slab. This method usually results in a deeper slab with complex reinforcement especially at the column positions.

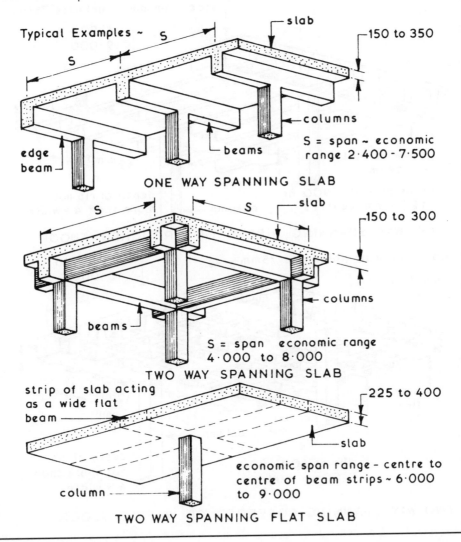

Typical Examples ~ S

slab

150 to 350

columns

edge beam

beams

S = span ~ economic range 2·400 - 7·500

ONE WAY SPANNING SLAB

slab

150 to 300

columns

beams

S = span economic range 4·000 to 8·000

TWO WAY SPANNING SLAB

strip of slab acting as a wide flat beam

225 to 400

slab

economic span range - centre to centre of beam strips ~ 6·000 to 9·000

column

TWO WAY SPANNING FLAT SLAB

Ribbed Floors ~ to reduce the overall depth of a traditional cast insitu reinforced concrete beam and slab suspended floor a ribbed floor could be used. The basic concept is to replace the wide spaced deep beams with narrow spaced shallow beams or ribs which will carry only a small amount of slab loading. These floors can be designed as one or two way spanning floors. One way spanning ribbed floors are sometimes called troughed floors whereas the two way spanning ribbed floors are called coffered or waffle floors. Ribbed floors are usually cast against metal or reinforced plastic formers in the form of preformed moulds which can be left in place to provide the soffit finish.

Typical Examples ~

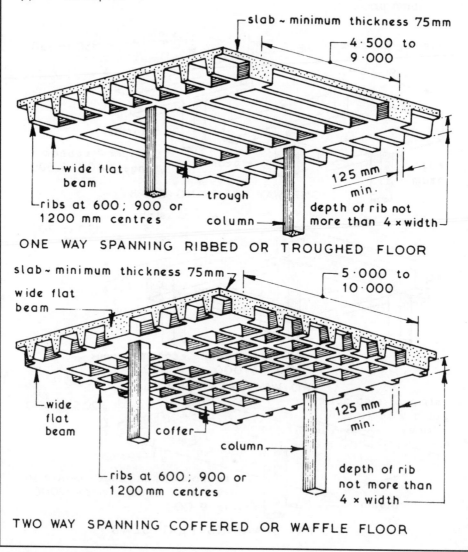

slab ~ minimum thickness 75mm

4·500 to 9·000

wide flat beam

trough

ribs at 600; 900 or 1200 mm centres

column

125 mm min.

depth of rib not more than 4 × width

ONE WAY SPANNING RIBBED OR TROUGHED FLOOR

slab ~ minimum thickness 75mm

wide flat beam

5·000 to 10·000

wide flat beam

coffer

column

125 mm min.

ribs at 600; 900 or 1200 mm centres

depth of rib not more than 4 × width

TWO WAY SPANNING COFFERED OR WAFFLE FLOOR

Hollow Pot Floors ~ these are in essence a ribbed floor with permanent formwork in the form of hollow clay or concrete pots. The main advantage of this type of cast insitu floor is that it has a flat soffit which is suitable for the direct application of a plaster finish or an attached dry lining. The voids in the pots can be utilised to house small diameter services within the overall depth of the slab. These floors can be designed as one or two way spanning slabs, the common format being the one way spanning floor.

Typical Example ~

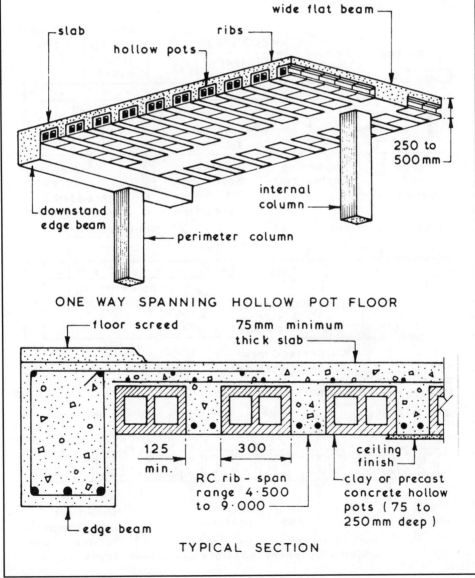

ONE WAY SPANNING HOLLOW POT FLOOR

TYPICAL SECTION

Soffit and Beam Fixings ~ concrete suspended floors can be designed to carry loads other than the direct upper surface loadings. Services can be housed within the voids created by the beams or ribs and suspended or attached ceilings can be supported by the floor. Services which run at right angles to the beams or ribs are usually housed in cast-in holes. There are many types of fixings available for use in conjunction with floor slabs, some are designed to be cast-in whilst others are fitted after the concrete has cured. All fixings must be positioned and installed so that they are not detrimental to the structural integrity of the floor.

Typical Examples ~

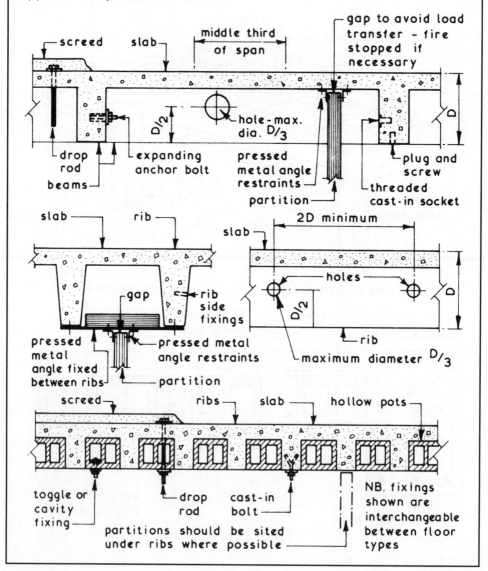

Precast Concrete Floors ~ these are available in several basic formats and provide an alternative form of floor construction to suspended timber floors and insitu reinforced concrete suspended floors. The main advantages of precast concrete floors are:-

1. Elimination of the need for formwork except for nominal propping which is required with some systems.
2. Curing time of concrete is eliminated therefore the floor is available for use as a working platform at an earlier stage.
3. Superior quality control of product is possible with factory produced components.

The main disadvantages of precast concrete floors when compared with insitu reinforced concrete floors are:-

1. Less flexible in design terms.
2. Formation of large openings in the floor for ducts, shafts and stairwells usually have to be formed by casting an insitu reinforced concrete floor strip around the opening position.
3. Higher degree of site accuracy is required to ensure that the precast concrete floor units can be accommodated without any alterations or making good

Typical Basic Formats ~

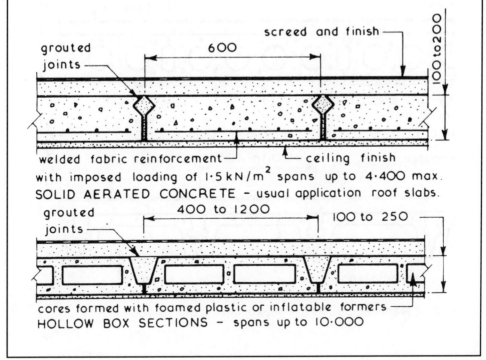

with imposed loading of $1 \cdot 5 \, kN/m^2$ spans up to 4·400 max.
SOLID AERATED CONCRETE - usual application roof slabs.

cores formed with foamed plastic or inflatable formers
HOLLOW BOX SECTIONS - spans up to 10·000

Typical Basic Formats ~

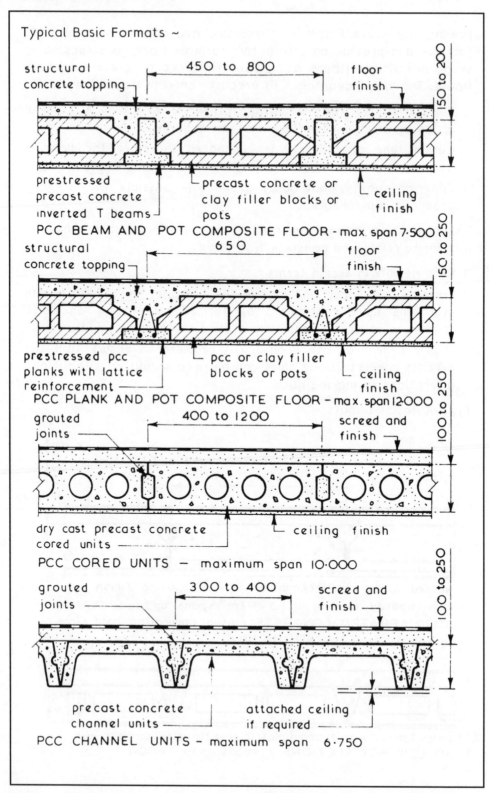

structural concrete topping

450 to 800

floor finish

150 to 200

prestressed precast concrete inverted T beams

precast concrete or clay filler blocks or pots

ceiling finish

PCC BEAM AND POT COMPOSITE FLOOR - max. span 7·500

structural concrete topping

650

floor finish

150 to 250

prestressed pcc planks with lattice reinforcement

pcc or clay filler blocks or pots

ceiling finish

PCC PLANK AND POT COMPOSITE FLOOR - max. span 12·000

grouted joints

400 to 1200

screed and finish

100 to 250

dry cast precast concrete cored units

ceiling finish

PCC CORED UNITS — maximum span 10·000

grouted joints

300 to 400

screed and finish

100 to 250

precast concrete channel units

attached ceiling if required

PCC CHANNEL UNITS — maximum span 6·750

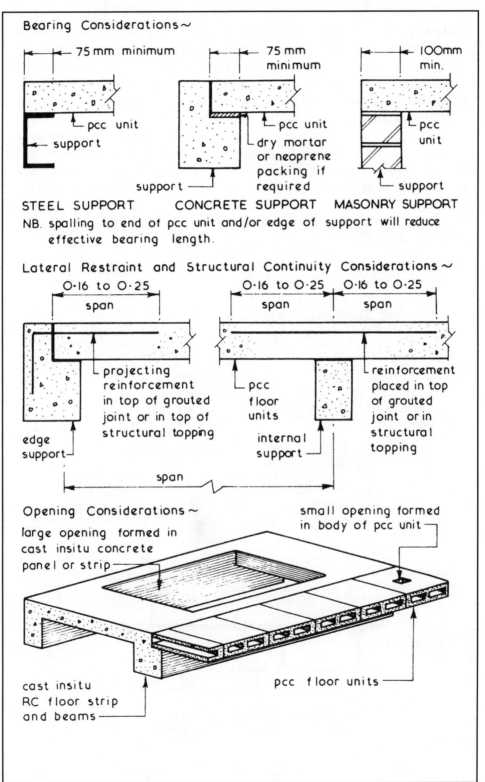

Bearing Considerations ~

75 mm minimum
pcc unit
support

75 mm minimum
pcc unit
dry mortar or neoprene packing if required
support

100mm min.
pcc unit
support

STEEL SUPPORT   CONCRETE SUPPORT   MASONRY SUPPORT

NB. spalling to end of pcc unit and/or edge of support will reduce effective bearing length.

Lateral Restraint and Structural Continuity Considerations ~

0·16 to 0·25 span
projecting reinforcement in top of grouted joint or in top of structural topping
edge support
span

0·16 to 0·25 span   0·16 to 0·25 span
pcc floor units
internal support
reinforcement placed in top of grouted joint or in structural topping

Opening Considerations ~

large opening formed in cast insitu concrete panel or strip

small opening formed in body of pcc unit

cast insitu RC floor strip and beams

pcc floor units

Raised Flooring ~ developed in response to the high-tech boom of the 1970s. It has proved expedient in accommodating computer and communications cabling as well as numerous other established services. The system is a combination of adjustable floor pedestals, supporting a variety of decking materials. Pedestal height ranges from as little as 30 mm up to about 600 mm, although greater heights are possible at the expense of structural floor levels. Decking is usually in loose fit squares of 600 mm, but may be sheet plywood or particleboard screwed direct to closer spaced pedestal support plates on to joists bearing on pedestals.

Application ~

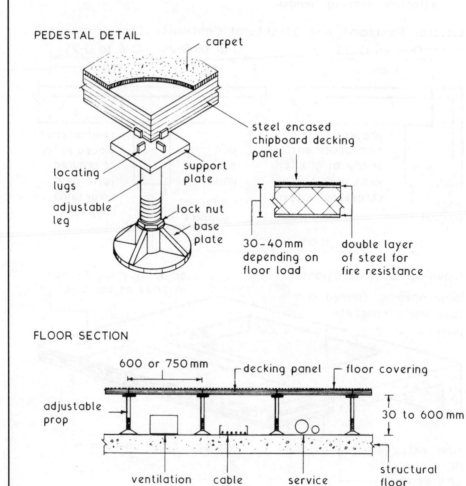

PEDESTAL DETAIL

carpet

steel encased chipboard decking panel

locating lugs

support plate

adjustable leg

lock nut

base plate

30–40 mm depending on floor load

double layer of steel for fire resistance

FLOOR SECTION

600 or 750 mm

decking panel

floor covering

adjustable prop

30 to 600 mm

ventilation duct

cable tray

service pipes

structural floor

Primary Functions ~

1. Provide a means of circulation between floor levels.

2. Establish a safe means of travel between floor levels.

3. Provide an easy means of travel between floor levels.

4. Provide a means of conveying fittings and furniture between floor levels.

Constituent Parts ~

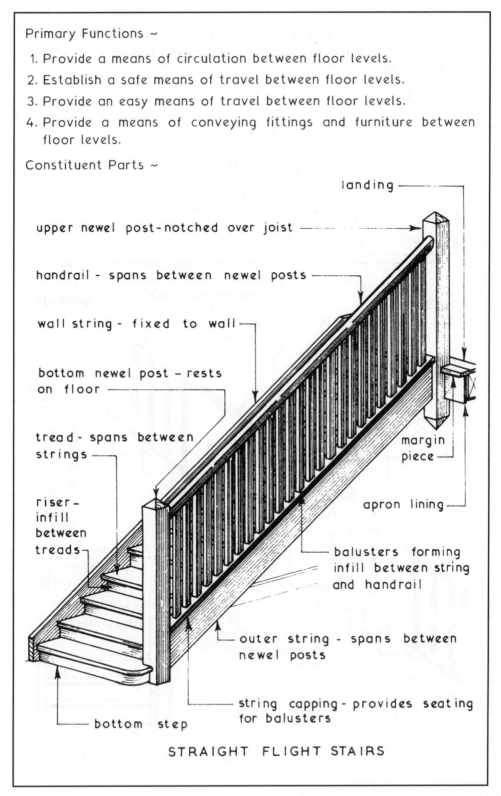

landing

upper newel post-notched over joist

handrail - spans between newel posts

wall string - fixed to wall

bottom newel post - rests on floor

tread - spans between strings

riser - infill between treads

margin piece

apron lining

balusters forming infill between string and handrail

outer string - spans between newel posts

string capping - provides seating for balusters

bottom step

STRAIGHT FLIGHT STAIRS

All dimensions quoted are the minimum required for domestic stairs exclusive to one dwelling as given in Approved Document K unless stated otherwise.

Terminology ~

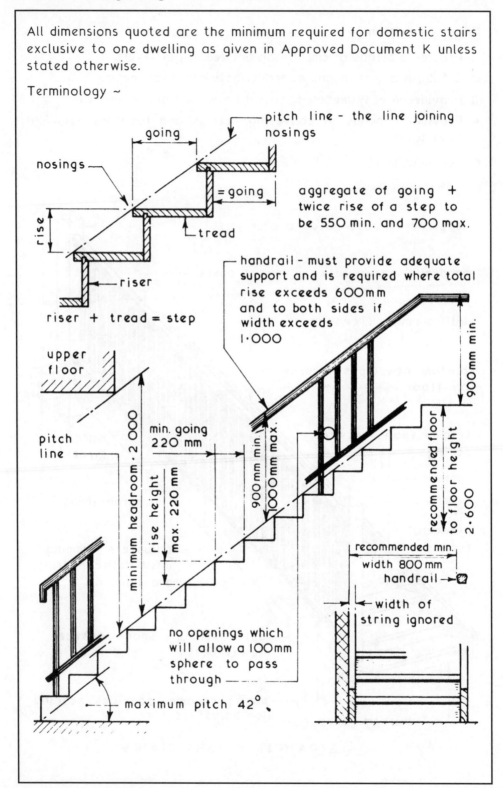

pitch line – the line joining nosings

going

nosings

= going

tread

rise

riser

riser + tread = step

aggregate of going + twice rise of a step to be 550 min. and 700 max.

handrail – must provide adequate support and is required where total rise exceeds 600mm and to both sides if width exceeds 1·000

upper floor

pitch line

minimum headroom · 2 000

min. going 220 mm

rise height max. 220 mm

900 mm min.
1000 mm max.

900mm min.

recommended floor to floor height 2·600

recommended min. width 800 mm

handrail

width of string ignored

no openings which will allow a 100mm sphere to pass through

maximum pitch 42°

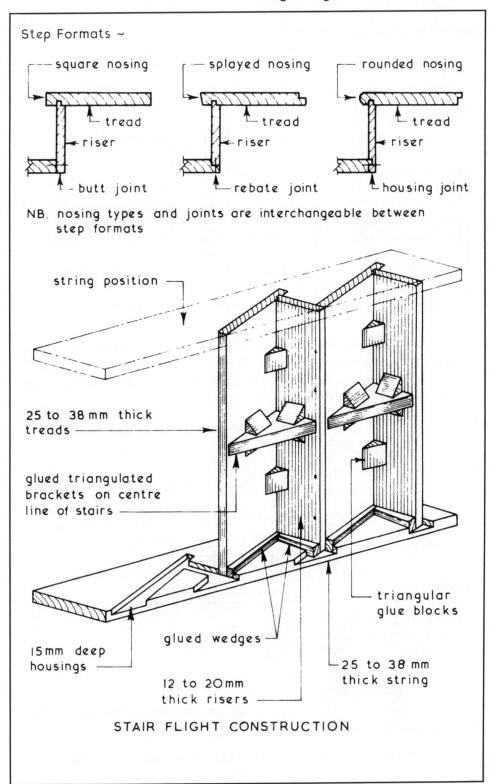

Step Formats ~

square nosing
tread
riser
butt joint

splayed nosing
tread
riser
rebate joint

rounded nosing
tread
riser
housing joint

NB. nosing types and joints are interchangeable between step formats

string position

25 to 38 mm thick treads

glued triangulated brackets on centre line of stairs

triangular glue blocks

glued wedges

15mm deep housings

12 to 20mm thick risers

25 to 38 mm thick string

STAIR FLIGHT CONSTRUCTION

Bottom Step Arrangements ~

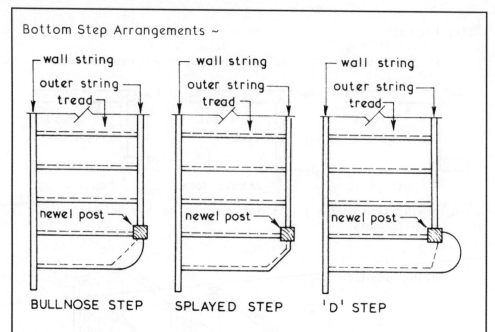

BULLNOSE STEP   SPLAYED STEP   'D' STEP

Projecting bottom steps are usually included to enable the outer string to be securely jointed to the back face of the newel post and to provide an easy line of travel when ascending or descending at the foot of the stairs.

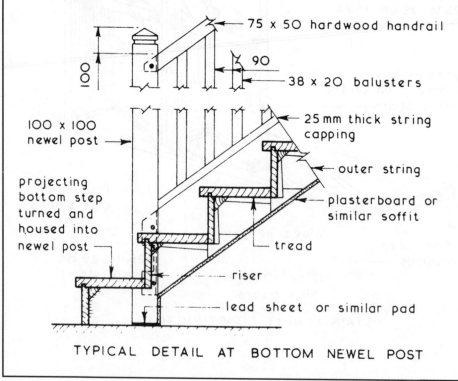

TYPICAL DETAIL AT BOTTOM NEWEL POST

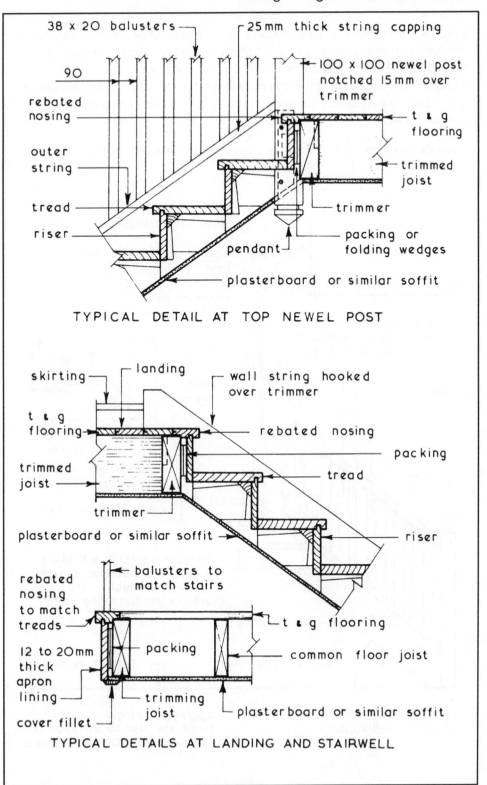

TYPICAL DETAIL AT TOP NEWEL POST

- 38 × 20 balusters
- 25mm thick string capping
- 90
- 100 × 100 newel post notched 15mm over trimmer
- rebated nosing
- t & g flooring
- outer string
- trimmed joist
- tread
- trimmer
- riser
- pendant
- packing or folding wedges
- plasterboard or similar soffit

TYPICAL DETAILS AT LANDING AND STAIRWELL

- skirting
- landing
- wall string hooked over trimmer
- t & g flooring
- rebated nosing
- packing
- trimmed joist
- tread
- trimmer
- plasterboard or similar soffit
- riser
- rebated nosing to match treads
- balusters to match stairs
- t & g flooring
- 12 to 20mm thick apron lining
- packing
- common floor joist
- cover fillet
- trimming joist
- plasterboard or similar soffit

Timber Open Riser Stairs

Open Riser Timber Stairs ~ these are timber stairs constructed to the same basic principles as standard timber stairs excluding the use of a riser. They have no real advantage over traditional stairs except for the generally accepted aesthetic appeal of elegance. Like the traditional timber stairs they must comply with the minimum requirements set out in Part K of the Building Regulations.

Typical Requirements for Stairs in a Small Residental Building ~

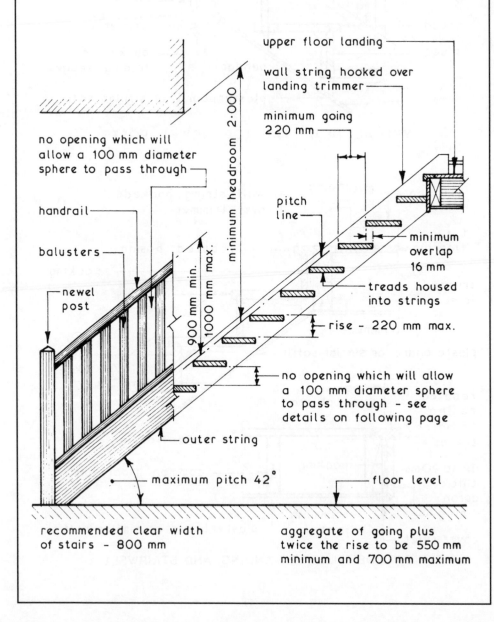

upper floor landing

wall string hooked over landing trimmer

minimum going 220 mm

pitch line

minimum overlap 16 mm

treads housed into strings

rise - 220 mm max.

no opening which will allow a 100 mm diameter sphere to pass through - see details on following page

floor level

no opening which will allow a 100 mm diameter sphere to pass through

handrail

balusters

newel post

minimum headroom 2·000

900 mm min.

1000 mm max.

outer string

maximum pitch 42°

recommended clear width of stairs - 800 mm

aggregate of going plus twice the rise to be 550 mm minimum and 700 mm maximum

Design and Construction ~ because of the legal requirement of not having a gap between any two consecutive treads through which a 100 mm diameter sphere can pass and the limitation relating to the going and rise, as shown on the previous page, it is generally not practicable to have a completely riserless stair for residential buildings since by using minimum dimensions a very low pitch of approximately $27^{1}/_{2}°$ would result and by choosing an acceptable pitch a very thick tread would have to be used to restrict the gap to 100 mm.

Possible Solutions ~

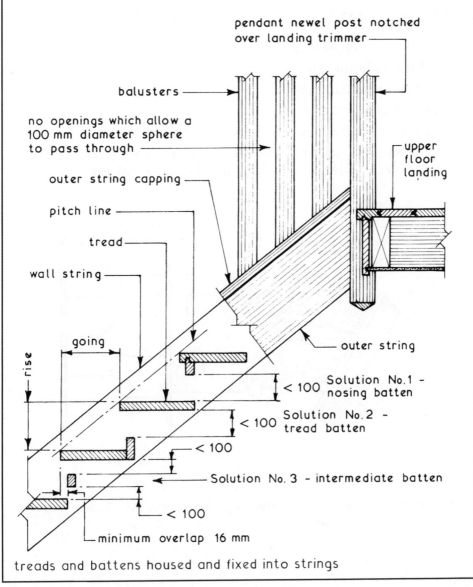

pendant newel post notched over landing trimmer

balusters

no openings which allow a 100 mm diameter sphere to pass through

upper floor landing

outer string capping

pitch line

tread

wall string

outer string

going

rise

< 100 Solution No.1 - nosing batten

< 100 Solution No.2 - tread batten

< 100

Solution No. 3 - intermediate batten

< 100

minimum overlap 16 mm

treads and battens housed and fixed into strings

Application — a straight flight for access to a domestic loft conversion only. This can provide one habitable room, plus a bathroom or WC. The WC must not be the only WC in the dwelling.

Practical issues — an economic use of space, achieved by a very steep pitch of about 60° and opposing overlapping treads.

Safety — pitch and tread profile differ considerably from other stairs, but they are acceptable to Building Regulations by virtue of "familiarity and regular use" by the building occupants.

Additional features are:

* a non-slip tread surface.
* handrails to both sides.
* minimum going 220 mm.
* maximum rise 220 mm.
* (2 × rise) + (going) between 550 and 700 mm.
* a stair used by children under 5 years old, must have the tread voids barred to leave a gap not greater than 100 mm.

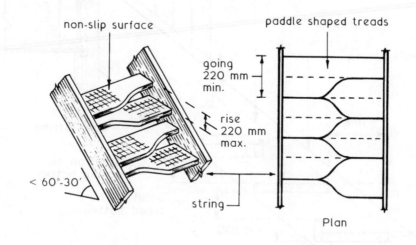

non-slip surface

paddle shaped treads

going 220 mm min.

rise 220 mm max.

< 60°·30′

string

Plan

Ref. Building Regulations, Approved Document K1: Stairs, ladders and ramps: Section 1.29

Timber Stairs ~ these must comply with the minimum requirements set out in Part K of the Building Regulations. Straight flight stairs are simple, easy to construct and install but by the introduction of intermediate landings stairs can be designed to change direction of travel and be more compact in plan than the straight flight stairs.

Landings ~ these are designed and constructed in the same manner as timber upper floors but due to the shorter spans they require smaller joist sections. Landings can be detailed for a 90° change of direction (quarter space landing) or a 180° change of direction (half space landing) and can be introduced at any position between the two floors being served by the stairs.

Typical Layouts ~

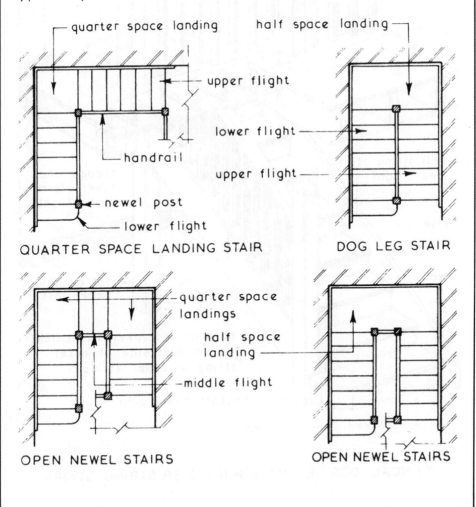

QUARTER SPACE LANDING STAIR

DOG LEG STAIR

OPEN NEWEL STAIRS

OPEN NEWEL STAIRS

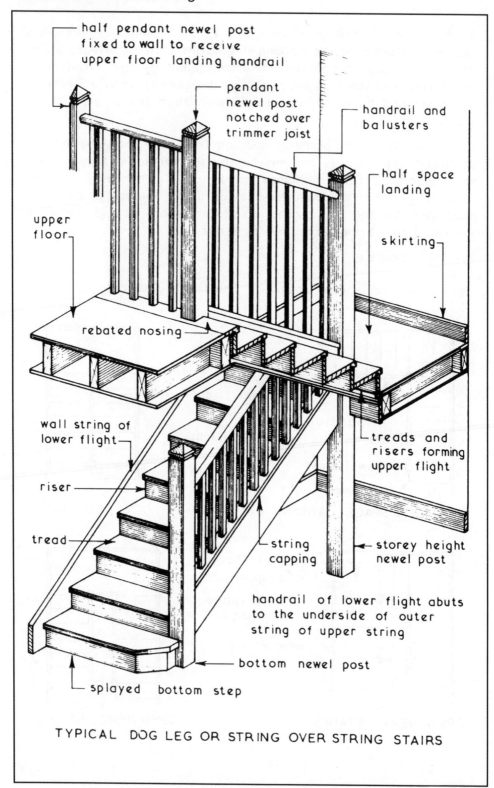

half pendant newel post
fixed to wall to receive
upper floor landing handrail

pendant
newel post
notched over
trimmer joist

handrail and
balusters

half space
landing

upper
floor

skirting

rebated nosing

treads and
risers forming
upper flight

wall string of
lower flight

riser

string
capping

storey height
newel post

tread

handrail of lower flight abuts
to the underside of outer
string of upper string

bottom newel post

splayed bottom step

TYPICAL DOG LEG OR STRING OVER STRING STAIRS

Insitu Reinforced Concrete Stairs ~ a variety of stair types and arrangements are possible each having its own appearance and design characteristics. In all cases these stairs must comply with the minimum requirements set out in Part K of the Building Regulations in accordance with the purpose group of the building in which the stairs are situated.

Typical Examples ~

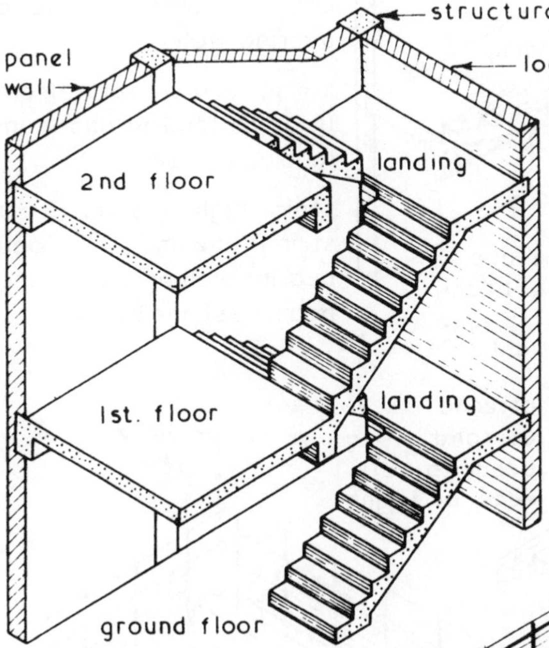

structural frame

panel wall

load bearing wall

INCLINED SLAB STAIR

landing

landings span from well edge to load bearing wall

stair flights span from floor to landing and from landing to floor

for detailed example see page 568

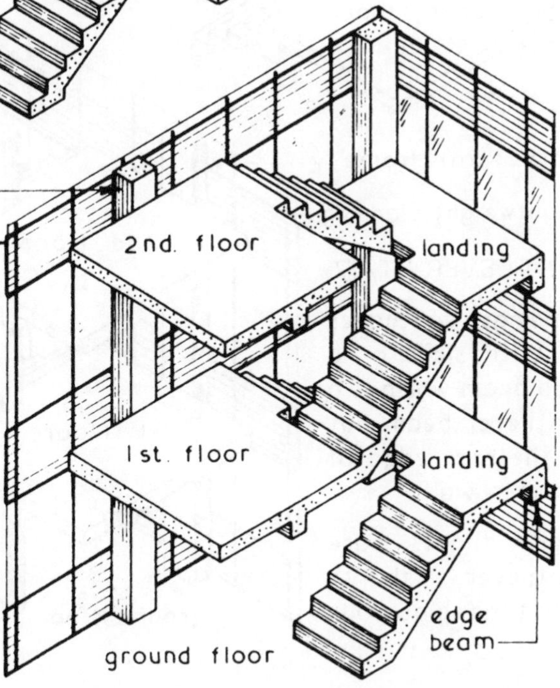

2nd floor

1st. floor

landing

ground floor

structural frame

lightweight cladding

CRANKED SLAB STAIR

Stair flights span as a cranked slab from floor to landing edge beam and from landing edge beam to floor

If no structural support is given at landing levels stairs are called a continuous slab or scissor stair

2nd. floor

landing

1st. floor

landing

edge beam

ground floor

Typical Examples ~

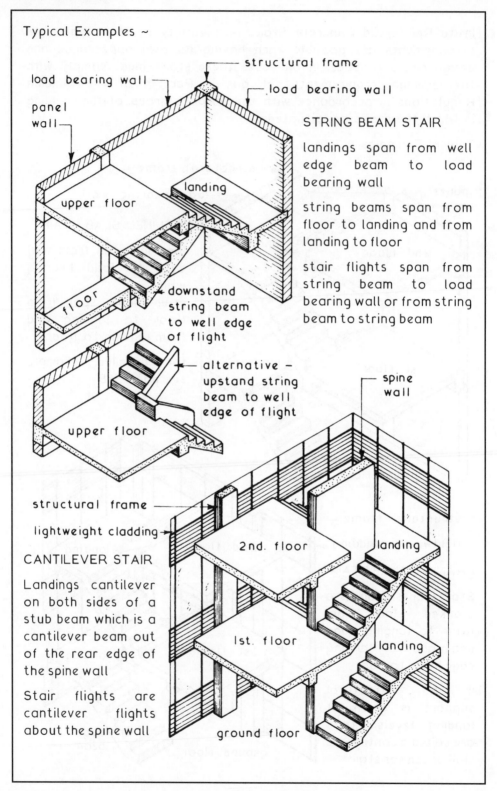

STRING BEAM STAIR

landings span from well edge beam to load bearing wall

string beams span from floor to landing and from landing to floor

stair flights span from string beam to load bearing wall or from string beam to string beam

load bearing wall

structural frame

load bearing wall

panel wall

landing

upper floor

floor

downstand string beam to well edge of flight

alternative – upstand string beam to well edge of flight

upper floor

spine wall

structural frame

lightweight cladding

2nd. floor

landing

CANTILEVER STAIR

Landings cantilever on both sides of a stub beam which is a cantilever beam out of the rear edge of the spine wall

Stair flights are cantilever flights about the spine wall

1st. floor

landing

ground floor

Spiral and Helical Stairs ~ these stairs constructed in insitu reinforced concrete are considered to be aesthetically pleasing but are expensive to construct. They are therefore mainly confined to prestige buildings usually as accommodation stairs linking floors within the same compartment. Like all other forms of stair they must conform to the requirements of Part K of the Building Regulations and if used as a means of escape in case of fire with the requirements of Part B. Spiral stairs can be defined as those describing a helix around a central column whereas a helical stair has an open well. The open well of a helical stair is usually circular or elliptical in plan and the formwork is built up around a vertical timber core.

Typical Example of a Helical Stair ~

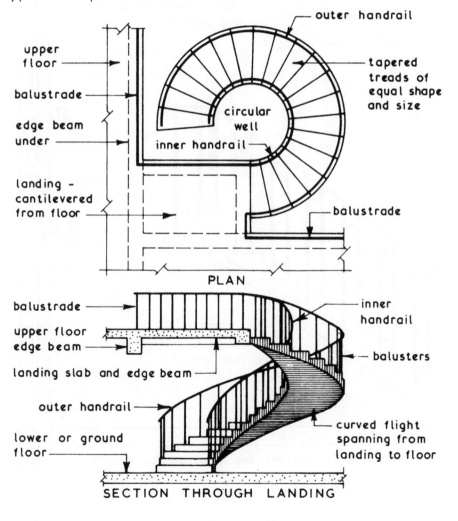

PLAN

SECTION THROUGH LANDING

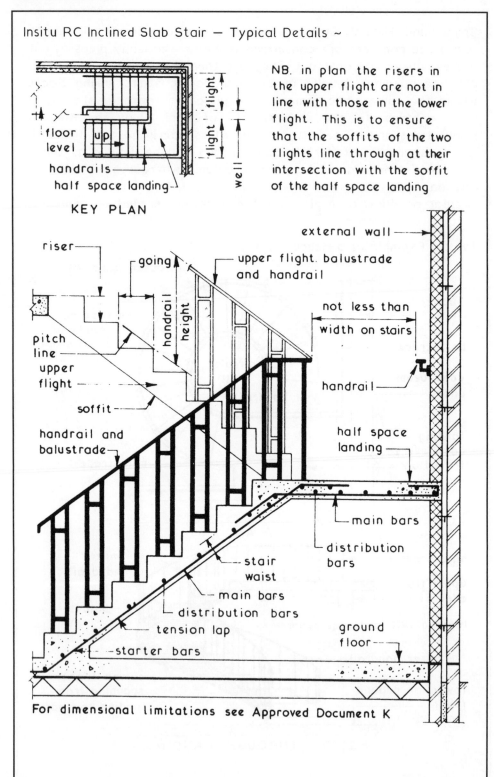

Insitu RC Inclined Slab Stair — Typical Details ~

NB. in plan the risers in the upper flight are not in line with those in the lower flight. This is to ensure that the soffits of the two flights line through at their intersection with the soffit of the half space landing

floor level
up
handrails
half space landing
KEY PLAN

riser
going
upper flight. balustrade and handrail
external wall
pitch line upper flight
handrail height
not less than
width on stairs
soffit
handrail
handrail and balustrade
half space landing
main bars
distribution bars
stair waist
main bars
distribution bars
tension lap
starter bars
ground floor

For dimensional limitations see Approved Document K

Insitu Reinforced Concrete Stair Formwork ~ in specific detail the formwork will vary for the different types of reinforced concrete stair but the basic principles for each format will remain constant.

Typical RC Stair Formwork Details ~ (see page 568 for Key Plan)

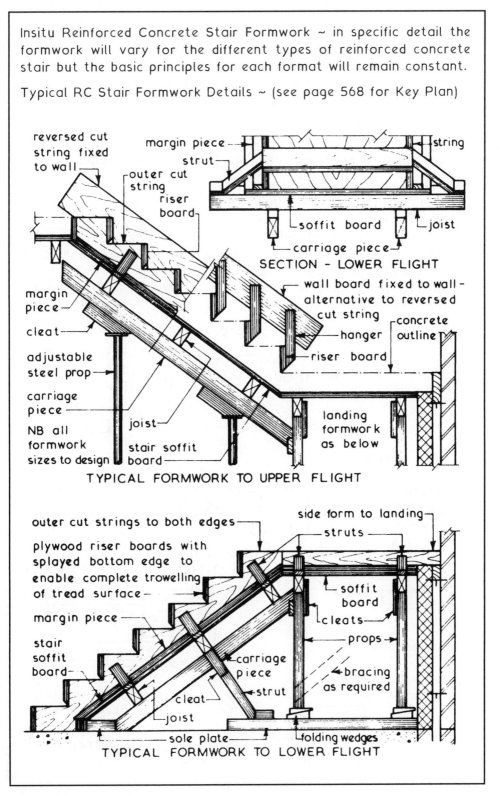

reversed cut string fixed to wall

margin piece

strut

outer cut string

riser board

string

SECTION - LOWER FLIGHT

soffit board

joist

carriage piece

wall board fixed to wall - alternative to reversed cut string

concrete outline

hanger

riser board

margin piece

cleat

adjustable steel prop

carriage piece

joist

stair soffit board

landing formwork as below

NB all formwork sizes to design

TYPICAL FORMWORK TO UPPER FLIGHT

outer cut strings to both edges

side form to landing

struts

plywood riser boards with splayed bottom edge to enable complete trowelling of tread surface

margin piece

stair soffit board

soffit board

cleats

props

carriage piece

bracing as required

cleat

strut

joist

sole plate

folding wedges

TYPICAL FORMWORK TO LOWER FLIGHT

Precast Concrete Stairs

Precast Concrete Stairs ~ these can be produced to most of the formats used for insitu concrete stairs and like those must comply with the appropriate requirements set out in Part K of the Building Regulations. To be economic the total production run must be sufficient to justify the costs of the moulds and therefore the designers choice may be limited to the stair types which are produced as a manufacturer's standard item.

Precast concrete stairs can have the following advantages:-

1. Good quality control of finished product.
2. Saving in site space since formwork fabrication and storage will not be required.
3. The stairs can be installed at any time after the floors have been completed thus giving full utilisation to the stair shaft as a lifting or hoisting space if required.
4. Hoisting, positioning and fixing can usually be carried out by semi-skilled labour.

Typical Example ~ Straight Flight Stairs

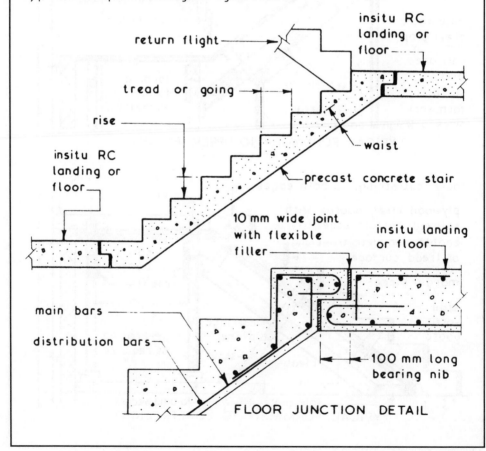

FLOOR JUNCTION DETAIL

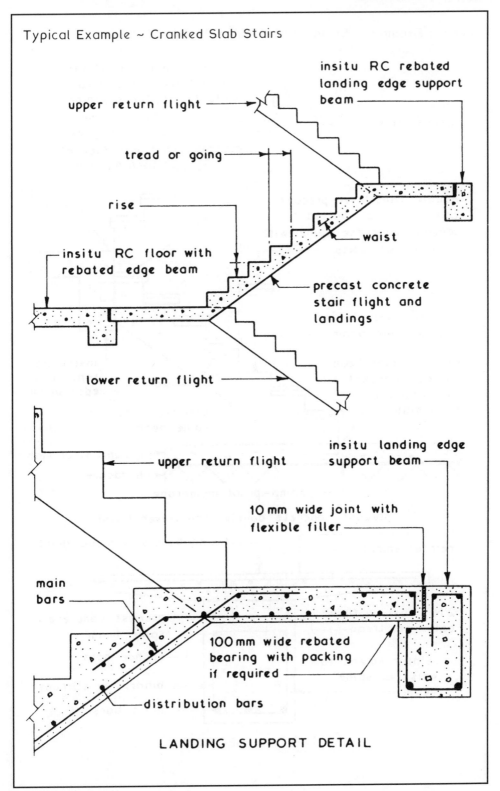

Typical Example ~ Cranked Slab Stairs

upper return flight

insitu RC rebated landing edge support beam

tread or going

rise

insitu RC floor with rebated edge beam

waist

precast concrete stair flight and landings

lower return flight

upper return flight

insitu landing edge support beam

10 mm wide joint with flexible filler

main bars

100 mm wide rebated bearing with packing if required

distribution bars

LANDING SUPPORT DETAIL

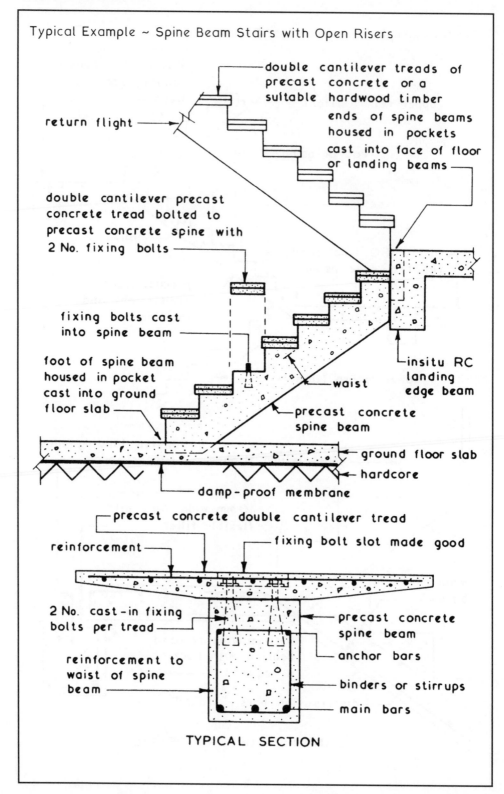

Typical Example ~ Spine Beam Stairs with Open Risers

double cantilever treads of precast concrete or a suitable hardwood timber

ends of spine beams housed in pockets cast into face of floor or landing beams

return flight

double cantilever precast concrete tread bolted to precast concrete spine with 2 No. fixing bolts

fixing bolts cast into spine beam

foot of spine beam housed in pocket cast into ground floor slab

insitu RC landing edge beam

waist

precast concrete spine beam

ground floor slab

hardcore

damp-proof membrane

precast concrete double cantilever tread

reinforcement

fixing bolt slot made good

2 No. cast-in fixing bolts per tread

precast concrete spine beam

anchor bars

reinforcement to waist of spine beam

binders or stirrups

main bars

TYPICAL SECTION

Precast Concrete Spiral Stairs ~ this form of stair is usually constructed with an open riser format using tapered treads which have a keyhole plan shape. Each tread has a hollow cylinder at the narrow end equal to the rise which is fitted over a central steel column usually filled with insitu concrete. The outer end of the tread has holes through which the balusters pass to be fixed on the underside of the tread below, a hollow spacer being used to maintain the distance between consecutive treads.

Typical Example ~

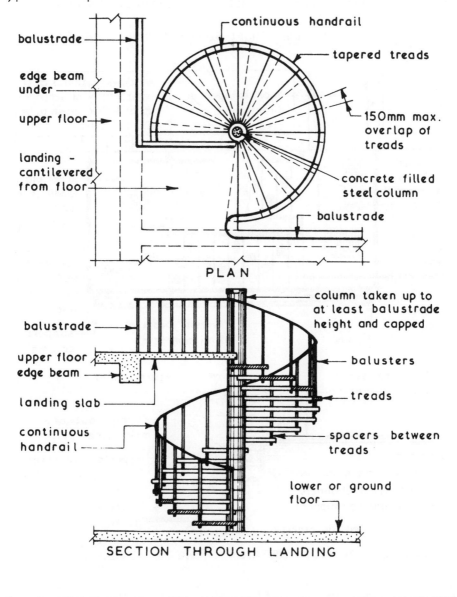

PLAN

SECTION THROUGH LANDING

Metal Stairs ~ these can be produced in cast iron, mild steel or aluminium alloy for use as escape stairs or for internal accommodation stairs. Most escape stairs are fabricated from cast iron or mild steel and must comply with the Building Regulation requirements for stairs in general and fire escape stairs in particular. Most metal stairs are purpose made and therefore tend to cost more than comparable concrete stairs. Their main advantage is the elimination of the need for formwork whilst the main disadvantage is the regular maintenance in the form of painting required for cast iron and mild steel stairs.

Typical Example ~ Straight Flight Steel External Escape Stair

door to open in direction of but clear of stairs

windows within 1·800 horizontally or vertically up or 9·000 down to be fixed and 1/2 hr. f.r.

minimum width as for internal stairs

landing width not less than stair width

support column

steel plated landings on framed steel channels

continuous handrails 900mm min. above pitch line

up

universal beam at each floor level

support column

PLAN

perforated steel plates to landing

return flight

strings out of mild steel plate

framed steel channels forming landing support

perforated steel treads with 25mm overlap of tread below

handrail omitted for clarity

LANDING JUNCTION DETAIL

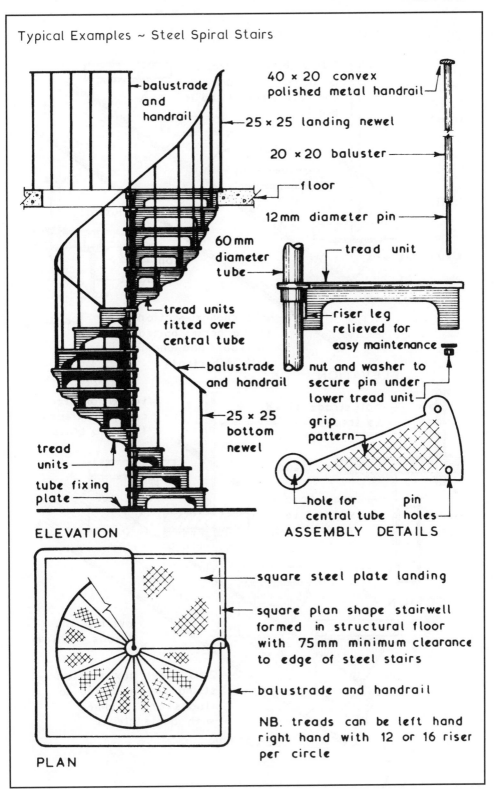

Typical Examples ~ Steel Spiral Stairs

balustrade and handrail

40 × 20 convex polished metal handrail

25 × 25 landing newel

20 × 20 baluster

floor

12mm diameter pin

60 mm diameter tube

tread unit

tread units fitted over central tube

riser leg relieved for easy maintenance

balustrade and handrail

nut and washer to secure pin under lower tread unit

25 × 25 bottom newel

grip pattern

tread units

tube fixing plate

hole for central tube

pin holes

ELEVATION

ASSEMBLY DETAILS

square steel plate landing

square plan shape stairwell formed in structural floor with 75mm minimum clearance to edge of steel stairs

balustrade and handrail

NB. treads can be left hand right hand with 12 or 16 riser per circle

PLAN

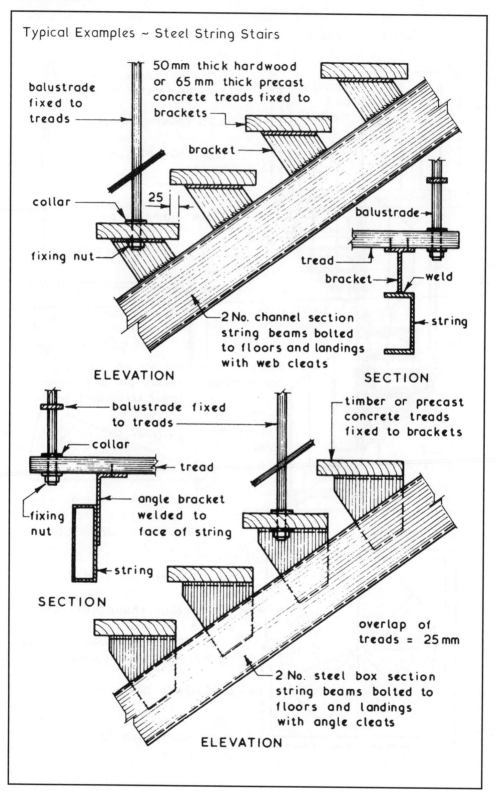

Typical Examples ~ Steel String Stairs

balustrade fixed to treads

50 mm thick hardwood or 65 mm thick precast concrete treads fixed to brackets

bracket

collar

25

fixing nut

balustrade

tread

bracket

weld

string

2 No. channel section string beams bolted to floors and landings with web cleats

ELEVATION

SECTION

balustrade fixed to treads

collar

tread

angle bracket welded to face of string

fixing nut

string

SECTION

timber or precast concrete treads fixed to brackets

overlap of treads = 25 mm

2 No. steel box section string beams bolted to floors and landings with angle cleats

ELEVATION

Balustrades and Handrails ~ these must comply in all respects with the requirements given in Part K of the Building Regulations and in the context of escape stairs are constructed of a non-combustible material with a handrail shaped to give a comfortable hand grip. The handrail may be covered or capped with a combustible material such as timber or plastic. Most balustrades are designed to be fixed after the stairs have been cast or installed by housing the balusters in a preformed pocket or by direct surface fixing.

Typical Details ~

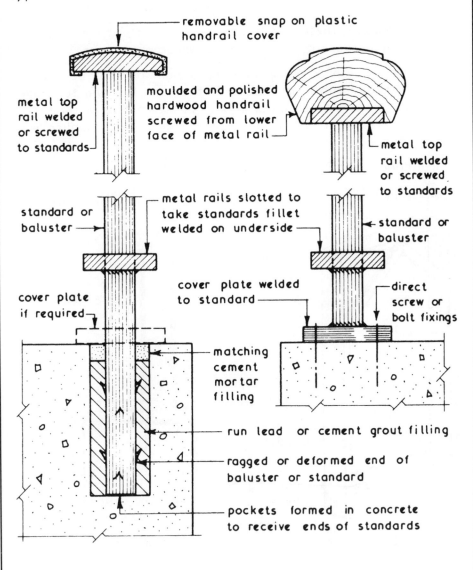

removable snap on plastic handrail cover

metal top rail welded or screwed to standards

moulded and polished hardwood handrail screwed from lower face of metal rail

metal top rail welded or screwed to standards

standard or baluster

metal rails slotted to take standards fillet welded on underside

standard or baluster

cover plate if required

cover plate welded to standard

direct screw or bolt fixings

matching cement mortar filling

run lead or cement grout filling

ragged or deformed end of baluster or standard

pockets formed in concrete to receive ends of standards

Doors and Door Linings

Functions ~ the main functions of any door are to:

1. Provide a means of access and egress.
2. Maintain continuity of wall function when closed.
3. Provide a degree of privacy and security.

Choice of door type can be determined by:-

1. Position — whether internal or external.
2. Properties required — fire resistant, glazed to provide for borrowed light or vision through, etc.
3. Appearance — flush or panelled, painted or polished, etc.

Door Schedules ~ these can be prepared in the same manner and for the same purpose as that given for windows on page 325.

Internal Doors ~ these are usually lightweight and can be fixed to a lining, if heavy doors are specified these can be hung to frames in a similar manner to external doors. An alternative method is to use door sets which are usually storey height and supplied with prehung doors.

Typical door Lining Details ~

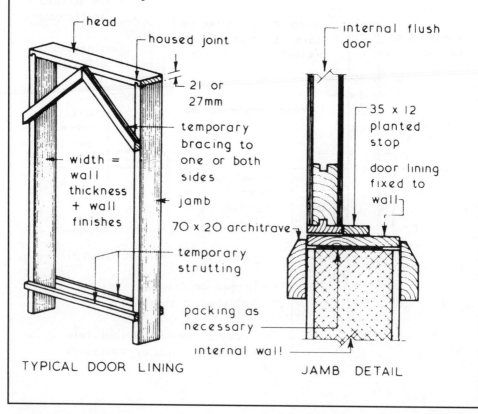

TYPICAL DOOR LINING

JAMB DETAIL

Internal Doors ~ these are similar in construction to the external doors but are usually thinner and therefore lighter in weight.

Typical Examples ~

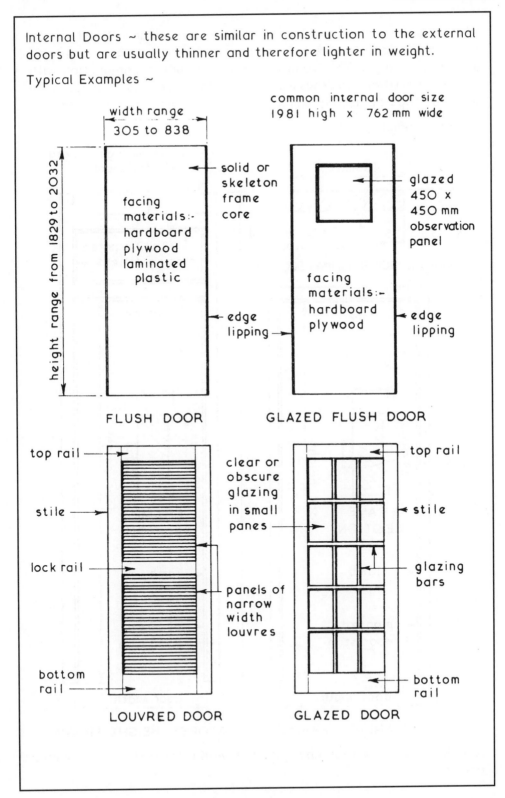

common internal door size
1981 high x 762 mm wide

width range
305 to 838

height range from 1829 to 2032

solid or skeleton frame core

facing materials:-
hardboard
plywood
laminated
plastic

edge lipping

glazed 450 x 450 mm observation panel

facing materials:-
hardboard
plywood

edge lipping

FLUSH DOOR

GLAZED FLUSH DOOR

top rail

stile

lock rail

bottom rail

clear or obscure glazing in small panes

panels of narrow width louvres

top rail

stile

glazing bars

bottom rail

LOUVRED DOOR

GLAZED DOOR

Internal Door Frames and linings ~ these are similar in construction to external door frames but usually have planted door stops and do not have a sill. The frames sized to be built in conjunction with various partition thicknesses and surface finishes. Linings with planted stops ae usually employed for lightweight domestic doors.

Typical Examples ~

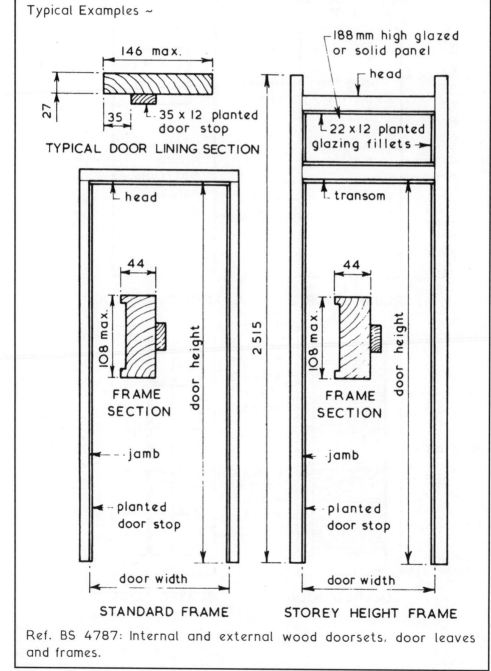

146 max.

27

35

35 x 12 planted door stop

TYPICAL DOOR LINING SECTION

head

44

108 max.

door height

FRAME SECTION

jamb

planted door stop

door width

STANDARD FRAME

188 mm high glazed or solid panel

head

22 x 12 planted glazing fillets

transom

2 515

44

108 max.

door height

FRAME SECTION

jamb

planted door stop

door width

STOREY HEIGHT FRAME

Ref. BS 4787: Internal and external wood doorsets, door leaves and frames.

Door sets ~ these are factory produced fully assembled prehung doors which are supplied complete with frame, architraves and ironmongery except for door furniture. The doors hung to the frames using pin butts for easy door removal. Prehung door sets are available in standard and storey height versions and are suitable for all internal door applications with normal wall and partition thicknesses.

Typical Examples ~

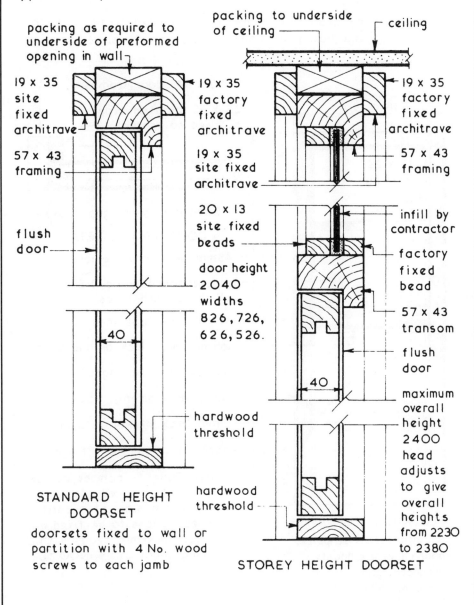

packing as required to underside of preformed opening in wall

19 x 35 site fixed architrave

57 x 43 framing

flush door

40

packing to underside of ceiling

ceiling

19 x 35 factory fixed architrave

19 x 35 site fixed architrave

20 x 13 site fixed beads

door height 2040

widths 826, 726, 626, 526.

hardwood threshold

STANDARD HEIGHT DOORSET

doorsets fixed to wall or partition with 4 No. wood screws to each jamb

hardwood threshold

19 x 35 factory fixed architrave

57 x 43 framing

infill by contractor

factory fixed bead

57 x 43 transom

flush door

40

maximum overall height 2400 head adjusts to give overall heights from 2230 to 2380

STOREY HEIGHT DOORSET

Half Hour Flush Fire Doors ~ these are usually based on the recommendations given in BS 8214: Code of practice for fire doors. A wide variety of door constructions are available from various manufacturers but generally they all have to be fitted to a similar frame.

A door's resistance to fire is measured by:-

1. Insulation — resistance to thermal transmittance, see BS 476—20 & 22.

2. Integrity — resistance in minutes to the penetration of flame and hot gases under simulated fire conditions.

Typical Details ~

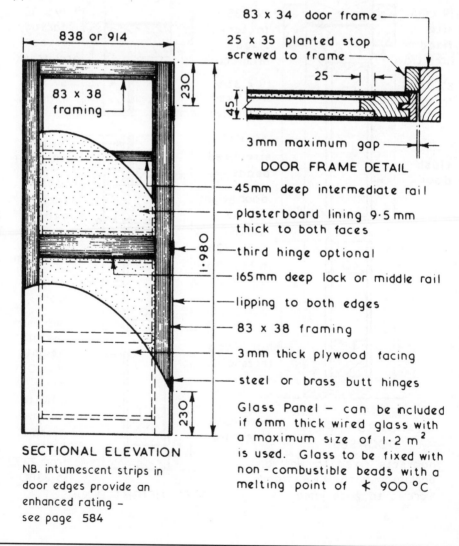

SECTIONAL ELEVATION

NB. intumescent strips in door edges provide an enhanced rating – see page 584

838 or 914

83 x 38 framing

230

1·980

230

83 x 34 door frame

25 x 35 planted stop screwed to frame

25

45

3mm maximum gap

DOOR FRAME DETAIL

45mm deep intermediate rail

plasterboard lining 9·5 mm thick to both faces

third hinge optional

165mm deep lock or middle rail

lipping to both edges

83 x 38 framing

3mm thick plywood facing

steel or brass butt hinges

Glass Panel — can be included if 6mm thick wired glass with a maximum size of $1·2\,m^2$ is used. Glass to be fixed with non-combustible beads with a melting point of $\ngtr 900\,°C$

One Hour Flush Fire Door ~ like the half hour flush fire door shown on page 582 these doors are based on the recommendations given in BS 8214 which covers both door and frame. A wide variety of door constructions are available from various manufacturers but most of these are classified as a one hour fire resistant door with both insulation and integrity ratings of 60 minutes.

Typical Details ~

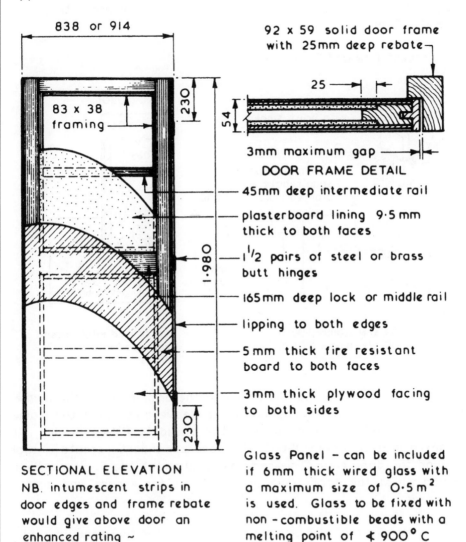

838 or 914

83 x 38 framing

230

54

92 x 59 solid door frame with 25mm deep rebate

25

3mm maximum gap
DOOR FRAME DETAIL

1·980

230

— 45mm deep intermediate rail

— plasterboard lining 9·5 mm thick to both faces

— 1½ pairs of steel or brass butt hinges

— 165mm deep lock or middle rail

— lipping to both edges

— 5 mm thick fire resistant board to both faces

— 3mm thick plywood facing to both sides

SECTIONAL ELEVATION
NB. intumescent strips in door edges and frame rebate would give above door an enhanced rating ~
see page 584.

Glass Panel – can be included if 6mm thick wired glass with a maximum size of 0·5 m² is used. Glass to be fixed with non-combustible beads with a melting point of ≮ 900° C

Fire and Smoke Resistance ~ Doors can be assessed for both integrity and smoke resistance. They are coded accordingly, for example FD30 or FD30s. FD indicates a fire door and 30 the integrity time in minutes. The letter 's' denotes that the door or frame contains a facility to resist the passage of smoke.

Manufacturers produce doors of standard ratings — 30, 60 and 90 minutes, with higher ratings available to order. A colour coded plug inserted in the door edge corresponds to the fire rating. See BS 8214, Table 1 for details.

Intumescent Fire and Smoke Seals ~

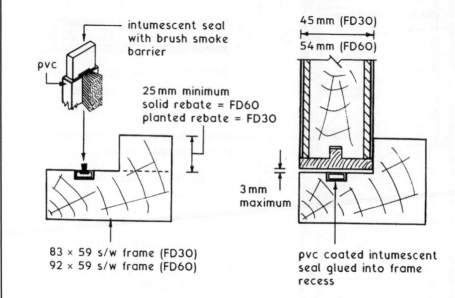

pvc

intumescent seal with brush smoke barrier

25 mm minimum
solid rebate = FD60
planted rebate = FD30

3 mm maximum

45 mm (FD30)
54 mm (FD60)

83 × 59 s/w frame (FD30)
92 × 59 s/w frame (FD60)

pvc coated intumescent seal glued into frame recess

The intumescent core may be fitted to the door edge or the frame. In practice, most joinery manufacturers leave a recess in the frame where the seal is secured with rubber based or PVA adhesive. At temperatures of about 150°C, the core expands to create a seal around the door edge. This remains throughout the fire resistance period whilst the door can still be opened for escape and access purposes.

NB. The smoke seal will also function as an effective draught seal.

Typical Details ~

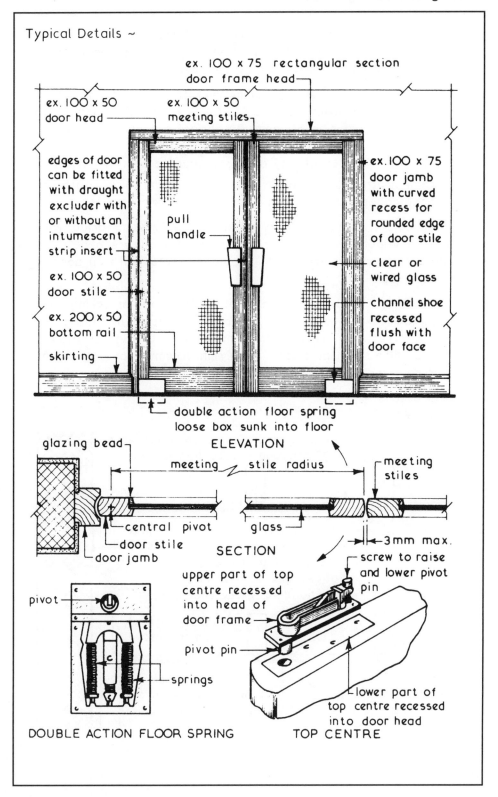

ex. 100 x 75 rectangular section door frame head

ex. 100 x 50 door head

ex. 100 x 50 meeting stiles

ex. 100 x 75 door jamb with curved recess for rounded edge of door stile

edges of door can be fitted with draught excluder with or without an intumescent strip insert

pull handle

clear or wired glass

ex. 100 x 50 door stile

ex. 200 x 50 bottom rail

channel shoe recessed flush with door face

skirting

double action floor spring loose box sunk into floor

ELEVATION

glazing bead

meeting stile radius

meeting stiles

central pivot

door stile

door jamb

glass

SECTION

3mm max.

screw to raise and lower pivot pin

upper part of top centre recessed into head of door frame

pivot pin

lower part of top centre recessed into door head

pivot

springs

DOUBLE ACTION FLOOR SPRING

TOP CENTRE

Plasterboard ~ this is a rigid board made with a core of gypsum sandwiched between face sheets of strong durable paper. In the context of ceilings two sizes can be considered —

1. Baseboard 2·400×1·200×9·5mm thick for supports at centres not exceeding 400mm; 2·400×1·200×12·5mm for supports at centres not exceeding 600mm. Baseboard has square edges and therefore the joints will need reinforcing with jute scrim at least 90mm wide or alternatively a special tape to prevent cracking.

2. Gypsum Lath 1·200×406×9·5 or 12·5mm thick. Lath has rounded edges which eliminates the need to reinforce the joints.

Baseboard is available with a metallized polyester facing which acts as a vapour control layer to prevent moisture penetrating the insulation and timber, joints should be sealed with an adhesive metallized tape.

The boards are fixed to the underside of the floor or ceiling joists with galvanised or sheradised plasterboard nails at not more than 150mm centres and are laid breaking the joint. Edge treatments consist of jute scrim or plastic mesh reinforcement or a preformed plaster cove moulding.

Typical details ~

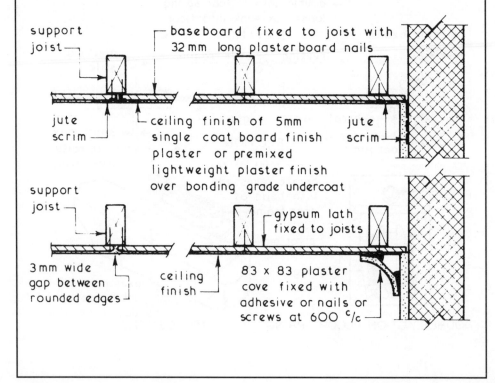

support joist —

baseboard fixed to joist with 32mm long plasterboard nails

jute scrim —

ceiling finish of 5mm single coat board finish plaster or premixed lightweight plaster finish over bonding grade undercoat

jute scrim —

support joist —

3mm wide gap between rounded edges —

ceiling finish —

gypsum lath fixed to joists

83 x 83 plaster cove fixed with adhesive or nails or screws at 600 ^{c}/$_c$

Suspended Ceilings ~ these can be defined as ceilings which are fixed to a framework suspended from main structure thus forming a void between the two components. The basic functional requirements of suspended ceilings are:-

1. They should be easy to construct, repair, maintain and clean.

2. So designed that an adequate means of access is provided to the void space for the maintenance of the suspension system, concealed services and/or light fittings.

3. Provide any required sound and/or thermal insulation.

4. Provide any required acoustic control in terms of absorption and reverberation.

5. Provide if required structural fire protection to structural steel beams supporting a concrete floor.

6. Conform with the minimum requirements set out in the Building Regulations and in particular the regulations governing the restriction of spread of flame over surfaces of ceilings and the exeptions permitting the use of certain plastic materials.

7. Design to be based on a planning module preferably a dimensional coordinated system with a first preference module of 300 mm.

Typical Suspended Ceiling Grid Framework Layout ~

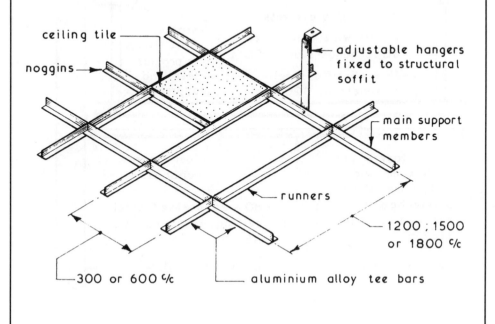

ceiling tile

noggins

adjustable hangers fixed to structural soffit

main support members

runners

1200 ; 1500 or 1800 c/c

300 or 600 c/c

aluminium alloy tee bars

Classification of Suspended Ceiling ~ there is no standard method of classification since some are classified by their function such as illuminated and acoustic suspended ceilings others are classified by the materials used and classification by method of construction is also very popular. The latter method is simple since most suspended ceiling types can be placed in one of three groups:-

1. Jointless suspended ceilings.

2. Panelled suspended ceilings — see page 589.

3. Decorative and open suspended ceilings — see page 590.

Jointless Suspended Ceilings ~ these forms of suspended ceilings provide a continuous and jointless surface with the internal appearance of a conventional ceiling. They may be selected to fulfil fire resistance requirements or to provide a robust form of suspended ceiling. The two common ways of construction are a plasterboard or expanded metal lathing soffit with hand applied plaster finish or a sprayed applied rendering with a cement base.

Typical Details ~

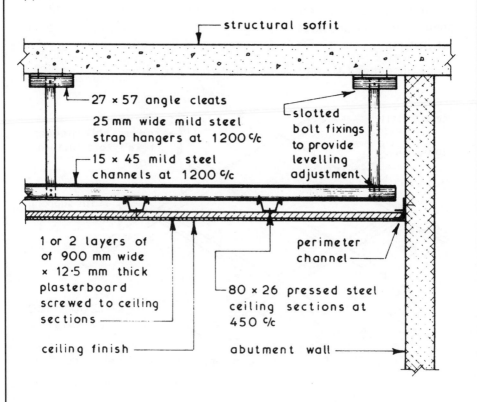

Panelled Suspended Ceilings ~ these are the most popular form of suspended ceiling consisting of a suspended grid framework to which the ceiling covering is attached. The covering can be of a tile, tray, board or strip format in a wide variety of materials with an exposed or concealed supporting framework. Serivces such as luminaries can usually be incorporated within the system. Generally panelled systems are easy to assemble and install using a water level or laser beam for initial and final levelling. Provision for maintenance access can be easily incorporated into most systems and layouts.

Typical Support Details ~

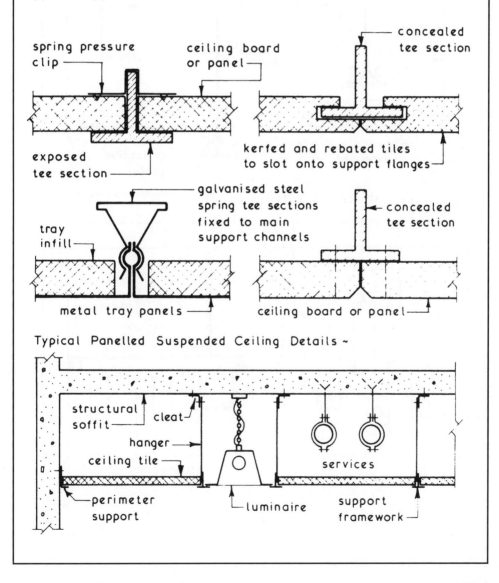

Typical Panelled Suspended Ceiling Details ~

Decorative and Open Suspended Ceilings ~ these ceilings usually consist of an openwork grid or suspended shapes onto which the lights fixed at, above or below ceiling level can be trained thus creating a decorative and illuminated effect. Many of these ceilings are purpose designed and built as opposed to the proprietary systems associated with jointless and panelled suspended ceilings.

Typical Examples ~

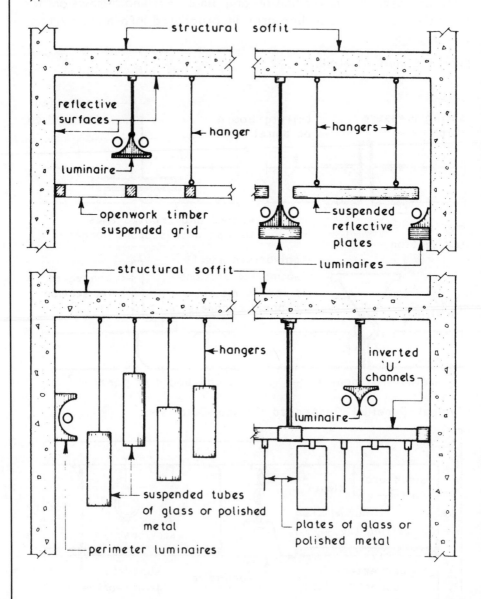

Functions ~ the main functions of paint are to provide:-

1. An economic method of surface protection to building materials and components.
2. An economic method of surface decoration to building materials and components.

Composition ~ the actual composition of any paint can be complex but the basic components are:-

1. Binder ~ this is the liquid vehicle or medium which dries to form the surface film and can be composed of linseed oil, drying oils, synthetic resins and water. The first function of a paint medium is to provide a means of spreading the paint over the surface and at the same time acting as a binder to the pigment.
2. Pigment ~ this provides the body, colour, durability and corrosion protection properties of the paint. White lead pigments are very durable and moisture resistant but are poisonous and their use is generally restricted to priming and undercoating paints. If a paint contains a lead pigment the fact must be stated on the container. The general pigment used in paint is titanium dioxide which is not poisonous and gives good obliteration of the undercoats.
3. Solvents and Thinners ~ these are materials which can be added to a paint to alter its viscosity.

Paint Types — there is a wide range available but for most general uses the following can be considered:-

1. Oil Based paints — these are available in priming, undercoat and finishing grades. The latter can be obtained in a wide range of colours and finishes such as matt, semi-matt, eggshell, satin, gloss and enamel. Polyurethane paints have a good hardness and resistance to water and cleaning. Oil based paints are suitable for most applications if used in conjunction with correct primer and undercoat.
2. Water Based Paints — most of these are called emulsion paints the various finishes available being obtained by adding to the water medium additives such as alkyd resin & polyvinyl acetate (PVA). Finishes include matt, eggshell, semi-gloss and gloss. Emulsion paints are easily applied, quick drying and can be obtained with a washable finish and are suitable for most applications.

Paints and Painting

Supply ~ paint is usually supplied in metal containers ranging from 250 millilitres to 5 litres capacity to the colour ranges recommended in BS 381C (colours for specific purposes) and BS 4800 (paint colours for building purposes).

Application ~ paint can be applied to almost any surface providing the surface preparation and sequence of paint coats are suitable. The manufacturers specification and/or the recommendations of BS 6150 (painting of buildings) should be followed. Preparation of the surface to receive the paint is of the utmost importance since poor preparation is one of the chief causes of paint failure. The preperation consists basically of removing all dirt, grease, dust and ensuring that the surface will provide an adequate key for the paint which is to be applied. In new work the basic build-up of paint coats consists of:-

1. Priming Coats — these are used on unpainted surfaces to obtain the necessary adhesion and to inhibit corrosion of ferrous metals. New timber should have the knots treated with a solution of shellac or other alcohol based resin called knotting prior to the application of the primer.

2. Undercoats — these are used on top of the primer after any defects have been made good with a suitable stopper or filler. The primary function of an undercoat is to give the opacity and build-up necessary for the application of the finishing coat(s).

3. Finish — applied directly over the undercoating in one or more coats to impart the required colour and finish.

Paint can applied by:-

1. Brush — the correct type, size and quality of brush such as those recommended in BS 2992 needs to be selected and used. To achieve a first class finish by means of brush application requires a high degree of skill.

2. Spray — as with brush application a high degree of skill is required to achieve a good finish. Generally compressed air sprays or airless sprays are used for building works.

3. Roller — simple and inexpensive method of quickly and cleanly applying a wide range of paints to flat and textured surfaces. Roller heads vary in size from 50 to 450mm wide with various covers such as sheepskin, synthetic pile fibres, mohair and foamed polystyrene. All paint applicators must be thoroughly cleaned after use.

Painting ~ the main objectives of applying coats of paint to a surface are preservation, protection and decoration to give a finish which is easy to clean and maintain. To achieve these objectives the surface preparation and paint application must be adequate. The preparation of new and previously painted surfaces should ensure that prior to painting the surface is smooth, clean, dry and stable.

Basic Surface Preparation Techniques ~

Timber — to ensure a good adhesion of the paint film all timber should have a moisture content of less than 18%. The timber surface should be prepared using an abrasive paper to produce a smooth surface brushed and wiped free of dust and any grease removed with a suitable spirit. Careful treatment of knots is essential either by sealing with two coats of knotting or in extreme cases cutting out the knot and replacing with sound timber. The stopping and filling of cracks and fixing holes with putty or an appropriate filler should be carried out after the application of the priming coat. Each coat of paint must be allowed to dry hard and be rubbed down with a fine abrasive paper before applying the next coat. On previously painted surfaces if the paint is in a reasonable condition the surface will only require cleaning and rubbing down before repainting, when the paint is in a poor condition it will be necessary to remove completely the layers of paint and then prepare the surface as described above for new timber.

Building Boards — most of these boards require no special preparation except for the application of a sealer as specified by the manufacturer.

Iron and Steel — good preparation is the key to painting iron and steel successfully and this will include removing all rust, mill scale, oil, grease and wax. This can be achieved by wire brushing, using mechanical means such as shot blasting, flame cleaning and chemical processes and any of these processes are often carried out in the steel fabrication works prior to shop applied priming.

Plaster — the essential requirement of the preparation is to ensure that the plaster surface is perfectly dry, smooth and free of defects before applying any coats of paint especially when using gloss paints. Plaster which contains lime can be alkaline and such surfaces should be treated with an alkali resistant primer when the surface is dry before applying the final coats of paint.

Paint Defects ~ these may be due to poor or incorrect preparation of the surface, poor application of the paint and/or chemical reactions. The general remedy is to remove all the affected paint and carry out the correct preparation of the surface before applying in the correct manner new coats of paint. Most paint defects are visual and therefore an accurate diagnosis of the cause must be established before any remedial treatment is undertaken.

Typical Paint Defects ~

1. Bleeding — staining and disruption of the paint surface by chemical action, usually caused by applying an incorrect paint over another. Remedy is to remove affected paint surface and repaint with correct type of overcoat paint.

2. Blistering — usually caused by poor presentation allowing resin or moisture to be entrapped, the subsequent expansion causing the defect. Remedy is to remove all the coats of paint and ensure that the surface is dry before repainting.

3. Blooming — mistiness usually on high gloss or varnished surfaces due to the presence of moisture during application. It can be avoided by not painting under these conditions. Remedy is to remove affected paint and repaint.

4. Chalking — powdering of the paint surface due to natural ageing or the use of poor quality paint. Remedy is to remove paint if necessary, prepare surface and repaint.

5. Cracking and Crazing — usually due to unequal elasticity of successive coats of paint. Remedy is to remove affected paint and repaint with compatible coats of paint.

6. Flaking and Peeling — can be due to poor adhesion, presence of moisture, painting over unclean areas or poor preparation. Remedy is to remove defective paint, prepare surface and repaint.

7. Grinning — due to poor opacity of paint film allowing paint coat below or background to show through, could be the result of poor application; incorrect thinning or the use of the wrong colour. Remedy is to apply further coats of paint to obtain a satisfactory surface.

8. Saponification — formation of soap from alkali present in or on surface painted. The paint is ultimately destroyed and a brown liquid appears on the surface. Remedy is to remove the paint films and seal the alkaline surface before repainting.

Joinery Production ~ this can vary from the flow production where one product such as flush doors is being made usually with the aid of purpose designed and built machines, to batch production where a limited number of similar items are being made with the aid of conventional woodworking machines. Purpose made joinery is very often largely hand made with a limited use of machines and is considered when special and/or high class joinery components are required.

Woodworking Machines ~ except for the portable electric tools such as drills, routers, jigsaws and sanders most woodworking machines need to be fixed to a solid base and connected to an extractor system to extract and collect the sawdust and chippings produced by the machines.

Saws — basically three formats are available, namely the circular cross cut and band saws. Circular are general purpose saws and usually have tungsten carbide tipped teeth with feed rates of up to 60·000 per minute. Cross cut saws usually have a long bench to support the timber, the saw being mounted on a radial arm enabling the circular saw to be drawn across the timber to be cut. Band saws consist of an endless thin band or blade with saw teeth and a table on which to support the timber and are generally used for curved work.

Planers — most of these machines are combined planers and thicknessers, the timber being passed over the table surface for planning and the table or bed for thicknessing. The planer has a guide fence which can be tilted for angle planing and usually the rear bed can be lowered for rebating operations. The same rotating cutter block is used for all operations. Planing speeds are dependent upon the operator since it is a hand fed operation whereas thicknessing is mechanically fed with a feed speed range of 6·000 to 20·000 per minute. Maximum planing depth is usually 10 mm per passing.

Morticing Machines — these are used to cut mortices up to 25 mm wide and can be either a chisel or chain morticer. The former consists of a hollow chisel containing a bit or auger whereas the latter has an endless chain cutter.

Tenoning Machines — these machines with their rotary cutter blocks can be set to form tenon and scribe. In most cases they can also be set for trenching, grooving and cross cutting.

Spindle Moulder — this machine has a horizontally rotating cutter block into which standard or purpose made cutters are fixed to reproduce a moulding on timber passed across the cutter.

Purpose Made Joinery ~ joinery items in the form of doors, windows, stairs and cupboard fitments can be purchased as stock items from manufacturers. There is also a need for purpose made joinery to fulfil client/designer/user requirement to suit a specific need, to fit into a non-standard space, as a specific decor requirement or to complement a particular internal environment. These purpose made joinery items can range from the simple to the complex which require high degrees of workshop and site skills.

Typical Purpose Made Counter Details ~

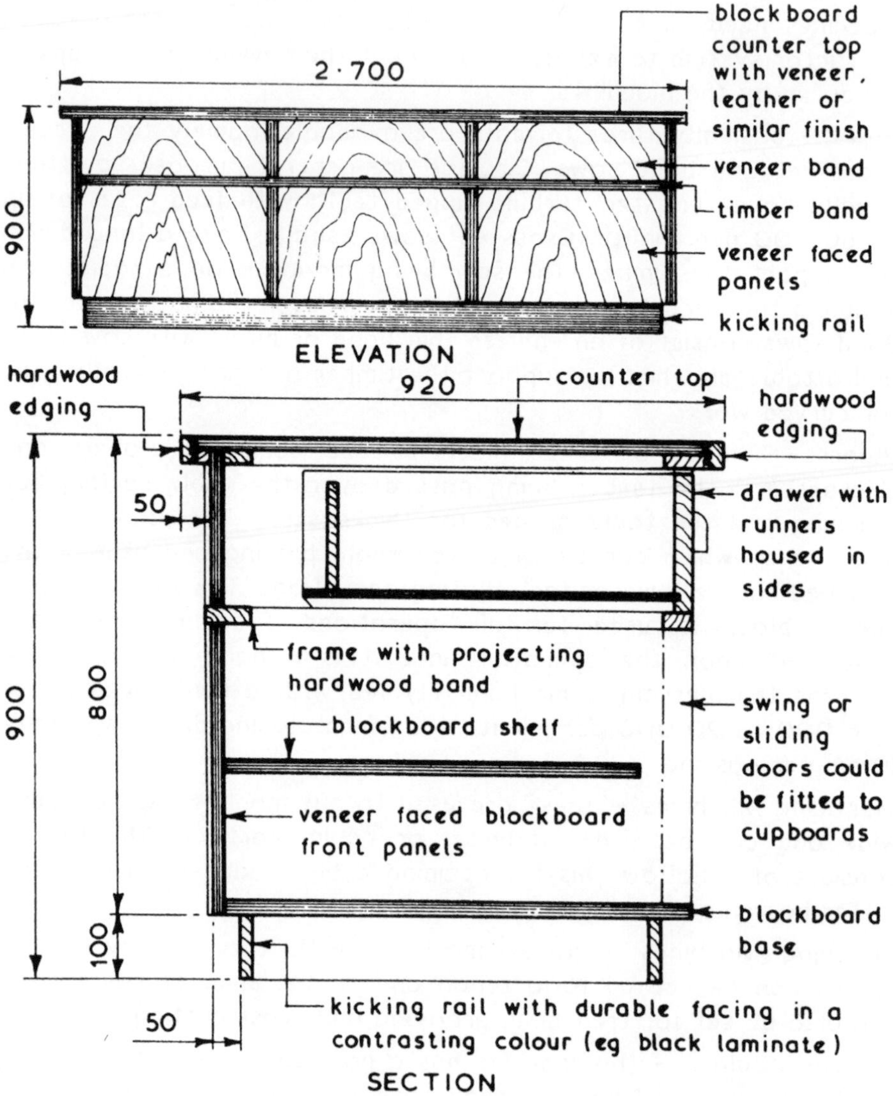

ELEVATION

SECTION

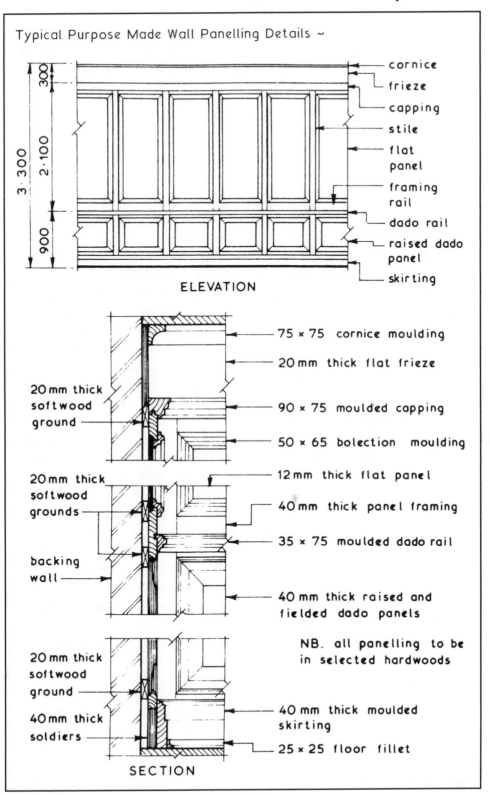

Typical Purpose Made Wall Panelling Details ~

300
2·100
900
3·300

cornice
frieze
capping
stile
flat panel
framing rail
dado rail
raised dado panel
skirting

ELEVATION

20 mm thick softwood ground

20 mm thick softwood grounds

backing wall

20 mm thick softwood ground

40 mm thick soldiers

75 × 75 cornice moulding

20 mm thick flat frieze

90 × 75 moulded capping

50 × 65 bolection moulding

12 mm thick flat panel

40 mm thick panel framing

35 × 75 moulded dado rail

40 mm thick raised and fielded dado panels

NB. all panelling to be in selected hardwoods

40 mm thick moulded skirting

25 × 25 floor fillet

SECTION

Joinery Timbers ~ both hardwoods and softwoods can be used for joinery works. Softwoods can be selected for their stability durability and/or workability if the finish is to be paint but if it is left in its natural colour with a sealing coat the grain texture and appearance should be taken into consideration. Hardwoods are usually left in their natural colour and treated with a protective clear sealer or polish therefore texture, colour and grain pattern are important when selecting hardwoods for high class joinery work.

Typical Softwoods Suitable for Joinery Work ~

1. Douglas Fir — sometimes referred to as Columbian Pine or Oregon Pine. It is available in long lengths and has a straight grain. Colour is reddish brown to pink. Suitable for general and high class joinery. Approximate density 530 kg/m³.

2. Redwood — also known as Scots Pine. Red Pine, Red Deal and Yellow Deal. It is a widely used softwood for general joinery work having good durability a straight grain and is reddish brown to straw in colour. Approximate density 430 kg/m³.

3. European Spruce — similar to redwood but with a lower durability. It is pale yellow to pinkish white in colour and is used mainly for basic framing work and simple internal joinery. Approximate density 650 kg/m³.

4. Pitch Pine — durable softwood suitable for general joinery work. It is light red to reddish yellow in colour and tends to have large knots which in some cases can be used as a decorative effect. Approximate density 650 kg/m³.

5. Parana Pine — moderately durable straight grained timber available in a good range of sizes. Suitable for general joinery work especially timber stairs. Light to dark brown in colour with the occasional pink stripe. Approximate density 560 kg/m³.

6. Western Hemlock — durable softwood suitable for interior joinery work such as panelling. Light yellow to reddish brown in colour. Approximate density 500 kg/m³.

7. Western Red Cedar — originates from British Columbia and Western USA. A straight grained timber suitable for flush doors and panel work. Approximate density 380 kg/m³.

Typical Hardwoods Suitable for Joinery Works ~

1. Beech — hard close grained timber with some silver grain in the predominately reddish yellow to light brown colour. Suitable for all internal joinery. Approximately density 700 kg/m^3.

2. Iroko — hard durable hardwood with a figured grain and is usually golden brown in colour. Suitable for all forms of good class joinery. Approximate density 660 kg/m^3.

3. Mahogany (African) — interlocking grained hardwood with good durability. It has an attractive light brown to deep red colour and is suitable for panelling and all high class joinery work. Approximate density 560 kg/m^3.

4. Mahogany (Honduras) — durable hardwood usually straight grained but can have a mottled or swirl pattern. It is light red to pale reddish brown in colour and is suitable for all good class joinery work. Approximate density 530 kg/m^3.

5. Mahogany (South American) — a well figured, stable and durable hardwood with a deep red or brown colour which is suitable for all high class joinery particularly where a high polish is required. Approximate density 550 kg/m^3.

6. Oak (English) — very durable hardwood with a wide variety of grain patterns. It is usually a light yellow brown to a warm brown in colour and is suitable for all forms of joinery but should not be used in conjunction with ferrous metals due to the risk of staining caused by an interaction of the two materials. (The gallic acid in oak causes corrosion in ferrous metals.) Approximate density 720 kg/m^3.

7. Sapele — close texture timber of good durability, dark reddish brown in colour with a varied grain pattern. It is suitable for most internal joinery work especially where a polished finish is required. Approximate density 640 kg/m^3.

8. Teak — very strong and durable timber but hard to work. It is light golden brown to dark golden yellow in colour which darkens with age and is suitable for high class joinery work and laboratory fittings. Approximate density 650 kg/m^3.

9. Jarrah (Western Australia) — hard, dense, straight grained timber. Dull red colour, suited to floor and stair construction subjected to heavy wear. Approximate density 820 kg/m^3.

Composite Boards ~ are factory manufactured, performed sheets with a wide range of properties and applications. The most common size is 2440 × 1220 mm or 2400 × 1200 mm in thicknesses from 3 to 50 mm.

1. Plywood (BS EN636) — produced in a range of laminated thicknesses from 3 to 25 mm, with the grain of each layer normally at right angles to that adjacent. 3,7,9 or 11 plies make up the overall thickness and inner layers may have lower strength and different dimensions to those in the outer layers. Adhesives vary considerably from natural vegetable and animal glues to synthetics such as urea, melamine, phenol and resorcinol formaldehydes. Quality of laminates and type of adhesive determine application. Surface finishes include plastics, decorative hardwood veneers, metals, rubber and mineral aggregates.

2. Block and Stripboards (BS 3444) — range from 12 to 43 mm thickness, made up from a solid core of glued softwood strips with a surface enhancing veneer. Appropriate for dense panelling and doors.

   Battenboard — strips over 30 mm wide (unsuitable for joinery).
   Blockboard — strips up to 25 mm wide.
   Laminboard — strips up to 7 mm wide.

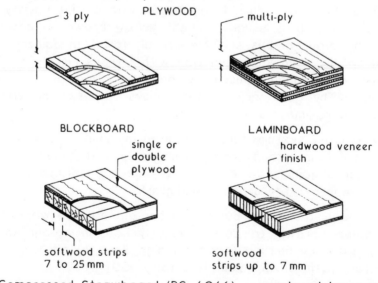

3. Compressed Strawboard (BS 4046) — produced by compacting straw under heat and pressure, and edge binding with paper. Used as panels with direct decoration or as partitioning with framed support. Also, for insulated roof decking with 58 mm slabs spanning 600 mm joist spacing.

4. Particle Board (BS 7916)

Chipboard — bonded waste wood or chip particles in thicknesses from 6 to 50mm, popularly used for floors in 18 and 22mm at 450 and 600mm maximum joist spacing, respectively. Sheets are produced by heat pressing the particles in thermosetting resins.

Wood Cement Board — approximately 25% wood particles mixed with water and cement, to produce a heavy and dense board often preferred to plasterboard and fibre cement for fire cladding.
Often 3 layer boards, from 6 to 40mm in thickness.

Oriented Strand Board — composed of wafer thin strands of wood, approximately 80mm long × 25m wide, resin bonded and directionally oriented before superimposed by further layers. Each layer is at right angles to adjacent layers, similar to the structure of plywood. A popular alternative for wall panels, floors and other chipboard and plywood applications, they are produced in a range of thicknesses from 6 to 25mm.

5. Fibreboards (BS EN 622) — basically wood in composition, reduced to a pulp and pressed to achieve 3 categories:

Hardboard — density at least $800kg/m^3$ in thicknesses from 3·2 to 8mm. Provides an excellent base for coatings and laminated finishes.

Mediumboard (low density) 350 to $560kg/m^3$ for pinboards and wall linings in thicknesses of 6·4,9, and 12·7mm.

Mediumboard (high density) 560 to $800kg/m^3$ for linings and partitions in thicknesses of 9 and 12mm.

Softboard, otherwise known as insulating board with density usually below $250kg/m^3$. Thicknesses from 9 to 25mm, often found impregnated with bitumen in existing flat roofing applications. Ideal as pinboard.

Medium Density Fibreboard, differs from other fibreboards with the addition of resin bonding agent. These boards have a very smooth surface, ideal for painting and are available moulded for a variety of joinery applications. Density exceeds $600kg/m^3$ and common board thicknesses are 9, 12, 18 and 25mm for internal and external applications.

6. Woodwool (BS 1105) — units of 600 mm width are available in 50, 75 and 100 mm thicknesses. They comprise long wood shavings coated with a cement slurry, compressed to leave a high proportion of voids. These voids provide good thermal insulation and sound absorption. The perforated surface is an ideal key for direct plastering and they are frequently specified as permanent formwork.

Plastics ~ the term plastic can be applied to any group of substances based on synthetic or modified natural polymers which during manufacture are moulded by heat and/or pressure into the required form. Plastics can be classified by their overall grouping such as polyvinyl chloride (PVC) or they can be classified as thermoplastic or thermosetting. The former soften on heating whereas the latter are formed into permanent non-softening materials. The range of plastics available give the designer and builder a group of materials which are strong, reasonably durable, easy to fit and maintain and since most are mass produced of relative low cost.

Typical Applications of Plastics in Buildings ~

Application	Plastics Used
Rainwater goods	unplasticised PVC (uPVC or PVC-U).
Soil, waste, water and gas pipes and fittings	uPVC; polyethylene (PE); acrylonitrile butadiene styrene (ABS), polypropylene (PP).
Hot and cold water pipes	chlorinated PVC; ABS; polypropylene; polyethylene; PVC (not for hot water).
Bathroom and kitchen fittings	glass fibre reinforced polyester (GRP); acrylic resins.
Cold water cisterns	polypropylene; polystyrene; polyethylene.
Rooflights and sheets	GRP; acrylic resins; uPVC.
DPC's and membranes, vapour control layers	low density polyethylene (LDPE); PVC film; polypropylene.
Doors and windows	GRP; uPVC.
Electrical conduit and fittings	plasticised PVC; uPVC; phenolic resins.
Thermal insulation	generally cellular plastics such as expanded polystyrene bead and boards; expanded PVC; foamed polyurethane; foamed phenol formaldehyde; foamed urea formaldehyde.
Floor finishes	plasticised PVC tiles and sheets; resin based floor paints; uPVC.
Wall claddings and internal linings	unplasticised PVC; polyvinyl fluoride film laminate; melamine resins; expanded polystyrene tiles & sheets.

7 DOMESTIC SERVICES

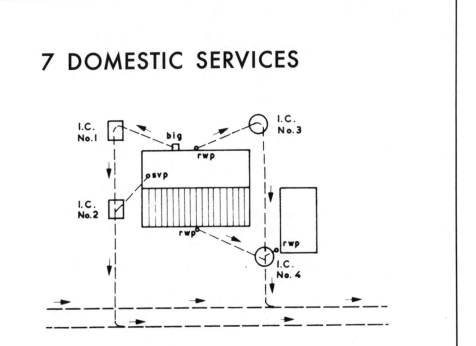

DRAINAGE EFFLUENTS
SUBSOIL DRAINAGE
SURFACE WATER REMOVAL
ROAD DRAINAGE
RAINWATER INSTALLATIONS
DRAINAGE SYSTEMS
DRAINAGE PIPE SIZES AND GRADIENTS
WATER SUPPLY
COLD WATER INSTALLATIONS
HOT WATER INSTALLATIONS
CISTERNS AND CYLINDERS
SANITARY FITTINGS
SINGLE AND VENTILATED STACK SYSTEMS
DOMESTIC HOT WATER HEATING SYSTEMS
ELECTRICAL SUPPLY AND INSTALLATION
GAS SUPPLY AND GAS FIRES
SERVICES FIRE STOPS AND SEALS
OPEN FIREPLACES AND FLUES
TELEPHONE INSTALLATIONS

Effluent ~ can be defined as that which flows out. In building drainage terms there are three main forms of effluent :-

1. Subsoil Water ~ water collected by means of special drains from the earth primarily to lower the water table level in the subsoil. It is considered to be clean and therefore requires no treatment and can be discharged direct into an approved water course.

2. Surface water ~ effluent collected from surfaces such as roofs and paved areas and like subsoil water is considered to be clean and can be discharged direct into an approved water course or soakaway

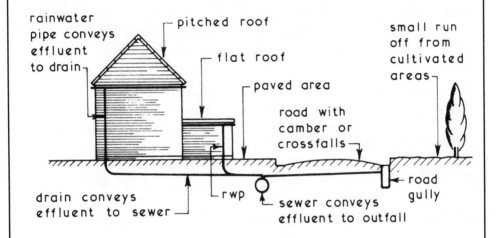

3. Foul or Soil Water ~ effluent contaminated by domestic or trade waste and will require treatment to render it clean before it can be discharged into an approved water course.

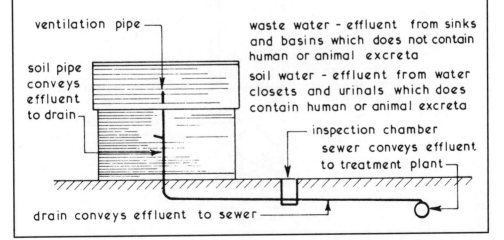

Subsoil Drainage ~ Building Regulation C3 requires that subsoil drainage shall be provided if it is needed to avoid :-

a) the passage of ground moisture into the interior of the building or

b) damage to the fabric of the building.

Subsoil drainage can also be used to improve the stability of the ground, lower the humidity of the site and enhance its horticultural properties. Subsoil drains consist of porous or perforated pipes laid dry jointed in a rubble filled trench. Porous pipes allow the subsoil water to pass through the body of the pipe whereas perforated pipes which have a series of holes in the lower half allow the subsoil water to rise into the pipe. This form of ground water control is only economic up to a depth of 1·500, if the water table needs to be lowered to a greater depth other methods of ground water control should be considered (see page 270 to 274).

The water collected by a subsoil drainage system has to be conveyed to a suitable outfall such as a river, lake or surface water drain or sewer. In all cases permission to discharge the subsoil water will be required from the authority or owner and in the case of streams, rivers and lakes, bank protection at the outfall may be required to prevent erosion. (see page 607)

Typical Subsoil Drain Details ~

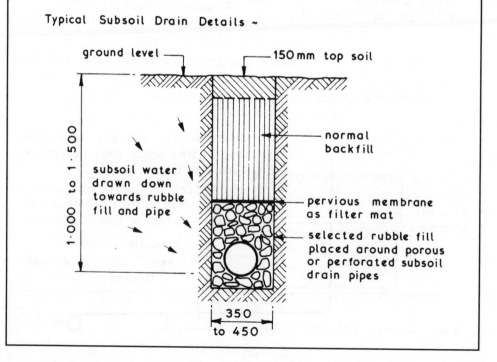

Subsoil Drainage Systems ~ the lay out of subsoil drains will depend on whether it is necessary to drain the whole site or if it is only the substructure of the building which needs to be protected. The latter is carried out by installing a cut off drain around the substructure to intercept the flow of water and divert it away from the site of the building. Junctions in a subsoil drainage system can be made using standard fittings or by placing the end of the branch drain onto the crown of the main drain.

Typical Examples ~

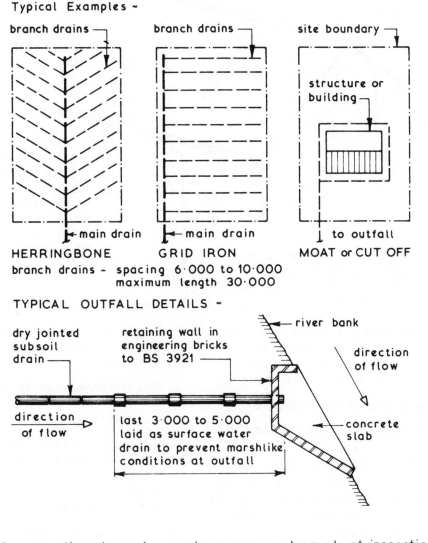

branch drains

branch drains

site boundary

structure or building

main drain

main drain

to outfall

HERRINGBONE

GRID IRON

MOAT or CUT OFF

branch drains - spacing 6·000 to 10·000
maximum length 30·000

TYPICAL OUTFALL DETAILS ~

dry jointed subsoil drain

retaining wall in engineering bricks to BS 3921

river bank

direction of flow

direction of flow

last 3·000 to 5·000 laid as surface water drain to prevent marshlike conditions at outfall

concrete slab

NB. connections to surface water sewer can be made at inspection chamber or direct to the sewer using a saddle connector- it may be necessary to have a catchpit to trap any silt (see page 611)

General Principles ~ a roof must be designed with a suitable fall towards the surface water collection channel or gutter which in turn is connected to vertical rainwater pipes which convey the collected discharge to the drainage system. The fall of the roof will be determined by the chosen roof covering or the chosen pitch will limit the range of coverings which can be selected.

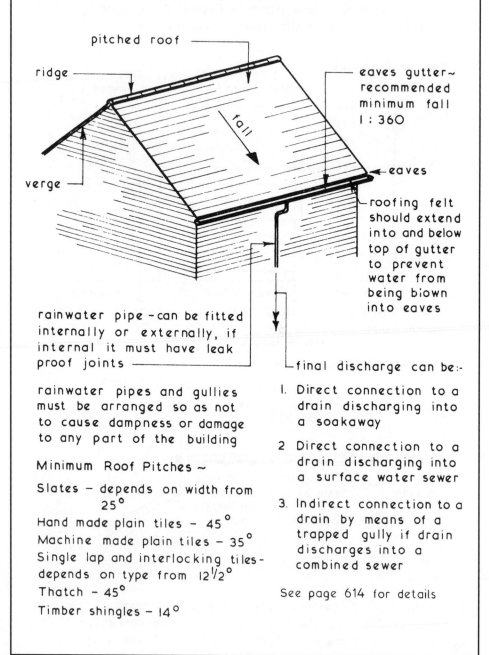

pitched roof

ridge

eaves gutter~ recommended minimum fall 1 : 360

fall

verge

eaves

roofing felt should extend into and below top of gutter to prevent water from being blown into eaves

rainwater pipe –can be fitted internally or externally, if internal it must have leak proof joints

final discharge can be:-

rainwater pipes and gullies must be arranged so as not to cause dampness or damage to any part of the building

Minimum Roof Pitches ~

Slates – depends on width from 25°
Hand made plain tiles – 45°
Machine made plain tiles – 35°
Single lap and interlocking tiles- depends on type from 12½°
Thatch – 45°
Timber shingles – 14°

1. Direct connection to a drain discharging into a soakaway

2. Direct connection to a drain discharging into a surface water sewer

3. Indirect connection to a drain by means of a trapped gully if drain discharges into a combined sewer

See page 614 for details

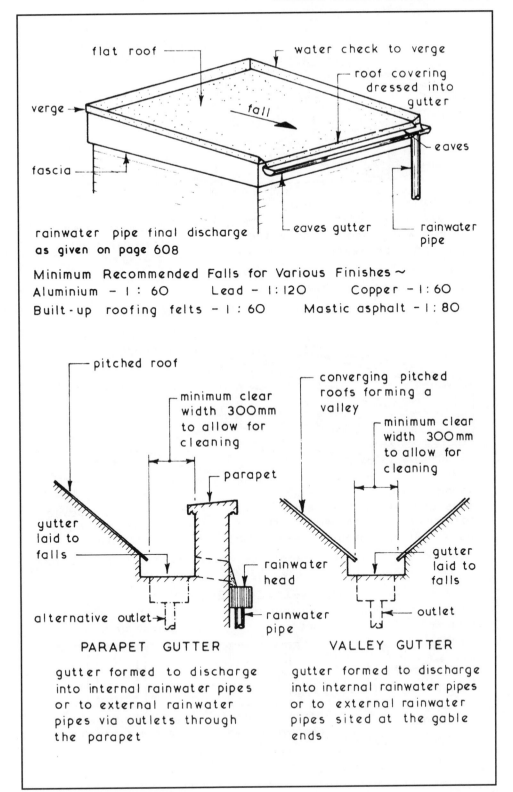

flat roof

water check to verge

roof covering
dressed into
gutter

verge

fall

eaves

fascia

eaves gutter

rainwater
pipe

rainwater pipe final discharge
as given on page 608

Minimum Recommended Falls for Various Finishes ∼
Aluminium − 1 : 60   Lead − 1:120   Copper −1:60
Built-up roofing felts − 1 : 60   Mastic asphalt −1:80

pitched roof

minimum clear
width 300mm
to allow for
cleaning

parapet

converging pitched
roofs forming a
valley

minimum clear
width 300mm
to allow for
cleaning

gutter
laid to
falls

gutter
laid to
falls

rainwater
head

alternative outlet

rainwater
pipe

outlet

PARAPET GUTTER

VALLEY GUTTER

gutter formed to discharge
into internal rainwater pipes
or to external rainwater
pipes via outlets through
the parapet

gutter formed to discharge
into internal rainwater pipes
or to external rainwater
pipes sited at the gable
ends

609

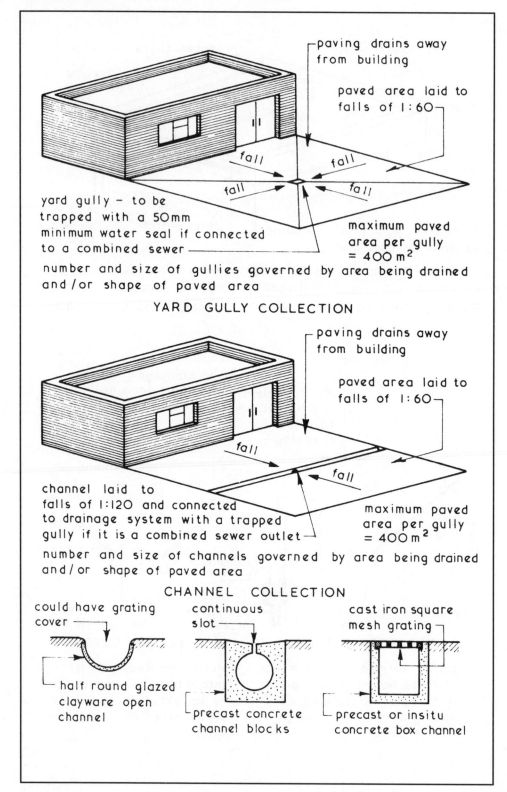

paving drains away from building

paved area laid to falls of 1:60

fall fall

fall fall

yard gully – to be trapped with a 50mm minimum water seal if connected to a combined sewer

maximum paved area per gully = 400 m²

number and size of gullies governed by area being drained and/or shape of paved area

YARD GULLY COLLECTION

paving drains away from building

paved area laid to falls of 1:60

fall

fall

channel laid to falls of 1:120 and connected to drainage system with a trapped gully if it is a combined sewer outlet

maximum paved area per gully = 400 m²

number and size of channels governed by area being drained and/or shape of paved area

CHANNEL COLLECTION

could have grating cover

half round glazed clayware open channel

continuous slot

precast concrete channel blocks

cast iron square mesh grating

precast or insitu concrete box channel

Highway Drainage ~ the stability of a highway or road relies on two factors —

1. Strength and durability of upper surface
2. Strength and durability of subgrade which is the subsoil on which the highway construction is laid.

The above can be adversely affected by water therefore it may be necessary to install two drainage systems. One system (subsoil drainage) to reduce the flow of subsoil water through the subgrade under the highway construction and a system of surface water drainage.

Typical Highway Subsoil Drainage Methods ~

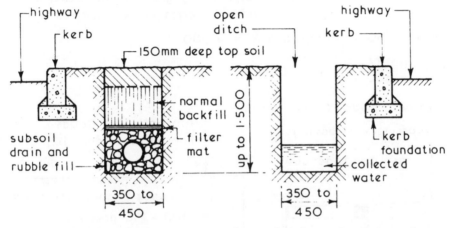

Subsoil Drain - acts as a cut off drain and can be formed using perforated or porous drain pipes. If filled with rubble only it is usually called a French or rubble drain.

Open Ditch - acts as a cut off drain and could also be used to collect surface water discharged from a rural road where there is no raised kerb or surface water drains.

Surface Water Drainage Systems~

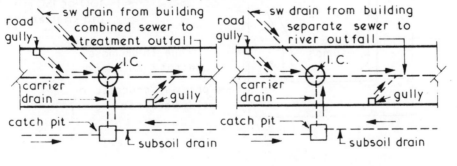

Road Drainage ~ this consists of laying the paved area or road to a suitable crossfall or gradient to direct the run-off of surface water towards the drainage channel or gutter. This is usually bounded by a kerb which helps to convey the water to the road gullies which are connected to a surface water sewer. For drains or sewers under 900 mm internal diameter inspection chambers will be required as set out in the Building Regulations. The actual spacing of road gullies is usually determined by the local highway authority based upon the carriageway gradient and the area to be drained into one road gully. Alternatively the following formula could be used :-

$$D = \frac{280\sqrt{s}}{W}$$

where D = gully spacing
S = carriageway gradient (per cent)
W = width of carriageway in metres

∴ If S = 1:60 = 1·66 % and W = 4·500

$$D = \frac{280\sqrt{1.66}}{4\cdot500} = \text{say } 80\cdot000$$

Typical Road Gully Detail ~

- carriageway paving
- cast iron road gully grating and frame
- footpath (see Roads– Footpaths on page 112)
- drainage channel (fall 1:200)
- 300 × 150 precast concrete kerb
- levelling brick course
- kerb foundation strip
- branch drain to surface water sewer
- stopper
- 100 or 150 mm dia. outlet
- 85 min.
- sump
- 150 mm thick mass concrete surround
- precast concrete or clayware road gully

Materials ~ the traditional material for domestic eaves gutters and rainwater pipes is cast iron but uPVC systems are very often specified today because of their simple installation and low maintenance costs. Other materials which could be considered are aluminium alloy, galvanized steel and stainless steel but whatever material is chosen it must be of adequate size, strength and durability.

Typical Eaves Details ~

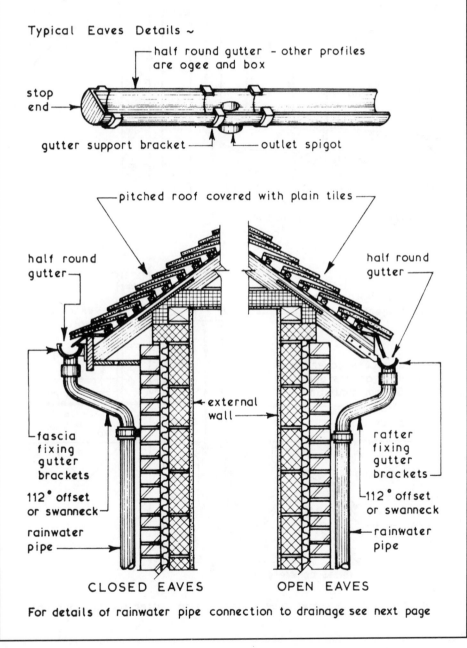

half round gutter - other profiles are ogee and box

stop end

gutter support bracket

outlet spigot

pitched roof covered with plain tiles

half round gutter

half round gutter

external wall

fascia fixing gutter brackets

rafter fixing gutter brackets

112° offset or swanneck

112° offset or swanneck

rainwater pipe

rainwater pipe

CLOSED EAVES

OPEN EAVES

For details of rainwater pipe connection to drainage see next page

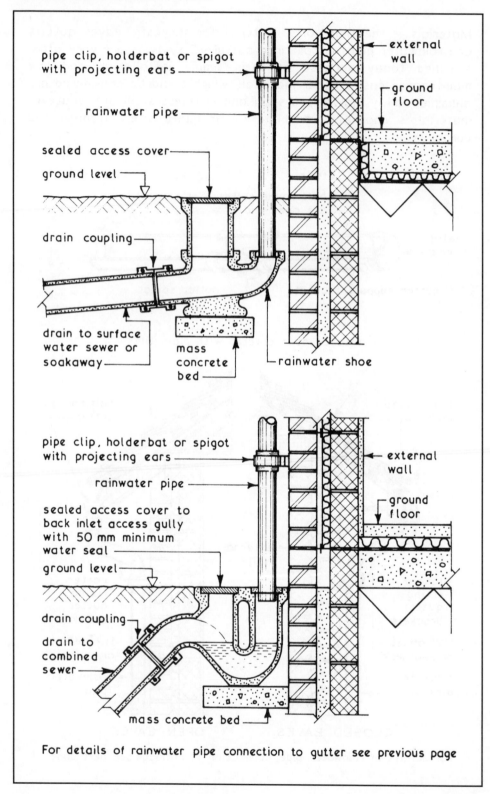

pipe clip, holderbat or spigot with projecting ears

rainwater pipe

sealed access cover

ground level

drain coupling

drain to surface water sewer or soakaway

mass concrete bed

external wall

ground floor

rainwater shoe

pipe clip, holderbat or spigot with projecting ears

rainwater pipe

sealed access cover to back inlet access gully with 50 mm minimum water seal

ground level

drain coupling

drain to combined sewer

mass concrete bed

external wall

ground floor

For details of rainwater pipe connection to gutter see previous page

Soakaways ~ provide a means for collecting and controlling the seapage of rainwater into surrounding granular subsoils. They are not suitable in clay subsoils. Siting is on land at least level and preferably lower than adjacent buildings and no closer than 5 m to a building. Concentration of a large volume of water any closer could undermine the foundations. The simplest soakaway is a rubble filled pit, which is normally adequate to serve a dwelling or other small building. Where several buildings share a soakaway, the pit should be lined with precast perforated concrete rings and surrounded in free—draining material.

BRE Digest 365 provides capacity calculations based on percolation tests. The following empirical formula will prove adequate for most situations:-

$$C = \frac{AR}{3}$$ where: C = capacity (m³)

A = area on plan to be drained (m²)

R = rainfall (m/h)

e.g. roof plan area 60 m² and rainfall of 50 mm/h (0·05 m/h)

$$C = \frac{60 \times 0 \cdot 05}{3} = 1 \cdot 0 \, m^3 \text{ (below invert of discharge pipe)}$$

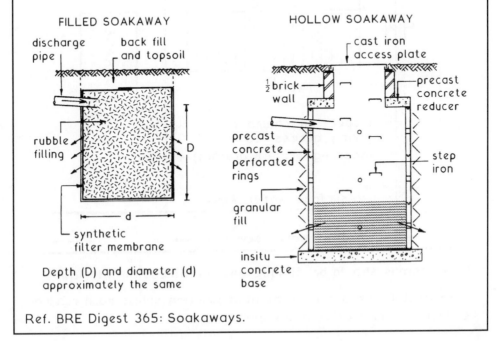

FILLED SOAKAWAY

discharge pipe

back fill and topsoil

rubble filling

D

d

synthetic filter membrane

Depth (D) and diameter (d) approximately the same

HOLLOW SOAKAWAY

cast iron access plate

½ brick wall

precast concrete perforated rings

granular fill

insitu concrete base

precast concrete reducer

step iron

Ref. BRE Digest 365: Soakaways.

Drains ~ these can be defined as a means of conveying surface water or foul water below ground level.

Sewers ~ these have the same functions as drains but collect the discharge from a number of drains and convey it to the final outfall. They can be a private or public sewer depending on who is responsible for the maintenance.

Basic Principles ~ to provide a drainage system which is simple efficient and economic by laying the drains to a gradient which will render them self cleansing and will convey the effluent to a sewer without danger to health or giving nuisance. To provide a drainage system which will comply with the minimum requirements given in Part H of the Building Regulations

Typical Basic Requirements ~

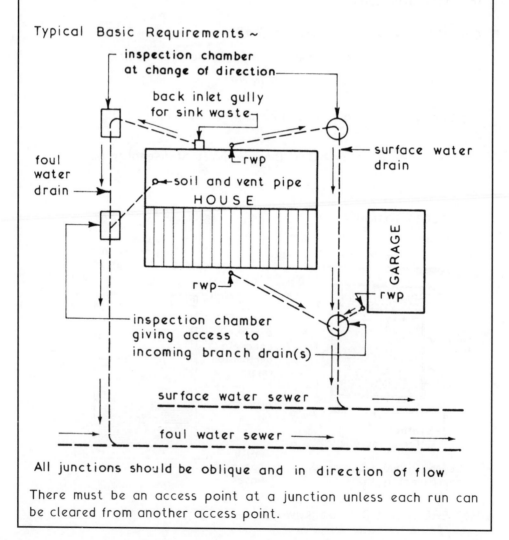

All junctions should be oblique and in direction of flow

There must be an access point at a junction unless each run can be cleared from another access point.

Separate System ~ the most common drainage system in use
where the surface water discharge is conveyed in separate drains
and sewers to that of foul water discharges and therefore receives
no treatment before the final outfall.

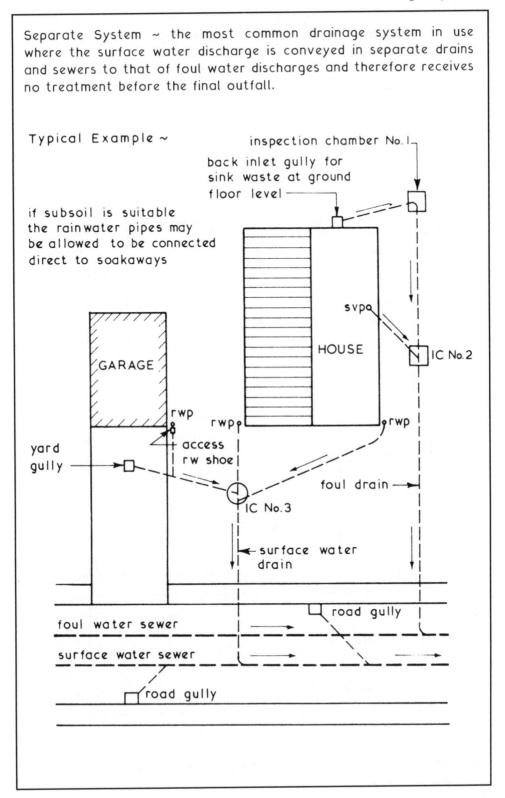

Typical Example ~

inspection chamber No. 1

back inlet gully for
sink waste at ground
floor level

if subsoil is suitable
the rainwater pipes may
be allowed to be connected
direct to soakaways

GARAGE

HOUSE

svp

IC No. 2

rwp

rwp

rwp

yard
gully

access
rw shoe

IC No. 3

foul drain

surface water
drain

road gully

foul water sewer

surface water sewer

road gully

Combined System ~ this is the simplest and least expensive system to design and install but since all forms of discharge are conveyed in the same sewer the whole effluent must be treated unless a sea outfall is used to discharge the untreated effluent.

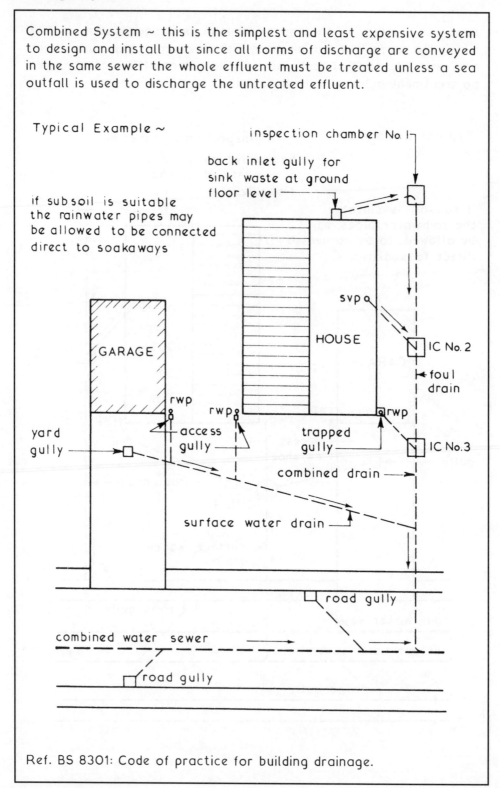

Typical Example ~

inspection chamber No. 1

back inlet gully for sink waste at ground floor level

if subsoil is suitable the rainwater pipes may be allowed to be connected direct to soakaways

svp

GARAGE

HOUSE

IC No. 2

foul drain

rwp

rwp

rwp

yard gully

access gully

trapped gully

IC No. 3

combined drain

surface water drain

road gully

combined water sewer

road gully

Ref. BS 8301: Code of practice for building drainage.

Partially Separate System ~ a compromise system — there are two drains, one to convey only surface water and a combined drain to convey the total foul discharge and a proportion of the surface water.

Typical Example ~

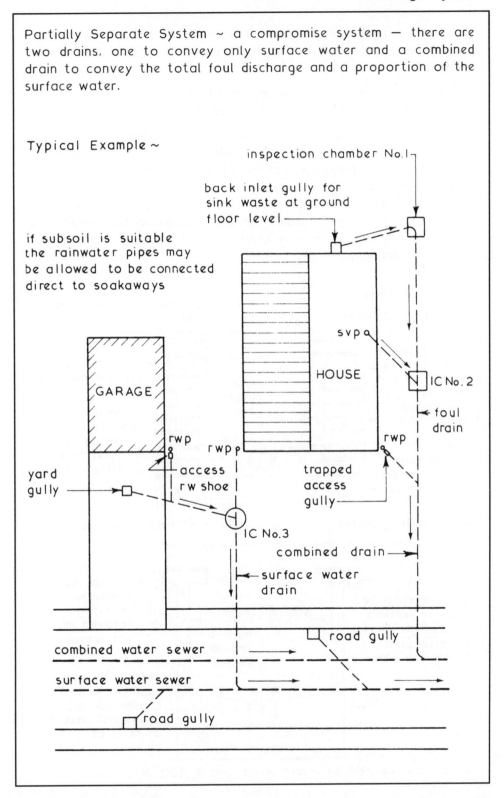

inspection chamber No.1

back inlet gully for sink waste at ground floor level

if subsoil is suitable the rainwater pipes may be allowed to be connected direct to soakaways

svp

HOUSE

GARAGE

IC No. 2

foul drain

rwp

rwp

rwp

yard gully

access rw shoe

trapped access gully

IC No.3

combined drain

surface water drain

road gully

combined water sewer

surface water sewer

road gully

619

Simple Drainage—Inspection Chambers

Inspection Chambers ~ these provide a means of access to drainage systems where the depth to invert level does not exceed 1·000.

Manholes ~ these are also a means of access to the drains and sewers, and are so called if the depth to invert level exceeds 1·000.

These means of access should be positioned in accordance with the requirements of part H of the Building Regulations. In domestic work inspection chambers can be of brick, precast concrete or preformed in plastic for use with patent drainage systems. The size of an inspection chamber depends on the depth to invert level, size and number of branch drains to be accommodated within the chamber-guidance to sizing is given in BS 8301: Code of practice for building drainage.

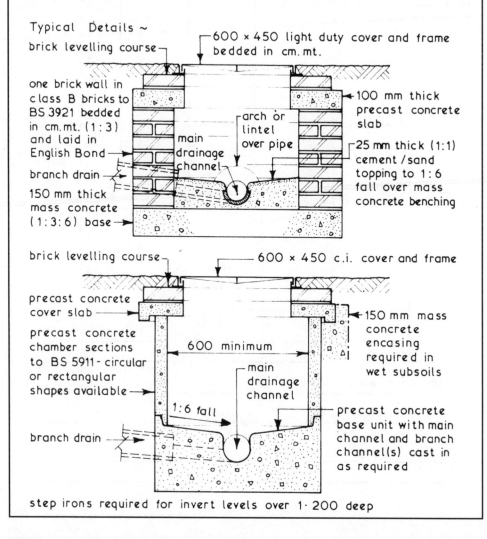

Typical Details ~

brick levelling course

600 × 450 light duty cover and frame bedded in cm. mt.

one brick wall in class B bricks to BS 3921 bedded in cm.mt. (1:3) and laid in English Bond

branch drain

150 mm thick mass concrete (1:3:6) base

main drainage channel

arch or lintel over pipe

100 mm thick precast concrete slab

25 mm thick (1:1) cement/sand topping to 1:6 fall over mass concrete benching

brick levelling course

600 × 450 c.i. cover and frame

precast concrete cover slab

precast concrete chamber sections to BS 5911 - circular or rectangular shapes available

branch drain

600 minimum

1:6 fall

main drainage channel

150 mm mass concrete encasing required in wet subsoils

precast concrete base unit with main channel and branch channel(s) cast in as required

step irons required for invert levels over 1·200 deep

620

Plastic Inspection Chambers ~ the raising piece can be sawn horizontally with a carpenter's saw to suit depth requirements with the cover and frame fitted at surface level. Bedding may be a 100 mm prepared shingle base or 150 mm wet concrete to ensure a uniform support.

The unit may need weighting to retain it in place in areas of high water table, until backfilled with granular material. Under roads a peripheral concrete collar is applied to the top of the chamber in addition to the 150 mm thickness of concrete surrounding the inspection chamber.

Typical Example ~

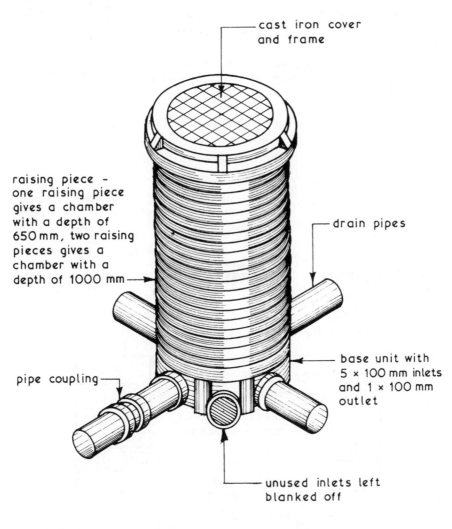

cast iron cover and frame

raising piece - one raising piece gives a chamber with a depth of 650 mm, two raising pieces gives a chamber with a depth of 1000 mm

drain pipes

pipe coupling

base unit with 5 × 100 mm inlets and 1 × 100 mm outlet

unused inlets left blanked off

Excavations ~ drains are laid in trenches which are set out, excavated and supported in a similar manner to foundation trenches except for the base of the trench which is cut to the required gradient or fall.

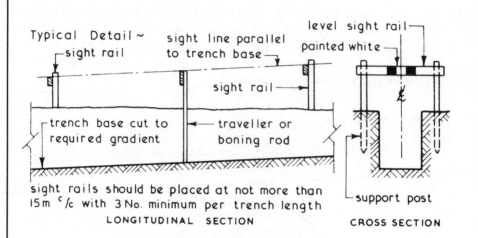

Typical Detail ~
sight rail
sight line parallel to trench base
level sight rail
painted white
sight rail
trench base cut to required gradient
traveller or boning rod
sight rails should be placed at not more than 15m ^{c}/c with 3 No. minimum per trench length

LONGITUDINAL SECTION

support post

CROSS SECTION

Joints ~ these must be watertight under all working and movement conditions and this can be achieved by using rigid and flexible joints in conjuntion with the appropriate bedding.

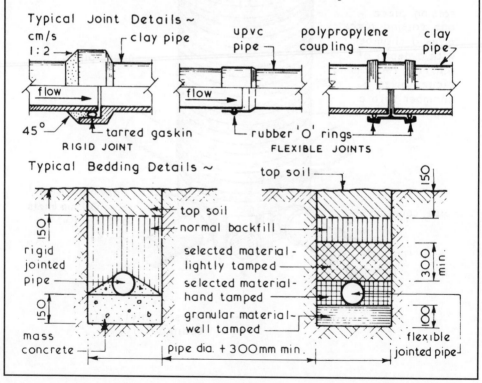

Typical Joint Details ~

cm/s 1:2
clay pipe
flow
45°
tarred gaskin
RIGID JOINT

upvc pipe
flow
rubber 'O' rings
FLEXIBLE JOINTS

polypropylene coupling
clay pipe

Typical Bedding Details ~

top soil
150

rigid jointed pipe
150
150
mass concrete

top soil
normal backfill
selected material - lightly tamped
selected material - hand tamped
granular material - well tamped
pipe dia. + 300mm min.

300 min
100
flexible jointed pipe

Watertightness ~ must be ensured to prevent water seapage and erosion of the subsoil. Also, in the interests of public health, foul water should not escape untreated. The Building Regulaions, Approved Document H1: Section 2 specifies either an air or water test to determine soundness of installation.

AIR TEST ~ equipment : manometer and accessories (see page 641) 2 drain stoppers, one with tube attachment

Application ~

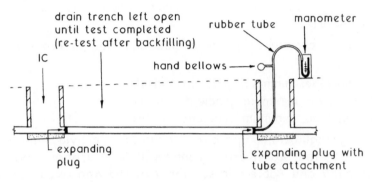

Test ~ 100 mm water gauge to fall no more than 25 mm in 5 mins. Or, 50 mm w.g. to fall no more than 12 mm in 5 mins.

WATER TEST ~ equipment : Drain stopper
                          Test bend
                          Extension pipe

Application ~

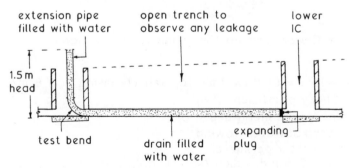

Test ~ 1·5 m head of water to stand for 2 hours and then topped up. Leakage over the next 30 minutes should be minimal, i.e.
100 mm pipe — 0·05 litres per metre, which equates to a drop of 6·4 mm/m in the extension pipe, and
150 mm pipe — 0·08 litres per metre, which equates to a drop of 4·5 mm/m in the extension pipe.

Drainage Pipes ~ sizes for normal domestic foul water applications:-

<20 dwellings   = 100 mm diameter
20—150 dwellings = 150 mm diameter

Exceptions: 75 mm diameter for waste or rainwater only (no WCs)
            150 mm diameter minimum for a public sewer

Other situations can be assessed by summating the Discharge Units from appliances and converting these to an appropriate diameter stack and drain, see BS5572: Code of practice for sanitary pipework. Gradient will also affect pipe capacity and when combined with discharge calculations, provides the basis for complex hydraulic theories.

The simplest correlation of pipe size and fall, is represented in Maguire's rule:-

4″ (100 mm) pipe, minimum gradient 1 in 40
6″ (150 mm) pipe, minimum gradient 1 in 60
9″ (225 mm) pipe, minimum gradient 1 in 90

The Building Regulations, approved Document H provides more scope and relates to foul water drains running at 0·75 proportional depth. See Diagram 7 and Table 6 in section 2 of the Approved Document.

Other situations outside of design tables and empirical practice can be calculated.

eg. A 150 mm diameter pipe flowing 0·5 proportional depth.

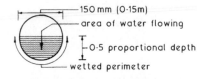

Applying the Chezy formula for gradient calculations:-

$$v = c\sqrt{m \times i}$$

where: v = velocity of flow, (min for self cleansing = 0·8 m/s)
       c = Chezy coefficient (58)
       m = hydraulic mean depth or;
$$\frac{\text{area of water flowing}}{\text{wetted perimeter}}$$
           for 0·5 p.d. = diam/4
       i = inclination or gradient as a fraction 1/x

Selecting a velocity of 1 m/s as a margin of safety over the minimum:-

$$1 = 58\sqrt{0·15/4 \times i}$$

i = 0·0079 where i = 1/x

So, x = 1/0·0079 = 126, i.e. a minimum gradient of 1 in 126

Water supply ~ an adequate supply of cold water of drinking quality should be provided to every residential building and a drinking water tap installed within the building. The installation should be designed to prevent waste, undue consumption, misuse, contamination of general supply, be protected against corrosion and frost damage and be accessible for maintenance activities. The intake of a cold water supply to a building is owned jointly by the water authority and the consumer who therefore have joint maintenance responsibilities.

Typical Water Supply Arrangement ~

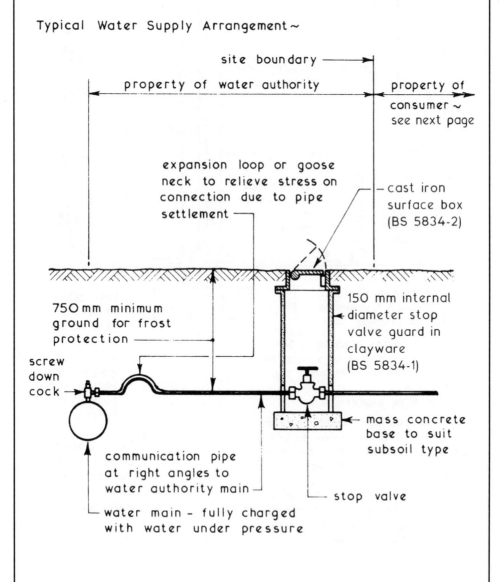

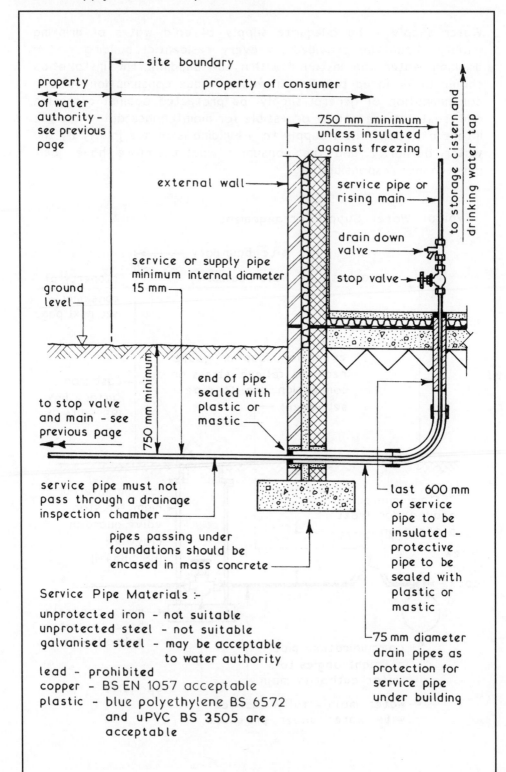

site boundary

property of water authority - see previous page

property of consumer

750 mm minimum unless insulated against freezing

external wall

to storage cistern and drinking water tap

service pipe or rising main

service or supply pipe minimum internal diameter 15 mm

drain down valve

stop valve

ground level

750 mm minimum

to stop valve and main - see previous page

end of pipe sealed with plastic or mastic

service pipe must not pass through a drainage inspection chamber

last 600 mm of service pipe to be insulated - protective pipe to be sealed with plastic or mastic

pipes passing under foundations should be encased in mass concrete

75 mm diameter drain pipes as protection for service pipe under building

Service Pipe Materials :-

unprotected iron - not suitable
unprotected steel - not suitable
galvanised steel - may be acceptable
                  to water authority
lead - prohibited
copper - BS EN 1057 acceptable
plastic - blue polyethylene BS 6572
          and uPVC BS 3505 are
          acceptable

General ~ when planning or designing any water installation the basic physical laws must be considered :-

1. Water is subject to the force of gravity and will find its own level.
2. To overcome friction within the conveying pipes water which is stored prior to distribution will require to be under pressure and this is normally achieved by storing the water at a level above the level of the outlets. The vertical distance between these levels is usually called the head.
3. Water becomes less dense as its temperature is raised therefore warm water will always displace colder water whether in a closed or open circuit.

Direct Cold Water Systems ~ the cold water is supplied to the outlets at mains pressure the only storage requirements is a small capacity cistern to feed the hot water storage tank. These systems are suitable for districts which have high level reservoirs with a good supply and pressure. The main advantage is that drinking water is available from all cold water outlets, disadvantages include lack of reserve in case of supply cut off, risk of back syphonage due to negative mains pressure and a risk of reduced pressure during peak demand periods.

Typical Direct Cold Water System ~

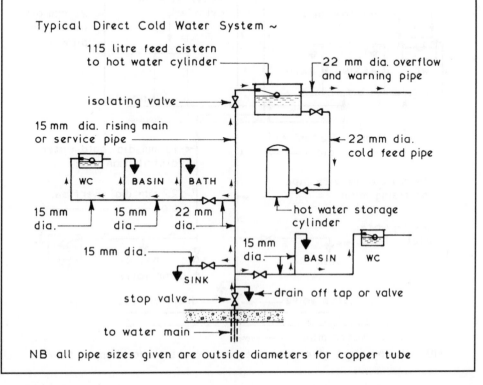

NB all pipe sizes given are outside diameters for copper tube

Indirect Systems ~ Cold water is supplied to all outlets from a cold water storage cistern except for the cold water supply to the sink(s) where the drinking water tap is connected directly to incoming supply from the main. This system requires more pipework than the direct system but it reduces the risk of back syphonage and provides a reserve of water should the mains supply fail or be cut off. The local water authority will stipulate the system to be used in their area.

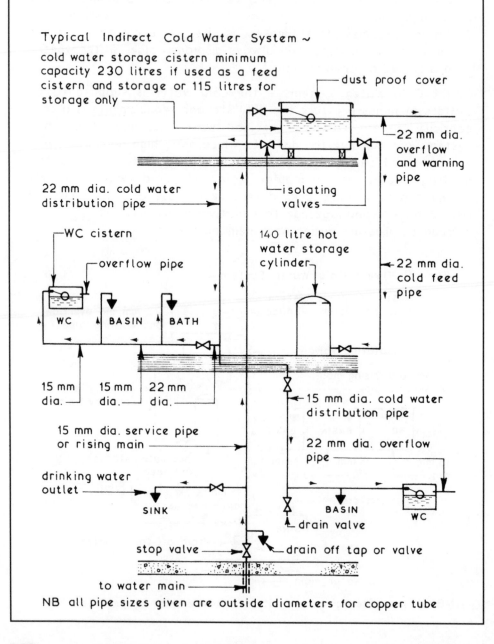

Typical Indirect Cold Water System ~

cold water storage cistern minimum capacity 230 litres if used as a feed cistern and storage or 115 litres for storage only

dust proof cover

22 mm dia. overflow and warning pipe

22 mm dia. cold water distribution pipe

isolating valves

140 litre hot water storage cylinder

22 mm dia. cold feed pipe

WC cistern

overflow pipe

WC   BASIN   BATH

15 mm dia.   15 mm dia.   22 mm dia.

15 mm dia. cold water distribution pipe

15 mm dia. service pipe or rising main

22 mm dia. overflow pipe

drinking water outlet

SINK

BASIN

drain valve

WC

stop valve

drain off tap or valve

to water main

NB all pipe sizes given are outside diameters for copper tube

Direct System ~ this is the simplest and least expensive system of hot water installation. The water is heated in the boiler and the hot water rises by convection to the hot water storage tank or cylinder to be replaced by the cooler water from the bottom of the storage vessel. Hot water drawn from storage is replaced with cold water from the cold water storage cistern. Direct systems are suitable for soft water areas and for installations which are not supplying a central heating circuit.

Typical Direct Hot Water System ~

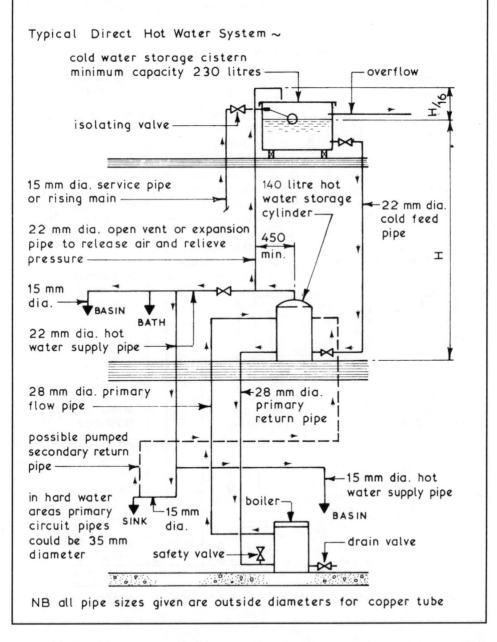

NB all pipe sizes given are outside diameters for copper tube

Indirect System ~ this is a more complex system than the direct system but it does overcome the problem of furring which can occur in direct hot water systems. This method is therefore suitable for hard water areas and in all systems where a central heating circuit is to be part of the hot water installation. Basically the pipe layouts of the two systems are similar but in the indirect system a separate small capacity feed cistern is required to charge and top up the primary circuit. In this system the hot water storage tank or cylinder is in fact a heat exchanger — see page 634

Typical Indirect Hot Water System ~

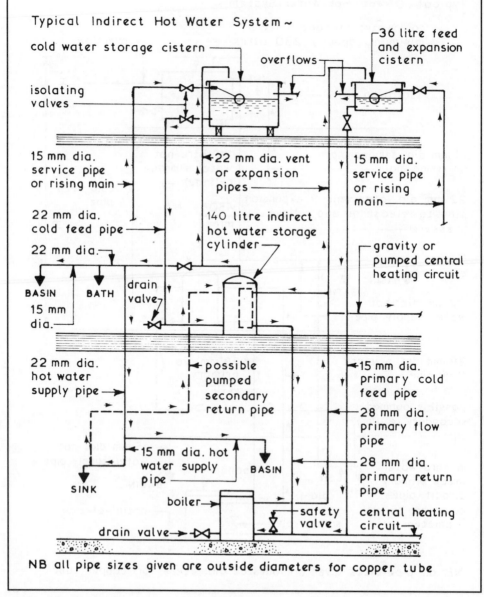

NB all pipe sizes given are outside diameters for copper tube

Mains Fed Indirect System ~ now widely used as an alternative to conventional systems. It eliminates the need for cold water storage and saves considerably on installation time. This system is established in Europe and the USA, but only acceptable in the UK at the local water authority's discretion. It complements electric heating systems, where a boiler is not required. An expansion vessel replaces the standard vent and expansion pipe and may be integrated with the hot water storage cylinder. It contains a neoprene diaphragm to separate water from air, the air providing a 'cushion' for the expansion of hot water. Air loss can be replenished by foot pump as required.

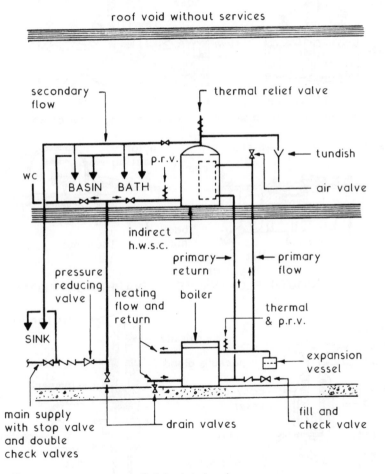

NB. p.r.v. = pressure relief (safety) valve

Flow Controls ~ these are valves inserted into a water installation to control the water flow along the pipes or to isolate a branch circuit or to control the draw-off of water from the system.

Typical Examples ~

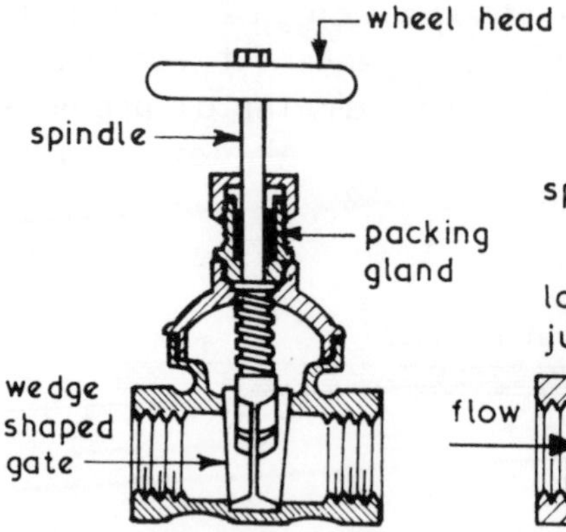

wheel head

spindle

packing gland

wedge shaped gate

GATE VALVE
used to control flow of water

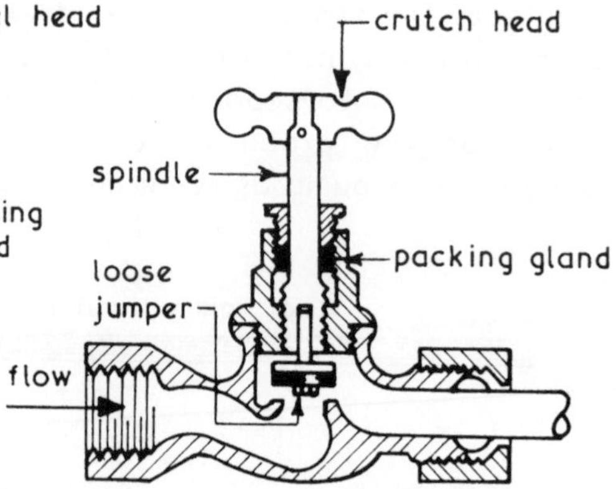

crutch head

spindle

loose jumper

packing gland

flow

STOP VALVE
used to stop flow of water

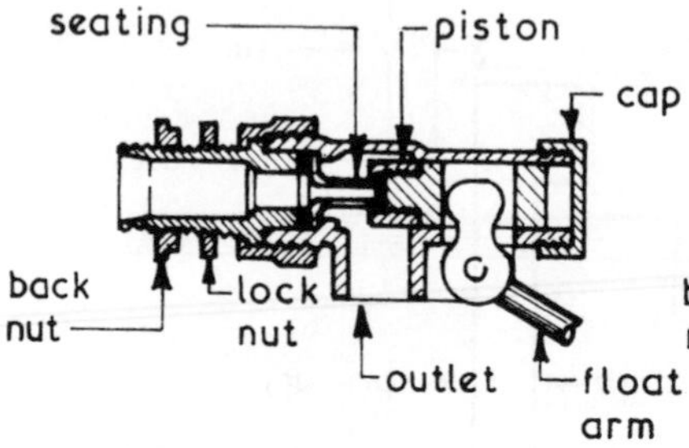

seating

piston

cap

back nut

lock nut

outlet

float arm

PORTSMOUTH FLOATVALVE

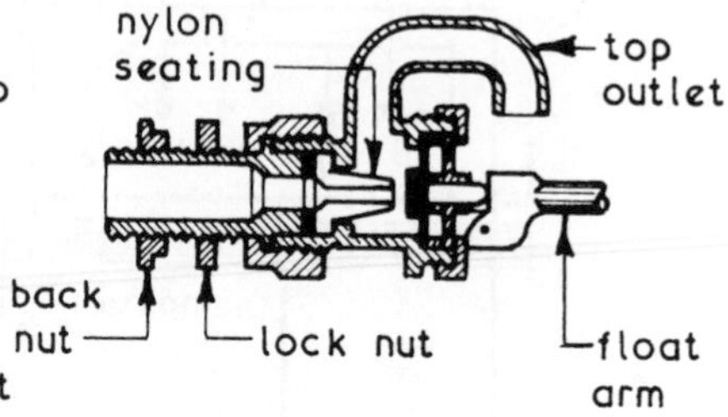

nylon seating

top outlet

back nut

lock nut

float arm

DIAPHRAGM FLOATVALVE

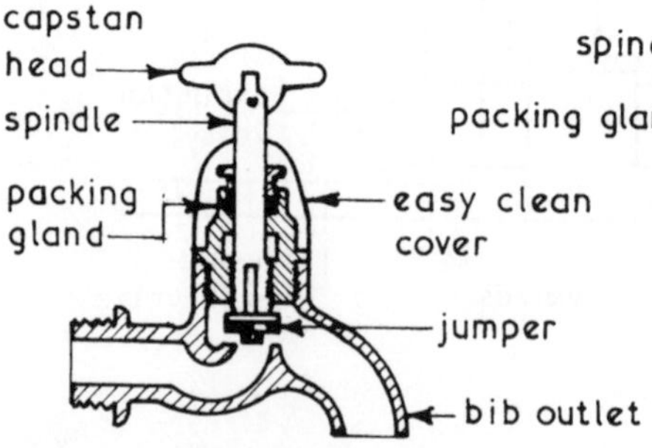

capstan head

spindle

packing gland

easy clean cover

jumper

bib outlet

BIB TAP
horizontal inlet - used over sinks and for hose pipe outlets

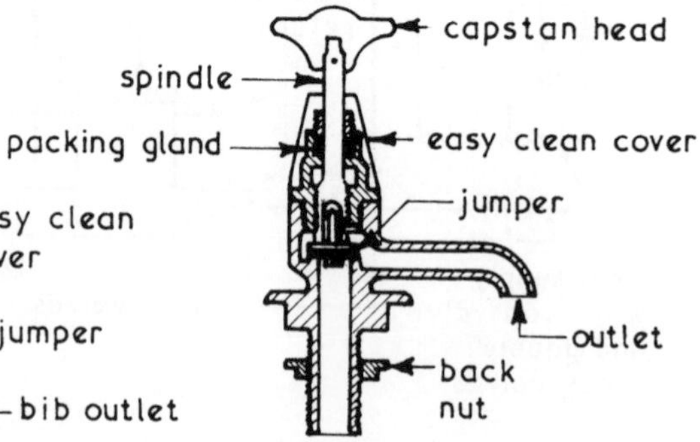

capstan head

spindle

packing gland

easy clean cover

jumper

outlet

back nut

PILLAR TAP
vertical inlet - used in conjunction with fittings

Cisterns ~ these are fixed containers used for storing water at atmospheric pressure. The inflow of water is controlled by a floatvalve which is adjusted to shut off the water supply when it has reached the designed level within the cistern. The capacity of the cistern depends on the draw off demand and whether the cistern feeds both hot and cold water systems. Domestic cold water cisterns should be placed at least 750 mm away from an external wall or roof surface and in such a position that it can be inspected, cleaned and maintained. A minimum clear space of 300 mm is required over the cistern for floatvalve maintenance. An overflow or warning pipe of not less than 22 mm diameter must be fitted to fall away to discharge in a conspicuous position. All draw off pipes must be fitted with a gate valve positioned as near to the cistern as possible.

Cisterns are available in a variety of sizes and materials such as galvanised mild steel (BS 417), moulded plastic (BS 4213) and reinforced plastic (BS 4994). If the cistern and its associated pipework are to be housed in a cold area such as a roof they should be insulated against freezing.

Typical Details ~

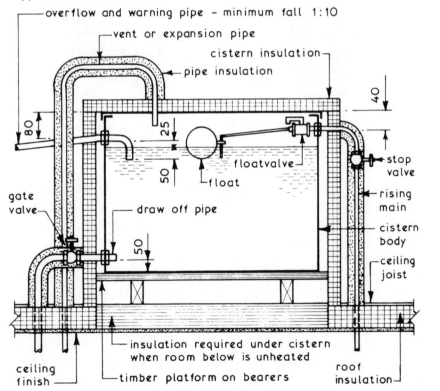

633

Indirect Hot Water Cylinders ~ these cylinders are a form of heat exchanger where the primary circuit of hot water from the boiler flows through a coil or annulus within the storage vessel and transfers the heat to the water stored within. An alternative hot water cylinder for small installations is the single feed or 'Primatic' cylinder which is self venting and relies on two air locks to separate the primary water from the secondary water. This form of cylinder is connected to pipework in the same manner as for a direct system (see page 629) and therefore gives savings in both pipework and fittings. Indirect cylinders usually conform to the recommendations of BS 1565 (galvanized mild steel) or BS1566 (copper).

Typical Examples ~

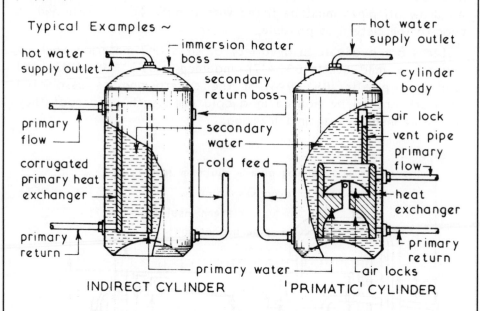

INDIRECT CYLINDER   'PRIMATIC' CYLINDER

Primatic Cylinders ~

1. Cylinder is filled in the normal way and the primary system is filled via the heat exchanger, as the initial filling continues air locks are formed in the upper and lower chambers of the heat exchanger and in the vent pipe.
2. The two air locks in the heat exchanger are permanently maintained and are self-recuperating in operation. These air locks isolate the primary water from the secondary water almost as effectively as a mechanical barrier.
3. The expansion volume of total primary water at a flow temperature of 82°C is approximately 1/25 and is accommodated in the upper expansion chamber by displacing air into the lower chamber, upon contraction reverse occurs.

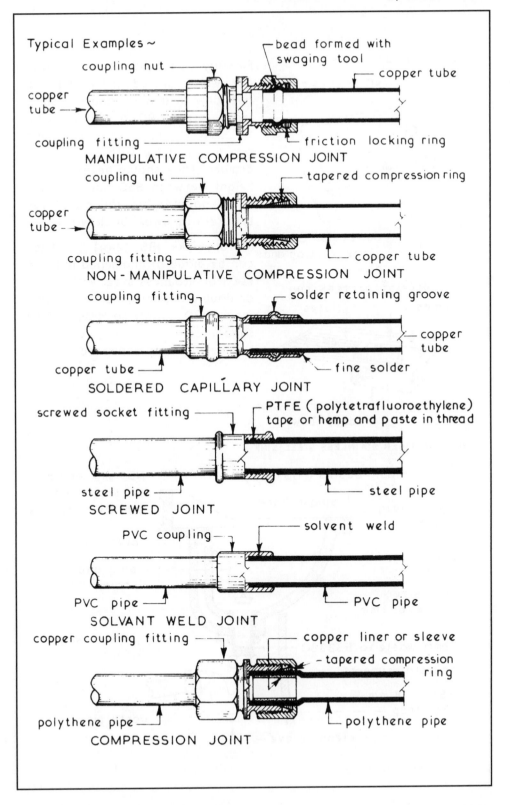

Typical Examples ~

bead formed with swaging tool

coupling nut

copper tube

copper tube

coupling fitting

friction locking ring

MANIPULATIVE COMPRESSION JOINT

coupling nut

tapered compression ring

copper tube

coupling fitting

copper tube

NON-MANIPULATIVE COMPRESSION JOINT

coupling fitting

solder retaining groove

copper tube

copper tube

fine solder

SOLDERED CAPILLARY JOINT

screwed socket fitting

PTFE (polytetrafluoroethylene) tape or hemp and paste in thread

steel pipe

steel pipe

SCREWED JOINT

PVC coupling

solvent weld

PVC pipe

PVC pipe

SOLVANT WELD JOINT

copper coupling fitting

copper liner or sleeve

tapered compression ring

polythene pipe

polythene pipe

COMPRESSION JOINT

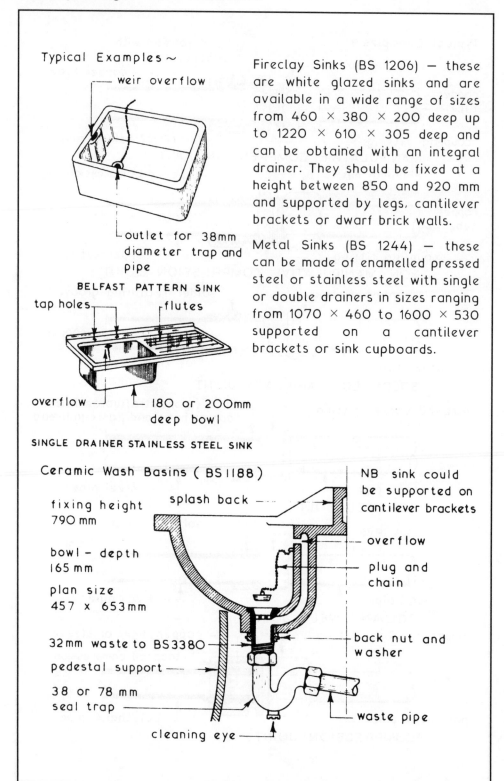

Typical Examples ~

weir overflow

outlet for 38mm diameter trap and pipe

BELFAST PATTERN SINK

tap holes

flutes

overflow

180 or 200mm deep bowl

SINGLE DRAINER STAINLESS STEEL SINK

Fireclay Sinks (BS 1206) — these are white glazed sinks and are available in a wide range of sizes from 460 × 380 × 200 deep up to 1220 × 610 × 305 deep and can be obtained with an integral drainer. They should be fixed at a height between 850 and 920 mm and supported by legs, cantilever brackets or dwarf brick walls.

Metal Sinks (BS 1244) — these can be made of enamelled pressed steel or stainless steel with single or double drainers in sizes ranging from 1070 × 460 to 1600 × 530 supported on a cantilever brackets or sink cupboards.

Ceramic Wash Basins (BS 1188)

fixing height 790 mm

splash back

bowl - depth 165 mm

plan size 457 x 653mm

32mm waste to BS 3380

pedestal support

38 or 78 mm seal trap

cleaning eye

NB sink could be supported on cantilever brackets

overflow

plug and chain

back nut and washer

waste pipe

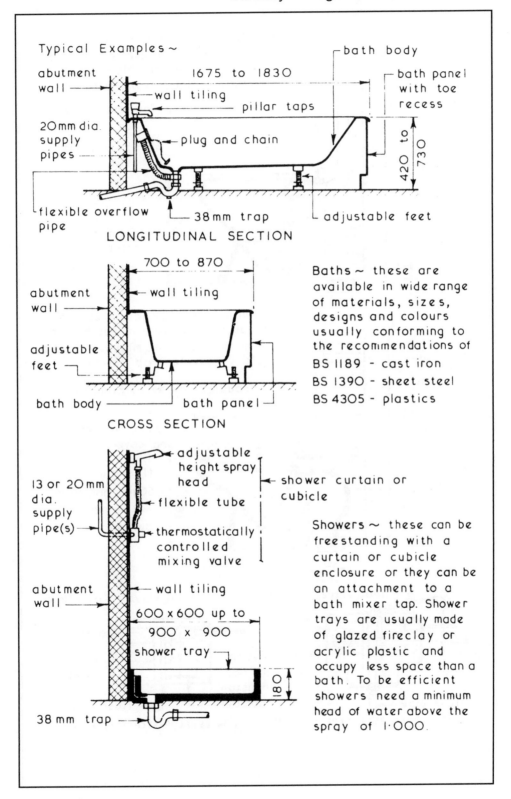

Typical Examples ~

abutment wall
1675 to 1830
bath body
wall tiling
bath panel with toe recess
pillar taps
20 mm dia. supply pipes
plug and chain
420 to 730
flexible overflow pipe
38 mm trap
adjustable feet

LONGITUDINAL SECTION

abutment wall
700 to 870
wall tiling
adjustable feet
bath body
bath panel

CROSS SECTION

Baths ~ these are available in wide range of materials, sizes, designs and colours usually conforming to the recommendations of
BS 1189 - cast iron
BS 1390 - sheet steel
BS 4305 - plastics

13 or 20 mm dia. supply pipe(s)
adjustable height spray head
shower curtain or cubicle
flexible tube
thermostatically controlled mixing valve
abutment wall
wall tiling
600 x 600 up to 900 x 900
shower tray
180
38 mm trap

Showers ~ these can be freestanding with a curtain or cubicle enclosure or they can be an attachment to a bath mixer tap. Shower trays are usually made of glazed fireclay or acrylic plastic and occupy less space than a bath. To be efficient showers need a minimum head of water above the spray of 1·000.

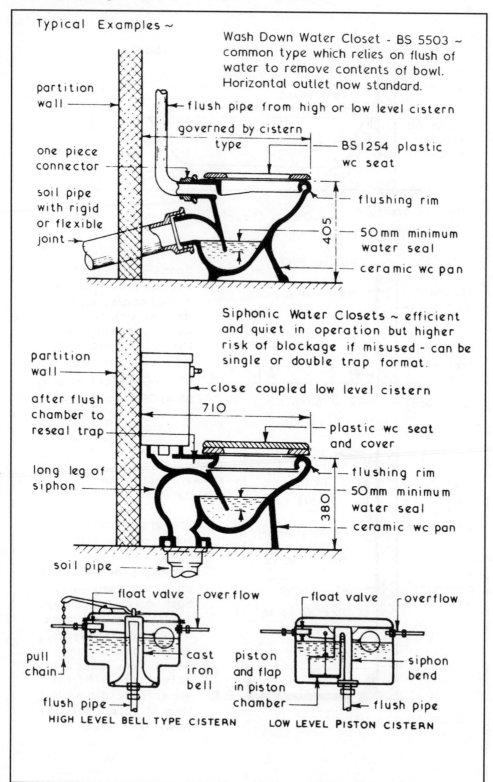

Typical Examples ~

Wash Down Water Closet - BS 5503 ~ common type which relies on flush of water to remove contents of bowl. Horizontal outlet now standard.

partition wall

one piece connector

soil pipe with rigid or flexible joint

flush pipe from high or low level cistern governed by cistern type

BS 1254 plastic wc seat

flushing rim

405

50 mm minimum water seal

ceramic wc pan

Siphonic Water Closets ~ efficient and quiet in operation but higher risk of blockage if misused - can be single or double trap format.

partition wall

after flush chamber to reseal trap

long leg of siphon

soil pipe

close coupled low level cistern

710

plastic wc seat and cover

flushing rim

50 mm minimum water seal

380

ceramic wc pan

float valve   overflow

pull chain

cast iron bell

flush pipe

HIGH LEVEL BELL TYPE CISTERN

float valve   overflow

piston and flap in piston chamber

siphon bend

flush pipe

LOW LEVEL PISTON CISTERN

638

Single Stack System ~ method developed by the Building Research Establishment to eliminate the need for ventilating pipework to maintain the water seals in traps to sanitary fittings. The slope and distance of the branch connections must be kept within the design limitations given below. This system is only possible when the sanitary appliances are closely grouped around the discharge stack.

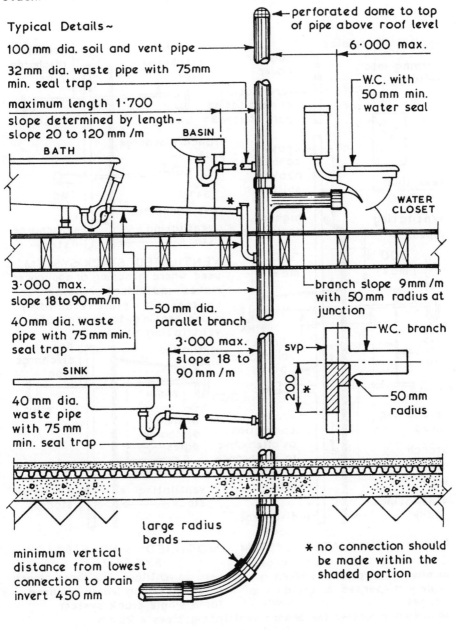

Typical Details ~

100 mm dia. soil and vent pipe

32mm dia. waste pipe with 75mm min. seal trap

maximum length 1·700
slope determined by length—
slope 20 to 120 mm /m

BATH

perforated dome to top of pipe above roof level

6·000 max.

W.C. with 50 mm min. water seal

BASIN

WATER CLOSET

branch slope 9mm /m with 50 mm radius at junction

3·000 max.
slope 18 to 90mm/m

50 mm dia. parallel branch

40mm dia. waste pipe with 75 mm min. seal trap

SINK

3·000 max.
slope 18 to 90 mm /m

40 mm dia. waste pipe with 75 mm min. seal trap

svp

200

W.C. branch

50 mm radius

large radius bends

minimum vertical distance from lowest connection to drain invert 450 mm

* no connection should be made within the shaded portion

639

Ventilated Stack Discharge Systems

Ventilated Stack Systems ~ where the layout of sanitary appliances is such that they do not conform to the requirements for the single stack system shown on page 639 ventilating pipes will be required to maintain the water seals in the traps. Three methods are available to overcome the problem, namely a fully ventilated system, a ventilated stack system and a modified single stack system which can be applied over any number of storeys.

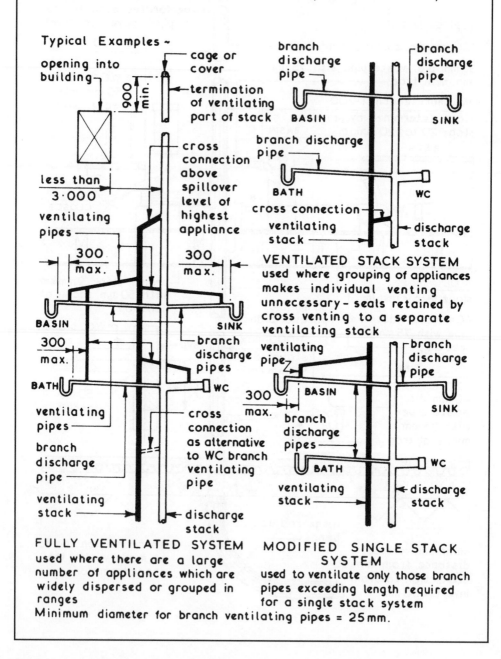

Typical Examples ~

VENTILATED STACK SYSTEM used where grouping of appliances makes individual venting unnecessary ~ seals retained by cross venting to a separate ventilating stack

FULLY VENTILATED SYSTEM used where there are a large number of appliances which are widely dispersed or grouped in ranges

MODIFIED SINGLE STACK SYSTEM used to ventilate only those branch pipes exceeding length required for a single stack system

Minimum diameter for branch ventilating pipes = 25 mm.

Airtightness ~ must be ensured to satisfy public health legislation. The Building Regulations, Approved Document H1:Section 1, provides minimum standards for test procedures. An air or smoke test on the stack must produce a pressure at least equal to 38 mm water gauge for not less than 3 minutes.

Application ~

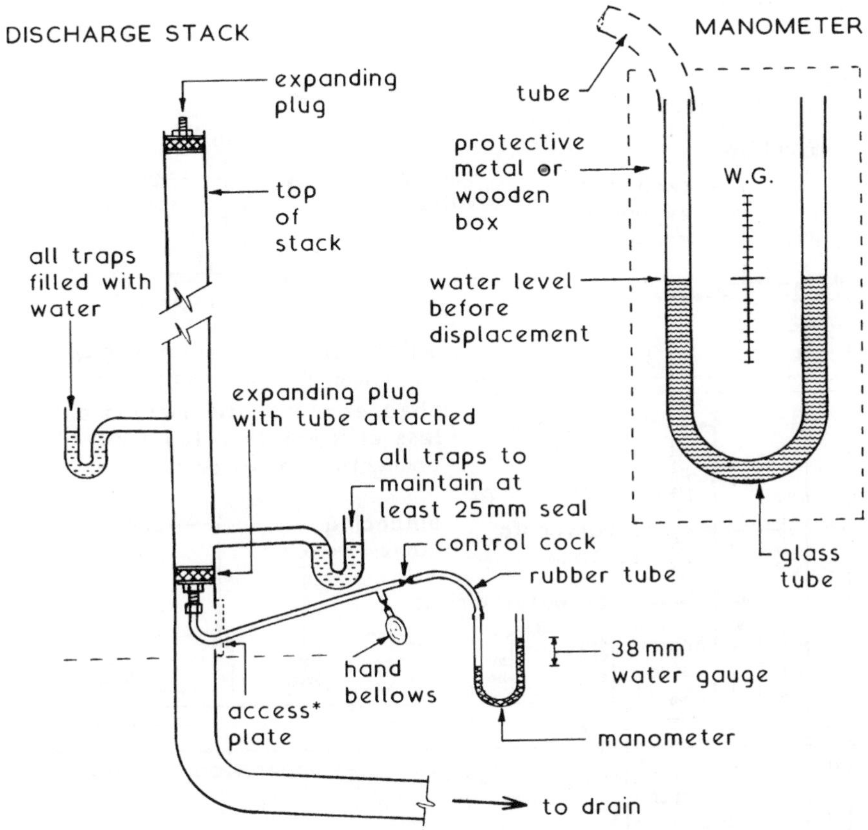

DISCHARGE STACK

MANOMETER

expanding plug

tube

top of stack

protective metal or wooden box

W.G.

all traps filled with water

water level before displacement

expanding plug with tube attached

all traps to maintain at least 25mm seal

control cock

rubber tube

glass tube

hand bellows

38 mm water gauge

access* plate

manometer

to drain

* if access plate is not provided, top connection to first IC may be plugged and rubber tube inserted through wc pan seal.

NB. Smoke tests are rarely applied now as the equipment is quite bulky and unsuited for use with uPVC pipes. Smoke producing pellets are ideal for leakage detection, but must not come into direct contact with plastic materials.

One Pipe System ~ the hot water is circulated around the system by means of a centrifugal pump. The flow pipe temperature being about 80°C and the return pipe temperature being about 60 to 70°C. The one pipe system is simple in concept and easy to install but has the main disadvantage that the hot water passing through each heat emitter flows onto the next heat emitter or radiator, therefore the average temperature of successive radiators is reduced unless the radiators are carefully balanced or the size of the radiators at the end of the circuit are increased to compensate for the temperature drop.

Typical Layout ~

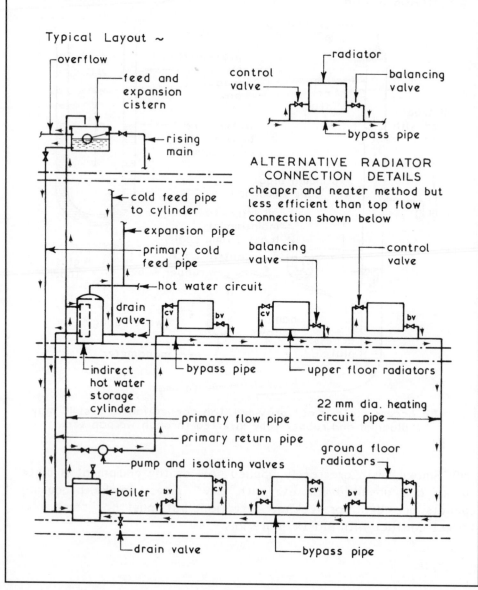

ALTERNATIVE RADIATOR CONNECTION DETAILS
cheaper and neater method but less efficient than top flow connection shown below

Two Pipe System ~ this is a dearer but much more efficient system than the one pipe system shown on the previous page. It is easier to balance since each radiator or heat emitter receives hot water at approximately the same temperature because the hot water leaving the radiator is returned to the boiler via the return pipe without passing through another radiator.

Typical Layout ~

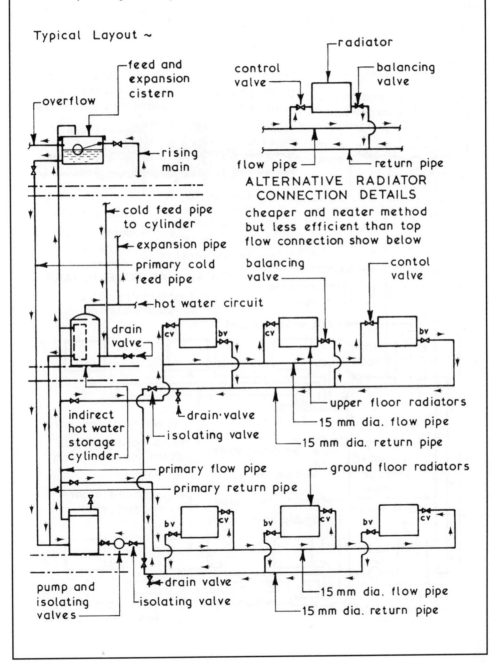

ALTERNATIVE RADIATOR CONNECTION DETAILS

cheaper and neater method but less efficient than top flow connection show below

Micro Bore System ~ this system uses 6 to 12mm diameter soft copper tubing with an individual flow and return pipe to each heat emitter or radiator from a 22mm diameter manifold. The flexible and unobstrusive pipework makes this system easy to install in awkward situations but it requires a more powerful pump than that used in the traditional small bore systems. The heat emitter or radiator valves can be as used for the one or two pipe small bore systems alternatively a double entry valve can be used.

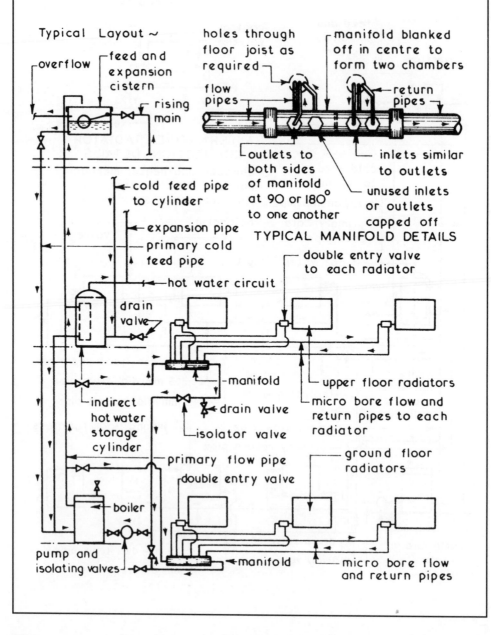

Typical Layout ~

overflow

feed and expansion cistern

rising main

holes through floor joist as required

flow pipes

manifold blanked off in centre to form two chambers

return pipes

outlets to both sides of manifold at 90 or 180° to one another

inlets similar to outlets

unused inlets or outlets capped off

TYPICAL MANIFOLD DETAILS

cold feed pipe to cylinder

expansion pipe

primary cold feed pipe

hot water circuit

double entry valve to each radiator

drain valve

manifold

upper floor radiators

micro bore flow and return pipes to each radiator

indirect hot water storage cylinder

drain valve

isolator valve

primary flow pipe

double entry valve

ground floor radiators

boiler

pump and isolating valves

manifold

micro bore flow and return pipes

Controls ~ the range of controls available to regulate the heat output and timing operations for a domestic hot water heating system is considerable, ranging from thermostatic radiator control valves to programmers and controllers.

Typical Example ~

Boiler — fitted with a thermostat to control the temperature of the hot water leaving the boiler.

Heat Emitters or Radiators — fitted with thermostatically controlled radiator valves to control flow of hot water to the radiators to keep room at desired temperature.

Programmer/Controller — this is basically a time switch which can usually be set for 24 hours, once daily or twice daily time periods and will generally give separate programme control for the hot water supply and central heating systems. The hot water cylinder and room thermostatic switches control the pump and motorised valve action.

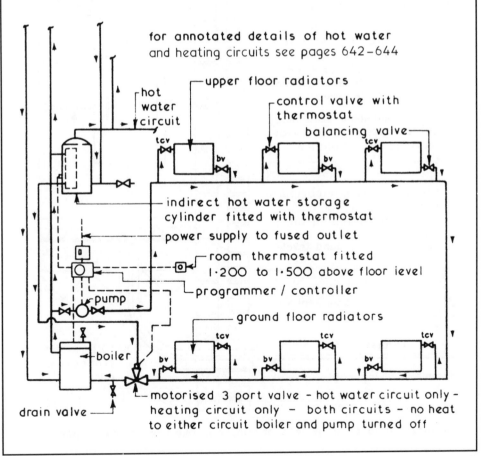

for annotated details of hot water and heating circuits see pages 642–644

upper floor radiators

hot water circuit

control valve with thermostat

balancing valve

indirect hot water storage cylinder fitted with thermostat

power supply to fused outlet

room thermostat fitted 1·200 to 1·500 above floor level

programmer / controller

pump

ground floor radiators

boiler

motorised 3 port valve - hot water circuit only - heating circuit only - both circuits - no heat to either circuit boiler and pump turned off

drain valve

Electrical Supply ~ in England and Wales electricity is generated and supplied by National Power, PowerGen and Nuclear Electric and distributed through regional supply companies, whereas in Scotland it is generated, supplied and distributed by Scottish Power and the Scottish Hydro-Electric Power Company. The electrical supply to a domestic installation is usually 230 volt single phase and is designed with the following basic aims :-

1. Proper earthing to avoid shocks to occupant.

2. Prevention of current leakage.

3. Prevention of outbreak of fire.

Typical Electrical Supply Intake Details ~

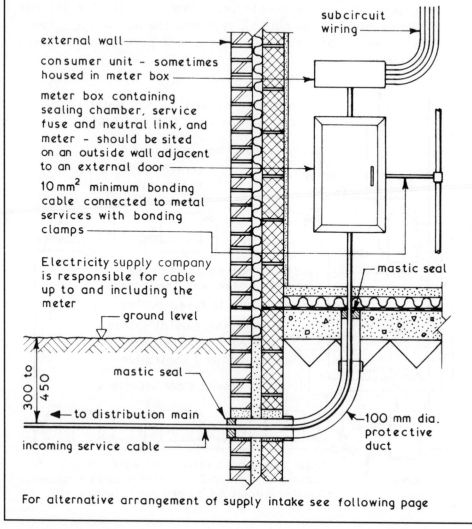

For alternative arrangement of supply intake see following page

Electrical Supply Intake ~ although the electrical supply intake can be terminated in a meter box situated within a dwelling most supply companies prefer to use the external meter box to enable the meter to be read without the need to enter the premises.

Typical Electrical Supply Intake Details ~

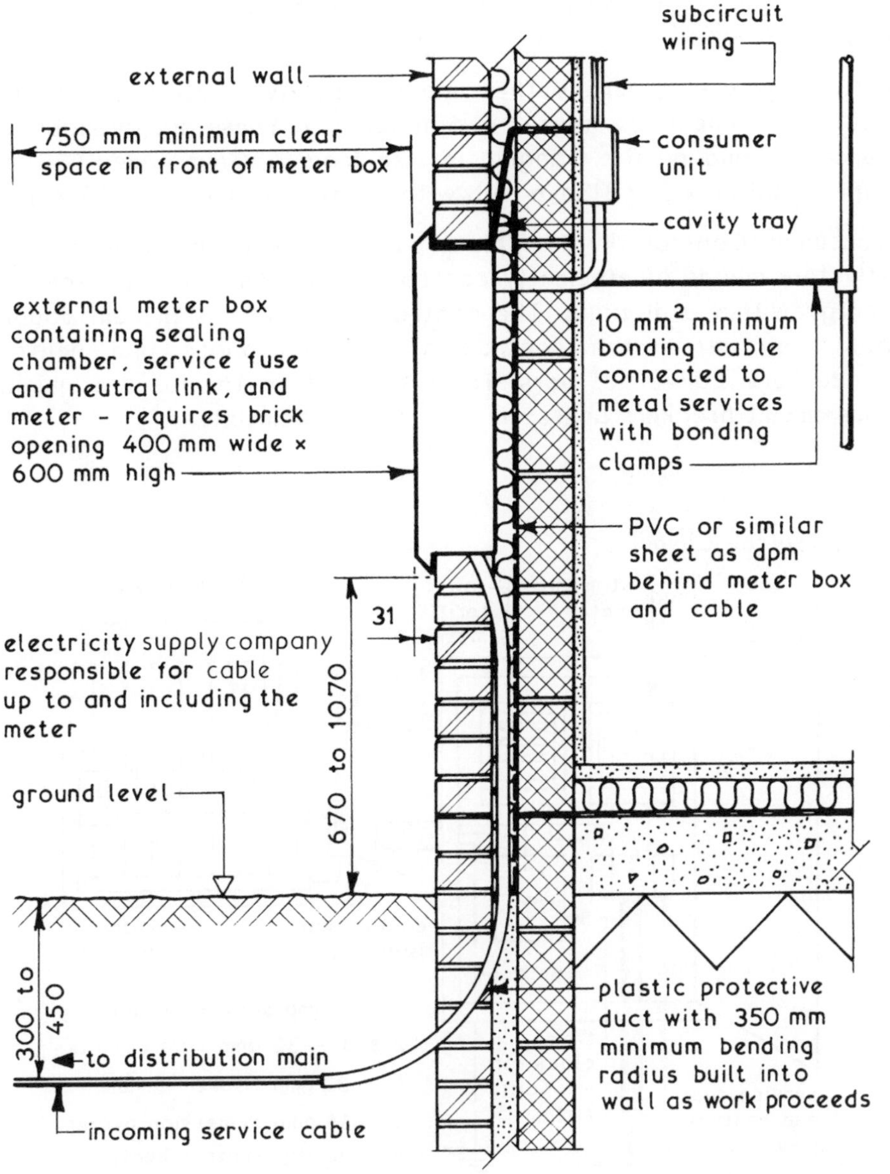

subcircuit wiring

external wall

750 mm minimum clear space in front of meter box

consumer unit

cavity tray

external meter box containing sealing chamber, service fuse and neutral link, and meter - requires brick opening 400 mm wide × 600 mm high

10 mm^2 minimum bonding cable connected to metal services with bonding clamps

electricity supply company responsible for cable up to and including the meter

PVC or similar sheet as dpm behind meter box and cable

31

670 to 1070

ground level

300 to 450

to distribution main

incoming service cable

plastic protective duct with 350 mm minimum bending radius built into wall as work proceeds

For alternative arrangement of supply intake see previous page

Electrical Supply—Basic Requirements

Entry and Intake of Electrical Service ~ the local electricity supply company is responsible for providing electricity up to and including the meter, but the consumer is responsible for safety and protection of the company's equipment. The supplier will install the service cable up to the meter position where their termination equipment is installed. This equipment may be located internally or fixed externally on a wall, the latter being preferred since it gives easy access for reading the meter — see details on the previous page.

Meter Boxes — generally the supply company's meters and termination equipment are housed in a meter box. These are available in fibreglass and plastic, ranging in size from 450 mm wide × 638 mm high to 585 m wide × 815 mm high with an overall depth of 177 mm.

Consumer Control Unit — this provides a uniform, compact and effective means of efficiently controlling and distributing electrical energy within a dwelling. The control unit contains a main double pole isolating switch controlling the live phase and neutral conductors, called bus bars. These connect to the fuses or miniature circuit breakers protecting the final subcircuits.

Typical Layout ~

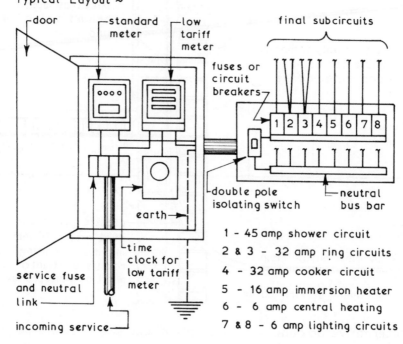

1 - 45 amp shower circuit
2 & 3 - 32 amp ring circuits
4 - 32 amp cooker circuit
5 - 16 amp immersion heater
6 - 6 amp central heating
7 & 8 - 6 amp lighting circuits

Electric Cables ~ these are made up of copper or aluminium wires called conductors surrounded by an insulating material such as PVC or rubber.

Typical Examples ~

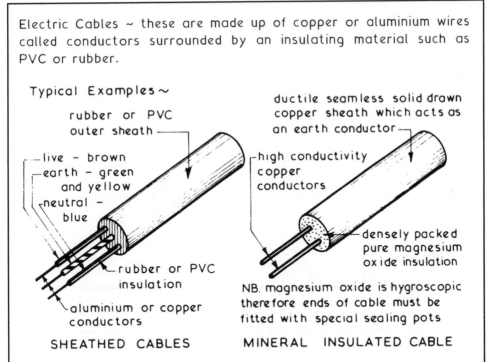

rubber or PVC outer sheath

live - brown
earth - green and yellow
neutral - blue

rubber or PVC insulation

aluminium or copper conductors

SHEATHED CABLES

ductile seamless solid drawn copper sheath which acts as an earth conductor

high conductivity copper conductors

densely packed pure magnesium oxide insulation

NB. magnesium oxide is hygroscopic therefore ends of cable must be fitted with special sealing pots

MINERAL INSULATED CABLE

Conduits ~ these are steel or plastic tubes which protect the cables. Steel conduits act as an earth conductor whereas plastic conduits will require a separate earth conductor drawn in. Conduits enable a system to be rewired without damage or interference of the fabric of the building. The cables used within conduits are usually insulated only, whereas in non-rewireable systems the cables have a protective outer sheath.

Typical Conduit Fittings ~

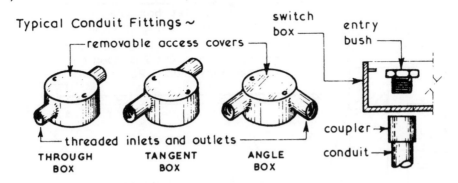

removable access covers

switch box

entry bush

threaded inlets and outlets

coupler
conduit

THROUGH BOX

TANGENT BOX

ANGLE BOX

Trunking — alternative to conduit and consists of a preformed cable carrier which is surface mounted and is fitted with a removable or 'snap on' cover which can have the dual function of protection and trim or surface finish.

Wiring systems ~ rewireable systems housed in horizontal conduits can be cast into the structural floor slab or sited within the depth of the floor screed. To ensure that such a system is rewireable, draw - in boxes must be incorporated at regular intervals and not more than two right angle boxes to be included between draw-in points. Vertical conduits can be surface mounted or housed in a chase cut in to a wall provided the depth of the chase is not more than one third of the wall thickness. A horizontal non-rewireable system can be housed within the depth of the timber joists to a suspended floor whereas vertical cables can be surface mounted or housed in a length of conduit as described for rewireable systems.

Typical Examples~

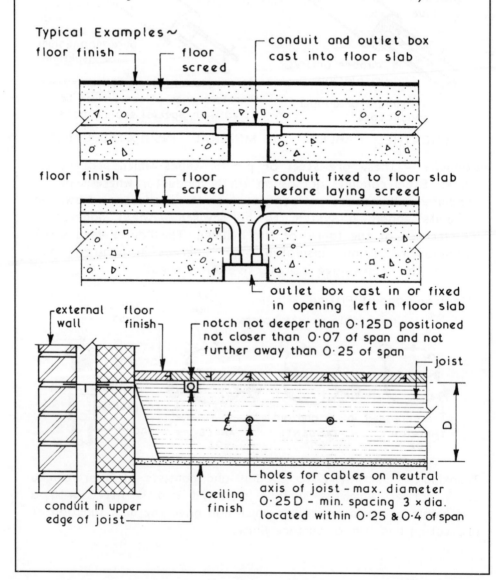

floor finish ── ── floor screed ── conduit and outlet box cast into floor slab

floor finish ── ── floor screed ── conduit fixed to floor slab before laying screed

outlet box cast in or fixed in opening left in floor slab

external wall ── floor finish ── notch not deeper than 0·125 D positioned not closer than 0·07 of span and not further away than 0·25 of span ── joist

conduit in upper edge of joist ── ceiling finish ── holes for cables on neutral axis of joist – max. diameter 0·25 D – min. spacing 3 × dia. located within 0·25 & 0·4 of span

Cable Sizing ~ the size of a conductor wire can be calculated taking into account the maximum current the conductor will have to carry (which is limited by the heating effect caused by the resistance to the flow of electricity through the conductor) and the voltage drop which will occur when the current is carried. For domestic electrical installations the following minimum cable specifications are usually suitable —

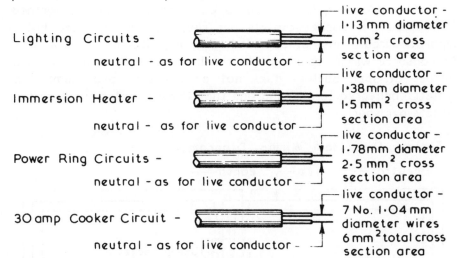

Lighting Circuits -
         neutral - as for live conductor —
live conductor - 1·13 mm diameter 1mm^2 cross section area

Immersion Heater -
         neutral - as for live conductor —
live conductor – 1·38mm diameter 1·5 mm^2 cross section area

Power Ring Circuits -
         neutral - as for live conductor —
live conductor – 1·78mm diameter 2·5 mm^2 cross section area

30 amp Cooker Circuit -
         neutral - as for live conductor -
live conductor - 7 No. 1·04 mm diameter wires 6 mm^2 total cross section area

All the above ratings are for one twin cable with or without an earth conductor.

Electrical Accessories ~ for power circuits these include cooker control units and fused connector units for fixed appliances such as immersion heaters, water heaters and refrigerators.

Socket Outlets ~ these may be single or double outlets, switched or unswitched, surface or flush mounted and may be fitted with indicator lights. Recommended fixing heights are —

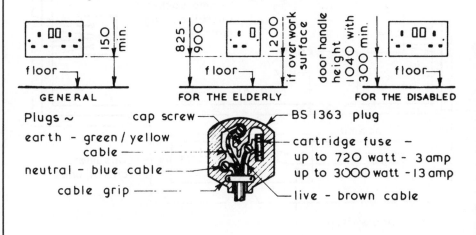

GENERAL   FOR THE ELDERLY   FOR THE DISABLED

Plugs ~   cap screw ——— BS 1363 plug

earth - green/yellow cable ———

neutral - blue cable ———

cable grip ——— · —

cartridge fuse — up to 720 watt - 3 amp up to 3000 watt - 13 amp

live - brown cable

651

Power Circuits ~ in new domestic electrical installations the ring main system is usually employed instead of the older system of having each socket outlet on its own individual fused circuit with unfused round pin plugs. Ring circuits consist of a fuse or miniature circuit breaker protected subcircuit with a 32 amp rating of a live conductor, neutral conductor and an earth looped from socket outlet to socket outlet. Metal conduit systems do not require an earth wire providing the conduit is electrically sound and earthed. The number of socket outlets per ring main is unlimited but a separate circuit must be provided for every 100 m^2 of floor area. To conserve wiring, spur outlets can be used as long as the total number of spur outlets does not exceed the total number of outlets connected to the ring and that there is not more than two outlets per spur.

Typical Ring Main Wiring Diagram ~

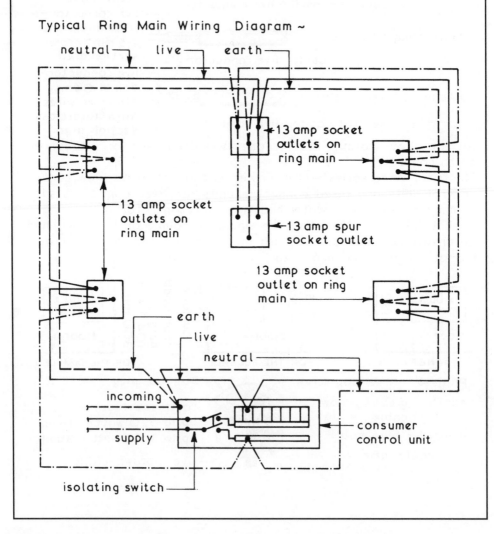

Lighting Circuits ~ these are usually wired by the loop-in method using an earthed twin cable with a 6 amp fuse or miniature circuit breaker protection. In calculating the rating of a lighting circuit an allowance of 100 watts per outlet should be used. More than one lighting circuit should be used for each installation so that in the event of a circuit failure some lighting will be in working order.

Typical Lighting Circuit Wiring Diagram ~

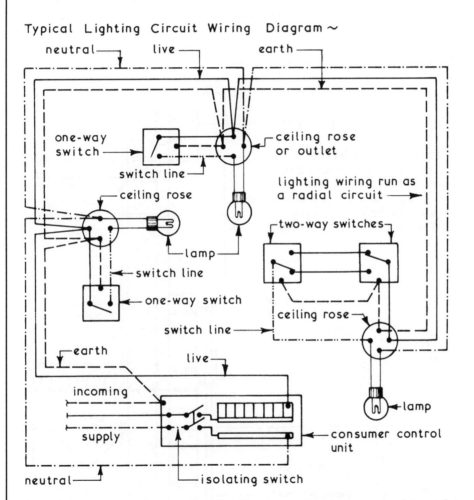

Electrical Accessories ~ for lighting circuits these consist mainly of switches and lampholders, the latter can be wall mounted, ceiling mounted or pendant in format with one or more bulb or tube holders. Switches are usually rated at 5 amps and are available in a variety of types such as double or 2 gang, dimmer and pull or pendant switches. The latter must always be used in bathrooms.

Gas Supply ~ potential consumers of mains gas may apply to the regional office of Transco (Lattice Group plc) for a connection. The cost is normally based on a fee per metre run. However, where the distance is considerable, the gas authority may absorb some of the cost if there is potential for more customers. The supply, appliances and installation must comply with the safety requirements made under the Gas Safety (Installation and Use) Regulations, 1994, and Part J of the Building Regulations.

Typical Gas Supply Arrangement ~

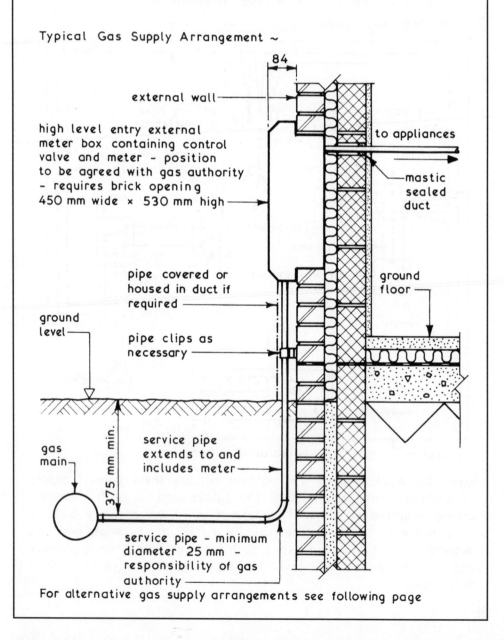

high level entry external meter box containing control valve and meter - position to be agreed with gas authority - requires brick opening 450 mm wide × 530 mm high

external wall

to appliances

mastic sealed duct

pipe covered or housed in duct if required

ground floor

ground level

pipe clips as necessary

gas main

service pipe extends to and includes meter

375 mm min.

service pipe - minimum diameter 25 mm - responsibility of gas authority

For alternative gas supply arrangements see following page

Gas Service Pipes ~

1. Whenever possible the service pipe should enter the building on the side nearest to the main.
2. A service pipe must not pass under the foundations of a building.
3. No service pipe must be run within a cavity but it may pass through a cavity by the shortest route.
4. Service pipes passing through a wall or solid floor must be enclosed by a sleeve or duct which is end sealed with mastic.
5. No service pipe shall be housed in an unventilated void.
6. Suitable materials for service pipes are copper (BS EN 1057) and steel (BS 1387 and BS 3601)

Typical Gas Supply Arrangement ~

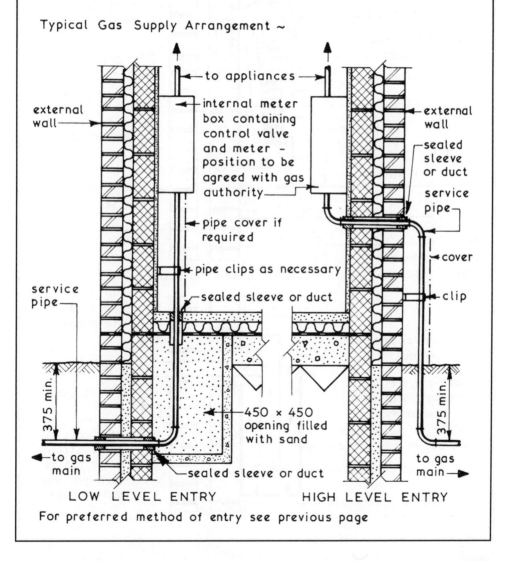

For preferred method of entry see previous page

Gas Fires ~ for domestic use these are classified as a gas burning appliance with a rated input of up to 60 kW and must be installed in accordance with minimum requirements set out in Part J of the Building Regulations. Most gas fires connected to a flue are designed to provide radiant and convected heating whereas the room sealed balanced flue appliances are primarily convector heaters.

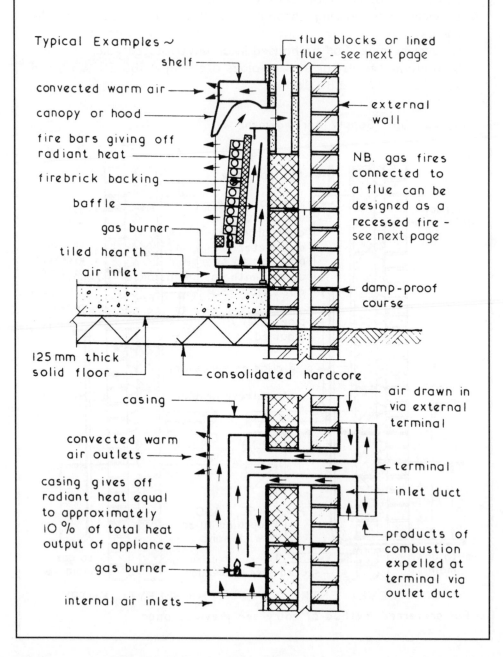

Typical Examples ~

shelf

flue blocks or lined flue - see next page

convected warm air

external wall

canopy or hood

fire bars giving off radiant heat

NB. gas fires connected to a flue can be designed as a recessed fire - see next page

firebrick backing

baffle

gas burner

tiled hearth

air inlet

damp-proof course

125 mm thick solid floor

consolidated hardcore

casing

air drawn in via external terminal

convected warm air outlets

terminal

casing gives off radiant heat equal to approximately 10 % of total heat output of appliance

inlet duct

products of combustion expelled at terminal via outlet duct

gas burner

internal air inlets

Gas Fire Flues ~ these can be defined as a passage for the discharge of the products of combustion to the outside air and can be formed by means of a chimney, special flue blocks or by using a flue pipe. In all cases the type and size of the flue as recommended in Approved Document J, BS 1289 and BS 5440 will meet the requirements of the Building Regulations.

Typical Single Gas Fire Flues ~

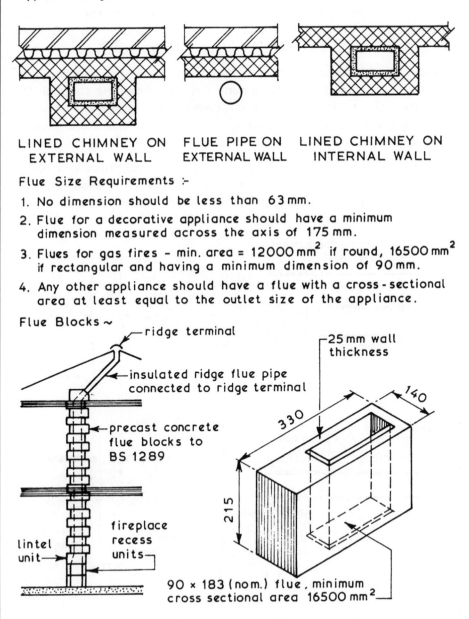

LINED CHIMNEY ON   FLUE PIPE ON   LINED CHIMNEY ON
EXTERNAL WALL   EXTERNAL WALL   INTERNAL WALL

Flue Size Requirements :-

1. No dimension should be less than 63 mm.

2. Flue for a decorative appliance should have a minimum dimension measured across the axis of 175 mm.

3. Flues for gas fires - min. area = $12000\,mm^2$ if round, $16500\,mm^2$ if rectangular and having a minimum dimension of 90 mm.

4. Any other appliance should have a flue with a cross-sectional area at least equal to the outlet size of the appliance.

Flue Blocks ~

— ridge terminal

— insulated ridge flue pipe connected to ridge terminal

— precast concrete flue blocks to BS 1289

fireplace recess units

lintel unit

—25 mm wall thickness

330

140

215

90 × 183 (nom.) flue, minimum cross sectional area $16500\,mm^2$

Fire Protection of Services Openings ~ penetration of compartment walls and floors (zones of restricted fire spread, eg. flats in one building), by service pipes and conduits is very difficult to avoid. An exception is where purpose built service ducts can be accommodated. The Building Regulations, Approved Document B3:Section 10 determines that where a pipe passes through a compartment interface, it must be provided with a proprietary seal. Seals are collars of intumescent material which expands rapidly when subjected to heat, to form a carbonaceous charring. The expansion is sufficient to compress warm plastic and successfully close a pipe void for up to 4 hours.

In some circumstances fire stopping around the pipe will be acceptable, provided the gap around the pipe and hole through the structure are filled with non-combustible material. Various materials are acceptable, including reinforced mineral fibre, cement and plasters, asbestos rope and intumescent, mastics.

Pipes of low heat resistance, such as PVC, lead, aluminium alloys and fibre cement may have a protective sleeve of non-combustible material extending at least 1 m either side of the structure.

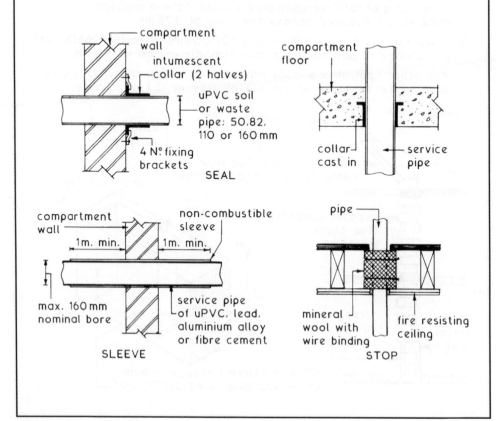

658

Open Fireplaces ~ for domestic purposes these are a means of providing a heat source by consuming solid fuels with an output rating of under 45 kW. Room-heaters can be defined in a similar manner but these are an enclosed appliance as opposed to the open recessed fireplace.

Components ~ the complete construction required for a domestic open fireplace installation is composed of the hearth, fireplace recess, chimney, flue and terminal.

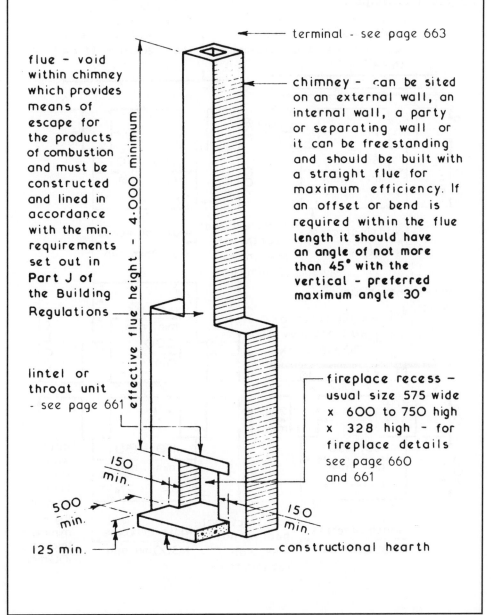

terminal - see page 663

flue - void within chimney which provides means of escape for the products of combustion and must be constructed and lined in accordance with the min. requirements set out in Part J of the Building Regulations

effective flue height - 4.000 minimum

chimney - can be sited on an external wall, an internal wall, a party or separating wall or it can be freestanding and should be built with a straight flue for maximum efficiency. If an offset or bend is required within the flue length it should have an angle of not more than 45° with the vertical - preferred maximum angle 30°

lintel or throat unit - see page 661

fireplace recess - usual size 575 wide x 600 to 750 high x 328 high - for fireplace details see page 660 and 661

150 min.

500 min.

150 min.

125 min.

constructional hearth

Open Fireplace Recesses ~ these must have a constructional hearth and can be constructed of bricks or blocks of concrete or burnt clay or they can be of cast insitu concrete. All fireplace recesses must have jambs on both sides of the opening and a backing wall of a minimum thickness in accordance with its position and such jambs and backing walls must extend to the full height of the fireplace recess.

Typical Examples ~

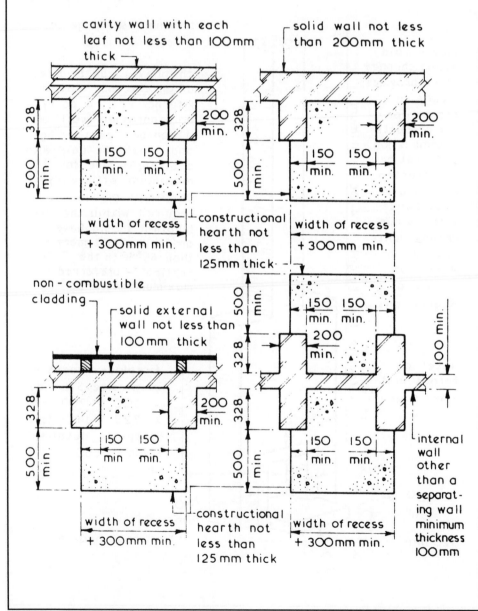

Traditional Fireplace Details ~

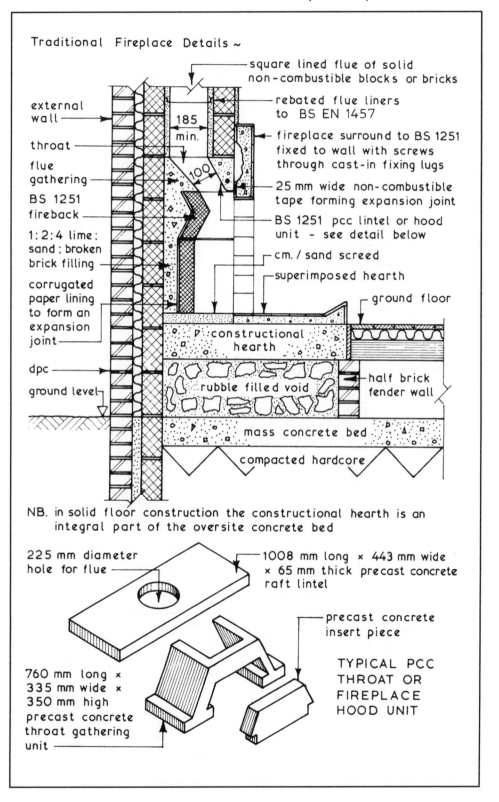

square lined flue of solid non-combustible blocks or bricks

rebated flue liners to BS EN 1457

external wall

throat

flue gathering

BS 1251 fireback

1:2:4 lime: sand: broken brick filling

corrugated paper lining to form an expansion joint

dpc

ground level

185 min.

100

fireplace surround to BS 1251 fixed to wall with screws through cast-in fixing lugs

25 mm wide non-combustible tape forming expansion joint

BS 1251 pcc lintel or hood unit - see detail below

cm./sand screed

superimposed hearth

ground floor

constructional hearth

rubble filled void

half brick fender wall

mass concrete bed

compacted hardcore

NB. in solid floor construction the constructional hearth is an integral part of the oversite concrete bed

225 mm diameter hole for flue

1008 mm long × 443 mm wide × 65 mm thick precast concrete raft lintel

precast concrete insert piece

760 mm long × 335 mm wide × 350 mm high precast concrete throat gathering unit

TYPICAL PCC THROAT OR FIREPLACE HOOD UNIT

Open Fireplace Chimneys and Flues ~ the main functions of a chimney and flue are to :-

1. Induce an adequate supply of air for the combustion of the fuel being used.
2. Remove the products of combustion.

In fulfilling the above functions a chimney will also encourage a flow of ventilating air promoting constant air changes within the room which will assist in the prevention of condensation.

Approved Document J recommends that all flues should be lined with approved materials so that the minimum size of the flue so formed will be 200mm diameter or a square section of equivalent area. Flues should also be terminated above the roof level as set out in the Approved Document.

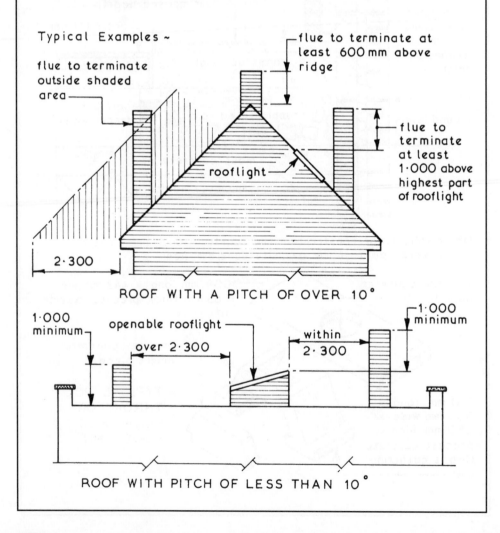

Typical Examples ~

flue to terminate outside shaded area

flue to terminate at least 600 mm above ridge

rooflight

flue to terminate at least 1·000 above highest part of rooflight

2·300

ROOF WITH A PITCH OF OVER 10°

1·000 minimum

openable rooflight

over 2·300

within 2·300

1·000 minimum

ROOF WITH PITCH OF LESS THAN 10°

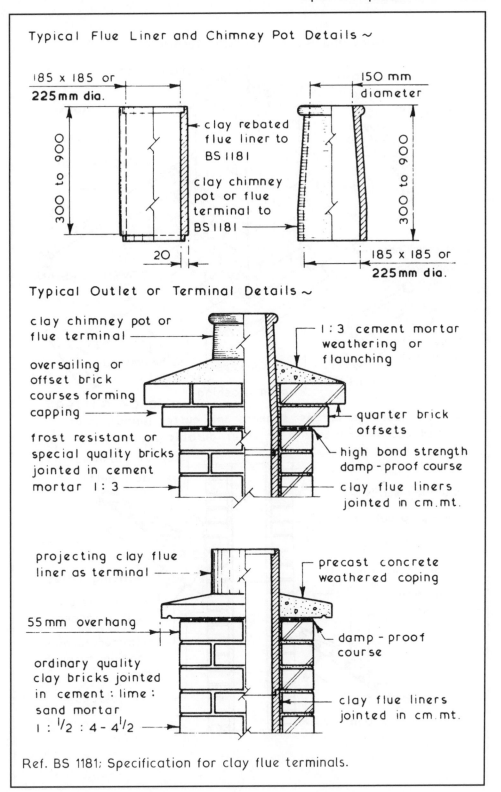

Typical Flue Liner and Chimney Pot Details ~

185 x 185 or **225mm dia.**

300 to 900

← clay rebated flue liner to BS 1181

clay chimney pot or flue terminal to BS 1181 →

20

150 mm diameter

300 to 900

185 x 185 or **225mm dia.**

Typical Outlet or Terminal Details ~

clay chimney pot or flue terminal →

oversailing or offset brick courses forming capping →

frost resistant or special quality bricks jointed in cement mortar 1:3 →

1:3 cement mortar weathering or flaunching

quarter brick offsets

high bond strength damp-proof course

clay flue liners jointed in cm.mt.

projecting clay flue liner as terminal →

55mm overhang

ordinary quality clay bricks jointed in cement : lime : sand mortar 1 : $\frac{1}{2}$: 4 - 4$\frac{1}{2}$ →

precast concrete weathered coping

damp-proof course

clay flue liners jointed in cm.mt.

Ref. BS 1181; Specification for clay flue terminals.

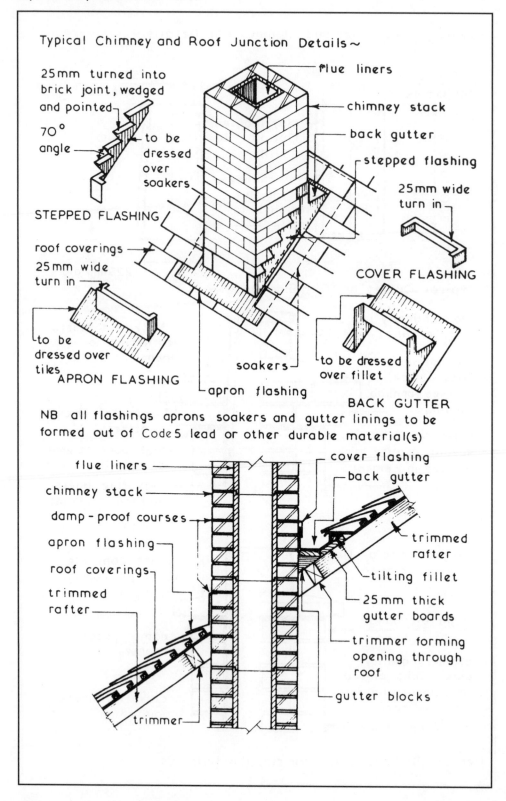

Typical Chimney and Roof Junction Details~

flue liners

25mm turned into brick joint, wedged and pointed

chimney stack

70° angle

back gutter

to be dressed over soakers

stepped flashing

STEPPED FLASHING

25mm wide turn in

COVER FLASHING

roof coverings 25mm wide turn in

to be dressed over tiles

APRON FLASHING

soakers

apron flashing

to be dressed over fillet

BACK GUTTER

NB all flashings aprons soakers and gutter linings to be formed out of Code 5 lead or other durable material(s)

flue liners

chimney stack

damp-proof courses

apron flashing

roof coverings

trimmed rafter

trimmer

cover flashing

back gutter

trimmed rafter

tilting fillet

25mm thick gutter boards

trimmer forming opening through roof

gutter blocks

Combustion Air ~ it is a Building Regulation requirement that in the case of open fireplaces provision must be made for the introduction of combustion air in sufficient quantity to ensure the efficient operation of the open fire. Traditionally such air is taken from the volume of the room in which the open fire is situated, this can create air movements resulting in draughts. An alternative method is to construct an ash pit below the hearth level fret and introduce the air necessary for combustion via the ash by means of a duct.

Typical Details ~

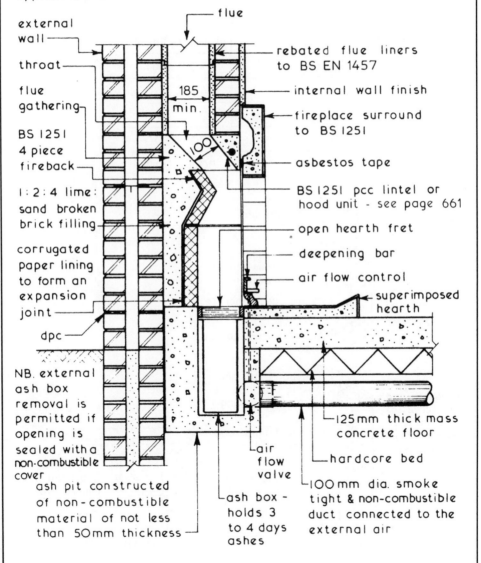

flue

external wall

throat

flue gathering

BS 1251 4 piece fireback

1:2:4 lime: sand broken brick filling

corrugated paper lining to form an expansion joint

dpc

185 min.

100

rebated flue liners to BS EN 1457

internal wall finish

fireplace surround to BS 1251

asbestos tape

BS 1251 pcc lintel or hood unit - see page 661

open hearth fret

deepening bar

air flow control

superimposed hearth

NB. external ash box removal is permitted if opening is sealed with a non-combustible cover

ash pit constructed of non-combustible material of not less than 50mm thickness

air flow valve

ash box - holds 3 to 4 days ashes

125mm thick mass concrete floor

hardcore bed

100 mm dia. smoke tight & non-combustible duct connected to the external air

Lightweight Pumice Chimney Blocks ~ these are suitable as a flue system for solid fuels, gas and oil. The highly insulative properties provide low condensation risk, easy installation as a supplement to existing or on-going construction and suitability for use with timber frame and thatched dwellings, where fire safety is of paramount importance. Also, the natural resistance of pumice to acid and sulphurous smoke corrosion requires no further treatment or special lining. A range of manufacturer's accessories allow for internal use with lintel support over an open fire or stove, or as an external structure supported on its own foundation. Whether internal or external, the units are not bonded in, but supported on purpose made ties at a maximum of 2 metre intervals.

flue (mm)	plan size (mm)
150 dia.	390 × 390
200 dia.	440 × 440
230 dia.	470 × 470
260 square	500 × 500
260 × 150 oblong	500 × 390

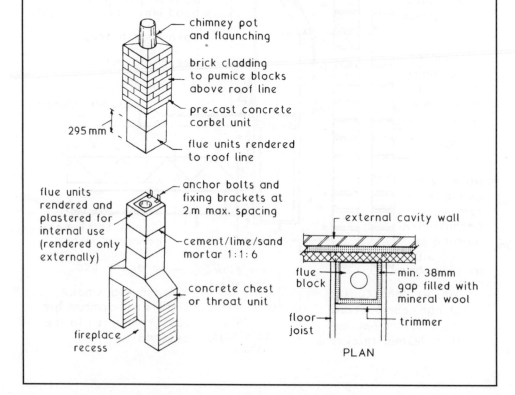

295 mm

chimney pot and flaunching

brick cladding to pumice blocks above roof line

pre-cast concrete corbel unit

flue units rendered to roof line

flue units rendered and plastered for internal use (rendered only externally)

anchor bolts and fixing brackets at 2 m max. spacing

cement/lime/sand mortar 1:1:6

concrete chest or throat unit

fireplace recess

external cavity wall

flue block

min. 38mm gap filled with mineral wool

floor joist

trimmer

PLAN

Telephone Installations ~ unlike other services such as water, gas and electricity, telephones cannot be connected to a common mains supply. Each telephone requires a pair of wires connecting it to the telephone exchange. The external supply service and connection to the lead-in socket is carried out by telecommunication engineers. Internal extensions can be installed by the site electrician.

Typical Supply Arrangements ~

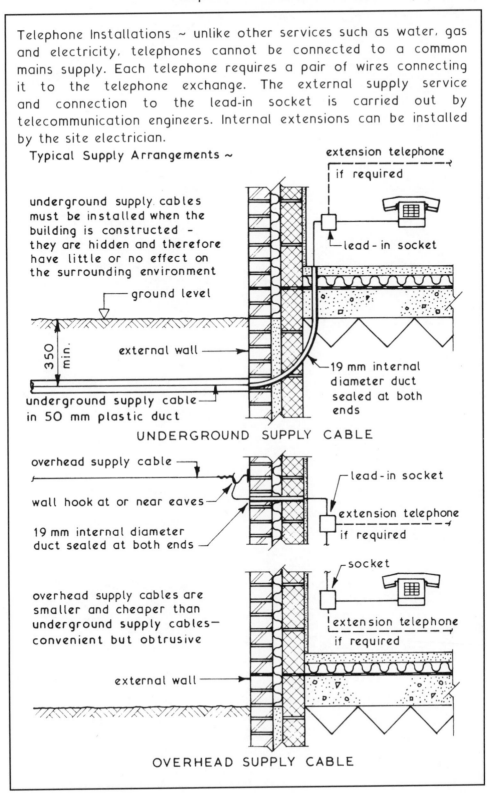

underground supply cables must be installed when the building is constructed - they are hidden and therefore have little or no effect on the surrounding environment

extension telephone
if required

lead-in socket

ground level

350 min.

external wall

19 mm internal diameter duct sealed at both ends

underground supply cable in 50 mm plastic duct

UNDERGROUND SUPPLY CABLE

overhead supply cable

wall hook at or near eaves

19 mm internal diameter duct sealed at both ends

lead-in socket

extension telephone
if required

overhead supply cables are smaller and cheaper than underground supply cables— convenient but obtrusive

socket

extension telephone
if required

external wall

OVERHEAD SUPPLY CABLE

INDEX

Access for disabled, 505–7
Access to sites, 10
Accommodation on sites, 76
Active earth pressures, 218
Adhesives, 395
Admixtures, 235
Aerated concrete floor units, 551
Aggregate samples, 95
Air lock hopper, 199
Air test, 623, 641
Alternating tread stair, 562
Aluminium alloy infill panels, 466
Anchorages, 373
Anchor bars, 351
Anchor straps, 433
Angledozers, 142
Angles, 376
Angles of repose, 244
Angle piling, 263
Apex hinges, 390
Approved Documents, 46–8
Approved inspector, 49
Apron cladding panels, 474
Apron flashing, 664
Arches, 308–10
Asphalt tanking, 237–8
Attached piers, 286–7, 289
Augers, 65
Automatic level, 106
Average U value, 490, 498
Axial grid, 41
Axonometric projection, 24

Backacter, 141, 148, 231
Backgutter, 664
Backshore, 130
Balloon frame, 345
Balustrade, 577
Bargeboard, 439
Barrel vaults, 450–2
Basement excavation, 228–31
Basement insulation, 240
Basements, 228–40
Basic forms of structure, 8, 11

Basic formwork, 360–1
Basic module grid, 40
Basic roof forms, 400–2
Basic types of structure, 8, 11
Basins, 636
Batch mixers, 170–1
Baths, 637
Bay window, 324
Beam and pot floor, 552
Beam and slab raft, 195
Beam formwork, 361–2
Beam to column connection, 368, 380
Beam design:
    concrete, 352–3
    steel, 384–6
    timber, 545–6
Beams, 351, 377–81
Bearing piles, 197
Beech, 599
Benchmark, 59, 102, 106–7
Bending moment, 384, 388, 545
Bentonite, 200, 272–3
Bib tap, 632
Binder, 591
Binders, 351
Binding, 130
Birdcage scaffold, 123
Blockboard, 600
Block plan, 27, 35
Block walls, 515–7
Bloom base, 194
Boarded web girder, 396
Bolection moulding, 597
Bolt box, 366, 380
Bonding bricks, 285–8
Bonnet tiles, 410, 414
Boot lintels, 311
Bore hole, 60, 67
Borrowed light, 521
Bottom shore, 130
Box beam, 396
Box caisson, 255
Box pile foundation, 207
Braced structures, 398

Index

Breather membrane, 409, 412–3, 461
Brick:
  infill panels, 465
  internal walls, 513
  panel walls, 462
  retaining walls, 216, 219
  strength, 306
  testing, 93
  underpinning, 259
Brick bonding:
  attached piers, 289
  English bond, 286
  Flemish bond, 287
  principles, 285
  special bonds, 288
Brickwork cladding support, 463
Bridging of dpc's, 301
British Standards, 52
Buckling factor, 357
Builders plant, 139–74
Building control, 49
Building Regulations, 46–9
Building Regulations application, 51
Building Regulations exemptions, 50
Built environment, 3–5
Built-up roofing felt, 429
Bulk density, 73
Bulking of sand, 94
Bulldozer, 142
Bus bar, 648

Cab crane control, 165
Cable sizing, 651
Caissons, 255–6
Calculated brickwork, 306–7
Calculation of storage space, 89
Camber arch, 309
Cantilever:
  beam, 351
  foundations, 196
  retaining wall, 221
  scaffold, 125
  stairs, 566
  structures, 397
Cap tendons, 372
Carbon emissions, 494
Carbon index, 487, 492, 494
Cartridge fuse, 651
Cased pile foundation, 208
Casein adhesive, 395
Casement windows:

  ironmongery, 319
  metal, 318
  timber, 316–7
Castellated beam, 378
Cast insitu diaphragm walls, 233, 272
Cast insitu pile foundation, 211
Cast-in wall ties, 465
Cast-on finishes, 477
Cavity barrier, 302, 346, 412
Cavity closer, 302, 411–2
Cavity tray, 310
Cavity walls, 291–2, 305
Cedar, 598
Ceilings:
  plasterboard, 528, 586
  suspended, 587–90
Cellular basement, 232
Cement grouts, 274
Central heating, 642–5
Centres, 309
Certificates of ownership, 35, 39
Channel floor units, 552
Channels, 376–7
Chemical dpc, 299–300
Chemical grouts, 274
Chezy formula, 624
Chimney pot, 22, 663
Chimneys, 659, 662, 664
Chipboard, 536, 601
Circular bay window, 324
Circular column, 355
Cisterns, 633, 638
CI/SfB system, 55
Cladding panels, 464
Clamp vibrator, 174
Classification of piled foundations, 197
Clay cutter, 199
Clear glass, 326
Client, 43
Climbing cranes, 164, 168
Climbing formwork, 223
Closed couple roof, 404
Codes of Practice, 52
Cofferdams, 253–4
Coil wall ties, 222
Cold bridging, 302, 497–8
Cold deck, 428
Cold water installations, 627–8
Collar roof, 404
Column design, 356–7, 387
Column formwork, 363–4

Column to column
    connection, 367, 381
Column underpinning, 264
Columns, 355, 376–7, 380–1
Combined column foundation, 194
Combined method, 484
Combustion air, 665
Communicating information:
  elevations, 26
  floor plans, 26
  isometric projection, 24
  orthographic projection, 23
  perspective projection, 25
  sketches, 22
Complete excavation, 230
Composite beams, 396
Composite boarding, 600–1
Composite floors, 350, 552
Composite lintels, 311
Composite panel, 466
Composite piled foundation, 208–9
Composite timber beams, 396
Compound sections, 377
Compressed strawboard, 600
Compressible joints, 404, 462,
    517, 571
Concrete:
  admixtures, 235
  beam design, 352–3
  claddings, 464, 473–5
  column design, 356–7
  floor screed, 539–40
  mixers, 170–1
  placing, 174
  production, 247–52
  pumps, 173
  slab design, 352–3
  stairs, 565–73
  surface finishes, 476
  test cubes, 96
  testing, 97–9
Concrete production:
  designated mix, 251
  designed mix, 251
  materials, 247
  prescribed mix, 251
  site storage, 248
  specification, 251
  standard mix, 251
  supply, 252
  volume batching, 249

weight batching, 250
Concreting, 169
Concreting plant, 169–74
Conductivity, 481–2
Conduit and fittings, 649–50
Conoids, 452
Consolidation of soil, 74
Construction activities, 19
Construction, Design and
    Management Regs., 43
Construction joints, 111, 236, 517
Construction Regulations, 42–3
Consumer control unit, 648
Contiguous piling, 271
Continuous column
    foundations, 193
Contraction joints, 111
Controlling dimensions, 40
Controlling grid, 40
Controlling lines, 40
Cooker circuit cable, 650
Coping stones, 215–6, 293
Core drilling, 66
Core structures, 398
Cored floor units, 552
Cor-ply beam, 396
Cornice, 597
Corrugated sheet, 438–41
Coulomb's line, 73
Coulomb's wedge theory, 227
Counter batten, 413
Couple roof, 404
Cove mouldings, 527, 586
Cover flashing, 664
CPI System of Coding, 54
Cradles, 124
Crane operation, 165
Crane rail track, 167
Cranes, 157–68
Crane skips, 169
Cranked slab stairs, 565, 571
Crawler crane, 161
Creep, 370
Crib retaining wall, 225
Crosswall construction, 343–4
Crown hinges, 390
Curtain walling, 470–2
Curved laminated timber, 394
Curved tendons, 372
Cut and fill, 241
Cylinders, 634

Dado panel, 597
Dado rail, 597
Damp proof course, 296–302
Damp proof course materials, 297–8
Damp proof membrane, 296, 303
Datum, 102–5
Datum post, 102, 104–5
Dead loads, 29–30, 134, 187
Dead shoring, 128–9, 134–5
Decorative suspended ceiling, 590
Deep basements, 233–4
Deep strip foundation, 193
Defects in painting, 594
Deflection, 386, 396, 546
Demountable partitions, 521–2
Dense monolithic concrete, 235
Designer, 43
Density of materials, 30, 481–2
Design of foundations, 187, 192
Dewatering, 265–9
Diaphragm floatvalve, 632
Diaphragm walls, 233, 272–3, 295
Diesel hammers, 213
Dimensional coordination, 40
Dimensional grids, 40
Dipper arm, 231
Direct cold water system, 627
Direct hot water system, 629
Displacement pile foundation, 197,
    203–14
Disturbed soil sample, 61, 64
Documents for construction, 20
Dog leg stair, 563–4
Domelights, 460
Domes, 448–9
Domestic drainage, 616–9
Domestic electrical installations,
    646–53
Domestic floor finishes, 535–6
Domestic ground floors, 531–4
Domestic heating systems, 642–5
Doors:
    external, 334–6
    fire, 581–3
    frames, 337, 580
    glazed double swing, 585
    industrial, 339–42
    internal, 579
    ironmongery, 338
    linings, 577
    performance, 334

    sliding, 339, 341
    types, 335–6
Door set, 581
Dormer window, 402, 427
Double acting hammers, 213
Double action floor spring, 585
Double flying shore, 132
Double glazing, 317, 323, 328
Double hung sash windows, 320–1
Double lap tiling, 410–1, 414–5
Double layer grids, 448
Double rafter roof, 405
Double swing doors, 585
Douglas fir, 598
Dovetail anchor slots, 462, 465
Draft for Development, 52
Draglines, 149
Drainage:
    effluent, 605
    gradient, 624
    paved areas, 610
    pipe sizes, 624
    proportional depth, 624
    roads, 611–2
    simple, 616–22
    systems, 616–9
    testing, 623
Drained cavities, 239
Draught proofing, 499
Drawings:
    axonometric projection, 24
    construction process, 21
    hatchings, symbols and notation,
        31–4
    isometric projection, 24
    orthographic projection, 23
    perspective projection, 25
    plans and elevations, 26
    sketches, 22
Drilling rigs, 199–202
Driven insitu piled foundations, 210
Drop arch, 309
Drop hammer, 199, 205, 208–10, 212
Dry linings, 525–7
Dumpers, 152, 172
Dumpling, 229
Dynamic compaction, 276, 278

Earth pressures, 218, 226
Earthworks for roads, 109
Eaves:

closure piece, 439–40
details, 411–3, 417–8, 495
filler piece, 439–40
ventilation, 412, 417
Effective height of walls, 307
Effective length of columns, 387
Effective thickness of walls, 305, 307
Effluents, 605
Electrical cables, 649
Electrical installations, 652–3
Electrical site lighting, 81–4
Electricity:
    domestic supply, 646–8
    supply to sites, 84
Elemental method, 487–8, 493
Elevations, 26
End bearing pile foundation, 197
Energy efficiency, 487
Energy roof system, 441
English bond, 286
Environment, 3–5
Equivalent area, 356–7
Espagnolette, 323
European spruce, 598
European Standards, 53
Excavating machines, 146–50, 231
Excavations:
    basement, 228–31
    dewatering, 265–9
    oversite, 241
    pier holes, 242
    reduced level, 241
    setting out, 102–7
    trench, 242
    temporary support, 243–6
Expansion joint, 111, 236, 451
Exposed aggregate, 477
External asphalt tanking, 237
External doors, 334–6
External envelope, 18, 283
External escape stairs, 574

Facade bracing, 116
Face grid, 41
Face shovel, 147, 231
Facings to panel walls, 464
Factories Acts, 42
Fencing, 76–7, 87
Festoon lighting, 83
Fibreboard, 601
Finger joints, 392

Finishes, 15
Fin walls, 294
Fire back, 661
Fire doors, 582–3
Fireplaces and flues, 659–66
Fire protection of steelwork, 382–3
Fire resistance of steelwork, 382
Fire resisting doors, 584
Fire stops and seals, 658
Fish plates, 381
Fixed portal frame, 388–9, 391
Flashings, 293, 415, 426, 429, 664
Flat roofs, 424–9
Flat sawn timber, 393
Flat slabs, 547
Flat top girder, 442, 445
Flemish bond, 287
Flexible paving, 109
Flitch beam, 396
Floating floor, 503
Floating pile foundation, 197
Float valves, 632
Floor plans, 26, 34
Floor springs, 585
Floors:
    domestic ground, 531–4
    finishes, 535–6
    fixings, 550
    flat slab, 547, 550
    hollow pot, 549–50
    insitu RC suspended, 547–50
    large cast insitu, 537–8
    precast concrete, 551–3
    ribbed, 548
    screeds, 539–40
    sound insulation, 504–5
    suspended timber ground, 531, 533
    suspended timber upper, 541–4
    waffle, 548
Flue blocks, 657, 666
Flue lining, 661–5
Flush bored pile foundation, 197
Fly jib, 160–1
Flying shore, 128, 131–3
Folded plate construction, 9
Footpaths, 112
Forklift truck, 153
Formwork:
    beams, 361–2
    columns, 363–4
    principles, 360

slab, 349
stairs, 569
Foundation hinges, 390
Foundations:
 basic sizing, 186
 beds, 185
 calculated sizing, 187
 functions, 177
 grillage, 194
 isolated, 194
 piled, 197–214
 raft, 191
 short bored piled, 190, 197–9
 simple reinforced, 189
 stepped, 188
 subsoil movement, 178–80
 types, 192–6
Four-in-one bucket, 145
Four wheeled grader, 144
Framed structures, 351, 355, 380–1, 397
Freyssinet anchorage, 373
Friction hinge, 323
Friction piling, 197
Frieze, 597
Frodingham box pile, 207
Frodingham sheet pile, 254
Full height casting, 222
Furniture beetle, 434

Gambrel roof, 402, 408
Gantry crane, 162
Gantry girder, 377
Gantry scaffolding, 127
Garden wall bonds, 288
Gas:
 fires, 656
 flues, 657, 665
 service intake, 654–5
 supply, 654
Gas resistant membranes, 203
Gate valve, 632–3
Gauge box, 249
Geodistic dome, 448
Girders, 444–5
Glass and glazing, 326–331
Glass block wall, 332–3
Glazed cladding, 469
Gluelam, 392–4
Grab bucket, 149
Graders, 144

Granular soil shell, 199
Grillage foundation, 194
Groins, 452
Ground anchors, 233, 273, 375
Ground freezing, 275
Ground vibration, 276–7
Ground water, 265
Ground water control, 266–75
Grouting-subsoil, 274
Grouting-tiles, 530
Guard rail, 117–9, 121–2, 125–6
Gusset plate, 391, 437, 440

Half hour fire door, 582
Hammer head junction, 108
Hand auger, 65
Hand auger holes, 60
Handrail, 577
Hardwoods, 599
Health and Safety at Work, etc. Act, 42, 44
Hearths, 660–1, 665
Helical binding, 355
Helical stairs, 567
Helmets, 205, 209
Hemispherical dome, 449
High performance window, 317
Highway dumpers, 152
Hinge or pin joints, 390
Hip tiles, 410, 414
Hoardings, 76–80
Hoists, 154–5
Holding down bolts, 366, 380
Hollow box floor units, 551
Hollow pot floor, 549
Hollow sections, 376
Horizontal shore, 131–3
Horizontal sliding sash window, 322
Hot water:
 cylinders, 634
 direct system, 629
 expansion vessel, 631
 heating systems, 642–5
 indirect system, 630
 mains fed, 631
House Longhorn beetle, 434
Hull core structures, 398
Hyperbolic paraboloid roof, 453–4

Immersion heater cable, 651
Inclined slab stair, 565, 568

Independent scaffold, 118
Indirect cold water system, 628
Indirect cylinder, 634
Indirect hot water system, 630
Industrial doors, 339–42
Infill panel walls, 464–7
Inspection chambers, 605, 611, 616–21
Insulating dpc, 302
Insulation of basements, 240
Insulation – sound, 500–4
Insulation – thermal, 478–96
Integrity, 582, 584
Interest on capital outlay costing, 140
Internal asphalt tanking, 238
Internal doors, 578
Internal drop hammer, 199, 208, 210
Internal elements, 511
Internal environment, 3
Internal partitions, 518–22
Internal walls:
   block, 515–7
   brick, 514
   functions, 512
   plaster finish, 523–4
   plasterboard lining, 525–7
   types, 513
International Standards, 53
Intumescent collar, 658
Intumescent strips, 582–4
Inverted warm deck, 428
Ironmongery, 319, 338
Isolated foundations, 184, 194
Isometric projection, 24

Jack pile underpinning, 260
Jambs, 312
Jarrah, 599
Jet grouting, 274, 276, 279
Jetted sumps, 267
Jetted well points, 268–9
Joinery production, 595–9
Joinery timbers, 598–9
Jointless suspended ceiling, 588
Joints:
   basements, 236
   blockwork, 517
   drainage, 622
   laminated timber, 392
   portal frame, 389–91
   roads, 111

Joists:
   timber, 531, 541–6
   steel, 376
Joist sizing – timber, 545–6
Joist sizing – steel, 384–6

Kelly bar, 200–2
Kentledge, 264
Kitemark, 52

Ladders, 117–8, 121
Laminated timber, 392–4
Land caissons, 255
Landings, 559–61, 563–75
Lantern lights, 459
Large diameter piled foundations, 198, 202
Larseen sheet piling, 254
Lateral restraint, 433, 541, 543
Lateral support – basements, 234
Lateral support – walls, 407, 433, 543
Lattice beam, 379, 396, 444–5
Lattice girder, 379
Lattice jib crane, 160
Leader – piling rig, 205
Lean-to roof, 404
Lens light, 459
Levelling, 106–7
Lift casting, 223
Lighting:
   cable, 651
   circuits, 653
   sites, 81–83
Lightweight decking, 447
Lightweight infill panels, 464, 466–7
Lintels, 308, 311
Litzka beam, 378
Load-bearing concrete panels, 473
Load-bearing partitions, 518
Load-bearing internal walls, 510, 513–5
Locating services, 101
Loft hatch, 499
Long span roofs, 442–7
Loop ties, 224
Lorries, 151
Lorry mounted cranes, 159–60
Luffing jib, 163

Maguire's rule, 624
Mahogany, 599

Main beams, 17, 381
Mandrel, 209
Manometer, 623, 641
Mansard roof, 402, 408
Masonry partitions, 518
Mass concrete retaining wall, 220
Mass retaining walls, 219–20
Mast cranes, 163
Mastic asphalt tanking, 237–8
Mastic asphalt to flat roofs, 429
Materials:
  conductivity, 481–2
  density, 30, 481–2
  hoist, 154
  storage, 87–92
  testing, 93–99
  weights, 29–30
Mechanical auger, 65
Meeting rails, 321
Meeting stiles, 322, 585
Metal casement windows, 318
Metal section decking, 350
Metal stairs, 574–6
Meter box, 646–8, 654–5
Methane, 303
Method statement, 28
Micro-bore heating, 644
Middle shore, 130
Middle third rule, 216
Mineral insulating cable, 649
Mixing concrete, 169
Mobile cranes, 157
Mobile scaffold, 121
Modular coordination, 40–1
Modular ratio, 356
Mohr's circle, 73
Moment of resistance, 135, 384
Monitor roof, 442, 444
Monogroup cable, 370
Monolithic caissons, 255
Monostrand anchorage, 373
Monostrand cable, 370
Mortar strength, 306
Morticing machine, 595
Movement joint, 111, 463, 517, 571
Mud-rotary drilling, 66
Multi-purpose excavators, 146, 150
Multi-span portal frames, 389, 391
Multi-stage wellpoints, 269
Multi-storey structures, 397–8
Needle and pile underpinning, 261

Needle design, 134–5
Needles, 128, 130–3
Newel post, 555, 558–61, 564
Northlight barrel vault, 452
Northlight roofs, 442–3
Non-load-bearing partitions, 510–3, 518–22

Oak, 599
Oedometer, 74
Oil based paint, 591
One hour fire door, 583
One pipe heating, 642
Open caissons, 255
Open excavations, 228
Open fireplaces, 659–666
Open riser stairs, 560–2, 572
Open suspended ceiling, 590
Open web beam, 378
Openings:
  arches, 308–10
  heads, 311
  jambs, 312
  sills, 313
  support, 308
Oriel window, 324
Oriented strand board, 601
Orthographic projection, 23
Out-of-service crane condition, 164
Output and cycle times, 141
Overcladding, 468
Overhead and forklift trucks, 153

Pad foundation, 184, 194, 355, 380, 389, 391
Pad foundation design, 186
Pad template, 104
Paint defects, 594
Paints and painting, 591–3
Panelled suspended ceiling, 587, 589
Parallel lay cable, 370
Parane pine, 597
Parapet wall, 293
Partially preformed pile, 209
Particle board, 536, 601
Partitions:
  demountable, 521–2
  load-bearing, 518
  metal stud, 520
  non-load-bearing, 518–22
  timber stud, 519

Passenger hoist, 155
Passenger vehicles, 151
Passive earth pressures, 218
Patent glazing, 443–4, 457–8
Patent scaffolding, 122
Paved areas, 610
Paving, 109
Paving flags, 114
Pedestal, 554
Pendentive dome, 449
Penetration test, 70, 98
Percussion bored piling, 198–9
Performance requirements:
  doors, 334
  roofs, 399
  windows, 314
Perimeter trench excavation, 229
Permanent exclusion of water, 265,
  271–4
Permanent formwork, 350
Permitted development, 35
Phenol formaldehyde, 395, 602
Pigment, 591
Pile:
  beams, 214
  caps, 201, 214
  classification, 197
  testing, 214
  types, 198
Piled basements, 232
Piled foundations, 197–214
Piling:
  contiguous, 271
  hammers, 212–3
  helmets, 205, 209
  rigs, 205, 208–9
  steel sheet, 254
Pillar tap, 632
Pin or hinge joint, 390
Pitch pine, 598
Pitched roofs, 403–8
Pitched trusses, 442–3
Pivot window, 323
Placing concrete, 169
Plain tiles and tiling, 410–1, 414–5
Planer, 595
Plank and pot floor, 552
Planning application, 35–9
Planning grid, 40
Planning supervisor, 43
Plant:

bulldozer, 142
concreting, 169–74
considerations, 139
costing, 140
cranes, 157–68
dumpers, 152, 172
excavators, 146–50, 231
forklift trucks, 153
graders, 144
hoists, 154–5
scrapers, 143
tractor shovel, 145
transport vehicles, 151–3
Plaster cove, 586
Plaster finish, 524
Plasterboard, 525–6
Plasterboard ceiling, 586
Plasterboard dry lining, 525–7
Plasters, 523
Platform floor, 505
Platform frame, 345
Plug wiring, 651
Plywood, 600
Pneumatic caisson, 256
Poker vibrator, 174
Poling boards, 245
Polyurethane paint, 591
Portal frames, 388–91
Portsmouth float valve, 632
Post-tensioning, 372–3
Power circuit, 652
Precast concrete:
  diaphragm wall, 273
  floors, 534, 551–3
  frames, 365–8
  portal frames, 389
  stairs, 570–3
Preformed concrete pile, 205–6
Preservative treatment, 395, 435
Pressed steel lintels, 311
Pressure bulbs, 62–3
Prestressed concrete, 369–75
Pretensioning, 371
Primary elements, 13
Primatic cylinder, 634
Principal contractor, 43
Profile boards, 103–4
Profiled surface, 114, 476
Programmer/controller, 645
Prop design, 135
Proportional area method, 484, 486

Propped structures, 397
Protection orders, 100
Public utility services, 101
Published Document, 52
Pump sizing, 266
Purlin fixings, 439
Purlin roof, 405
Purpose designed excavators, 146
Purpose made joinery, 595
Putlog scaffold, 117
Putty, 327
Pynford stool underpinning, 362

Quarter sawn timber, 393

Radiators, 642–5
Radius of gyration, 387
Radon, 303
Raft basements, 232
Raft foundations, 191, 195
Rail tracks for cranes, 167
Rainscreen cladding, 464, 468
Rainwater drainage, 615
Rainwater installations, 613–4
Raised access floor, 554
Raking shore, 128, 130
Raking struts, 131–2, 230
Rankine's formula, 226
Rat trap bond, 288
Ready mixed concrete, 252
Ready mixed concrete truck, 172
Redwood, 598
Reinforced concrete:
  beams, 351–3
  column design, 356–7
  columns, 355–7
  floors, 547–50
  formwork, 349, 360–4, 569
  foundations, 189–91
  pile caps and beams, 214
  raft foundation, 191, 195
  reinforcement, 349–59
  retaining walls, 221
  slabs, 347–9
  stairs, 565–9
  strip foundations, 189, 193
Reinforcement:
  bar coding, 358
  bar schedule, 359
  grip length, 354
Remedial dpc, 298–300

Remote crane control, 165
Rendhex box pile, 207
Replacement piling, 197–8
Resin grout, 274
Resorcinol formaldehyde, 395, 600
Restraint straps, 433, 541, 543
Retaining walls, 215–27
Retaining walls – design, 226–7
Reversing drum mixer, 171
Ribbed floor, 548–50
Rider, 130
Ridge detail, 411–2, 417–8
Ridge piece, 439–40
Ridge roll, 423
Ridge tiles, 410–2, 414, 417–8
Ridge vent, 411, 417
Rigid pavings, 110
Rigid portal frames, 388–9, 391
Ring main wiring, 652
Roads:
  construction, 108–11
  drainage, 611–2
  earthworks, 109
  edgings, 113
  footpaths, 112
  forms, 110
  gullies, 610–12
  joints, 111
  kerbs, 113
  landscaping, 115
  pavings, 114
  services, 115
  setting out, 108
  signs, 115
Rolled steel joist, 376
Roll over crane skip, 169
Roller shutters, 339, 342
Roofs and roof covering:
  basic forms, 400–2
  built-up felt, 425–6
  flat top girder, 442, 445
  long span, 442–7
  mastic asphalt, 429
  monitor, 442, 444
  northlight, 442–3
  performance, 399
  sheet coverings, 438–41
  shells, 449–56
  slating, 418–22
  space deck, 447
  space frame, 448

surface water removal, 608–9, 613–4
thatching, 423
thermal insulation, 486, 488–9, 492–5
tiling, 410–1, 414–8
timber flat, 400, 424–9
timber pitched, 400–8
trussed rafter, 407
trusses, 406, 436
underlay, 409
Rooflights, 457–60
Room sealed appliance, 656
Rotary bored piling, 198, 201–2
Rotational dome, 449
Rubble chutes and skips, 156
Runners, 246

Saddle vault, 453
Safe bearing, 546
Safety signs, 44–5
Sampling shells, 65
Sand bulking test, 94
Sand pugging, 504
Sanitary fittings:
  basin, 636
  bath, 637
  discharge systems, 639–40
  shower, 637
  sink, 636
  wc pan, 638
Sanitation systems, 639–41
Sanitation system testing, 641
Sapele, 599
Sash weights, 320
Saws, 595
Scaffolding:
  birdcage, 12
  boards, 117–9
  cantilever, 125
  component parts, 116
  gantry, 127
  independent, 118
  ladders, 117–8, 121
  mobile, 121
  patent, 122
  putlog, 117
  slung, 123
  suspended, 124
  truss-out, 126
  tying-in, 120

working platform, 119
Scarf joint, 392
Scrapers, 143
Screed, 539–40
Secondary beams, 8, 381
Secondary elements, 14
Secondary glazing, 329
Section factor, 383
Section modulus, 135, 384
SEDBUK, 488, 491
Segmental arch, 309
Self propelled crane, 158
Self supporting static crane, 164–5
Services fire stops and seals, 658
Setting out:
  bases, 104
  basic outline, 102
  drainage, 622
  grids, 104
  levelling, 106–7
  reduced levels, 105
  roads, 108
  trenches, 103
Shear, 351, 385, 546
Shear bars, 351
Shear box, 74
Shear leg rig, 65–6, 199
Shear plate connector, 446
Shear strength of soils, 73–4
Shear wall structures, 398
Sheathed cables, 649
Sheet coverings, 438–41
Shell roofs, 449–56
Shoring, 128–35
Short bored pile foundation, 190
Shower, 637
Shutters, 342
Sight rails, 104–5, 108, 622
Sills, 313
Silt test for sand, 95
Simple drainage:
  bedding, 622
  inspection chambers, 620–1
  jointing, 622
  roads, 611–2
  setting out, 622
  systems, 616–9
Simply supported RC slabs, 347–9
Single acting hammer, 212
Single barrel vault, 450
Single flying shore, 131

Single lap tiles and tiling, 416–7
Single span portal frames, 388–9, 391
Single stack drainage, 639
Sink, 636
Site:
    construction activities, 19
    electrical supply, 84
    health and welfare, 86
    investigations, 59, 61, 276
    layout, 75–7
    lighting, 81–3
    materials testing, 93–9
    offices, 76–7, 85, 90
    plan, 27, 35
    security, 76, 78
    setting out, 102–7
    soil investigations, 60–6
    storage, 11, 87–92
Six wheeled grader, 144
Sketches, 22
Skimmer, 146
Skips, 156
Slates and slating, 418–22
Slenderness ratio, 135, 307
Sliding doors, 339–41
Sliding sash windows, 320–2
Slump test, 96
Slung scaffold, 123
Small diameter piled foundation,
    198–201
Smoke seal, 584
Smoke test, 641
Soakers, 415, 644
Soakaway, 615
Socket outlets, 651–2
Softwoods, 598
Soil:
    assessment, 68–74
    classification, 68–70
    improvement, 276–9
    investigation, 60–6, 276
    particle size, 68
    samples, 61
    stabilization, 276–9
    testing, 68–74
Solid block walls, 290
Solid brick walls, 284
Solid slab raft, 195
Sound insulating:
    floors, 503–4
    walls, 501–2

Sound insulation, 500–4
Sound reduction, 328, 500
Space deck, 447
Space frame, 448
Spindle moulder, 595
Spine beam stairs, 572
Splice joint, 391
Splicing collar, 205–6
Split barrel sampler, 70
Split ring connector, 446
Stairs:
    alternating tread, 562
    balusters, 555–61, 564, 567–8
    balustrade, 573, 575–7
    formwork, 569
    handrail, 555–8, 560–4, 567–8,
        573–7
    insitu RC, 565–8
    metal, 574–6
    open riser, 561
    precast concrete, 570–3
    timber, 555–64
Standard Assessment Procedure, 487
Standard dumper, 152
Standard crane skip, 169
Steel:
    beam, 376–7
    beam design, 384–6
    column, 376–7
    column design, 387
    compound sections, 377
    gantry girder, 377
    lattice beams, 379
    portal frames, 391
    roof trusses, 436–7
    screw pile, 207
    sheet piling, 253–4
    standard sections, 376
    string stairs, 576
    tube pile, 208
Stepped barrel vault, 452
Stepped flashing, 415, 644
Stepped foundation, 188
Stock holding policy, 90
Stop valve, 632–3
Storage of materials, 77, 87–92
Storey height cladding, 475
Straight flight stairs, 555–9
Straight line costing, 140
Straight mast forklift truck, 153
Strand, 370

Stress reduction in walls, 307
String beam stair, 565
Stripboard, 600
Strip foundations, 183, 186–9
Structural grid, 40
Structure:
 basic forms, 8–11
 basic types, 6–7
 functions, 16–17
 protection orders, 100
Strutting of floors, 542
Stud partitions, 519
Subsoil:
 drainage, 606–7
 movements, 178–81
 water, 265
Substructure, 12
Sump pumping, 266
Supply and storage of concrete, 169
Supported static tower crane, 163,
 166
Surcharging, 276
Surface water, 265
Surface water removal, 608–10
Suspended ceilings, 587–90
Suspended scaffols, 124
Suspended structures, 397
Swivel skip dumper, 152

Tactile pavings, 114
Tamping board vibrator, 174
Target U value, 487, 490–1, 498
Teak, 599
Telephone installation, 667
Telescopic boom forklift truck, 153
Telescopic crane, 159
Temporary bench mark, 102, 104–6
Temporary exclusion of water, 265–70
Temporary services, 76
Tendons, 370
Tenoning machine, 595
Test cubes, 96
Testing of materials, 93–9
Textured surfaces, 476
Thatched roof, 423
Thermal bridging, 497–8
Thermal conductivity, 479, 481–4, 486
Thermal insulation, 478–96
Thermal resistance, 478–80, 484, 486
Thin grouted membranes, 270
Thinners, 591

Three axle scraper, 143
Three centre arch, 309
Three pin portal frame, 388
Tile hanging, 461
Tilting drum mixer, 170–1
Tilting level, 106
Timber:
 beam design, 545–6
 casement windows, 316–8
 connectors, 446
 doors, 334–6, 340–1
 flat roofs, 424–9
 frame construction, 345–6
 girders, 445
 hardwoods, 599
 joinery production, 595–9
 pile foundation, 204
 pitched roofs, 403–8
 preservation, 434–5
 softwoods, 598
 stairs, 555–64
 storage, 92
 stud partition, 519
 trestle, 208
Timbering, 244–6
Tooled surface, 476
Toothed plate connector, 445–6
Top shore, 130, 133
Towed scraper, 143
Tower cranes, 157, 164–7
Tower scaffold, 121
Track mounted crane, 167
Tractor shovel, 145
Traditional strip foundation, 183
Traditional underpinning, 259
Translational dome, 449
Translucent glass, 326
Transport vehicles, 151
Transporting concrete, 172
Traveller, 103–5, 622
Travelling tower crane, 164, 167
Tree protection, 90, 100, 181
Trees – foundation damage, 179–80
Tremie pipe, 200, 272
Trench fill foundation, 193
Trench setting out, 103
Trench sheeting, 253–4
Trial pits, 60, 64
Triangle of forces, 216
Triangular chart, 69
Triaxial compression test, 73

Index

Tripod rig, 65–6, 199
Truss out scaffold, 126
Trussed rafter roof, 407
Tubular scaffolding, 116–23, 125–7
Two axle scraper, 143
Two pin portal frame, 388
Two pipe heating, 643
Tying-in scaffolding, 120

U value calculations, 478–9, 484–5
U values, 441, 478–9, 482–6,
    488–96
Unconfined compression test, 72
Underlay, 409
Underpinning, 257–64
Under-reaming, 202
Undersill cladding panels, 474
Universal beam, 376
Universal bearing pile, 207
Universal column, 376
Universal excavator, 146
Urea formaldehyde, 395, 602

Vacuum dewatering, 538
Valley beam, 443
Valley gutter, 443
Valley tiles, 410, 414
Vane test, 71
Vans, 151
Vapour control layer, 428, 430–2, 441,
    444, 528, 586
Vaults, 450–2
Ventilated stack discharge
    system, 640
Ventilation of roof
    space, 411–2, 430
Ventilation spacer, 411, 417–8
Verge details, 415, 426
Vertical casting, 477
Vertical laminations, 393
Vestibule frame, 330
Vibrating tamping beam, 174
Vibration of soil, 276–7
Vibro cast insitu piling, 211
Volume batching, 249

Wall hooks, 130
Wall plates, 130–3
Wallboard, 526–8
Wall profiles, 516
Walls:

cavity, 291–2, 305
curtain, 470–2
diaphragm, 233, 272–3
design strength, 306
diaphragm, 233, 272–3, 295
fin, 294
formwork, 222–4
glass block, 332–3
infill panel, 464–7
internal, 511–22
panelling, 597
retaining, 215–27
rising dampness, 298–301
slenderness ratio, 135, 307
solid block, 515
solid brick, 284–8, 514
sound insulation, 501–2
thermal insulation, 479–80, 484–5,
    488–90, 492–6
thickness, 305
tiling, 461, 529–30
underpinning, 258–63
waterproofing, 235–9
Warm deck, 428
Wash boring, 66
Water bar, 236, 337
Water based paint, 591
Water/cement ratio, 235, 250
Water closet pan, 638
Water installations:
cisterns, 633, 638
cold, 627–8
hot, 629–31
pipe joints, 635
supply, 625–6
Water table, 67, 265
Water test, 623
Waterproofing basements, 235–9
Weatherboarding, 461
Web cleats, 377–8
Weep holes, 215–21
Weight batching, 250
Weights of materials, 29–30
Wellpoint systems, 268–9
Western hemlock, 598
Western red cedar, 598
Windows:
bay, 324
double glazing, 317, 323, 328
glass and glazing, 326–31
high performance, 317

ironmongery, 319
metal casement, 318
oriel, 324
performance requirements, 314
pivot, 315, 323
schedules, 325
sliding sash, 320–2
timber casement, 316–7
types, 315
Wired glass, 326
Wiring systems, 650–3

Wood cement board, 601
Wood wool, 601
Woodworm, 434
Woodworking machinery, 595
Working platform, 119
Wrought iron dogs, 129

Yokes, 364

Zed beam, 439, 441
Zones, 41